Hyperbolic Equations and Waves

Battelle Seattle 1968 Recontres

Edited by M. Froissart

With 49 Figures

Springer-Verlag Berlin · Heidelberg · New York 1970

Professor M. FROISSART
CEN Saclay, Gif-sur-Yvette, France

We would like to thank

American Mathematical Society, Providence, R. I., USA,
Institute of Mathematics and its Applications, Southend-on-Sea, Essex, Great Britain,
Cambridge University Press, London NWI, Great Britain,
Royal Society, London SWI, Great Britain.
Librairie Universitaire Uystpruyst, Louvain, Belgium,

for their kind permission to reproduce the following contributions
Phase Jumps (M. S. HOWE),
On Waves Generated in Dispersive Systems by Travelling Forcing Effects, with
Applications to the Dynamics of Rotating Fluids (M. J. LIGHTHILL),
Group Velocity (M. J. LIGHTHILL),
Contributions to the Theory of Waves in Non-linear Dispersive Systems
(M. J. LIGHTHILL),
Variational Methods and Applications to Water Waves (G. B. WHITHAM),
Some Special Cases Treated by the Whitham Theory (M. J. LIGHTHILL),
Wavetrains in Inhomogeneous Moving Media
(F. B. BRETHERTON and M. J. LIGHTHILL),
Relativistic Hydrodynamics (A. H. TAUB).
Systèmes Linéaires, Hyperboliques Non-Stricts (J. LERAY and Y. OHYA)
Norme Formelle d'une Fonction Composée (J. LERAY and L. WAELBROECK)

ISBN 978-3-642-87027-9   ISBN 978-3-642-87025-5 (eBook)
DOI 10.1007/978-3-642-87025-5

Title No. 1607

Preface

The success of the 1967 Battelle Rencontres was so much appreciated by the participants and organizers of this experimental set-up that it was soon decided to go on with the experiment. Mathematicians and physicists had found a very suitable frame to overcome their natural shyness, to get occasionally interested into each others' work, to talk about it, and eventually to know each other. The 1968 Rencontres have been organized with the same idea in mind, and even somewhat enlarged in the following sense: the topic chosen — hyperbolic equations and waves — has proved a cornerstone of physics for more than a century and extends over most fields of contemporary physics. It follows immediately that the wide range of physicists concerned could not be represented by more than a couple of specialists in any single field. Thus, aside from bridging the gap between mathematicians and physicists, the 1968 Recontres provided a rather unique occasion to plug many intra-disciplinary gaps among physicists.

This made the Rencontres quite unpredictable as to how people would — and could — interact, and created a very stimulating environment for an unprecedented intellectual venture. From the outside, it may very well look like a hodge-podge of quite unrelated ideas. But it was much less so at the level of day-to-day discussions and informal gatherings where all slowly acquired a comprehensive synthetic view of the subject. Such a view is impossible to put in words, and has to be abstracted from the diversity of the points of view, of the approaches, of the analytic descriptions.

This diversity appeared noticeably at the level of the lectures and seminars, which ranged from basic lectures, just to define a common vocabulary, to very sophisticated accounts of current research. In view of this, it was thought preferable not to adopt any systematic policy concerning the form of the written reports contained in the present book. It was clearly useless to include basic lectures, which would duplicate other well-written texts, but also to include material on rapidly evolving subjects which would be superseded very soon. Also, it was recognized that lectures given about published work could be replaced with advantage in the present book by an introduction to the existing literature, and reproduction of the main papers, so that the participants could go farther on the way pointed to by the lecturers.

It is our hope that the readers who could not attend the Rencontres will benefit from these hints and will be able to obtain a more synthetic

feeling for this wide field through an acquaintance with subjects somewhat out of their day-to-day preoccupations.

I would like to take this occasion to thank on behalf of all participants and on my own the officers and staff of the Battelle Memorial
Institute, who made it possible to get together and think in this unique
fashion.

I also acknowledge the kind permission of the publishers to reproduce
the papers contained in this book: Cambridge University Press, the
American Mathematical Society, Librairie Universitaire Uystpruyst,
Academic Press, the Institute of Mathematics and its Applications, the
Royal Society, Springer-Verlag.

Saclay, July, 1969   M. FROISSART

Contents

List of Participants

Adler, Wm. F., Battelle Memorial Inst. 505 King Avenue, Columbus, Ohio 43201, USA

Ahluwalia, Daljit S., Courant Institute of Mathematical Sciences, 251 Mercer Street, New York, N. Y. 10012, USA

Baroody, Eugene M., Battelle Memorial Institute, 505 King Avenue, Columbus, Ohio 43201, USA

Bejda, Jozef, Institute of Basic Technical Research of Polish Academy of Sciences, Warsaw, Swietokrzyska 21, Poland

Bros, Jacques, School of Natural Sciences, Institute of Advanced Study, Princeton, N. J. 08540. USA

Chandrasekhar, S., Lab. of Astrophysics & Space Research, University of Chicago, Chicago, Illinois, USA

Dwork, Bernard M., Dept. of Mathematics, Princeton University, Princeton, N. J. 88540, USA

Dyson, Freeman J., Institute for Advanced Study, School of Natural Sciences, Princeton, N. J. 08540, USA

Ehrenpreis, Leon, Courant Institute of Mathematical Science, 251 Mercer Street, New York, N. Y. 10012, USA

Froissart, M., C. E. N. Saclay, 91 Gif-sur-Yvette, France

Gårding, L., Lunds Universitets Mathematiska Institution, Lund, Sweden

Harounian, Houshang, Dept. of Applied Math & Theoretical Physics, Silver Street, Cambridge, Great Britain

Hayes, Wallace D., Princeton University, School of Engineering & Applied Science, Princeton, N. J. 08540, USA

Hersh, Reuben, Dept. of Math, University of New Mexico, Albuquerque, New Mexico 87106, USA

Keller, Joseph, B., Courant Institute of Mathematical Science, Division of Electromagnetics Research, 251 Mercer Street, New York, N. Y. 10012, USA

Kruskal, Martin, Princeton University Observatory, Peyton Hall, Princeton, N. J..08540, USA

Lax, Peter D., Courant Institute of Math Science, AEC Computing & Mathematics Center, 251 Mercer Street, New York, N. Y. 10012, USA

Leray, Jean, Collège de France, Chaire de Théorie des Équations. Différentielles et Fonctionnelles, Paris, France

Lighthill, M. J., Imperial College of Science & Technology, University of London, South Kensington, London, S. W. 7, Great Britain

Murman, Earll M., Boeing Scientific Research Laboratories, P. O. Box 3981, Seattle, Washington 98124, USA

Nestell, M. K., Battelle-Northwest, P. O. Box 999, Richland, Washington 99352, USA

Osher, Stanley J., Department of Mathematics, Universitiy of California, Berkeley, California 94720, USA

Segal, Irving, 25 Moon Hill Road, Lexington, Mass. 02173, USA

Shih, Weishu, Institut des Hautes Etudes Scientifiques, Bures-Sur-Yvette (91), France

Smith, Donald, University of California Dept. of Mathematics, P. O. Box 109, La Jolla, California 92037, USA

Smoller, Joel, Dept. of Mathematics, University of Michigan, Ann Arbor, Mich. 48104, USA

Taub, A. H., Mathematics Dept., 301 Campbell Hall, University of California, Berkeley, California 94720, USA

Thorne, Kip, California Institute of Technology, W. K. Kellogg Radiation Laboratory, Pasadena, Calif. 91109, USA

Van Winter, Clasine, Department of Mathematics, University of Kentucky, Lexington, Kentucky 40506, USA

Zimmerman, Robert L., Institute of Theoretical Science, University of Oregon, Eugene, Oregon 97403, USA

Wave Mathematics

M. J. LIGHTHILL

1. General Introduction

It is appropriate at a meeting like this to present some substantial body of rather general theory, and I tried therefore to select material of this kind, that may be of interest to everyone in the audience, in the form of a general mathematical approach to treating a wide class of physical problems that involve waves. Accordingly, my first five lectures embody such a presentation. It is concerned with both linear and non-linear aspects of waves, with the emphasis on dispersive waves, and mainly on waves in fluids.

The waves are not necessarily governed by strictly hyperbolic equations, since non-relativistic formulations are often used. It follows that wave velocities may, for example, tend to infinity in the limit of zero wave number (in contradiction to strict hyperbolicity). Even when the equations are strictly hyperbolic, however, the highest-order terms may not be the most important, because excitation may take place at wave numbers for which lower-order terms are dominant. Variation of wave velocity with wave number, and (in the non-linear cases) also with amplitude, is found to be of crucial importance.

A physicist is, nevertheless, unhappy with a presentation that is devoted exclusively to general theory. There is furthermore, a danger that mathematicians may receive a misleading impression if physicists present to them a picture of physicists' activities as consisting purely of theoretical studies of a markedly generalized kind. For this reason, the last three lectures are devoted to a quite specific application of wave mathematics, directed towards predicting the dynamic response of ocean currents to meteorological variations.

The material of the eight lectures is set out in detail in eight papers (five of my own and three by other authors), here referred to as A to H. Papers A of G are included in the present book in the form of photographic reprints as follows:

A   Group Velocity   p. 96

These give a general account of dispersive waves. The present descriptive guide indicates a possible order in which various parts of them may be read. It sketches also the relations between them, as well as relations to various other papers cited in the bibliographies of papers A to G. Finally, paper H (not appended, but to be published shortly in *Phil. Trans. Roy. Soc. London*) is a self-contained account of the particular topic of dynamic response of ocean currents, entitled "Dynamic response of the Indian Ocean to onset of the Southwest Monsoon".

2. Introduction to Dispersive Wave Theory

It is essential to begin by reviewing the linear theory of dispersive waves. Theories using either the kinematics of surfaces of constant phase, or the method of superposition of plane-wave solutions, are reviewed in paper A, sections 1 to 7. The properties of group velocity, and the importance of wavenumber surfaces, emerge clearly, but there is at this stage no demonstration that the group velocity is the velocity of energy propagation. Nevertheless, the results derived for wave amplitude distributions are shown to be consistent with that assumption.

This linear theory is further extended in paper B, sections 1 to 3, which investigates the case of waves generated by travelling forcing effects. Such waves are greatly influenced by the often very large distortion of the wavenumber surface that Doppler effect produces. Sections 6 to 9 give illustrative examples that may be helpful at this stage, concerned with ship waves, with waves in stratified fluids and with TAYLOR columns. (Section 4 and 5, on the other hand, are more relevant to the second part of the lecture course, on dynamic response of ocean currents.)

After these preliminaries, it is appropriate to consider energy flux in relation to group velocity. The method of doing this described in paper A, section 8, produces the unexpected bonus that it leads to definite formulas for energy flux in plane periodic waves in a very general non-linear conservative system. These are given in terms of the averaged Lagrangian density $\mathscr{L}$, expressed as a function of the frequency ω and wavenumber k. Energy flux is $-\omega \mathscr{L}_k$ and energy density is $\omega \mathscr{L}_\omega - \mathscr{L}$.

Their ratio, the energy propagation velocity, can be written $d\omega/dk$ for changes keeping constant the measure of wave amplitude $\mathscr{L}/\omega$.

In 1964, I obtained this result for non-linear waves as an interesting extension of the result for waves of infinitesimal amplitude (which satisfy the dispersion relation $\mathscr{L}(\omega, k) = 0$). Correspondence on the matter with G. B. WHITHAM then revealed that he had obtained essentially the same result as a by-product of his attempts to extend the methods of geometrical optics to non-linear dispersive waves. At first, he tried to use that result directly; for systems homogeneous in space and time, a "smoothed" energy equation in the form

$$\frac{\partial}{\partial t}(\omega \mathscr{L}_\omega - \mathscr{L}) + \frac{\partial}{\partial x}(-\omega \mathscr{L}_k) = 0 \tag{1}$$

was accordingly employed. (Here x and k are vectors with the number of spatial dimensions appropriate to the problem.)

However, equation (1) simplifies, if ω and k are related to a generalized phase function θ by equations

$$\theta_t = -\omega, \quad \theta_x = k, \tag{2}$$

to a form

$$\frac{\partial \mathscr{L}_\omega}{\partial t} = \frac{\partial \mathscr{L}_k}{\partial x}. \tag{3}$$

It may similarly be shown that a smoothed equation for momentum conservation,

$$\frac{\partial}{\partial t}(k \mathscr{L}_\omega) + \frac{\partial}{\partial x}(\mathscr{L} I - k \mathscr{L}_k) = 0 \tag{4}$$

(where the second bracketed expression is a matrix representing momentum flux), simplifies to the form (3). WHITHAM then realized that (3) could be more simply derived, and derived even for systems not homogeneous in space and time, from Hamilton's principle in a smoothed form.

WHITHAM in a 1965 paper proposed this approach, and has given a more fully developed account of it in paper C. The smoothed form of Hamilton's principle,

$$\delta \iint \mathscr{L}(\omega, k, t, x) dt dx = 0, \tag{5}$$

is the basis of this work. Any expectation, based on experience with infinitesimal-amplitude (linear) systems, that energy-propagation relations would be of key importance in non-linear systems, would, as he showed, have been ill-founded. Essentially this is because non-linear effects allow continuous energy exchange between different modes with different wavenumbers.

Before describing the WHITHAM method in detail, we may note that paper D, by BRETHERTON and GARRETT, extended the analysis of paper A, section 8, to infinitesimal-amplitude waves in a non-uniformly moving fluid medium. If $\mathscr{L}^r(\omega, k)$ denotes the averaged Lagrangian density for plane periodic waves in a medium at rest, the energy density and energy flux are calculated as $(\omega - U \cdot k) \mathscr{L}^r_\omega$ and $(\omega - U \cdot k) \mathscr{L}^r_k$ respectively, where the fluid velocity is U and the waves satisfy the infinitesimal-amplitude dispersion relation $\mathscr{L}^r(\omega, k) = 0$. It follows that the energy of a wave packet varies in proportion to $\omega - U \cdot k$ as it moves through the system.

This is interesting, because for a steady system (one not changing with time) the energy of a wave packet might be supposed constant (just as its frequency ω would remain constant). This would be correct for a medium at rest, but not for a moving medium. Wave packets gain energy at the expense of the energy of the fluid motion if $\omega - U \cdot k$ increases; that is (under steady conditions with ω constant) if, along the path of a wave packet, the component of fluid velocity in a direction opposite to the phase velocity of the waves is increasing.

3. Detailed Aspects of the Whitham Theory

Paper E, sections 1 to 3, and also paper F, sections 1 and 2, discuss several aspects of the WHITHAM theory, but mainly in relation to problems with two independent variables, for which they give the criterion governing whether equation (3) is elliptic or hyperbolic, both in general and in a special form for moderate amplitude. The latter may be extended to the case of several independent variables as follows.

Eq. (3), written out by means of (2) in terms of generalized phase function θ, always takes the form

$$\mathscr{L}_{\omega\omega} \theta_{tt} - 2 \mathscr{L}_{\omega k} \theta_{tx} + \mathscr{L}_{kk} \theta_{xx} = R, \tag{6}$$

where R is zero for systems homogeneous in space and time, and furthermore for other systems consists only of terms of lower than second order. Eq. (6) is elliptic if the quadratic form

$$\mathscr{L}_{\omega\omega} T^2 + 2 \mathscr{L}_{\omega k} T X + \mathscr{L}_{kk} X^2 \tag{7}$$

(where X is a vector and because $\mathscr{L}_{kk}$ is a symmetric matrix $X^* \mathscr{L}_{kk} X$ has been written $\mathscr{L}_{kk} X^2$ without serious risk of ambiguity) is definite; it is hyperbolic if (7) is nonsingular but indefinite.

For moderate amplitude, $\mathscr{L}$ can be expanded as a series

$$\mathscr{L} = (\tfrac{1}{2}) g(k) \mathscr{T}^2 + \tfrac{1}{6} h(k) \mathscr{T}^3 + \dots, \tag{8}$$

where $\mathscr{T} = \omega - f(k)$, and then the quadratic form (7) can be written

$$(g + h\mathscr{T})[(T - f_k X + g_k g^{-1} \mathscr{T} X)^2 - \mathscr{T} f_{kk} X^2] + 0(\mathscr{T}^2). \qquad (9)$$

For small enough $\mathscr{T}$, then, it follows in general (specifically, when the matrix f_{kk} is non-singular), that eq. (6) is elliptic if $\mathscr{T} f_{kk}$ is negative definite, and otherwise hyperbolic. It is "strictly hyperbolic" if $\mathscr{T} f_{kk}$ is positive definite. (Note that for given k, the sign of $\mathscr{T}$ for small $\mathscr{T}$ is determinate, since the departure of ω from $f(k)$ is an even function of some measure of amplitude.)

Paper F develops in particular (see section 5) the implication of the elliptic-hyperbolic criterion for stability of a plane periodic wave of finite amplitude. A parallel criterion for waves of arbitrary amplitude (not just moderate amplitude) is developed in section 4 for waves on deep water, using an expression for the smoothed Lagrangian expected to be rather accurate for all amplitudes.

WHITHAM has considered cases when the form of plane periodic waves of given wavenumber may depend not only on the wavenumber and amplitude (or, what is the same thing, wavenumber and frequency), but also on at least one other variable. This is developed for example in paper C, section 4, where waves on shallow water are investigated in detail as a particular case. In these, the velocity potential may contain a slowly varying non-periodic mean portion ψ whose gradient $\psi_x = \beta$ (where x is a twodimensional vector) represents a mean horizontal current induced in the shallow water by the waves. We may then have $\mathscr{L} = \mathscr{L}(\omega, k, \gamma, \beta)$, where $\gamma = -\psi_t$, leading to a double system of equations

$$\frac{\partial \mathscr{L}_\omega}{\partial t} = \frac{\partial \mathscr{L}_k}{\partial x}, \quad \frac{\partial \mathscr{L}_\gamma}{\partial t} = \frac{\partial \mathscr{L}_\beta}{\partial x}. \qquad (10)$$

WHITHAM shows that for moderate amplitude this system is elliptic (corresponding to instability) if and only if the wavelength to depth ratio is less than 4.6.

Solutions involving, not just a slight perturbation of uniform initial data, but large changes, have been obtained so far only for waves in homogeneous systems with two independent variables. Paper E has obtained detailed solutions for waves of moderate amplitude in both the hyperbolic case (section 7 to 9) and the elliptic case (section 12 to 15). A particularly noteworthy feature of these solutions, discussed at length in both paper E and paper F, is the appearance of a discontinuity. Beyond a certain time no continuous solution of the equations exists.

There is nothing very unexpected about this in the hyperbolic case. It is rather well-known that a quasi-linear hyperbolic equation in two independent variables, such as eq. (6) with just one space dimension, normally exhibits a running together of characteristics. They are found to have a cusped envelope and, for times beyond that corresponding to the cusp, can possess only solutions in a "weak" sense, that is, discontinuous solutions satisfying the original equations in an integrated form.

The Riemann surface on which a solution satisfying regular Cauchy initial data is found to exist in such a case is single-sheeted except between the two branches of the cusped curve, where it has three sheets. These take the form of the original sheet continued across the first branch of the cusped curve, folded back on the second and back again on reaching the first. All these expectations in the hyperbolic case are borne out in general terms by examples calculated in paper E, sections 7 to 9. The detailed form of the solutions is of some interest, but nothing really unexpected appears.

The appearance of similar discontinuities in the elliptic case was not really expected. However, paper E, section 6 shows that the usual argument, why this cusped singular curve, the "limiting line", cannot occur in the elliptic case, does not in any way rule out an isolated singularity. In section 14, this is shown to occur in an illustrative elliptic case. The Riemann surface on which the solution exists is then an ordinary two-sheeted Riemann surface with a winding-point of order 2 at the singularity.

Paper F shows that such a singularity occurs after a finite time for a wide class of elliptic initial-value problems. These are concerned with the development of a wave group whose amplitude distribution is symmetrical about its maximum. This property is predicted to remain true at later times, but energy is redistributed within the group so that the maximum amplitude increases. Its rate of increase, and also the spatial gradient of wave number, have become infinite when the singularity appears. A physical interpretation of this process is given (see especially paper F, section 8).

What happens beyond the critical time? Paper E showed that "weak" solutions could actually be found for these problems (with symmetrical initial conditions); solutions which conserved number of wave crests but allowed a certain amount of dissipation of energy by some unspecified means: in these solutions the group remains symmetrical as it moves on. However, paper F pointed out that although experiments of FEIR demonstrate the maintenance of symmetry up to a critical time close to that predicted, a complete breakdown of symmetry occurs immediately beyond that time.

One possible hypothesis deserving further investigation is that a different kind of weak solution of an asymmetric character may appear, with energy conserved but without conservation of number of wave crests. HOWE, in paper G, has appropriately christened such discontinuities "phase jumps". He studies them for a problem where a fairly precise comparison with experiment may be possible. This is the problem of stationary deep-water waves on a uniform stream, analyzed in a preliminary fashion in paper F, section 3.

HOWE, in an earlier paper, had obtained solutions to this problem in the case when the waves are generated by the stream flowing past a wavy wall of particular shape. He used the numerical method developed by GARABEDIAN for solving elliptic equations with Cauchy initial data. He made no assumption restricting the amplitude of the waves studied. He observed the same tendency for a manyvalued solution to appear as was found in papers E and F.

In paper G he showed how a weak solution involving a "phase jump" but exactly satisfying conservation of energy (as well as and momentum) could be found in this case. This is an intriguing prediction which will be investigated experimentally and, no doubt, also in other ways. Until it is confirmed or otherwise, we can only say that Whitham's method appears to give valuable results up to the time when it predicts a breakdown of continuous solutions, and that the predicted breakdown does correspond to a marked change in character of the phenomenon observed, but that it remains uncertain whether that change in character can be predicted in detail.

Non Strictly Hyperbolic Operators

J. Leray

The theory contained in the last three papers [2−4], hereafter reprinted, deals with a type of system that is so special that the properties of the strictly hyperbolic operators can be applied. That type of system occurs in Relativity theory.

The theory of strictly hyperbolic operators has to consider functions and distributions belonging to Sobolev spaces. But, as the first paper [1], hereafter reprinted, shows, by a counter example due to De Giorgi, the theory of the not strictly hyperbolic systems has to make use of data and unknowns belonging to a perfectly well defined Gevrey's class, which in one extreme case is the class of the holomorph functions and which, in the other extreme case, has to be replaced by Sobolev spaces.

That necessary use of Gevrey's class − hence of infinitely differentiable functions − is quite unpleasant for applications of the theory to relativistic magnetohydrodynamics (A. Lichnerowicz and Y. Choquet-Bruhat: see [4]). But what the Relativity theory essentially requires is a well known influence domain and we prove that it does exist.

References

1. Leray, J.: Équations hyperboliques non-strictes: contre-exemples, du type De Giorgi, aux théorèmes d'existence et d'unicité. Math. Annalen 162, 228−236 (1966).
2. −, et Y. Ohya: Systèmes linéaires, hyperboliques non stricts. Colloque sur L'Analyse fonctionnelle. Liège: CBRM 105−144 (1964).
3. −, et L. Waelbroeck: Norme formelle d'une fonction composée. Colloque sur L'Analyse fonctionnelle. Liége: CBRM 145−152 (1964).
4. −, et Y. Ohya: Équations et systèmes non-linéaires, hyperboliques non-stricts. Math. Annalen 170, 167−205 (1967).

Cauchy's Problem

1. Cauchy's problem with holomorphic data can be studied at the characteristic points of the hypersurface S carrying Cauchy's data; (for linear equations, see [4], which improves [3; I]; for non linear equations, see Y. CHOQUET-BRUHAT [1]). In general its solution u is algebroid at those points (but an equation with constant coefficients and constant data on a hyperplane constitutes, from this point of view, an exceptional case!). The solution u can be uniformed by a convenient map.

In the linear case that map is defined by an ordinary differential system of Hamiltonian type; the solution u ramifies on the characteristic hypersurface K tangent to the hypersurface S carrying Cauchy's data; the behavior of u on K can be described by an asymptotic expansion, whose terms are given by integrations along the bicharacteristics issued from the characteristic points of S; those bicharacteristics generate K and satisfy the Hamiltonian system; the invariant differential form of the Hamiltonian system gives the first term of that asymptotic expansion of u on K.

Note. In mechanics, the solutions of an Hamiltonian system constitute trajectories of particles and its invariant differential form defines the density of a mass with conservation law. Thus u behaves on K as the asymptotic solutions [2] of the equation behave on their domain of definition.

2. Cauchy's problems with analytic data having singularities can be studied with precision by applying convenient transforms to the preceding results. Those transforms operate on germs of uniformisable analytic functions but have the formal properties of classical transforms of functions defined on $\mathbb{R}^n$.

Such a transform, having the properties of the classical Laplace transform, gives the expression and the behavior of the elementary solution of a linear hyperbolic operator [3; IV] (the assumption of hyperbolicity could be easily removed, when the dimension of the space is odd).

Such another transform, having the classical properties of the convolution by a Laplace transform, gives the expression of the solution of

Cauchy's problem with analytic data and makes possible the study of the singularities of that solution, which are issued from the singularities of the data [3; V].

References

1. CHOQUET-BRUHAT, Y.: Uniformisation de la solution d'un problème de Cauchy non linéaire à données holomorphes. Bull. Soc. Math. France **94**, 25 – 48 (1966).
2. LAX, P., and M. J. LIGHTHILL: see their lectures herewith.
3. LERAY, J.: Problème de Cauchy: (I) Uniformisation de la solution du problème linéaire analytique de Cauchy près de la variété qui porte les données de Cauchy. Bull. Soc. Math. France **85**, 389 – 429 (1957); – (II) La solution unitaire d'un opérateur différentiel linéaire. Bull. Soc. Math. France **86**, 75 – 96 (1958); – (III) Le calcul différentiel et intégral sur une variété analytique complexe. Bull. Soc. Math. France **87**, 81 – 180 (1959); – (IV) Un prolongement de la transformation de Laplace qui transforme la solution unitaire d'un opérateur hyperbolique en sa solution élémentaire. Bull. Soc. Math. France **90**, 39 – 156 (1969); – (V) Données analytiques non holomorphes. To be published. – Un aperçu des méthodes qu'emploie (V) se trouve dans: Le problème de Cauchy pour une équation linéaire à coefficients polynômiaux. C. R. Acad. Sc. Paris **242**, 953 – 959 (1956).
4. GÅRDING, L., T. KOTAKE et J. LERAY: Uniformisation et développement asymptotique de la solution du problème de Cauchy linéaire, à données holomorphes; analogie avec la théorie des ondes asymptotiques et approchées. Bull. Soc. Math. France **92**, 263 – 361 (1964).

On Feynman's Integrals

J. LERAY

The following result confirms an hypothesis made in theoretical physics (see J. LASCOUX and F. PHAM [2]), whose main part was recently proved by D. FOTIADI [1], (namely the existence of the algebraic singular support).

Theorem

Let P be a finite set of polynomials in
$$x = (x_1, \dots, x_l) \in X = \mathbb{R}^l;$$
let them depend algebraicly on a parameter
$$t \in T = \mathbb{C}^m;$$
denote by $p(t, x)$ the value of $p \in P$ at (t, x). Assume there are a set of numbers
$$a_p \in \mathbb{C} \quad \text{(where } p \in P)$$

and a domain D_T of the real part of T such that:
$$p(t, x) \geq 0 \quad \text{for all} \quad t \in D_T \quad \text{and all} \quad x \in X;$$
$$\int_X \prod_{p \in \varphi} |p(t, x)|^{\operatorname{Re} a_p} d x_1 \wedge \dots \wedge d x_l$$

is a bounded function of $t \in D_T$.

Then the integral
$$F(t) = \int_X [p(t, x)]^{a_p} d x_1 \wedge \dots \wedge d x_l$$

defines on D_T the germ of a function of t which belongs to the Nilsson class Nils(T).

The Nilsson Class

Nils(T) is the class of the analytic functions F of $t \in T$ having the three properties:

1. There is an algebraic hypersurface of T, called the singular support of F and denoted by $\mathrm{Ss}[F]$, such that F is holomorph on the simply connected covering space of the complement of $\mathrm{Ss}[F]$ in T; (in other words, F has an *algebraic singular support*); hence F is a multivalued function of t;

2. There is a finite number of branches of $F(t)$ such that any other branch is a linear combination (with constant coefficients) of them; (in other words, F is *ζ-fuchsian*);

3. F has a *slow growth* along lines or algebraic paths of real dimension 1; (see N. NILSSON [5] for the precise definition).

Proof

The proof of the preceding theorem [4] results from a fundamental theorem due to N. NILSSON [5], from variants of his definition of slow growth and from properties of the restriction to a subspace of T of a function belonging to Nils(T).

Note. An algebraic construction of $Ss[F]$ can be given by means of [3] and of the properties of the restriction [4]. But till now this construction is not very explicit for cases as special as Feynman's one: each $p(t,x)$, which is of degree 2, depends only on the image x_p of x by a map:

$$X \ni x \mapsto x_p \in X_p = \mathbb{C}^4$$

References

1. FOTIADI, D.: Thesis (to be published in Journal de math. pures et appl.).
2. LASCOUX, J.: Perturbation theory in quantun field theory and homology. —
   PHAM, F.: Landau singuarities in the physical region. Benjamin: Battelle Rencontres, 1967, Lectures in Mathematics and Physics (1968).
3. LERAY, J.: Un Complément au théorème de N. Nilsson sur les intégrales de formes differentielles à support singulier algébrique. Bull. Soc. Math. France **95**, 313–374 (1967).
4. — In preparation.
5. NILSSON, N.: Some growth and ramification properties of certain integrals on algebraic manifolds. Arkiv för Math. **5**, 463–476 (1963–65).
6. — Asymptotic estimates for spectral functions connected with hypoelliptic differential operators. Arkiv för Math. **5**, 527–540 (1963–65).

The Theory of Lacunas

L. GÅRDING

This is a report on joint work with M. F. ATIYAH and R. BOTT. Our aim is to analyze and generalize an article by I. G. PETROVSKY [1].

1. Generalities

Let $a(\xi) = a(\xi_1, \ldots, \xi_n)$ be a polynomial of degree m and let $a(D)$, $D_k = \partial/\partial x_k$, be the corresponding differential operator. The polynomial $a(\xi)$ and the operator $a(D)$ are said to be hyperbolic if $a(D)$ has an elementary solution (necessarily unique) with support in a proper cone K with its vertex at the origin 0. If $\check{K} = K - 0$ is contained in a half-space $x\vartheta = x_1\vartheta_1 + \cdots + x_n\vartheta_n > 0$ and if a is homogeneous, the hyperbolicity is equivalent to the condition that $a(\xi + t\vartheta) = 0$ has m real roots t for every real ξ. Let $\mathrm{Hyp}(\vartheta, m)$ be the set of these a's and let $\mathrm{Hyp}^*(\vartheta, m)$ be the set of $a \in \mathrm{Hyp}(\vartheta, m)$ which are strongly hyperbolic in the sense that the roots are all different when ξ is not proportional to ϑ. When m is not fixed, we write $\mathrm{Hyp}(\vartheta)$ and $\mathrm{Hyp}^*(\vartheta)$. Modulo a complex factor, every $a \in \mathrm{Hyp}(\vartheta)$ is real and we limit ourselves to real polynominals in the sequel. $\mathrm{Hyp}(\vartheta)$ is closed under multiplication and factorization. Let $A \subset \mathbb{C}^n$ be the complex hypersurface $a(\xi) = 0$. Then $\mathrm{Re}\,A$ is generated by the m points $\xi + t\vartheta$ where ξ is real and $a(\xi + t\vartheta) = 0$. It is a conoid with m possibly intersecting sheets which do not contain ϑ. Let $\Gamma = \Gamma(A, \vartheta)$ be the component of $R^n - \mathrm{Re}\,A$ which contains ϑ. It is a convex open cone and $a(D)$ has the fundamental solution

$$E(a, x) = (2\pi)^{-n} \int e^{ix\zeta} a(i\zeta)^{-1} d\zeta \tag{1}$$

where $\mathrm{Im}\,\zeta \in -\Gamma$ is constant and the integral does not depend on $\mathrm{Im}\,\zeta$. By the Paley-Wiener theorem, the convex hull of the support of E is the dual cone $K(A, \vartheta) = \{x; x\Gamma \geq 0\}$ of Γ.

The singularities of E have been studied by HERGLOTZ [2] and PETROVSKY [1] for $a \in \mathrm{Hyp}^*(\vartheta)$ and by LERAY [3] for some strongly hyperbolic operators with analytic coefficients. It turns out that E is holomorphic in $\mathring{K}$ except on the dual $^\circ\mathrm{Re}\,A$ of $\mathrm{Re}\,A$, i.e., the set of $x \in R^n$ such that $\mathrm{Re}\,X$ is tangent to $\mathrm{Re}\,A$. Here X is the complex hyperplane $x\xi = 0, \xi \in \mathbb{C}^n$. Physically, E represents the propagation of δ-shock at the origin and at time $x\vartheta = 0$ in a generalized elastic medium of $n-1$ dimensions, the singularities corresponding to the wave fronts. We are going to study the general case $a \in \mathrm{Hyp}(\vartheta)$. Let $L \subset \mathring{K}$ be a maximal connected open set where E is holomorphic. We say that L is a weak (strong) lacuna if E is the restriction to L of an entire function $(E = 0$ in $L)$. The interior of the light cone is a strong lacuna for the wave operator $D_1^2 - D_2^2 - D_3^2 - D_4^2$ and a weak lacuna for all its powers. PETROVSKY [1] found sufficient and almost necessary conditions for L to be a lacuna which involve the topology of (A, X) when $x \in L$. We are going to generalize his results. In the sequel, we shall use two spaces $Z = \{\xi, \eta, \ldots\} = \mathbb{C}^n$ and $\mathbf{Z} = \{x, y, \ldots\} = \mathbb{C}^n$ coupled by a duality $x\xi = \sum x_k \xi_k$.

2. The Geometry of Hyperbolic Surfaces the Wave Front Surface

Let $H(m)$ be the space of all real homogeneous polynomials $a(\xi) \not\equiv 0$ of degree m. It has been shown (W. NUIJ [4]) that $\mathrm{Hyp}^*(\vartheta, m)$ is open in $H(m)$, that $\mathrm{Hyp}(\vartheta, m)$ is the part of closure of $\mathrm{Hyp}^*(\vartheta, m)$ where $a(\vartheta) \neq 0$ and that both these spaces have two simply connected components characterized by $\mathrm{sgn}\,a(\vartheta)$. When a is a polynomial, let $\Lambda(a)$ be its lineality, i.e., the space of ζ such that $a(\xi + t\zeta) = a(\xi)$ for all ξ and all t. It is a linear space and one has $\Lambda(ab) = \Lambda(a) \cap \Lambda(b)$ so that $\Lambda(a) = \Lambda(A)$ only depends on A. If $a \in \mathrm{Hyp}(\vartheta)$, $\mathrm{Re}\,\Lambda(A)$ is also the edge $\{\zeta; \zeta \in \mathrm{Re}\,Z, \Gamma = \Gamma + \zeta\}$ of $\Gamma = \Gamma(A, \vartheta)$ and $K = K(A, \vartheta)$ is orthogonal to $\mathrm{Re}\,\Lambda(A)$. Special case: a constant $\neq 0$, $\Gamma = \mathrm{Re}\,Z$, $K = 0$. When $\xi \in \mathrm{Re}\,Z$, let $p = p(a, \xi)$ be the minimal multiplicity at $t = 0$ of the polynomial $a(\xi + t\zeta)$ as ζ varies and define the polynomial $a_\xi(\zeta)$ by $a(\xi + t\zeta) = t^p a_\xi(\zeta) + O(t^{p+1})$. In particular, A_ξ is the tangent conoid of A at ξ, transported to the origin. If $a \in \mathrm{Hyp}(\vartheta)$, the multiplicity p is attained if $\zeta \in \Gamma$ and one has $a_\xi \in \mathrm{Hyp}(\vartheta)$ for all real ξ.

Examples. If $a(\xi) \neq 0$, then $p(a, \xi) = 0$ and $a_\xi(\zeta) = a(\xi)$ is constant. If $a(\xi) = 0$, $\mathrm{grad}\,a(\xi) \neq 0$, then $p(a, \xi) = 1$ and $a_\xi(\zeta) = \sum(\partial a(\xi)/\partial \xi_k)\zeta_k$ is linear. If $a \in \mathrm{Hyp}^*(\vartheta)$, all a_ξ with $\xi \neq 0$ are constant or linear. If $a \in \mathrm{Hyp}(\vartheta)$, $\mathrm{Re}\,A$ may have singular points $\xi \neq 0$ at which the degree of a_ξ exceeds 1.

Definition. When $a \in \mathrm{Hyp}(\vartheta)$, $\xi \in \mathrm{Re}\,Z$, put $\Gamma_\xi = \Gamma_\xi(A, \vartheta) = \Gamma(A_\xi, \vartheta)$ and $K_\xi = K_\xi(A, \vartheta) = K(A_\xi, \vartheta)$ and let the wave front surface $W(A, \vartheta)$ be the union of all $K_\xi(A, \vartheta)$ for $0 \neq \xi \in \mathrm{Re}\,Z$.

Examples. If $a(\xi) \neq 0$, then $\Gamma_\xi = \mathrm{Re}\ \mathbb{Z}$ and $K_\xi = 0$. If $a(\xi) = 0$, grad $a(\xi) \neq 0$, then Γ_ξ is the halfspace $a_\xi(\vartheta)\,a_\xi(\zeta) > 0$ and K_ξ is the half-ray $\lambda a_\xi(\vartheta)\,\mathrm{grad}\,a(\xi)$, $\lambda \geq 0$. If $a \in \mathrm{Hyp}^*(\vartheta)$, W is the part of $^\circ\mathrm{Re}\,A$ contained in K, but if $a \in \mathrm{Hyp}(\vartheta)$, W may be smaller (see item 7 in the first set of figures). We shall see that $E(a,x)$ is holomorphic in $\mathring{K}$ outside W.

Since $a_\xi \in \mathrm{Hyp}(\zeta)$ for all $\zeta \in \Gamma(A,\vartheta)$, $\Gamma(A_\zeta,\vartheta) \supset \Gamma(A,\vartheta)$ and, consequently, $K_\xi \subset K = K(A,\vartheta)$. Moreover, the Γ_ξ and K_ξ have the following important continuity property: if $\mathrm{Re}\,\mathbb{Z} \ni \zeta \to 0$ and $\mathrm{Hyp}(\vartheta,m) \ni b \to a \in \mathrm{Hyp}(\vartheta,m)$, then $\Gamma_{\xi+\zeta}(B,\vartheta) \cap \Gamma_\xi \to \Gamma_\xi$ so that also $K_{\xi+\zeta}(B,\vartheta) \cup K_\xi \to K_\xi$ and $W(B,\vartheta) \cup W(A,\vartheta) \to W(A,\vartheta)$. Note that $^\circ\mathrm{Re}\,A \cap K$ is not a continuous function of a.

Re A   W(A,ϑ)

1. m = 2

2. m = 3

3. m = 3

4. m = 4

5. m = 4

6. m = 4

7. m = 6

3. Vector Fields and Cycles

It is possible to modify the chain of integration in (1) if one stays away from A and keeps the real part of the exponent bounded from above. Also, the homogeneity makes it possible to make an explicit radial integration. These, in principle, are the procedures of HERGLOTZ and PETROVSKY who succeeded in writing $E(a, x)$ as an integral of a rational form over a cycle in projective space $\mathbb{Z}^* = \dot{\mathbb{Z}}/\mathbb{C}$. We shall get the same result in the general case. The chains of integration will be parametrized as $\operatorname{Re}\mathbb{Z} \ni \xi \to \xi - iv(\xi)$ where the $v(\xi) \in \operatorname{Re} Z$ are suitable vector fields. Let U be the set of such fields having the properties that $v(\xi) \in \mathbb{C}^\infty(\operatorname{Re}\dot{\mathbb{Z}})$ and $v(\lambda\xi) = |\lambda| v(\xi)$ when $\lambda \in \dot{R}$. Let $V(A, \vartheta)$ and $V(A, X, \vartheta)$ be the $v \in U$ such that $v(\xi) \in \Gamma_\xi(A, \vartheta)$ and $v(\xi) \in \Gamma_\xi(A, \vartheta) \bigcap \operatorname{Re} X$ respectively and, in addition, $a(\xi \pm itv(\xi)) \neq 0$ when $0 < t \le 1$.

Example. If $\eta \in \Gamma$, $v(\xi) = |\xi| \eta \in V(A, \vartheta)$; if also $\eta x = 0$, then $v \in V(A, X, \vartheta)$. Hence, if $x \bar{\in} \pm K$ then $V(A, X, \vartheta)$ contains an element with constant direction. By virtue of the continuity properties of the Γ_ξ, $V(A, \vartheta)$ is not empty and, by the definition of $W = W(A, \vartheta)$, $V(A, X, \vartheta)$ is not empty when $x \bar{\in} \pm W$. Both $V(A, \vartheta)$ and $V(A, X, \vartheta)$ constitute one homotopy class. The essential reason for this is the convexity of the Γ_ξ.

Our vector fields give rise to cycles in projective space $\mathbb{Z}^* = \dot{\mathbb{Z}}/\mathbb{C}$. Let us put $\mathbb{Z}^+ = \dot{\mathbb{Z}}/R^+$ and let B^+ and B^* be the images of $B \subset \dot{\mathbb{Z}}$ in $\mathbb{Z}^+$ and $\mathbb{Z}^*$ respectively. Let $\tau_j(\xi)$ be the right cofactor of $d\xi_j$ in $\tau(\xi) = d\xi_1 \wedge \ldots \wedge d\xi_n$ and put $\omega(\xi) = \sum \xi_j \tau_j(\xi)$. This differential form will serve to orient our cycles. The corresponding form $\omega_x(\xi)$ on X is defined by $\omega(\xi) = d(x\xi) \wedge \omega_x(\xi) + O(x\xi)$. In our generalization of the HERGLOTZ-PETROVSKY-LERAY formulas we shall employ the following homology classes, used by LERAY when $a \in \operatorname{Hyp}^*(\vartheta)$.

Definition. Let $a \in \operatorname{Hyp}(\vartheta)$, $x \bar{\in} \pm W(A, \vartheta)$. When $v \in V(A, X, \vartheta)$, put

$$\alpha = \alpha_v = 2^{-1}\{\xi - iv(\xi); \xi \in \operatorname{Re}\dot{\mathbb{Z}}\}$$

and let $\bar{\alpha}$ be its complex conjugate. Orient α^+, $\bar{\alpha}^+$ by $x\xi\omega(\xi) > 0$, making them $(n-1)$ - chains of $\mathbb{Z}^+ - A^+$ and cycles of $(\mathbb{Z}^+ - A^+, X^+)$. Their boundaries are then

$$\partial\alpha^+, \quad \partial\bar{\alpha}^+ = \{\xi \pm iv(\xi); \xi \in \operatorname{Re}\dot{X}\}^+$$

oriented by $\omega_x(\xi) > 0$. Let

$$\alpha^* = \alpha(A, x, \vartheta)^* \in H_{n-1}(\mathbb{Z}^* - A^*, X^*)$$

be the homology class of α_v^* and

$$\partial\alpha^* = \partial\alpha(A, x, \vartheta)^* \in H_{n-2}(X^* - A^* \bigcap X^*)$$

its boundary.

Note. We use homology with compact supports and cohomology with arbitrary supports.

It is possible to show that α^* contains every α_v^* where $v \in V(A, \vartheta)$ has the property that $v(\xi) \in \Gamma_\xi(A, \vartheta) \bigcap \operatorname{Re} X$ when ξ is conically close to $\mathrm{Re}\,X$. The class $\alpha^*(A,y,\vartheta)$ contains cycles varying continuously with y when y is close to $x \stackrel{\varepsilon}{\in} \pm W(A,\vartheta)$. If $a \in \mathrm{Hyp}(\vartheta)$, $x \stackrel{\varepsilon}{\in} \pm W(A,\vartheta)$, $v \in V(A,X,\vartheta)$ then $\alpha^*_v \in \alpha(B,x,\vartheta)^*$ when $b \in \mathrm{Hyp}(\vartheta)$ is sufficiently close to a. The map $\xi \to -\xi$ sometimes reverses orientation. We have

$$\bar{\alpha}^* = (-1)^{n-1}\alpha^*, 2\alpha^* \ni \alpha^+ + (-1)^{n-1}\bar{\alpha}^+$$

and the same for the boundaries. We see here the influence of the parity of n.

Example. $n = 2$, picture of α^*. Crosses indicate A^*, $\mathbb{Z}^*$ is the complex plane.

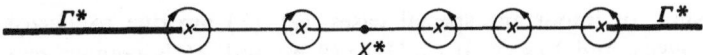

Example. $n = 3$, picture of $2\partial\alpha^*$. Crosses indicate $A^* \cap X^*$ and X^* is the complex plane.

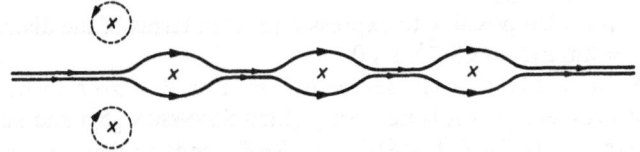

Note that the two dotted circles also represent $2\partial\alpha^*$.

4. Generalization of the Herglotz-Petrovsky-Leray Formulas

Let $a \in \mathrm{Hyp}(\vartheta,m)$, $x \stackrel{\varepsilon}{\in} W(A,\vartheta)$, $x \stackrel{\varepsilon}{\in} -K(A,\vartheta)$. Then $E(a,\cdot)$ is holomorphic close to x and

$$D^v E(a,x) = (2\pi i)^{1-n} \int_{\alpha^*} \chi^*_q(x\,\xi)\, \xi^v a(\xi)^{-1}\, \omega(\xi) \tag{2}$$

if $q = m - n - |v| \geq 0$ and

$$D^v E(a,x) = (2\pi i)^{-n} \int_{t_x \partial \alpha^*} \chi_q(x\,\xi)\, \xi^v a(\xi)^{-1}\, \omega(\xi) \tag{2'}$$

if $q < 0$. Here

$$\chi^*(t) = t^r/r!, r \geq 0, \chi^*(t) = (-1)^{r+1}(-r-1)!\,t^r, r < 0$$

and $\alpha^* = \alpha(A,x,\vartheta)^*$ and t_x is the tube operation mapping every point in $X^* - X^* \cap A^*$ into the boundary of a small complex disk with its center at the point and apart from its center contained in $Z^* - X^* \cup A^*$. If $b \in \mathrm{Hyp}\,(\vartheta, m)$ is sufficiently close to a, these formulas are true for b with α^* replaced by $\alpha_v^* \in \alpha^*$ where $v \in V(A, X, \vartheta)$ is fixed.

Note. The differential forms (2), (2′) are $(n-1)$-forms homogeneous of degree zero and hence closed. They are rational and holomorphic in $Z - A \cup X$. The proof of (2), (2′) runs as follows: replace (1) by $E_s(a, x) = (2\pi)^{-n} \int e^{ix\xi} a(i\zeta)^{-s} d\zeta$ where $s \in \mathbb{C}$. The right side is a distribution in x and entire in s. With $m \, \mathrm{Re}\, s - n < 0$, replace the chain of integration $\mathrm{Im}\, \zeta \in -\Gamma$ by a chain close to $\mathrm{Im}\, \zeta = -v(\xi) \in V(A, X, \vartheta)$, introduce polar coordinates, integrate radially and make an analytical continuation to $s = 1$.

Note. In important special cases, (2), (2′) are due to HERGLOTZ, PETROVSKY and LERAY. If $a \in \mathrm{Hyp}^*\,(\vartheta, m)$ and A^* is regular, one gets PETROVSKY's formulas by taking a residue onto A^* in (2) and two successive residues onto $A^* \cap X^*$ in (2′). For a class of strongly hyperbolic a with analytic coefficients, formulas analogous to (2), (2′) have been obtained by LERAY.

Note. It is also possible to express $E(a, x)$ in terms of the distribution $a_-(\xi)^{-1} = \lim a(\xi - is\,\vartheta)^{-1}$, $s \downarrow 0$.

Note. Let $a \in \mathrm{Hyp}\,(\vartheta, m)$, degree $b < m$. For $P = a - b$ to be hyperbolic with respect to ϑ it is necessary (LEIF SVENSSON [5]) and sufficient that $\sup |b(\xi \pm it\vartheta) a(\xi \pm it\vartheta)|^{-1}$, $\xi \in \mathrm{Re}\,\mathbb{Z}$, tends to zero as $\mathbb{R} \ni t \to \infty$. In this case $P(D) = a(D) - b(D)$ has the fundamental solution

$$E(P, x) = \sum_0^\infty b(D)^k E(a^{k+1}, x) \tag{3}$$

with support in $K(A, \vartheta)$.

5. Support of the Fundamental Solution

The exact determination of the support or singular support of fundamental solutions $E(P, \cdot)$ is beyond our present analysis. In general, however, the wave front surface W should be the entire singular support. To circumvent this difficulty we modify our first definition of a lacuna to a less sweeping one where we only consider components of $Z - W$. It now runs as follows.

By definition, a weak (strong) lacuna of P is a connected component of $\mathrm{Re}\, Z - W(A, \vartheta)$ where $E(P, \cdot)$ is the restriction of an entire function (vanishes). Trivial lacuna: the exterior of $K(A, \vartheta)$. If $P = a \in \mathrm{Hyp}\,(\vartheta, m)$, then $E(a, x)$ is homogeneous of degree $m - n$ so that $x \in L = $ weak lacuna $\Leftrightarrow E$ is a polynomial in L. If $m \prec n$, a weak lacuna is also strong. Stable lacuna $L: E(b, \cdot)$ remains locally a polynomial at every point $x \in L$ if $b \in \text{Hyp}(\vartheta, m)$ is sufficiently close to a. The trivial lacuna is stable. The following result is an immediate consequence of (2), (2').

Theorem. If $a \in \text{Hyp}(\vartheta, m)$, $x \in K(A, \vartheta) - W(A, \vartheta)$ and if $\partial\alpha(A, x, \vartheta)^* = 0$, then x is in a weak stable lacuna for all powers of a. If, in addition $\alpha(A, x, \vartheta)^* = 0$ and $m \geq n$, x is in strong stable lacuna for a.

Note. We shall see that the first part of the theorem has a converse and that there are no non-trivial stable strong lacunas for a when $m \geq n$.

Note. It follows from (3) that if $P = a - b$ is hyperbolic with respect to ϑ, then the weak lacunas for a that are characterized by $\partial\alpha(A, x, \vartheta)^* = 0$ are also weak lacunas for P.

6. The Topology of Algebraic Hypersurfaces

Let $\theta(\xi)$ be a rational function or differential form on Z. We say that θ is homogeneous of degree $h(\theta) = p$ if $\theta(l(\xi)\xi) = l(\xi)^p \theta(\xi)$ for every linear form $l(\xi)$. It is clear that $h(\theta) = 0$ if and only if θ is a rational form on projective space Z^*. We have the identity

$$(\partial f/\partial \xi_k)\omega(\xi) = -d(f(\xi)\tau_k(\xi;\xi)) \qquad (4)$$

where f is a rational function, $h(f) = 1 - n$ and $\tau_k(\xi;\xi) = \sum \xi_j \tau_{kj}(\xi)$, τ_{jk} being the right cofactor of $d\xi_j \wedge d\xi_k$ in $\tau(\xi) = d\xi_1 \wedge \ldots \wedge d\xi_n$. Note that every $(n-1)$-form θ with $h(\theta) = 0$ is closed and has the form $\theta(\xi) = g(\xi)\omega(\xi)$ where g is a rational function such that $h(g) = -n$.

Let $a(\xi) \not\equiv 0$ be a homogeneous polynomial, let $A: a(\xi) = 0$ be the associated hypersurface in Z and A^* its image in Z^*. Let $G^p = G^p(Z^*, A^*)$ be the space of rational p-forms on Z^*, holomorphic in $Z^* - A^*$. By a well-known theorem by GROTHENDIECK [6] which generalizes an earlier result by ATIYAH and HODGE we have

Theorem. $G^{n-1}(Z^*, A^*)$ generates $H^{n-1}(Z^* - A^*)$.

Let $a_0, \ldots, a_q$ be the irreducible factors of a. We say that $\theta \in G^{n-1}(Z^*, A^*)$ has singularity $\leq (p_0, \ldots, p_q)$ if $a_0^{p_0} \ldots a_q^{p_q} \theta$ is holomorphic on $\dot{Z}$. If A^* is globally simple, i. e., if $A_0^*, \ldots, A_q^*$ are non-singular and meet in general position, there is a more precise result: the $\theta \in G^{n-1}$ of singularity $\leq (\infty, 1, \ldots, 1)$ generate $H^{n-1}(Z^* - A^*)$. It is also possible to require that all the θ's be divisible by a fixed polynominal g that does not belong to the ideal generated by $a_1, \ldots, a_q$. The proof uses the preceding theorem, the formula (4) and simple computations.

When $a \in \text{Hyp}(\vartheta)$, let Γ^* be the image in Z^* of the closure of $\Gamma(A, \vartheta)$ in $\dot{Z}$. If $x \in K(A, \vartheta) - W(A, \vartheta)$, it is possible to show that

$\Gamma^* \in H_{n-1}(\mathbb{Z}^* - X^*, A^*)$ meets $\alpha(A, x, \vartheta)^* \in H_{n-1}(\mathbb{Z}^* - A^*, X^*)$ exactly once (see the first figure, p. 15). Hence, in this case

$$\alpha(A, x, \vartheta)^* \neq 0. \tag{5}$$

Finally, let us note that the tube operation

$$t_x\colon H_{n-2}(X^* - X^* \cap A^*) \to H_{n-1}(\mathbb{Z}^* - X^* \cup A^*)$$

is injective (LERAY, [3]).

7. Necessary Conditions for Lacunas

Theorem. Let $a \in \mathrm{Hyp}(\vartheta)$ and let $x \in K(A, \vartheta) - W(A, \vartheta)$ be in a weak lacuna for all powers of a. Then $\partial\alpha(A, x, \vartheta)^* = 0$ and the lacuna is stable. If A^* is globally simple, we have the same result when x belongs to a weak lacuna for a alone.

Proof. The differential forms that appear in (2′) with a replaced by a^k, $k = 1, 2, \ldots$, span $H^{n-1}(\mathbb{Z}^* - A^* \cup X^*)$ so that $t_x \partial\alpha^* = 0$ and hence also $\partial\alpha^* = 0$. Open problem: is this condition also necessary in the general case? A positive answer would imply that all weak lacunas are stable.

If $x \in K(A, \vartheta) - W(A, \vartheta)$ and if $\partial\alpha(A, x, \vartheta)^* = 0$, then $\alpha^* = \alpha(A, x, \vartheta)^* \in H_{n-1}(\mathbb{Z}^* - A^*)$ and if, in addition, $m \geq n$, then (2) holds with $v = 0$. Suppose that x belongs to a stable strong lacuna for a. Then, differentiating with respect to the coefficients of a, one gets $\int_{\alpha^*} \theta = 0$ for all $\theta \in G^{n-1}(\mathbb{Z}^*, A^*)$ so that $\alpha^* = 0$. But this is impossible in view of (5). The same reasoning goes through using only a subset of perturbations of a. Problem: show that the use of perturbations is not necessary. A positive answer would imply that if $m \geq n$, the exterior of K is the only strong lacuna.

Note. There are analogous results for systems.

8. References

1. PETROVSKY, I. G.: On the Diffusion of Waves and Lacunas for Hyperbolic Equations. Mat. Sbornik (Rec. Math.), N. S. **17**, 289–368 (1945).
2. HERGLOTZ, G.: Über die Integration Linearer, Partieller Differentialgleichungen mit Konstanten Koeffizienten. I (Anwendung Abelscher Integrale). II and III (Anwendung Fourierscher Integrale). Leipzig: Ber. Sächs. Akad. Wiss., Math.-Phys. Kl. **78**, 93–126 (1926); **80**, 6–114 (1928).

3. LERAY, J.: Un Prolongement de la Transformation de Laplace ... (Problème de Cauchy. IV). Bull. Soc. Math. de France **90**, 39–156 (1962).
4. NUIJ, W.: To be published in Math. Scand.
5. SVENSON, LEIF: To be published in Ark. f. Mat.
6. GROTHENDIECK, A.: On the de Rham cohomology of algebraic varieties. Publ. IHES **29**, 351—359 (1966).
7. ATIYAH, MICHAEL F., and HODGE, WILLIAM V. D.: Integrals of the Second Kind of an Algebraic Variety. Ann. of Math. **62**, 56–91 (1955).

General Relativistic Hydrodynamics

A. H. Taub

The material covered in the lecture given under this title is contained in the papers: "Relativistic Hydrodynamics" reprinted from the American Mathematical Society's *Lectures in Applied Mathematics*, Volume 8 [1], and "Stability of Fluid Motions and Variational Principles", a paper presented at the *International Colloque de Fluid Mechanique et Gravitation in Paris, June 1967* [2]. In the first paper the equations of motion for special relativistic hydrodynamics are derived from a Lorentz invariant formulation of the Boltzmann equation and various properties of the solution of these equations which hold for an arbitrary space-time are discussed. The Einstein field whose source is a fluid mass are described. Three methods of dealing equations for a gravitationed field: (1) The determination of field equations involving the metric alone from the algebraic properties of the fluid stress energy tensor. (2) The supplementation of the field equations by the conservation of particle number. (3) The supplementation of the field equation by an equation of state, that is by the requirement that the pressure p be a function of the energy density W alone, (or by an equivalent statement).

In the second paper it is shown that in case (3) above, the equations of motion

$$T^{\mu\nu}_{;\nu} = 0$$

may be integrated in a comoving coordinate system to give

$$U^{\mu} = e^{-\varphi} \delta^{\mu}_4$$

$$g_{44} = e^{2\varphi}$$

$$g_{4i} = e^{2\varphi} c_i(x_j) \qquad i,j, = 1,2,3$$

$$e^{\varphi} = \frac{\sigma}{W + p/c^2}$$

where

$$\frac{d\sigma}{\sigma} = \frac{dW}{W + p/c^2} .$$

---

[1] This paper is reprinted on p. 370.
[2] This paper is reproduced on p. 24.

It further follows that

$$(\sigma U^\mu)_{;\mu} = 0.$$

It is also shown that the field equation may be derived from the variational principle

$$\delta \int (R + 2K p) \sqrt{-g} \; d^4 X = 0,$$

where p is regarded as a function of g_{44} via the above equation.

Stability of Fluid Motions and
Variational Principles *

A. H. Taub

1. Introduction

It is the purpose of this paper to show that the equations satisfied by the difference of two "nearly equal" solutions of the Einstein field equations are derivable from a variational principle and indicate how this principle may be used to study the time dependence of the difference. The source of the gravitational fields will be assumed to be a perfect fluid obeying an equation of state. That is, the pressure p of the fluid will be assumed to be a function of only the energy density w. It will be further assumed that the perturbations in the fluid motions will be adiabatic. That is if the entropy for one solution is S and that for the nearby solution is $S + e S'$, then $S' = 0$. It will not be assumed that $S = $ constant for either solution although this condition does provide one of the possible equations of state that the fluid is required to obey.

We are considering the situation that arises when the metric tensor $g_{\mu\nu}$ of space time which is required to satisfy the Einstein field equations

$$G^{\mu\nu} = R^{\mu\nu} - \tfrac{1}{2} g^{\mu\nu} R = -\kappa T^{\mu\nu} = -\kappa \{(w + p) U^\mu U^\nu - p g^{\mu\nu}\} \quad (1.1)$$

is a function not only of the coordinate x^μ of space-time but also of a parameter e. Thus

$$g_{\mu\nu} = g_{\mu\nu}(x; e).$$

We are then concerned with the equations satisfied by

$$g'_{\mu\nu}(x) = \left(\frac{d g_{\mu\nu}}{d e}\right)_{e=0}. \quad (1.2)$$

* This work was supported in part by the United States Atomic Energy Commission under contract number AT(11−1)−34, Project Agreement No. 125. It was performed while the author was on sabbatical leave from the University of California, Berkeley and in residence at the Collège de France, Paris.

The $g'_{\mu\nu}$ is the quantity which represents the difference of two "nearly equal" solutions of the Einstein field equations.

Of course, the pressure p, the energy density w and the four velocity vector field U^μ which enter into the right hand side of eq. (1.1) must also be considered as functions of e. We will then have to consider the quantities p', w', U'^μ which are defined in an analogous fashion. For example

$$p' = \left(\frac{dp}{de}\right)_{e=0}. \tag{1.3}$$

Since

$$U^\mu U^\nu g_{\mu\nu} = 1 \tag{1.4}$$

it is evident that there will be relations between the primed quantities describing the fluid motion and the primed gravitational fields. One of our objectives will be to obtain these relations in explicit form. We shall do this by formulating eq. (1.1) and their immediate consequences

$$T^{\mu\nu}_{;\nu} = 0 \tag{1.5}$$

in comoving coordinates and thus removing the dependent variables U^μ from the problem.

That the equations

$$G'^{\mu\nu} = -\kappa T'^{\mu\nu} \tag{1.6}$$

are derivable from a variational principle should be expected from the fact that eq. (1.1) are themselves the Euler equations for a variational principle (cf [1] and [2]), based on the integral

$$I = \int (R - 2\kappa \mathscr{L}) \sqrt{-g}\, d^4 x.$$

Thus eqs. (1.1) insure that in the expansion

$$I(e) = I(0) + e\, I'(0) + \frac{e^2}{2} I''(0)$$

the quantity $I'(0)$ vanishes for arbitrary $g'_{\mu\nu}$ and corresponding primed variables for the other dependent variables in the problems.

The variational problem

$$\delta I''(0) = 0$$

the so called second variation problem or the Jacobi problem should have as its Euler equations, eqs. (1.6) or equivalent equations. We shall not use this method for deriving our results for it involves using too many sets of dependent variables. Instead we shall first discuss eqs. (1.5) in comoving coordinates and show how the thermodynamic variables are determined from the components of the metric tensor in case the fluid obeys an equation of state. It will then be possible to write eqs. (1.1) in terms of the components of the metric tensor alone. Eqs. (1.6) will then be expressible in terms of the $g_{\mu\nu}$ and the $g'_{\mu\nu}$. The associated variational principle will then be readily discernable.

2. The Equations of Motion

When $T^{\mu\nu}$ is given by the last of eqs. (1.1) it follows that eqs. (1.5) are equivalent to

$$(w + p) U^{\nu}_{;\nu} + w_{;\nu} U^{\nu} = 0 \tag{2.1}$$

and

$$(w + p) U^{\mu}_{;\nu} U^{\nu} = p_{;\nu} (g^{\mu\nu} - U^{\mu} U^{\nu}), \tag{2.2}$$

where the semicolon denotes the covariant derivative and the comma the ordinary derivative. The energy density w is related to the mass density ρ and the specific internal energy ε by the equation

$$w = \rho (c^2 + \varepsilon) \tag{2.3}$$

where c is the special relativity theory velocity of light.

In order to obtain a determinate problem eqs. (2.1) and (2.2) must be supplemented by an additional equation. We take this to be a relation of the form

$$p = p(w). \tag{2.4}$$

In case the conservation of mass obtains, that is in case

$$(\rho U^{\mu})_{;\mu} = 0. \tag{2.5}$$

Eq. (2.4) is equivalent to the statement that

$$S = \text{constant}$$

where S the entropy and the temperature Θ are defined by the equation

$$\Theta \, dS = d\varepsilon + p \, d\left(\frac{1}{\rho}\right). \tag{2.6}$$

When eq. (2.4) obtains, there exists a function σ such that

$$\frac{\sigma_{,\mu}}{\sigma} = \frac{w_{,\mu}}{w + p}. \tag{2.7}$$

and eq. (2.1) is equivalent to

$$(\sigma U^{\mu})_{;\mu} = 0. \tag{2.8}$$

Comoving coordinates are those in which the four velocity vector field has only one non-vanishing component, the time or x^4 component. In view of eq. (1.4) we then must have

$$U^\mu = \frac{1}{\sqrt{g_{44}}}\, \delta_4^\mu \tag{2.9}$$

in co-moving coordinates. Eq. (2.8) then becomes

$$\left(\frac{\sqrt{-g}\,\sigma}{\sqrt{g_{44}}}\right)_{,4} = 0\,.$$

Hence

$$\sqrt{-g}\,\sigma = \sqrt{g_{44}}\, f(x^i) \tag{2.10}$$

where f is an arbitrary function of x^1, x^2 and x^3 but independent of x^4.

In the comoving coordinate system, eq. (2.2) may be shown to be equivalent to the equations.

$$\left(\frac{p,\nu}{(w+p)} + \frac{1}{2g_{44}}\, g_{44,\nu}\right)(g^{\mu\nu} - U^\mu U^\nu) = g^{\mu\nu}\left(\frac{g_{4\nu}}{g_{44}}\right)_{,4}\,.$$

In deriving this equation the covariant derivative is written out in terms of the ordinary derivatives and the Christoffel symbols and these symbols are in turn expressed in terms of the $g_{\mu\nu}$. These equations may in turn be written as

$$\left(\frac{p,i}{w+p} + \frac{1}{2g_{44}}\, g_{44,i}\right) = \frac{1}{\sqrt{g_{44}}}\left(\frac{p,_4}{w+p}\, V_i + V_{i,4}\right) \tag{2.11}$$

where

$$V_i = \frac{g_{i4}}{\sqrt{g_{44}}}\,. \tag{2.12}$$

When eq. (2.7) holds, that is when an equation of state exists eq. (2.11) may be written as

$$\left(\frac{(w+p)\sqrt{g_{44}}}{\sigma}\right)_{,i} = \left(\frac{w+p}{\sigma}\, V_i\right)_{,4}\,. \tag{2.13}$$

The integrability conditions of these equations then lead to the condition that

$$\left(\frac{w+p}{\sigma}\, V_i\right)_{,j} - \left(\frac{w+p}{\sigma}\, V_j\right)_{,i} = F_{ij}(x^i)$$

where

$$F_{ij,k} + F_{jk,i} + F_{ki,j} = 0\,.$$

That is, we must have

$$V_i = \frac{\sigma}{w + p} C_i(x^j) + \frac{\sigma}{w + p} \varphi_{,i} \qquad (2.14)$$

where φ may be function of the x^i and x^4. Hence the solution of eq. (2.13) is given by

$$\frac{(w + p)}{\sigma} \sqrt{g_{44}} = \varphi_{,4} + k(x^4).$$

It is no restriction to take

$$\varphi = \text{constant}$$
$$k = 1$$

for if these conditions are not satisfied we may make coordinate transformation

$$\bar{x}^4 = \bar{x}^4(x^i, x^4)$$
$$\bar{x}^i = x^i$$

where

$$d\bar{x}^4 = (\varphi_{,4} + k(x^4)) dx^4 + \varphi_{,i} dx^i.$$

The $\bar{x}^\mu$ coordinate system is also a co-moving one and in it

$$\frac{\bar{g}_{i4}}{\sqrt{\bar{g}_{44}}} = \frac{\sigma}{w + p} C_i(x^j).$$

Hence in a co-moving coordinate system

$$V_i = \frac{\sigma}{w + p} C_i(x^j) \qquad (2.15)$$

and

$$(w + p) \sqrt{g_{44}} = \sigma. \qquad (2.16)$$

The function $f(x^i)$ and the vector field $C_i(x^i)$ may be said to characterize a particular solution of the equations of motion in co-moving coordinates. These functions are to be determined from the initial conditions of the problem. If these conditions in turn depend on a parameter e, so will the solutions. The vector field V_i is simply related to the vorticity vector v^μ and hence the rotation as may be seen by evaluating

$$v^\mu = \frac{1}{\sqrt{-g}} \varepsilon^{\mu\nu\sigma\tau} U_\nu U_{\sigma,\tau} \qquad (2.17)$$

in the co-moving coordinate system.

3. The Variational Principle for the Field Equation

We now consider the quantities entering into eq. (2.16) as functions of e and differentiate that equation with respect to this variable and set e equal to zero. We obtain

$$\frac{w' + p'}{w + p} + \frac{1}{2}\frac{g'_{44}}{g_{44}} = \frac{\sigma'}{\sigma}.$$

But from eq. (2.7) it follows that

$$\frac{\sigma'}{\sigma} = \frac{w'}{w + p}.$$

Hence we must have

$$p' = -\frac{(w + p)g'_{44}}{2g_{44}}.\tag{3.1}$$

In general we may consider p to be a function of w and S, the entropy. Hence

$$p' = \left(\frac{\partial p}{\partial w}\right)_s w' + \left(\frac{\partial p}{\partial S}\right)S'.$$

This may be written as

$$p' = \alpha^2 w' + \left(\frac{\partial p}{\partial S}\right)S'$$

where α is the ratio of the velocity of sound to the special theory of relativity velocity of light. We impose the condition that $S' = 0$. Hence

$$p' = \alpha^2 w'.\tag{3.2}$$

The effect of condition $S' = 0$ is to enable us to express w' in terms of w and p' alone without any recourse to the details of a particular class of solutions of the equations of motion.

In the co-moving coordinate system eqs. (1.1) become

$$G^{44} = -\kappa\frac{(w + p)}{g_{44}} + pg^{44}$$
$$G^{4i} = \kappa pg^{4i}\tag{3.3}$$
$$G^{ij} = \kappa pg^{ij}.$$

In these equations w and p are considered as functions of g_{44} determined by eq. (2.16) and the g_{4i} are given by eq. (2.15). Eq. (2.10) relates σ and $\sqrt{-g}$.

In this coordinate system

$$G^{\mu\nu} g'_{\mu\nu} = -\kappa \frac{(w + p)}{g_{44}} g'_{44} + \kappa p g^{\mu\nu} g'_{\mu\nu}.$$

It follows from eq. (3.1) that we may write this as

$$G^{\mu\nu} g'_{\mu\nu} = 2\kappa p' + \kappa p g^{\mu\nu} g'_{\mu\nu} = \frac{2\kappa}{\sqrt{-g}} (p\sqrt{g})'$$

that is,

$$\sqrt{-g} \, G^{\mu\nu} g'_{\mu\nu} = 2\kappa (p\sqrt{g})'.$$

But aside from a pure divergence term this equation may be written as

$$[\sqrt{-g}(R + 2\kappa p)]' = 0.$$

That is the field equations are derivable from the variational principle

$$\delta \int \sqrt{-g}(R + 2\kappa p) d^4 x = 0 \tag{3.4}$$

in the co-moving coordinate system where p, w, σ are functions of g_{44} and eqs. (2.10), (2.15) and (2.16) hold.

4. The Variation of the Field Equations

It then follows from differentiation of eq. (3.3) that

$$G'^{44} = \frac{\kappa}{g_{44}^2} g'_{44}(w + p) + \kappa p g'^{44} - \frac{\kappa}{g_{44}}(w' + p') + \kappa p' g^{44}. \tag{4.1}$$

In view of eqs. (3.1) and (3.2), this may be written as

$$G'^{44} = \kappa(w + p)\frac{g'_{44}}{2g_{44}^2}\left[3 + \frac{1}{\alpha^2} - g_{44}g^{44}\right] + \kappa p g'^{44} \tag{4.2}$$

$$G'^{4i} = \kappa p g'^{4i} + \kappa p' g^{4i}$$

or

$$G'^{4i} = \kappa p g'^{4i} - \frac{\kappa(w + p)}{2g_{44}} g'_{44} g^{4i}. \tag{4.3}$$

Similarly, we obtain

$$G'^{ij} = \kappa p g'^{ij} - \frac{\kappa(w + p)}{2g_{44}} g'_{44} g^{ij}. \tag{4.4}$$

Let $\delta g'_{\mu\nu}$ be arbitrary functions of the x^μ which will later be identified with variations of the $g_{\mu\nu}$. Then it follows from the above equations that

$$G^{\mu\nu}\delta g'_{\mu\nu} = -\kappa(w+p)\frac{\delta g'_{44}}{g_{44}} + \kappa p g^{\mu\nu}\delta g'_{\mu\nu} \qquad (4.5)$$

and that

$$G'^{\mu\nu}\delta g'_{\mu\nu} = \kappa\frac{(w+p)}{2g_{44}^2}g'_{44}\delta g'_{44}\left(3+\frac{1}{\alpha^2}\right) + \kappa p g'^{\mu\nu}\delta g'_{\mu\nu}$$

$$- \frac{\kappa(w+p)}{2g_{44}}g'_{44}g^{\mu\nu}\delta g'_{\mu\nu}.$$

Hence

$$\left(G'^{\mu\nu} + \frac{1}{2}g^{\sigma\tau}g'_{\sigma\tau}G^{\mu\nu}\right)\delta g'_{\mu\nu} = \frac{\kappa}{2}\frac{(w+p)}{g_{44}^2}g'_{44}\delta g'_{44}\left(3+\frac{1}{\alpha^2}\right)$$

$$+ \kappa p\left(g'^{\mu\nu}\delta g'_{\mu\nu} + \frac{1}{2}g^{\mu\nu}\delta g'_{\mu\nu}g^{\sigma\tau}g'_{\sigma\tau}\right)$$

$$- \frac{\kappa}{2}\frac{(w+p)}{g_{44}}(g'_{44}g^{\mu\nu}\delta g'_{\mu\nu} + \delta g'_{44}g^{\mu\nu}g'_{\mu\nu}).$$

We may write

$$\left(G'^{\mu\nu} + \frac{1}{2}g^{\sigma\tau}g'_{\sigma\tau}G^{\mu\nu}\right)\delta g'_{\mu\nu} = \frac{\kappa}{2}\delta\mathcal{M}(g_{\mu\nu};g'_{\sigma\tau}). \qquad (4.6)$$

Wherein the co-moving coordinate system

$$\mathcal{M} = \frac{1}{2}\frac{(w+p)}{g_{44}^2}\left(3+\frac{1}{\alpha^2}\right)g_{44}'^2 + p\left(g'^{\mu\nu}g'_{\mu\nu} + \frac{1}{2}(g^{\mu\nu}g'_{\mu\nu})^2\right)$$

$$- \frac{(w+p)}{g_{44}}g'_{44}g^{\mu\nu}g'_{\mu\nu}$$

and $\delta\mathcal{M}$ is the quantity obtained by varying the $g'_{\mu\nu}$ in this expression. The functions $g_{\mu\nu}$, p and w are to be kept constant in this variation.

Since in the co-moving coordinate system

$$\frac{g'_{44}}{g_{44}} = g'_{\mu\nu}U^\mu U^\nu$$

we may write

$$\mathcal{M} = (w+p)g'_{\mu\nu}U^\mu U^\nu(\beta^2 g'_{\sigma\tau}U^\sigma U^\tau - (g^{\sigma\tau}-U^\sigma U^\tau)g'_{\sigma\tau})$$

$$+ p\left(g'^{\mu\nu}g'_{\mu\nu} + \frac{1}{2}(g^{\mu\nu}g'_{\mu\nu})^2\right) \qquad (4.7)$$

where

$$\beta^2 = \frac{1}{2}\left(1 + \frac{1}{\alpha^2}\right) \geq 1 \tag{4.8}$$

since for any physical fluid we must have its velocity of sound less than that of light, that is we must have $\alpha^2 \leq 1$.

Since $\mathcal{M}$ is a scalar function, this expression is valid in all coordinate systems. If we define the tensor field

$$k_{\mu\nu} = g'_{\mu\nu} - \frac{1}{2}g_{\mu\nu}g^{\sigma\tau}g'_{\sigma\tau} \tag{4.9}$$

then

$$k = g^{\mu\nu}k_{\mu\nu} = -g^{\sigma\tau}g'_{\sigma\tau}$$

and hence

$$g'_{\mu\nu} = k_{\mu\nu} - \frac{k}{2}g_{\mu\nu} \tag{4.10}$$

$$U^\mu U^\nu g'_{\mu\nu} = K - \frac{k}{2}$$

where

$$K = U^\sigma U^\tau k_{\sigma\tau}. \tag{4.11}$$

Further we have

$$(g^{\sigma\tau} - U^\sigma U^\tau)g'_{\sigma\tau} = -\left(K + \frac{k}{2}\right). \tag{4.12}$$

Eq. (4.7) may then be written as

$$\mathcal{M} = (w + p)\left(K - \frac{k}{2}\right)\left(\beta^2\left(K - \frac{k}{2}\right) + K + \frac{k}{2}\right)$$

$$+ p\left(k^{\sigma\tau}k_{\sigma\tau} - \frac{k^2}{4}\right). \tag{4.13}$$

5. The Variational Principle

It can be shown that the left hand side of eq. (4.6) is given by

$$\left(G'^{\mu\nu} + \frac{1}{2}g^{\sigma\tau}g'_{\sigma\tau}G^{\mu\nu}\right)\delta g'_{\mu\nu} = \frac{1}{\sqrt{-g}}(\sqrt{-g}\,G^{\mu\nu})'\,\delta g'_{\mu\nu}$$

$$= -\frac{1}{2}\delta\mathcal{J}(g_{\mu\nu}; g'_{\sigma\tau})$$

where

$$\mathscr{I} = k_{\alpha\beta}k_{\sigma\tau}\left[2g^{\sigma\alpha}G^{\tau\beta} - G^{\sigma\tau}g^{\alpha\beta} + \frac{R}{2}\left(g^{\sigma\alpha}g^{\tau\beta} - \frac{1}{2}g^{\sigma\tau}g^{\alpha\beta}\right)\right]$$

$$- \frac{1}{2}\left[2k_{;\tau}^{\sigma\alpha}k_{\alpha\beta;\sigma}g^{\beta\tau} - k_{\alpha\beta;\sigma}k_{;\tau}^{\alpha\beta}g^{\sigma\tau} + \frac{1}{2}k_{,\sigma}k_{,\tau}g^{\sigma\tau}\right]. \tag{5.1}$$

Hence eqs. (4.6) are the Euler equations of the variational principle

$$\delta \int \sqrt{-g}(\mathscr{I} + \kappa\mathscr{M})d^4 x = 0 \tag{5.2}$$

where the tensor field to be varied is $k_{\alpha\beta}$ or equivalently $g'_{\alpha\beta}$. The first parenthesis on the right hand side of eq. (5.1) may be written in terms of w, p and U^μ since we are assuming that the field equations are satisfied by the $g_{\mu\nu}$ and these quantities.

It is convenient to write

$$k_{\alpha\beta} = KU_\alpha U_\beta + K_\alpha U_\beta + K_\beta U_\alpha + K_{\alpha\beta} + \frac{1}{3}(k - K)(g_{\alpha\beta} - U_\alpha U_\beta) \tag{5.3}$$

where K has been defined above and

$$K_\alpha = U^\mu k_{\mu\beta}(\delta_\alpha^\beta - U^\beta U^\alpha) \tag{5.4}$$

$$K_{\alpha\beta} = k_{\sigma\tau}(\delta_\alpha^\sigma - U^\sigma U_\alpha)(\delta_\beta^\tau - U^\tau U_\beta) - \frac{1}{3}(k - K)(g_{\alpha\beta} - U_\alpha U_\beta) \tag{5.5}$$

and hence

$$K_{\alpha\beta}g^{\alpha\beta} = 0. \tag{5.6}$$

Then if we define

$$\mathscr{A} = k_{\alpha\beta}k_{\sigma\tau}\left[2g^{\sigma\alpha}G^{\tau\beta} - G^{\sigma\tau}g^{\alpha\beta} + \frac{R}{2}\left(g^{\sigma\alpha}g^{\tau\beta} - \frac{1}{2}g^{\sigma\tau}g^{\alpha\beta}\right)\right]$$

we have

$$\mathscr{A} = \kappa(w + p)\left[\frac{1}{2}K^{\alpha\beta}K^{\alpha\beta} - K^\alpha K_\alpha - \left(K - \frac{k}{2}\right)^2 - \frac{2}{3}kK + \frac{1}{6}k^2\right]$$

and

$$\mathscr{A} + \kappa\mathscr{M} = \kappa(w + p)\left[(\beta^2 - 1)\left(K - \frac{k}{2}\right)^2 + \frac{1}{2}K^{\alpha\beta}K_{\alpha\beta} - K^\alpha K_\alpha \right.$$

$$\left. + \left(K - \frac{1}{3}k\right) - \frac{7}{36}k^2\right] + p\left(k^{\sigma\tau}k_{\sigma\tau} - \frac{k^2}{4}\right). \tag{5.7}$$

Some of the variables $k_{\alpha\beta}$ may be eliminated from the variational problem for we know their values. In the co-moving coordinate system

3 Hyperbolic Equations and Waves

eqs. (2.10) and (2.15) hold both for the metric $g_{\mu\nu}(x)$ and $g_{\mu\nu} + e g'_{\mu\nu}(x)$. That is, we may differentiate these equations with respect to e and set $e = 0$. From eqs. (2.10) we then obtain

$$\frac{1}{2} g^{\sigma\tau} g'_{\sigma\tau} = \beta^2 \frac{g'_{44}}{g_{44}} + \frac{f'}{f} = \beta^2 U^\sigma U^\tau g'_{\sigma\tau} + \frac{f'}{f}.$$

The last form of this equation holds in an arbitrary coordinate system. If we substitute for $g'_{\sigma\tau}$ in terms of $k_{\sigma\tau}$ we obtain

$$\beta^2 K - \frac{1}{2}(\beta^2 - 1)k = F \tag{5.8}$$

where we have written F for $-f'/f$ and this function is characterized by the fact that

$$U^\sigma F_{,\sigma} = 0 \tag{5.9}$$

in an arbitrary coordinate system. The function F represents the perturbation in the initial conditions for the solution of the equation of conservation of σ.

Eqs. (2.15) are treated similarly. On differentiating them with respect to e and setting $e = 0$ we obtain by use of eq. (2.16)

$$g'_{4i} = \frac{g'_{44}}{g_{44}} g_{4i} + g_{44} C'_i$$

in the co-moving coordinate system. Hence in this coordinate system we have

$$K_4 = 0, \quad K_i = \frac{\sigma}{w + p} \left(\frac{k}{2} C_i + C'_i \right).$$

We may write this equation as

$$K_\alpha = \frac{\sigma}{w + p} \left(\frac{k}{2} C_\alpha + C'_\alpha \right) \tag{5.10}$$

where C_α and C'_α are arbitrary vectors which satisfy the equations

$$C_\alpha W^\alpha = C'_\alpha W^\alpha = 0 \tag{5.11}$$

$$C_{\alpha;\sigma} W^\sigma + C_\sigma W^\alpha_{;\sigma} = C'_{\alpha;\sigma} W^\sigma + C'_\sigma W^\sigma_{;\alpha} = 0 \tag{5.12}$$

with

$$W^\sigma = \frac{\sigma}{w + p} U^\sigma.$$

The vectors C_α and C'_α are related to the initial rotation and the perturbed rotation.

6. Stability

We have thus seen that the Einstein field equations for a perfect fluid whose motion is such that an equation of state $p = p(w)$ is satisfied, may be derived from the variational principle described by eq. (3.4). In this equation p is a function of the components of the metric tensor. In the co-moving coordinate system this functional dependence is given by eq. (2.16) and then eqs. (2.15) and (2.10) relate g_{4i} and $\sqrt{g}$ to the initial conditions of the motion. We have further seen that the equations satisfied by small perturbations of the solutions of the field equations are also derivable from a variational principle if the perturbed fluid motion is an adiabatic one, that is has the same entropy as the original one. The latter variational principle is described by eq. (5.2) and the $g'_{\mu\nu}$ (or the $k_{\mu\nu}$) are restricted by equations derived from (2.10), (2.15) and (2.16).

The variational principle (5.2) is the one usually called the second variational problem associated with the variational principle (3.4). In fact if we call

$$I(e) = \int \sqrt{g}(R + 2\kappa p)d^4 x \qquad (6.1)$$

when the integrand is evaluated for the family of metric tensors $g_{\mu\nu}(x;e)$, then

$$I(e) = I(0) + \frac{e}{2}\int \sqrt{-g}(G^{\mu\nu} + \kappa T^{\mu\nu})(g'_{\mu\nu} + e g''_{\mu\nu})d^4 x$$
$$+ \frac{e^2}{2}\int \sqrt{-g}(\mathscr{I} + \kappa\mathscr{M})d^4 x + \cdots . \qquad (6.2)$$

The value $e = 0$ will be an extreme value for $I(e)$ if the field eqs. (1.1) are satisfied by $g_{\mu\nu}(x,0)$. These values of $g_{\mu\nu}$ will correspond to a minimum for $I(e)$ if $\mathscr{I} + \kappa\mathscr{M}$ which is a function of $g_{\mu\nu}(x,0)$ and $k_{\mu\nu}$, is positive for all allowed values of the $k_{\mu\nu}$. Such $g_{\mu\nu}(x,0)$ may be called stable solutions of the field equations.

For any solutions $g_{\mu\nu}(x,0)$ of the field equations the $k_{\mu\nu}(x)$ which satisfy eqs. (4.1) to (4.3), the Euler equations of the variational principle (5.2), the coefficient of e^2 in the above expansion takes on an extreme value.

The relation between the above definition of stability and the one based on the behavior of the time dependence of the variables $k_{\mu\nu}(x^\mu)$ may be readily discussed in the case where the $g_{\mu\nu}(x,0)$ are static stationary in the co-moving coordinate system. In that case the coefficients of the differential operators entering into eqs. (4.1) to (4.3) are independent of x^4 as are the coefficients of the $k_{\mu\nu}$ and $k_{\mu\nu;\rho}$ in the expressions for $\mathscr{I}$ and $\mathscr{M}$. We may then assume that

$$k_{\mu\nu}(x^\mu) = e^{i\sigma x^4}\mathring{k}_{\mu\nu}(x^i)$$

3*

and seek the conditions that ensure that the value of σ determined from the equations of motion is real. The relation between this condition and the condition which ensures that the integral of $\mathscr{J} + \kappa \mathscr{M}$ be positive will be reported on in a subsequent paper.

This paper will conclude with some additional remarks about the variational principle given by eq. (5.2). The integrand occurring in that equation depends on the tensor field $k_{\mu\nu}$ and may be looked upon as the Lagrangean function for this field. As such it is the unquantized Lagrangean for a field theory of a particle of spin 2. Associated with the $k_{\mu\nu}$ field there is a symmetric energy tensor $E^{\mu\nu}$ satisfying the conservation equations

$$E^{\mu\nu}_{;\nu} = 0. \tag{6.3}$$

The tensor $E^{\mu\nu}$ may be derived from the Lagrangean function by varying the $g_{\mu\nu}$ in it. That part of $E^{\mu\nu}$ arising from such a variation of $\mathscr{J}$ in the Lagrangean is the tensor I derived earlier [3] (cf [4]) and called the approximate stress energy tensor of the gravitational field. It is more appropriately called the gravitational part of the energy tensor of a perturbation of the gravitational field.

Eq. (6.3) may be used to study the behavior of various aspects of the perturbed motion of the fluid and its corresponding gravitational field. For example if the unperturbed motion is static then the vector field,

$$W^\mu = \frac{\sigma}{w + p} U^\mu = \delta^\mu_4$$

in the co-moving coordinate system is a Killing vector field. That is

$$W_{\mu;\nu} + W_{\nu;\mu} = 0.$$

In that case it follows from eq. (6.3) that

$$(W_\mu E^{\mu\nu})_{;\nu} = 0$$

and hence the closed hypersurface integral

$$\int W_\mu E^{\mu\nu} N_\mu d\Sigma = 0.$$

By choosing the closed hypersurface to contain the hypersurfaces $t = t_1$ and $t = t_2$ we may relate the perturbed energy contained in the fluid at these times with the flow of momentum from the boundaries of the fluid. This topic too will be discussed in a subsequent report.

References

1. TAUB, A. H.: General relativistic variational principle for perfect fluids. Phys. Rev. **94**, 1468 – 1470 (1954).
2. – Singular Hypersurfaces in General Relativity. Illinois Journal of Mathematics **1**, 370 – 388 (1957).
3. – Approximate Stress Energy Tensor for Gravitational Fields. Journ. Math. Phys. **2**, 787 – 793 (1961).
4. CAPELLA, A.: Thèse. Université de Paris (1963).

Pulsars

F. J. DYSON

Guide to the Literature

The subject of pulsars is so new and so rapidly developing that no summary of the literature can have more than ephemeral value. Since the original announcement in February 1968 by the Cambridge group, it has become traditional to publish most of the new discoveries in letters to "Nature". A fairly complete coverage of published work can be achieved by scanning the 1968 issues of "Nature", "Science", and "Astrophysical Journal Letters".

A more detailed account of the work done up to May 1968, including much unpublished material, will be contained in the Proceedings of a Pulsar Conference held at the Goddard Institute for Space Studies in New York, sponsored by Goddard and Yeshiva University. The proceedings will be published by Benjamin sometime about Fall of 1968. A good survey article based on the same conference is published by S. P. Maran and A. G. W. Cameron in "Physics To-day", August 1968, 41–49. This article has a complete bibliography of the subject up to May 1968.

My own work on the seismic detection of gravitational waves from pulsars is unpublished. Similar work is reported by J. Weber, Phys. Rev. Letters, 21, 395 (August 5, 1968).

A Difference Method for Plane Problems in Dynamic Viscoplasticity

J. Bejda*

I. Introduction

The equations governing the small dynamic deformations of an isotropic elastic/viscoplastic solid under condition of plane strain can be written in the following form:

$$u_t - p_x - q_x - \tau_y = 0, \tag{1a}$$

$$v_t - p_y + q_y - \tau_x = 0, \tag{1b}$$

$$-u_x - v_y + \frac{\Gamma^4}{3\Gamma^2 - 4} p_t + \frac{\Gamma^2(2-\Gamma^2)}{3\Gamma^2 - 4} r_t = -\frac{2}{3} < D > (p-r), \tag{1c}$$

$$-u_x + v_y + \Gamma^2 q_t = -2q < D > (r-p), \tag{1d}$$

$$\Gamma^2 \frac{2-\Gamma^2}{3\Gamma^2 - 4} p_t + \frac{\Gamma^2 - 1}{3\Gamma^2 - 4} r_t = -\frac{2}{3} < D > (r-p), \tag{1e}$$

$$-u_y - v_x + \Gamma^2 \tau_t = -2\tau < D >, \tag{1f}$$

where

$$< D > = \gamma \, \Phi \left(\frac{\sqrt{J_2}}{k_0} - 1 \right) \frac{1}{\sqrt{J_2}},$$

$$\sqrt{J_2} = \frac{1}{\sqrt{3}} \sqrt{(p-r)^2 + 3(q^2 + \tau^2)}.$$

The dimensionless velocities u and v, time t, dimensionless Cartesian coordinates x and y, plastic limit in simple shear k_0, viscosity coefficient γ, and dimensionless stresses p, q, τ, and r are determined in the following way:

---
* Institute of Basic Technical Research, Polish Academy of Sciences, Warsaw, Poland.

$$u = \frac{\hat{u}}{c_1}, \quad v = \frac{\hat{v}}{c_1}, \quad t = \frac{\hat{t}c_1}{b}, \quad x = \frac{\hat{x}}{b},$$

$$y = \frac{\hat{y}}{b}, \quad k_0 = \frac{\hat{k}_0}{\rho c_1^2}, \quad \gamma = \frac{b}{c_1}\hat{\gamma}, \quad \Gamma = \frac{c_1}{c_2}. \tag{2a}$$

$$p = \frac{1}{2}\frac{\sigma_x + \sigma_y}{\rho c_1^2}, \quad q = \frac{1}{2}\frac{\sigma_x - \sigma_y}{\rho c_1^2}, \quad \tau = \frac{\tau_{xy}}{\rho c_1^2}, \quad r = \frac{\sigma_z}{\rho c_1^2}. \tag{2b}$$

Here a hat "^" denotes the corresponding dimensional quantity, c_1 is the velocity of propagation of elastic, dilatational waves, c_2 the velocity of elastic, shear waves, Γ their ratio and b an arbitrary characteristic length. The speeds c_1 and c_2 are given by

$$c_1 = \left[\frac{3K + 4\mu}{3\rho}\right]^{1/2}, \quad c_2 = \left[\frac{\mu}{\rho}\right]^{1/2} \tag{3}$$

where K, μ, ρ denote Lamé constants and density, respectively.

Eqs. (1a–f) may be written in a more compact matrix form:

$$L[W] = A^t W_t + A^x W_x + A^y W_y - B = 0, \tag{4}$$

where the vectors W, B and tensors A^t, A^x, A^y are:

$$W = \begin{bmatrix} u \\ v \\ p \\ q \\ r \\ \tau \end{bmatrix}; \quad B = \begin{bmatrix} 0 \\ 0 \\ -\frac{2}{3} < D > (p-r) \\ -2 < D > q \\ -\frac{2}{3} < D > (r-p) \\ -2 < D > \tau \end{bmatrix}; \tag{5}$$

$$A^t = \begin{bmatrix} 1 & 0 & 0 & 0 & 0 & 0 \\ 0 & 1 & 0 & 0 & 0 & 0 \\ 0 & 0 & \dfrac{\Gamma^4}{3\Gamma^2 - 4} & 0 & \dfrac{\Gamma^2(2-\Gamma^2)}{3\Gamma^2 - 4} & 0 \\ 0 & 0 & 0 & \Gamma^2 & 0 & 0 \\ 0 & 0 & \dfrac{\Gamma^2(2-\Gamma^2)}{3\Gamma^2 - 4} & 0 & \dfrac{\Gamma^2(\Gamma^2 - 1)}{3\Gamma^2 - 4} & 0 \\ 0 & 0 & 0 & 0 & 0 & \Gamma^2 \end{bmatrix};$$

$$A^x = \begin{bmatrix} 0 & 0 & -1 & -1 & 0 & 0 \\ 0 & 0 & 0 & 0 & 0 & -1 \\ -1 & 0 & 0 & 0 & 0 & 0 \\ -1 & 0 & 0 & 0 & 0 & 0 \\ 0 & 0 & 0 & 0 & 0 & 0 \\ 0 & -1 & 0 & 0 & 0 & 0 \end{bmatrix};$$

$$A^y = \begin{bmatrix} 0 & 0 & 0 & 0 & 0 & -1 \\ 0 & 0 & -1 & 1 & 0 & 0 \\ 0 & -1 & 0 & 0 & 0 & 0 \\ 0 & 1 & 0 & 0 & 0 & 0 \\ 0 & 0 & 0 & 0 & 0 & 0 \\ -1 & 0 & 0 & 0 & 0 & 0 \end{bmatrix};$$

The matrices A^x and A^y are symmetric; the matrix A^t is symmetric positive definite. Thus, eqs. (4) constitute a symmetric hyperbolic system of partial differential equations.

II. Characteristic Properties of the Governing Equations

The system of eq. (4) is a system of six first-order semilinear hyperbolic partial differential equations in three independent variables with constant coefficients for the terms involving derivatives. A theory of such equations is given, for instance, in [1]. The method of finite differences along bicharacteristics will be used for the solution of eq. (4). Thus the geometry of the characteristic surfaces associated with eq. (4) will now be investigated.

The condition that a surface $\varphi(t, x, y) = \text{const}$, be a characteristic surface of eq. (4) is equivalent to the condition that the determinant of the characteristic matrix A be zero:

$$\text{Det } A = 0, \tag{6}$$

where

$$A = A^t \varphi_t + A^x \varphi_x + A^y \varphi_y. \tag{7}$$

Eq. (6) is equivalent to

$$\left[\varphi_t^2 - (\varphi_x^2 + \varphi_y^2) \right] \left[\varphi_t^2 - \frac{1}{\Gamma^2} (\varphi_x^2 + \varphi_y^2) \right] \varphi_t = 0. \tag{8}$$

The two terms in square brackets in eq. (8) describe the propagation of dilatational and shear waves with dimensionless wave velocities c equal to ± 1 and $\pm 1/\Gamma$ respectively; these correspond to the dimensional velocities c_1 and c_2.

It must be emphasized here that discontinuities are propagated in an elastic/viscoplastic medium with elastic wave velocities. This is explained mathematically by the fact that in eq. (4) the nonlinear term B, describing the plastic and viscous effects, does not enter into the analysis of the characteristic surfaces.

The bicharacteristics of eq. (4) are the generators of the following characteristic cones passing through the point (t_0, x_0, y_0):

$$c^2 (t-t_0)^2 = (x-x_0)^2 + (y-y_0)^2, \quad c = 1, \quad \frac{1}{\Gamma}. \tag{9}$$

It is convenient to introduce the following parametrization of the characteristic cones in terms of the two parameters α and τ:

$$x - x_0 = c\tau \cos \alpha, c = 1, \quad \frac{1}{\Gamma}, \tag{10a}$$

$$y - y_0 = c\tau \sin \alpha, \tag{10b}$$

$$t - t_0 = \tau. \tag{10c}$$

Relations (10) give the desired equations of bicharacteristics as the generators of the characteristic cones. The bicharacteristic strips associated with the bicharacteristic lines (10) are:

$$\varphi_t = c, \quad \varphi_x = -\cos \alpha, \quad \varphi_y = -\sin \alpha. \tag{11}$$

In order to determine the equations along bicharacteristics, the null vectors associated with the system (1) must first be determined. The null vectors $l = [l_1, l_2, l_3, l_4, l_5, l_6]$ are the solutions of the following homogeneous system of equations:

$$lA = 0. \tag{12}$$

Substituting for A from eqs. (7) and (11), solutions of eq. (12) corresponding to the dimensionless velocities $c = 1$ and $c = \frac{1}{\Gamma}$ are:

$$l = [-\Gamma^2 \cos \alpha, \ -\Gamma^2 \sin \alpha, \ \Gamma^2 - 1, \ \cos 2\alpha, \ \Gamma^2 - 2, \ \sin 2\alpha] \tag{13a}$$

for $c = 1$, and

$$l = [\Gamma \sin \alpha, \ -\Gamma \cos \alpha, \ 0, \ -\sin 2\alpha, \ 0, \ \cos 2\alpha] \tag{13b}$$

for $c = 1/\Gamma$.

The desired differential equations along bicharacteristics are obtained from the equation

$$l \cdot L[W] = 0 \tag{14}$$

where the dot denotes the inner product. In eq. (14) the partial derivatives with respect to t, can be eliminated by use of

$$W_t = \frac{dW}{dt} - W_x \frac{dx}{dt} - W_y \frac{dy}{dt} \tag{15}$$

where dW/dt is the total time derivative of W taken along a bicharacteristic and dx/dt, dy/dt are obtained by differentiation of eq. (10). In order to have a backward drawn bicharacteristic in the positive x-direction correspond to $\alpha = 0$ as indicated in Figure 1 we replace α by $\alpha + \pi$.

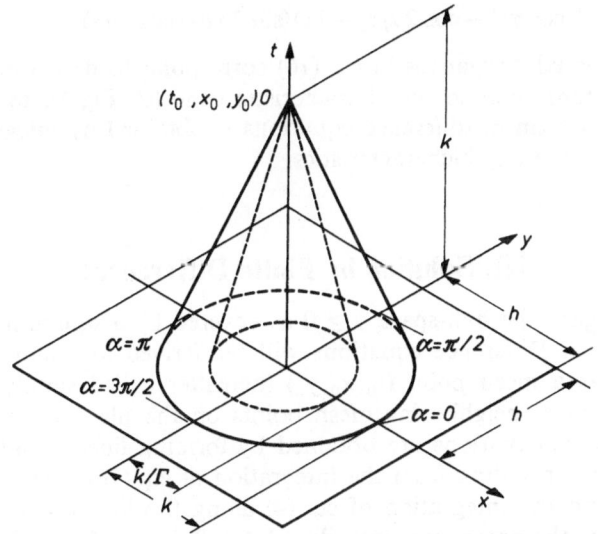

Fig. 1. Characteristic cones for the dynamic elastic, viscoplastic equations

Then, substituting eqs. (13) and (15) in eq. (14) gives

$$\cos \alpha \, du + \sin \alpha \, dv + dp + \cos 2\alpha \, dq + \sin 2\alpha \, d\tau = -S_1(\alpha)\, dt \quad (16\text{a})$$

for $c = 1$.

$$-\Gamma \sin \alpha \, du + \Gamma \cos \alpha \, dv - \Gamma^2 \sin 2\alpha \, dq + \Gamma^2 \cos 2\alpha \, dt = -S_2(\alpha)\, dt \quad (16\text{b})$$

for $c = 1/\Gamma$, where $S_1(\alpha)$ and $S_2(\alpha)$ are

$$
\begin{aligned}
S_1(\alpha) = & \left(-\sin^2 \alpha + \frac{1}{\Gamma^2}(1 - \cos 2\alpha)\right)u_x + \left(\frac{1}{2}\sin 2\alpha - \frac{1}{\Gamma^2}\sin 2\alpha\right)u_y \\
& + (-1 + \cos 2\alpha)q_x \cos \alpha + (1 + \cos 2\alpha)q_y \sin \alpha \\
& + \left(\frac{1}{2} - \frac{1}{\Gamma^2}\right)v_x \sin 2\alpha + \left(-\cos^2 \alpha + \frac{1}{\Gamma^2}(1 + \cos 2\alpha)\right)v_y \\
& + (\sin 2\alpha \cos \alpha - \sin \alpha)\tau_x + (\sin 2\alpha \sin \alpha - \cos \alpha)\tau_y \\
& + \frac{2D}{\Gamma^2}\left(\frac{1}{3}(p - r) + \tau \sin 2\alpha + q \cos 2\alpha\right)
\end{aligned}
\quad (16\text{c})
$$

and

$$S_2(\alpha) = \tfrac{1}{2}\sin 2\alpha(u_x - v_y) - \cos^2 \alpha u_y + \Gamma \sin \alpha p_x$$
$$+ \Gamma(\sin \alpha - \sin 2\alpha \cos \alpha)q_x + \Gamma \sin \alpha(1 + \cos 2\alpha)\tau_y \qquad (16d)$$
$$+ \sin^2 \alpha v_x - \tfrac{1}{2}\sin 2\alpha v_y - \Gamma \cos \alpha p_y + (1 - 2\sin 2\alpha)q_y$$
$$- \Gamma \cos \alpha(1 - \cos 2\alpha)\tau_x - 2D(\sin 2\alpha q - \cos 2\alpha \tau).$$

The spatial derivatives in eq. (16) correspond to derivatives in the direction tangential to the characteristic cone (cf. Fig. 1). In the next section a system of difference equations is obtained by integration of eqs. (16a – b) along bicharacteristics.

III. Solution by Finite Differences

We regard the half-space $y \geq 0$ as covered by a square mesh with mesh size h. Difference equations will be derived for computing the solution at a mesh point (t_0, x_0, y_0) (hereafter called simply 0) from known data at neighboring mesh points on the plane $t = t_0 - k$ (see Fig. 1). These equations are obtained by forming linear combinations of equations resulting from the integration of eq. (16) along bicharacteristics and the integration of eq. (4) along the line $x = x_0$, $y = y_0$. We follow the procedure introduced by BUTLER [2], and used by CLIFTON [3] for the elastic case.

Integrating eq. (16) along the bicharacteristic for which $\alpha = \alpha_i$, from the point 0 to the point $(t_0 - k, x_i, y_i)$ where this bicharacteristic intersects the plane $t = t_0 - k$, gives an equation involving the difference $\Delta W_i = W(t_0, x_0, y_0) - W(t_0 - k, x_i, y_i)$. It is convenient to eliminate ΔW_i by use of the identity

$$\Delta W_i = \delta W + W(t_0 - k, x_0, y_0) - W(t_0 - k, x_i, y_i) \qquad (17a)$$

where

$$\delta W = W(t_0, x_0, y_0) - W(t_0 - k, x_0, y_0), \qquad (17b)$$

When eq. (17) are used, the difference equations resulting from integrating eq. (16) along bicharacteristics for which $\alpha = \alpha_i$ are

$$\cos \alpha_i \delta u + \Gamma \cos \alpha_i \delta v - \Gamma^2 \sin 2\alpha_i \delta q + \Gamma^2 \cos 2\alpha_i \delta \tau$$
$$= -\tfrac{1}{2}k[S_1^0(\alpha_i) + S_1(\alpha_i)_i] - W_1 ; \qquad (18a)$$

and

$$-\Gamma \sin \alpha_i \delta u + \Gamma \cos \alpha_i \delta v - \Gamma^2 \sin 2\alpha_i \delta q + \Gamma^2 \cos 2\alpha_i \delta \tau$$
$$= -\tfrac{k}{2}[S_2(\alpha_i)^0 + S_2(\alpha_i)_i] - W_2(\alpha_i) \qquad (18b)$$

for the exterior and interior cones respectively, where

$$W_1(\alpha_i) = \cos \alpha_i (u_0 - u_i) + \sin \alpha_i (v_0 - v_i) + \cos 2\alpha_i (q_0 - q_i)$$
$$+ \sin 2\alpha_i (\tau_0 - \tau_i) + p_0 - p_i \,,$$
$$W_2(\alpha_i) = \Gamma \sin \alpha_i (u_0 - u_i) + \Gamma \cos \alpha_i (v_0 - v_i) - \Gamma^2 \sin (q_0 - q_i)$$
$$+ \Gamma^2 \cos 2\alpha_i (\tau_0 - \tau_i) \,.$$

In eq. (18) the superscript ^{0} denotes evaluation of the function at the point 0; the subscript $_0$ denotes evaluation of the function at the point $(t_0 - k, x_0, y_0)$; the subscript $_i$ denotes evaluation of the function at the point where the bicharacteristic α_i on the appropriate characteristic cone intersects the plane $t = t_0 - k$.

Six additional difference equations, obtained by integrating eq. (4) along the line $x = x_0$, $y = y_0$, are as follows

$$\delta u = \frac{k}{2} \left[(p_x + q_x + \tau_y)^0 + (p_x + q_x + \tau_y)_0 \right] \tag{19a}$$

$$\delta v = \frac{k}{2} \left[(p_y - q_y + \tau_x)^0 + (p_y - q_y + \tau_x)_0 \right] \tag{19b}$$

$$\delta p = \frac{k}{2} \left[\left(\frac{\Gamma^2 - 1}{\Gamma^2} (u_x + v_y) - \frac{2D}{3\Gamma^2} (p - r) \right)^0 \right.$$
$$\left. + \left(\frac{\Gamma^2 - 1}{\Gamma^2} (u_x + v_y) - \frac{2D}{3\Gamma^2} (p - r) \right)_0 \right] \tag{19c}$$

$$\Gamma^2 \delta q = \frac{k}{2} \left[(u_x - v_y - 2qD)^0 + (u_x - v_y - 2qD)_0 \right] \tag{19d}$$

$$\delta r = \frac{k}{2} \left[\left(\frac{\Gamma^2 - 2}{\Gamma^2} (u_x + v_y) - \frac{4D}{3\Gamma^2} (r - p) \right)^0 \right.$$
$$\left. + \left(\frac{\Gamma^2 - 2}{\Gamma^2} (u_x + v_y) - \frac{4D}{3\Gamma^2} (r - p) \right)_0 \right] \tag{19e}$$

$$\Gamma^2 \delta \tau = \frac{k}{2} \left[(u_y + v_x - 2\tau D)^0 + (u_y + v_x - 2\tau D)_0 \right] \tag{19f}$$

where the o superscripts and subscripts have the same meaning as in eq. (18).

All the terms on the right hand side of eqs. (18) and (19) can be evaluated from data on the plane $t = t_0 - k$ except those terms having a superscript ^{0}. Just as in the elastic case [3], all the terms involving partial derivatives at 0 can be eliminated by forming linear combinations of eq. (19), and the eight equations obtained by writing eq. (18) for

$$\alpha_i = \frac{(i-1)\pi}{2}, \quad (i = 1, 2, 3, 4).$$

In this way we obtain a system of six equations which determine the six unknown increments δu, δv, δp, δq, δr and $\delta \tau$. These equations involve terms, to be evaluated on the plane $t = t_0 - k$, which have the form

$$W(x_0 + ck, y_0) - W(x_0 - ck, y_0) \tag{20a}$$

$$W(x_0 + ck, y_0) + W(x_0 - ck, y_0) - 2W(x_0 y_0) \tag{20b}$$

$$k[W_x(x_0, y_0 + ck) - W_x(x_0 y_0 - ck)] \tag{20c}$$

$$k[W_x(x_0, y_0 + ck) + W_x(x_0, y_0 - ck) - 2W_x(x_0, y_0)] \tag{20d}$$

The expressions (20) can be replaced, respectively, by

$$2ck\,W_x(x_0, y_0) \tag{21a}$$

$$(ck)^2\,W_{xx}(x_0, y_0) \tag{21b}$$

$$2ck^2\,W_{xy}(x_0, y_0) \tag{21c}$$

$$0 \tag{21d}$$

with an accuracy of order k^3. In eqs. (20) and (21) c is equal to 1 for the exterior cones and to $1/\Gamma$ for the interior cones. The same relations hold if we change the roles of x and y in eqs. (20) and (21). Making use of the substitution of expressions (21) for those appearing in (20), the six equations for the six unknown increments become

$$2\delta u = \frac{k^2}{\Gamma^2}\left[(\Gamma^2 - 1)v_{yx} + \Gamma^2 u_{xx} + u_{yy} - 2\left(D\left(\frac{p-r}{3} + q\right)\right)_x - (2D\tau)_y\right]_0$$
$$+ 2k(q_x + p_x + \tau_y)_0\,, \tag{22a}$$

$$2\delta v = \frac{k^2}{\Gamma^2}\left[(\Gamma^2 - 1)u_{xy} + \Gamma^2 v_{yy} + v_{xx} - 2\left(D\left(\frac{p-r}{3} - q\right)\right)_y\right.$$
$$\left. - (2D\tau)_x\right]_0 + 2k(p_y - q_y + \tau_x)_0\,, \tag{22b}$$

$$2\Gamma^2\delta\tau = 2k^2(p_{xy})_0 + 2k(v_x + u_y)_0 + k^2(\tau_{xx} + \tau_{yy})_0$$
$$- \frac{k^3}{\Gamma^2}((D\tau)_{xx} + (D\tau)_{yy})_0\,, \tag{22c}$$

$$2\Gamma^2\delta q = -\frac{k}{2}[(4Dq)^0 + (4Dq)_0] - \frac{k^3}{\Gamma^2}\left[\left(D\left(\frac{p-r}{3} + q\right)\right)_{xx}\right.$$
$$\left. - \left(D\left(\frac{p-r}{3} - q\right)\right)_{yy}\right]_0 + k^2(q_{xx} + q_{yy} + p_{xx} - p_{yy})_0$$
$$+ 2k(u_x - v_y)_0\,; \tag{22d}$$

$$2\frac{\Gamma^2}{\Gamma^2-1}\,\delta p = k^2\big[q_{xx}-q_{yy}+p_{xx}+p_{yy}+2\tau_{xy}\big]_0 + 2k(u_x+v_y)_0$$

$$-\frac{k}{2}\left[\left(\frac{4D(p-r)}{3(\Gamma^2-1)}\right)^0 + \left(\frac{4D(p-r)}{3(\Gamma^2-1)}\right)_0\right]$$

$$-\frac{k^3}{\Gamma^2}\left[\left(D\left(\frac{p-r}{3}+q\right)\right)_{xx} + \left(D\left(\frac{p-r}{3}-q\right)\right)_{yy}\right]_0,$$

(22e)

$$\delta r = \frac{\Gamma^2-2}{\Gamma^2-1}\,\delta p + \frac{k}{3\Gamma^2}\frac{3\Gamma^2-4}{\Gamma^2-1}\big\{[D(p-r)]^0$$

$$+ [D(p-r)]_0\big\}.$$

(22f)

In order to convert eq. (22) to difference equations, the partial derivatives with respect to x and y are replaced by the corresponding centered difference equations. Thus, for example, the second partial derivative with respect to x is replaced by

$$W_{xx}(x_0,y_0) = \frac{1}{h^2}(W(x_0+h,y_0) + W(x_0-h,y_0) - 2W(x_0,y_0)). \quad (23)$$

The final difference equations correspond to a second order accurate difference method. That is, the error in the computed increments $\delta u, \delta v, \ldots, \delta r$ for a single step is $0(k^3)$.

Because stresses at 0 are unknowns, the terms involving superscripts $^\circ$ in eq. (22) are not known initially. As a result, an iterative procedure must be used for solving these equations.

The same basic formulae, eq. (22), are used in both elastic and viscoplastic regions. In elastic regions the coefficient D is equal to zero. The unknown function values on the plane $t = t_0$ are determined from known values at neighboring points on the plane $t = t_0 - k$ and, where applicable, from values given on the boundary. These points may be all in the plastic region, all in the elastic region, or some in each region. The value of the plastic yielding function F is calculated at every mesh point. Depending on whether $F < 0$ or $F \geq 0$, the viscosity coefficient γ is put equal to zero or to γ. In this way, the computations can be carried out without explicitly locating boundaries between elastic and plastic regions.

So far, we have only considered difference equations for interior points of the half-space $y \geq 0$. We now derive appropriate equations for mesh points on the boundary $y = 0$. These equations are obtained by eliminating equations along bicharacteristics for which $\alpha = \frac{3\pi}{2}$ on both exterior and interior cones since these bicharacteristics intersect the plane $t = t_0 - k$ at points outside the region of interest (cf. [2], [3]). Combining eqs. (18) and (19) as for interior points and then eliminating relations along the bicharacteristics corresponding to $\alpha = \dfrac{3\pi}{2}$ leads to the following equations for use at mesh points on the boundary $y = 0$.

$$2\delta u = (22a) + \frac{1}{\Gamma}(22c), \tag{24a}$$

$$2\delta v + \frac{2\Gamma^2}{\Gamma^2 - 1}\delta\rho = (22b) + (22e), \tag{24b}$$

$$-2\delta v + 2\Gamma^2\delta q = -(22b) + (22d), \tag{24c}$$

$$\delta r = (22f). \tag{24d}$$

The terms on the right-hand side indicate the right-hand side of the corresponding equation of eq. (22). Eq. (24) constitute four equations in six unknowns. The remaining two equations come from the boundary conditions on the boundary $y = 0$. The derivatives with respect to y that appear in eq. (24) are replaced by forward differences of sufficient accuracy so that the $0(k^3)$ accuracy is preserved. Thus, the following difference approximations are used.

$2ckW_y(x_0,0)$

$$= 2c\frac{k}{h}\left[2W(x_0,h) - \frac{1}{2}W(x_0,h) - \frac{3}{2}W(x_0,0)\right], \tag{25a}$$

$(ck)^2W_{yy}(x_0,0)$

$$= \left(\frac{ck}{h}\right)^2\left[W(x_0,0) - 2W(x_0,h) + W(x_0,2h)\right], \tag{25b}$$

$2ck^2W_{yx}(x_0,0)$

$$= \left(\frac{ck}{h}\right)^2\left[2W(x_0 + h,h) - \frac{1}{2}W(x_0 + h,2h) - \frac{3}{2}W(x_0 + h,0)\right.$$
$$\left. - 2W(x_0 - h,h) + \frac{1}{2}W(x_0 - h,2h) + \frac{2}{3}W(x_0 - h,0)\right]. \tag{25c}$$

It should be stressed here that in a case when the term B in eq. (1) is equal to zero, the difference schemes obtained in this paper are the same as those given by LAX and WENDROFF [4]. The more particular analysis of the problem under consideration is given in [5], where the propagation of two-dimensional stress waves in strain-rate sensitive elastic/viscoplastic half-space $y \geq 0$, which is loaded on the boundary $y = 0$ by an arbitrary normal pressure $P(x,t)$, is examined. Numerical calculations, when the material is assumed as mild steel, are presented.

The basic difficulty with the method considered above is the stability and convergence of the difference solution. Papers concerning this problem have recently begun to appear in mathematical literature [4–31]; however, for mixed initial and boundary value problems for systems of partial differential equations of the type obtained here, usable stability and convergence criteria are not available. A practical and useful estimation of the convergence of the difference solution for the elastic case was presented by CLIFTON [3], based on energy considerations.

IV. References

1. COURANT, R., and D. HILBERT: Methods of Mathematical Physics, Vol II, Partial Differential Equations. New York: Interscience 1962.
2. BUTLER, D. S.: The Numerical Solution of Hyperbolic Systems of Partial Differential Equations in Three Independent Variables. Proc. Roy. Soc. A 255, 232–252 (1962).
3. CLIFTON, R. J.: A Difference Method for Plane Problems in Dynamic Elasticity. Quarterly of Applied Mathematics 25, 97 (April 1967).
4. LAX, P. D., and B. WENDROFF: Difference Schemes with High Order of Accuracy for Solving Hyperbolic Equations. Comm. Pure Appl. Math. 17, 381–389 (1964).
5. BEJDA, J.: Propagation of Two-dimensional Stress Waves in an Elastic/Visco-Plastic Material. Proceedings of 12th International Congress of Applied Mechanics. Stanford: (August 26–30, 1968), Editors M. Hetényi and W. G. Wincenti, Springer-Verlag 1969.
6. ZHUKOV, A. I.: The Application of the Method of Characteristics to the Numerical Solution of One-dimensional Problems of Gas Dynamics. Trudy, Mathematitcheskogo Instituta im. V. A. Steklova 58 (1960), (in Russian).
7. PROUSE, G.: Sulla Risoluzione del Problema Misto per le Equazioni Iperboliche non Linerari Mediante le Differenze Finite. Ann. di Mat. 46, 313–341 (1958).
8. THOMÉE, V.: A Difference Method for a Mixed Boundary Problem for Symmetric Hyperbolic Systems. Arch. Rat. Mech. Anal. 13, 122–136 (1963).
9. KELLER, H. B., and V. THOMÉE: Unconditionally Stable Difference Methods for Mixed Problems for Quasi Linear Hyperbolic Systems in Two-dimensions. Comm. Pure Appl. Math. 15, 63–73 (1962).
10. DUFF, G. F. D.: Mixed Problems for Linear Systems of First Order Equations. Canad. J. Math. 10, 127–160 (1958).
11. WENDROFF, B.: Well-posed Problems and Stable Difference Operators: I. The Initial Value Problem with Constant Coefficients. SIAM J. Ser. B, Numer. Anal. 5, 71 (1968).
12. GODUNOV, S. K., and V. S. RYABENKII: The Theory of Difference Schemes. Interscience (1964).
13. RYABENKII, V. S.: Necessary and Sufficient Conditions for Good Definition of Boundary Value Problems for Systems of Ordinary Difference Equations. USSR Comp. Math. and Math. Phys. 4 (1), 43–61 (1964).

14. Richtmeyer, R. D., and K. W. Morton: Difference Methods for Initial-Value Problems. 2nd ed., Interscience (1967).
15. Bramble, J. H. (ed.): Numerical Solution of Partial Differential Equations. Academic Press (1966).
16. Kreiss, H. O.: Difference Approximations for the Initial-boundary Value Problems for Hyperbolic Differential Equations. Publ. 17, MRC, U.S. Army, Univ. of Wisconsin. Numerical Solution of Nonlinear Differential Equations. p. 141. Edited by Donald Greenspan, John Wiley and Sons 1966.
17. Sarason, L.: On Hyperbolic Mixed Problems. Arch. Rat. Mech. and Anal. 18, 310−334 (1965).
18. Strang, G.: Weiner-Hopf Difference Equations. J. Math. Mech. 13 (1), 85−96 (1964).
19. Courant, R., E. Isaacson, and Rees: On the Solution of Nonlinear Hyperbolic Differential Equations by Finite Differences. Comm. Pure Appl. Math. 5, 243−255 (1952).
20. Lax, P. D., and R. D. Richtmeyer: Survey of the Stability of Linear Finite Difference Equations. Comm. Pure Appl. Math. 9, 267−297 (1956).
21.−, and L. Nirenberg: On Stability for Difference Schemes. Comm. Pure Appl. Math. 14 (4), 473−492 (1966).
22. Morton, K. W., and S. Schechter: On the Stability of Finite Difference Matrices. SIAM J. Ser. B, Numer. Anal. 2 (2), 119−128 (1965).
23. Stetter, H. J.: Some Numerical Experiments Concerning Maximum Bounds for the Solutions of Partial Difference Equations. UCLA (August 1, 1963).
24. Lax, P. D.: Stability of Linear and Nonlinear Difference Schemes. In: J. Bramble (ed.): Numerical Solution of Partial Differential Equations. Academic Press (1966).
25. Osher, S.: Systems of Difference Equations with General Homogeneous Boundary Conditions. BNL 11857 (Brookhaven Ntl. Lab. Appl. Math. Dept.), (1968).
26. − Stability of Difference Approximations of Dissipative Type for Mixed Initial-boundary Value Problems. I. BNL 12414 (Brookhaven Ntl. Lab. Appl. Math. Dept.), (1968).
27. Hersh, R.: On the Theory of Difference Schemes for Mixed Initial-boundary Value Problems. (TR No. 149, Department of Mathematics and Statistics, The University of New Mexico, November 1967), to appear in SIAM J. Ser. B, Numer. Anal. (1968).
28. Glimm, J.: Solution in the Large for Nonlinear Hyperbolic Systems of Equations. Comm. Pure Appl. Math. 18, 697−715 (1965).
29. Conway, E., and J. Smoller: Global Solution of the Cauchy Problem for Quasi-linear First Order Equations in Several Space Variables. Comm. Pure Appl. Math. 19, 95−105 (1966).
30. Kreiss, H. O.: Stability Theory for Difference Approximations for Mixed Initial Boundary Value Problems, I. Math. Comp. (to appear).
31. −On Difference Approximations of the Dissipative Type for Hyperbolic Differential Equations. Comm. Pure and Appl. Math. 17, 335−353 (1964).

A Survey of Hyperbolic Systems of Conservation Laws in Two Dependent Variables

J. A. SMOLLER

I. Introduction

This paper is concerned with quasi-linear systems of equations in conservation form of the type

$$u_t + f(u,v)_x = 0, \quad v_t + g(u,v)_x = 0, \tag{1}$$

where f and g are smooth functions of two real variables u and v, and u and v are functions of x and t, $-\infty < x < \infty$, $t \geq 0$. We assume that the system (1) is hyperbolic, i.e., if we denote by F, the vector function (f,g), then $dF(u,v)$ has real and distinct eigenvalues $\lambda_1(u,v) < \lambda_2(u,v)$, for all values of the argument (u,v) in question. We consider the Cauchy problem for the system (1). Thus we seek a solution of (1) defined in $t \geq 0$, and having the given prescribed initial data

$$(u(0,x), v(0,x)) = (u_0(x), v_0(x)), \quad -\infty < x < \infty \tag{2}$$

where we assume that u_0 and v_0 are (at least) bounded and measurable functions. Systems of the type (1) arise in several physical situations. For example, they can be used to describe the equations of gas dynamics in a single space variable, with constant entropy.

It is well-known that classical solutions of $(1)-(2)$ cannot, in general exist for all $t \geq 0$, (see [10]). This is due to the non-linearity of f and g, which forces the characteristic speeds to depend on u and v. Therefore we seek a weak solution of $(1)-(2)$ which is a bounded and measurable function (u,v) which satisfies

$$\iint_{t \geq 0} [u\varphi_t + f(u,v)\varphi_x] \, dx \, dt + \int_{t=0} [u_0 \varphi] \, dx = 0$$

$$\iint_{t \geq 0} [v\varphi_t + g(u,v)\varphi_x] \, dx \, dt + \int_{t=0} [v_0 \varphi] \, dx = 0$$

for every smooth function $\varphi = \varphi(t,x)$ having compact support. The next five sections of this paper are devoted to giving the status of this problem up to the present time. The last section contains some conjectures and open problems.

II. Existence Theorems for Data Close to a Constant

For initial data sufficiently close to a constant, the existence theory is in fairly good shape, due to the work of LAX, [10], GLIMM, [3][1], and GLIMM and LAX, [4]. In [10], the Riemann problem for (1) is considered; that is, find a solution of (1) in $t \geq 0$ where the initial data is of the form

$$(u(0, x), v(0, x)) = \begin{cases} (u_l, v_l), & x < 0 \\ (u_r, v_r), & x > 0, \end{cases} \tag{3}$$

and (u_l, v_l), (u_r, v_r) are constant vectors. In order to solve this problem, LAX introduced the notion of genuine nonlinearity of the system (1), which is defined by requiring that $d\lambda_i(r_i(u, v)) \neq 0$, for $i = 1, 2$, where r_i is the right eigenvector of dF for the eigenvalue λ_i. Under the assumption that the system is genuinely non-linear, LAX showed that the Riemann problem (1)−(3) has a solution if (u_l, v_l) and (u_r, v_r) are sufficiently close. The solution is a function of x/t, consisting of (at most) three constant states separated by centered rarefaction waves or shock waves.

In [3], GLIMM gave a far reaching extension of the above result, by considering bounded and measurable data having variation sufficiently close to a constant. The ingenious method of GLIMM was to use a finite difference scheme, where the difference approximations involve a random choice. Let us briefly sketch this method. Suppose that the difference approximation (u, v) has been determined at the points $(x_0 - \Delta x, t_0)$ and $(x_0 + \Delta x, t_0)$. Then consider the Riemann problem for (1) with initial data

$$(u(0, x), v(0, x)) = \begin{cases} (u(x_0 - \Delta x, t_0), v(x_0 - \Delta x, t_0)), & x < x_0 \\ (u(x_0 + \Delta x, t_0), v(x_0 + \Delta x, t_0)), & x > x_0, \end{cases}$$

on the line $t = t_0$. By the results of [10], this problem has a solution $(\bar{u}, \bar{v})$ in the strip $t_0 \leq t \leq t_0 + \Delta t$. (This only holds if the initial points are sufficiently close − indeed, it is one of the main tasks of this paper to prove that the approximate solutions do not grow too large, so that the difference scheme can be defined for all x and t.) GLIMM then puts $(u(x_0, t_0 + \Delta t), v(x_0, t_0 + \Delta t) = (\bar{u}(a, t_0 + \Delta t), \bar{v}(a, t_0 + \Delta t)$, where a is

---

[1] These first two papers also contain existence theorem for hyperbolic systems of conservation laws in $n \geq 2$ dependent variables.

chosen at random in the interval $\lfloor x_0 - \Delta x, x_0 + \Delta x \rfloor$. He then proves that for almost all choices of the random point, a subsequence of these difference approximations converges to a solution of $(1)-(2)$ as the mesh lengths tend to zero. The proof depends first on a delicate and deep analysis of the interaction of approaching waves, and second, in order to keep control of the variation of the approximate solutions, GLIMM introduces some very interesting non-linear functionals.

In the paper [4], the results of GLIMM are extended in order to study the decay of the solution as $t \rightarrow +\infty$. The main result is that if the initial data is constant outside of a compact set, then the total variation decays like $0(t^{-1/2})$, while if the initial data is periodic, the total variation decays like $0(t^{-1})$. The proof of these results requires an extremely complex and detailed study of the difference approximations, and requires some difficult estimates for the rate of interaction and cancellation of shock and rarefaction waves.

III. Another Class of Initial Data

In a different direction, the papers [5-7, 13], are concerned with genuinely non-linear systems (1) in an open subset U of E^2, where the initial data (2) is not necessarily close to a constant, but is, however, restricted in a different way.

We assume that the initial data satisfies an order condition, condition (C), which was introduced in [5], and which was used in [6, 7] to also obtain existence theorems. This condition states that the initial data gives rise to a specific form of Riemann problem. Thus, if we let $(u_i, v_i) = (u_0(x_i), v_0(x_i))$, $i = 1, 2$, for $x_1 < x_2$, then the Riemann problem for (1) with initial data

$$(u(0,x), v(0,x)) = \begin{cases} (u_1, v_1), x < 0 \\ (u_2, v_2), x > 0 \end{cases}$$

is solvable by a shock wave of one characteristic family, and a rarefaction wave of the opposite characteristic family, the directions of these families being fixed, independent of the x_i's. In addition to these assumptions on the initial data, we assume that $f_v g_u > 0$ in U, a condition which implies that the system (1) is hyperbolic, and we also assume a condition which implies that the system (1) locally satisfies a shock interaction condition. This condition was introduced in [4], characterized in [7], and states that the interaction of two shock waves of the same characteristic family produces a shock wave of this family plus a rarefaction wave of the opposite family. It was shown in [7], that this condition holds locally in U if $l_i d^2 F(r_j, r_j) > 0$, $i, j = 1, 2$, $i \neq j$, where l_i is the left eigenvector of dF for the eigenvalue λ_i, normalized by $l_i r_i > 0$, and where we have also normalized r_i by $d\lambda_i(r_i) > 0$. It was further shown in [7], that $d\lambda_i(r_i) = l_i d^2 F(r_i, r_i)$, so that we can write our assumptions of the system (1) being genuinely non-linear and satisfying the shock interaction conditions in U as $l_i d^2 F(r_j, r_j) > 0$, $i, j = 1, 2$, (with the above normalizations for l_i and r_i, $i = 1, 2$). Finally, we assume that all shock waves under consideration arise from the intersection of two characteristics from precisely one characteristic family. This condition, which we call condition (L), holds for sufficiently weak shocks in U and also holds for arbitrary shocks in U under several natural conditions. For example (L) holds if $\lambda_2 \geq 0 \geq \lambda_1$ in U. We remark that condition (L) is part of the definition of a shock wave, given in [10]. It is also noteworthy to remark that these conditions hold for the systems considered in [5, 6]. (In fact we take the system of gas dynamics $u_t - v_x = 0$, $v_t + f(u)_x = 0$, where $f' < 0$, $f'' > 0$, as a test system, and we use this system as a check as to whether the hypotheses we introduce are valid for this system. If they fail to hold for this system, we reject them).

We can now state our main existence theorem.

Theorem 1. Let U be an open set in E^2 and suppose that $l_i d^2 F(r_j, r_j) > 0$, $i, j = 1, 2, f_v g_u > 0$ and condition (L) holds in U. If the bounded and measurable initial data satisfies condition (C) in U, then the problem $(1)-(2)$ has a global solution.

The solution is obtained as a limit of solutions of (1) having step data, and in the next sections we shall outline the proof.

IV. Properties of the Shock Curves

In order to prove theorem 1, we need to make a careful study of the geometry of the shock and rarefaction wave curves. This is facilitated by the fact that our hypotheses on f and g force certain convexity and monotonicity conditions on these curves. The rarefaction wave curves are easy. All the information we need to know about them is contained in the following theorem.

Theorem 2. Let the system (1) satisfy $f_v g_u > 0$ and $l_i d^2 F(r_j, r_j) > 0$, $i, j = 1, 2$ in U. Then through any point $P_0 \in U$, there exists an increasing (decreasing) convex downward (upward) curve, $v = w(u; P_0)$ ($v = w_1(u; P_0)$) defined for all $u \geq u_0$ ($u \geq u_0$) for which $(u, w(u; P_0)) \in U$ ($(u, w_1(u; P_0)) \in U$), having the property that each point on this curve is a state which can be connected to P_0 by a rarefaction wave of the first (second) characteristic family.

We turn now to the shock curves. These curves satisfy the Rankine-Hugoniot conditions for shocks which means that they satisfy

$$(u - u_0)(g(u, v) - g(u_0, v_0)) = (v - v_0)(f(u, v) - f(u_0, v_0))$$

where (u_0, v_0) is the state on the left of the shock and (u, v) is the state on the right. In addition, we define an i-shock, $i = 1, 2$, to be a discontinuity $x = x(t)$ of the function $(u(t, x), v(t, x))$ which satisfies the Rankine-Hugoniot condition and the inequality

$$\lambda_i(u(x + 0, t), v(x + 0, t)) < x(t) < \lambda_i(u(x - 0, t), v(x - 0, t))$$

where $\dot{x}(t)$ is the shock speed. This definition differs from the standard definition ([10]) since we do not require the condition $\lambda_i(u(x - 0, t), v(x - 0, t)) < \dot{x}(t)$ to hold for 2-shocks, nor the condition $\lambda_2(u(x + 0, t), v(x + 0, t)) > \dot{x}(t)$ to hold for 1-shocks. With this definition, we can state the following analogue of theorem 2 for shock curves.

Theorem 3. Under the assumptions of theorem 2, there exists through each point $P_0 \in U$, a decreasing (increasing), convex upwards (downwards) curve $v = s_2(u; P_0)$ ($v = s_1(u; P_0)$), having the property that each point on this curve is a state which can be connected to P_0 by a shock wave of the second (first) characteristic family.

In the rest of this section we shall fix our attention only on the 2-shock curves $v = s(u; P_0)$ (we are dropping the subscripts) and the 1-rarefaction wave curves $v = w(u; P_0)$ where $P_0 = (u_0, v_0)$. Analogous results will be seen to hold in the curves of the opposite families.

An important consequence of our assumptions is that we can show that a 2-shock curve originating on $v = s(u; P_0)$ has the slope of its tangent vector greater than the slope of the original shock curve at this point. Thus the new shock curve starts out "breaking to the right" (toward increasing u), and this implies that the shock interaction condition holds locally all along $v = s(u; P_0)$, and not just in a neighborhood of P_0. Note too that since $v = s(u; P_0)$ is decreasing and convex upward, it is defined for all $u \geq u_0$ for which $(u, s(u; P_0)) \in U$.

We shall next obtain a certain ordering principle which is the main tool in our proof of theorem 1. We first need some notation. Let $P_0 = (u_0, v_0) \in U$ and define

$$C(P_0) = \{(u, v) \in U : s(u; P_0) \leq v \leq w(u; P_0), u \geq u_0\}$$

Theorem 4 (Ordering Principle).

Let the assumptions of theorem 2 hold in U. Suppose too that condition (L) holds in U. If $P_1 \in C(P_0)$, then $C(P_1) \subseteq C(P_0)$. Conversely, if for all $P_0, P_1, P_1 \in C(P_0)$ implies that $C(P_1) \subseteq C(P_0)$, then condition (L) holds in U.

Let us now consider some interesting consequences of this theorem. First, we remark that if P_1 is not on $v = w(u; P_0)$, then $C(P_0) \cap C(P_1)$ contains at most one point (it is empty unless P_1 is on $v = s(u; P_0)$). Hence if P_1 is on $v = s(u; P_0)$, then the shock curve $v = s(u; P_1)$ meets $v = s(u; P_0)$ only at P_1. Thus the intersection of two shocks of the same family always produces a shock of that family plus a rarefaction wave of the opposite family; that is the shock interaction condition is valid globally in U. (This result is proved independently of assumption (L).) This is a mathematical proof of the theorem in gas dynamics, which states that if a gas is taken from state one to state two by a shock wave, and thence to state three by a second shock wave, it cannot go from state one to state three by a single shock wave (see [9], p. 321).

We can also show that the ordering principle will hold if we have a uniqueness theorem for Riemann problems in $C(P_0)$. Thus in view of theorem 4, we see that if the problem $(1)-(2)$ has a unique solution, then condition (L) holds. An old conjecture is that condition (L) should also be sufficient to isolate a unique solution, (see VII, where we consider this question).

V. Sketch of the Existence Theorem

We have now all of the machinery necessary to prove theorem 1. We first approximate the data by a sequence of step functions which also satisfy condition (C). We consider the Cauchy problem for (1) with initial data consisting of such a step function. The solution, for small t, consists of a finite number of constant states, separated by centered rarefaction waves of one characteristic family and shock waves of the other characteristic family. For sufficiently large t, these shocks and rarefaction waves interact with each other. The ordering principle allows us to solve such interactions, so that we can obtain an exact solution of these Cauchy problems. Furthermore, the ordering principle enables us to obtain estimates on the bounds and variation of these approximating solutions. We can thus show that these approximating solutions are uniformly bounded and have uniform bounded variation in the sense of TONELLI-CESARI, with respect to two independent (not necessarily orthogonal) directions[2]. It then follows that this sequence is compact in the topology of L_1-convergence on compact sets, so that a subsequence converges to a solution of $(1)-(2)$. We can also show that the solution is continuous almost everywhere in $t \geq 0$, in spite of

---

[2] Tonelli-Cesari variation was introduced in [1], and characterized in [8]. A function is in this class if and only if its distribution derivatives are measures. The technique of using this variation for Cauchy problems was introduced in [2].

the fact that the Tonelli-Cesari class contains functions which are discontinuous almost everywhere.

VI. Arbitary Riemann Problems

The ordering principle also enables us to obtain a solution to (1)−(2), under the same hypothesis as above, using the Glimm difference scheme. Thus the Glimm scheme can be used to prove existence theorems where the initial data is not necessarily close to a constant. It appears probable therefore, that the Glimm scheme, or some variant of it, will yield an existence theorem in the general case of bounded and measurable initial data. In order to obtain such a general theorem, it is first necessary to extend the results of [10], to solve Riemann problems in the large; i.e., to solve Riemann problems where the distance between (u_l, v_l) and (u_r, v_r) is arbitrary. This is the next topic that we wish to consider (see [13] for details).

From theorems 2 and 3, we see that the $u - v$ plane is divided into 4 regions by the two shock wave and two rarefaction wave curves. Thus, for each point $U_l = (u_l, v_l) \in E^2$, we define

$$S_1 = \{(u,v): v = s_1(u; U_l), u \le u_l\}$$
$$S_2 = \{(u,v): v = s_2(u; U_l), u \ge u_l\}$$
$$W_1 = \{(u,v): v = w_1(u; U_l), u \ge u_l\}$$
$$W_2 = \{(u,v): v = w_2(u; U_l), u \le u_l\}$$

where for $i = 1, 2$, the $S_i's$ and $W_i's$ are shock and rarefaction wave curves of the ith characteristic family. These sets divide the $u - v$ plane into the four regions illustrated in the figure below.

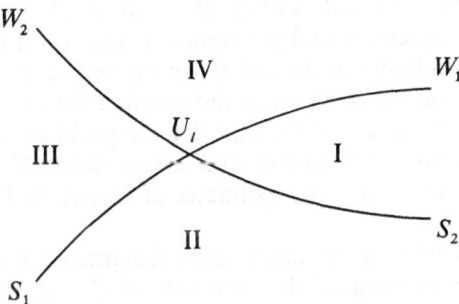

To solve the Riemann problem in the large can be restated as given $U_l = (u_l, v_l) \in E^2$, solve (1)−(3) for any $U_r = (u_r, v_r)$. By extending the arguments in [7], we can prove, somewhat surprisingly, that if U_r is in any of the regions I, II, or III, then the problem (1)–(4) is solvable. Thus our assumptions $f_v g_u > 0$ and $l_i d^2 F(r_j, r_j) > 0$, $i,j = 1,2$, are sufficient to solve all Riemann problems if U_r is in regions I, II, or III. For example, if U_r is in region II, we can show that there is a point U on S_1 for which the curve $v = s_2(u; U)$ passes through U_r. The solution of the Riemann problem consists of going from the constant state U_l to the constant state U by a 1-shock wave and then going from U to U_r by a 2-shock wave. The only difficulty comes in region IV and we shall show by an example that these hypotheses are not sufficient if U_r is in region IV. Thus consider the system

$$u_t - v_x = 0, \quad v_t + (e^{-2u/2})_x = 0,$$

with initial data $U_l = (0,0)$, $U_r = (0,2k)$, $k > 1$. It is easy to check that this system satisfies all of our hypotheses. Moreover, since the equation for W_1 is given by $v = 1 - e^{-u}$, and the equation for the 2-rarefaction wave through U_r is given by $v = 2k - 1 + e^{-u}$, these curves never meet. This simple example shows what can go wrong in the general case. Namely the trouble is that W_1 (and W_2 for $u \geq u_l$) has a horizontal asymptote. Thus we must add a hypothesis which rules out this behavior. In [13], we give several such conditions, most of which are easily checked. This settles the existence problem. For the uniqueness question, see [14], where we prove that there is at most one solution in the class of constant states separated by centered shocks and rarefaction waves which satisfies the Lax shock conditions; i.e., which satisfies the condition that a shock comes from the intersection of characteristics from precisely one characteristic family.

VII. Some Unsolved Problems

We shall end this brief survey by giving a few additional remarks, stating some conjectures and pointing out some open problems. First it appears very probable to us that using the results of [13], the Glimm difference scheme, or some variation of this scheme, will make it possible to prove an existence theorem for the problem (1)–(2) where the initial data is merely bounded and measurable. Of course, one will certainly have to extend the estimates of Glimm in [3], to this more general case.

Next, we would like to make some remarks on the very important question of uniqueness of the solutions of (1)–(2). In the preceding sections we have shown that we can obtain an existence theorem for the problem (1)–(2) in two completely different ways, namely by approximating the data by step data, and by using the Glimm difference scheme. However, it is an open question as to whether these two solutions agree almost everywhere in $t \geq 0$. We conjecture an affirmative answer to this question. However, it is well-known, [10], that even for a single equation in two independent variables, weak solutions are not uniquely determined by their initial data. One must impose additional assumptions on the solution in order to rule out certain extraneous unstable solutions. For the case of systems there is a theorem of OLEINIK [11], in which she considers the system $u_t - v_x = 0, v_t + f(u)_x = 0$, and proves a uniqueness theorem in a class of solutions containing at most a finite number of centered rarefaction waves. There is also another uniqueness theorem proved in [12], which applies to a very restrictive class of equations. In [14] the uniqueness problem is solved for Riemann problems. It is clear from Oleinik's paper, which is an extension of the classical technique of Holmgren, that the question of uniqueness for general data is intimately connected with the question of the existence of smooth solutions of linear hyperbolic systems with discontinuous coefficients. Of course as the basis of what we have said in IV, any uniqueness theorem must also contain as a hypothesis a condition which implies condition (L) cited above.

Finally, it is generally believed that another possible approach to an existence theorem for $(1)-(2)$ is by the "viscosity" method. Thus one considers the (parabolic) system

$$u_t + f(u,v)_x = \varepsilon_1 u_{xx}, \quad v_t + g(u,v)_x = \varepsilon_2 v_{xx}, \tag{4}$$

$\varepsilon_i > 0, i = 1, 2$, with the same initial data (2). Then plan is to first find a well-behaved solution $(u_\varepsilon, v_\varepsilon), \varepsilon = (\varepsilon_1 \varepsilon_2)$, of this problem for each ε, and then show that some subsequence of $\{(u_\varepsilon, v_\varepsilon)\}$, converges to a solution of the original problem $(1)-(2)$. The first difficulty one has to overcome is proving an existence theorem for $(2)-(4)$ — nothing along these lines is known. To carry out this program one has to obtain two sets of estimates, parabolic estimates in order to obtain solutions of $(2)-(4)$ and then hyperbolic estimates in order to obtain the desired convergent subsequence.

References

1. CESARI, L.: Sulle Funzioni a Variazione Limita. Ann. Scuola Norm. Pisa **5** (2), 299−313 (1936).
2. CONWAY, E., and J. SMOLLER: Global Solutions of the Cauchy Problem for Quasi-linear First Order Equations In Several Space Variables. Comm. Pur. Appl. Math. **19**, 95−105 (1966).
3. GLIMM, J.: Solutions in the Large for Nonlinear Hyperbolic Systems of Equations. Comm. Pur. Appl. Math. **18**, 697−715 (1965).
4. −, and P. D. LAX: Decay of Solutions of Systems of Hyperbolic Conservations Laws. (to appear).

5. YU-FA, GUA, and ZHANG TONG: A class of Initial-value Problems for Systems of Aerodynamic Equations. Acta. Math. Sinica, 386−396 (1965). English translation in Chin. Math. **7**, 90−101 (1965).

6. JOHNSON, J. L., and J. A. SMOLLER: Global Solutions for Certain Systems of Quasi-linear Hyperbolic Equations. J. Math. Mech. **17**, 561−576 (1967).

7. − − Global Solutions for an Extended Class of Hyperbolic Systems of Conservation Laws. Arch. Rat. Mech. Math. (to appear).

8. KRICKEBERG, K.: Distributionen, Funktionen Beschränkter Variation und Lebesguescher Inhalt Nicht Parametrischer Flächen. Ann. Mat. Pura Appl., Ser. 4, **44**, 105−133 (1957).

9. LANDAU, L. D., and E. M. LIFSHITZ: Fluid Mechanics. Reading, Mass.: Addison Wesley 1959.

10. LAX, P. D.: Hyperbolic Systems of Conservation Laws, II. Comm. Pur. Appl. Math. **10**, 537−566 (1957).

11. OLEINIK, O. A.: On the Uniqueness of the Generalized Solution of the Cauchy Problem for a Non-linear System of Equations Occurring in Mechanics. Usp. Mat. Nank. **78**, 169−176 (1957).

12. ROZDESTVENSKII, B.: Discontinuous Solutions of Hyperbolic Systems of Quasi-linear Equations. Ups. Mat. Nank. **15**, 59−117 (1960). English translation in Russ. Math. Surv. **15**, 55−111 (1960).

13. SMOLLER, J. A.: On the Solution of the Riemann Problem with General Step Data For an Extended Class of Hyperbolic Systems. Michigan Math. J. (to appear).

14. − A Uniqueness Theorem for Riemann Problems. Arch. Rat. Math. Mech. (to appear).

Maximum Norm Stability for Parabolic Difference Schemes in Half-Space

S. J. OSHER *

I. Introduction

Consider a partial differential operator of order $2m$:

$$P(\partial) = \sum_{|r|=2m} A_r \partial_1^{r_1}, \ldots, \partial_s^{r_s}$$

where $r = (r_1, \ldots, r_s)$, $|r| = \sum r_i$ and $\partial_j = \partial/\partial x_j$. The A_r are complex valued constant $n \cdot n$ matrices. I $\frac{\partial}{\partial t} - P(\partial)$ is said to be uniformly parabolic in the sense of PETROVSKII if there is a constant $\delta > 0$, such that all roots σ_i of

$$\det \left\{ \sum_{|r|=2m} A_r (i w_1)^{r_1}, \ldots, (i w_s)^{r_s} - \sigma I \right\} = 0$$

satisfy:

$$\mathrm{Re}\,\sigma_i \leq -\delta$$

for all $|w|_2 = 1$. Here $w = (w_1, \ldots, w_s)$, w_i real, and $|w|_2^2 = \sum_{i=1}^{s} |w_i|^2$. We also restrict ourselves to operators $P(\partial)$ having the property that the eigenvalues of the matrix $A_{(2m,0,0,\ldots,0)}$ are distinct.

Consider the following initial-boundary value problem:

$$\left[\frac{\partial}{\partial t} - P(\partial) \right] u(x,t) = 0 \tag{1.1}$$

$$u(x,0) = f(x)$$

$$[K u(x_1, \ldots, x_s, t)] = 0$$

$$x_1 = 0$$

* Department of Mathematics, University of California, Berkeley, California 94720.

for K some appropriate matrix differential operator with constant coefficients. This is to be solved for $0 \le x_1 < \infty$, $-\infty < x_i < \infty$, $i = 2_1 \dots, s, f(x)$ a given vector function with bounded components.

In this paper we shall not concern ourselves with well-posedness of this problem, rather we shall only engage in the study of stability conditions for difference approximations. We expect that the generalization to variable coefficients, and hence parabolic systems with lower order space derivatives, will not present many difficulties.

The author would like to thank Olof B. Widlund for several instructive discussions on this subject. His papers [1] and [2] are the principal references for this work, and much of the notation is his.

II. Preliminaries and Statement of Main Results

We introduce rectangular lattices of mesh points

$$R(h) = \{x; x_1 = n_i h, n_1 = -a_1, \dots, -1, 0, 1, \dots, n_i = 0, \pm 1, \pm 2, \dots\},$$
$$i = 2, \dots, s$$

a_1 is some positive integer. $R(\lambda, h) = \{(x, t); x \in R(h), t = nk = n\lambda h^{2m},$ n is an integer with $0 \le n \le T/k = \text{integer} > 0\}$. We shall always assume that $\lambda = k/h^{2m}$ is a fixed positive constant.

We shall study difference approximations to (1.1) of the following type:

$$\left(\sum_{j=-a}^{b} B_j S^j\right) u(x, t + k, h) = \left(\sum_{j=-a}^{b} C_j S^j\right) u(x, t, h) \tag{2.1}$$

where $j = (j_1, \dots, j_s)$, $S^j = (S_1^{j_1}, \dots, S_s^{j_s})$, $S_i^{j_i} u(x, t + k, h) = u(x_1, \dots, x_i + j_i h, x_{i+1}, \dots, x_s, t + k, h)$, $(x, t) \in R(\lambda, h)$, $0 \le t \le T - k$, $x_1 \ge 0$, $u(x, 0, h) = f(x)$, C_j, B_j are constant matrices depending only on λ.

We must now approximate the boundary conditions. The necessity for this becomes evident in the explicit case, i. e., when each $B_j = \delta(j) I$. Then for example, we have:

$$u(0, x_2, \dots, x_s, t + k, h)$$
$$= C_{-a_1, -a_2, \dots, -a_s} u(-a_1 h, x_2 - a_2 h, \dots, x_s - a_s h, t, h) + \cdots$$

We thus need some recipe for obtaining $u(x, t, h)$ if $x_1 = -h, -2h, \dots,$ $-a_1 h$. We do this in the following manner:

$$\sum_{j=-a}^{l} P_j^{(i)} S^j u(0, x_2, \dots, x_s, t + k, h) = \sum_{j=-a}^{l} Q_j^{(i)} S^j u(0, x_2, \dots, x_s, t, h),$$
$$i = -1, -2, \dots, -a_1. \tag{2.2}$$

Again, the $P_j^{(i)}$ and $Q_j^{(i)}$ are constant matrices. We shall use the following notation: If $a = (a_1, \dots, a_s)$, then $|a|_q^q = \left(\sum_{i=1}^{s} |a_i|^q\right) |a|_q$, $q \ge 1$. Let $v(x)$

be a vector defined on $R(h)$. Then

$$\|v\| = \sup_{x \in R(h)} |v(x)|_1 .$$

If $w(x,t)$ is a vector defined on $R(\lambda,h)$, we shall write

$$\||w|\| = \sup_{(x,t) \in R(\lambda,h)} |w(x,t)|_1 .$$

The norms of linear operators will be defined in the usual way. Thus, if A is an $n \cdot n$ matrix, then

$$|A|_q = \sup_{|a|_q = 1} |A a|_q \quad \text{etc.}$$

We shall Fourier transform the tangential variables, i. e., let

$$\hat{u}(x_1,\xi,t,h) = \sum_{j_2 = -\infty}^{\infty} , \dots , \sum_{j_s = -\infty}^{\infty} u(x_1,j_2 h, \dots, j_s h, t, h) e^{-i(j_2 \xi_2 + \dots + j_s \xi_s)}$$

for each ξ_i real, $|\xi_i| \le \pi$. The shift operator $S_p^{j_p}$ then transforms into multiplication by $e^{+ij_p\xi_p}$, $p = 2, \dots, s$. The eqs. (2.1) and (2.2) then transform into:

$$\sum_{j_1 = -a_1}^{b} B_{j_1}(\xi) S_1^{j_1} \hat{u}(x_1,\xi,t+k,h) = \sum_{j_1 = -a_1}^{b_1} C_{j_1}(\xi) S_1^{j_1} \hat{u}(x_1,\xi,t,h) \qquad (2.1)'$$

$$\sum_{j_1 = -a_1}^{l_1} P_{j_1}^{(p)}(\xi) S_1^{j_1} \hat{u}(0,\xi,t+k,h)$$
$$= \sum_{j_1 = -a_1}^{l_1} Q_{j_1}^{(p)}(\xi) S_1^{j_1} \hat{u}(0,\xi,t+k,h), \quad p = -1, \dots, -a_1 . \qquad (2.2)'$$

Each of the coefficient matrices is a 2π periodic function of the ξ_j.

We make the important *Symmetry Assumption:* for any real ξ, it is true that the roots k_i defined in 2.3 are real. Moreover, if we replace each ξ_j by $\xi_j + ihn_j$, $j = 2, \dots, s$, hn_j real, ξ_j, hn_j sufficiently small, then the roots k_i have imaginary parts which depend only on $\xi_j + ihn_j$, $j = 2, \dots, s$.

We may now make the following definition (see [1], [2]):

Definition 2.1. For ξ real, $|\xi_i| \le \pi$, we denote by k_i, $1 \le l \le n$, the roots of

$$\det \left\{ \left(\sum_{j_1 = -a_1}^{b_1} B_{j_1}(\xi) e^{ij_1\xi_1} \right) k - \left(\sum_{j = -a_1}^{b_1} C_{j_1}(\xi) e^{ij_1\xi_1} \right) \right\} . \qquad (2.3)$$

(2.1) is a parabolic difference scheme if there is a constant $\delta' > 0$, independent of ξ such that

$$|k_i| \le 1 - \delta' |\xi|_2^{2m}$$

and
$$-I + \sum_{j_1=-a_1}^{b_1} B_{j_1}(\xi) e^{ij_1\xi_1} = 0(|\xi|_2^{2m})$$

$$-I + \sum_{j_1=-a_1}^{b_1} C_{j_1}(\xi) e^{ij_1\xi_1} = 0(|\xi|_2^{2m}).$$

Widlund has shown [2] that to each parabolic system (1.1) there exists an explicit consistent difference scheme which fulfills both the parabolicity conditions.

We may now state our first main result:

Theorem 2.1. Eqs. (2.1) and (2.2) can be uniquely inverted for arbitrary bounded right hand sides if and only if:

(a) $\det\left\{ \sum_{j_1=-a_1}^{b_1} B_{j_1}(\xi) e^{ij_1\xi_1} \right\} \neq 0$ for any $|\xi_i| \leq \pi$, ξ_i real, $i = 1,2,\ldots,s$.

(b) $\dfrac{1}{2\pi} \Delta_{-\pi \leq \xi_1 \leq \pi} \arg \det\left\{ \sum_{j_1=-a_1}^{b_1} B_{j_1}(\xi) e^{ij_1\xi_1} \right\} = 0$ for any ξ_i real, $|\xi_i| < \pi$,

$i = 2,\ldots,s$.

(c) Eqs. (2.1) and (2.2) with right hand sides zero have only the unique solution zero.

Assumption 2.1. For any constant r, $|r| \geq 1$, and all real ξ_i, $|\xi_i| \leq \pi$, $\det\{C_{+b_1}(\xi) - r B_{+b_1}(\xi)\} \neq 0$.

Definition 2.2. Let
$$R(z,\xi,r) = \sum_{j=-a_1}^{b_1} (C_j(i) - r B_j(\xi)) z^{j+a_1} \tag{2.4}$$

for real ξ, $|r| > 1$, z such that $\det[R(z,\xi,r)] \neq 0$. Consider the equation

$$\frac{1}{2\pi i} \oint_{|z|=1} \left(\sum_{j=-a_1}^{l} (Q_j^{(-i)}(\xi) - r P_j^{(-i)}(\xi)) R^{-1}(z,\xi,r) z^{j+a_1} \right) \tag{2.5}$$

$$\left(\sum_{p=1}^{a_1+b_1} \sum_{t=b_1+1-p}^{b_1} (C_t(\xi) - r B_t(\xi)) z^{t+p-1-b_1} G_{pk}(r,\xi) \right) dz = \delta_{ik} I$$

$i = 1,2,\ldots,a_1;\quad k = 1,2,\ldots,a_1;\quad \delta_{ik}I \begin{array}{ll} = I & \text{if} \quad i = k, \\ = 0 & \text{if} \quad i \neq k, \end{array}$

to be solved for the $(a_1 + b_1)a_1$ ($n \cdot n$) matrices, G_{pk}, $p = 1,\ldots,a_1 + b_1$; $k = 1,2,\ldots,a_1$.

We say the difference scheme (2.1), (2.2), satisfies the *parabolic boundary criterion* if for $|r| \geq 1$ and all $\xi_2,\ldots,\xi_s$ real, with $|\xi_i| < \pi$ except $r = 1$ and $0 = \xi_2 = \cdots = \xi_s$, the eq. (2.4) has a unique solution set $G_{pk}(r,\xi)$ which are analytic in r, $e^{i\xi_2},\ldots,e^{i\xi_s}$, for $|r| \geq 1 - \delta^{III}$, $||e^{i\xi_j}| - 1| \leq \delta^{IV}$; $\delta^{III} > 0$, $\delta^{IV} > 0$.

Main Theorem. Suppose (2.1) is a parabolic difference scheme, consistent with (1.1), which obeys Assumption (2.1) and the Symmetry Assumption. Moreover, assume that the criteria of Theorem (2.1) are valid. Finally suppose that (2.1) and (2.2) obey the parabolic boundary criterion. Then there exists a constant C_0 independent of h and $f(x)$ such that:

$$\sup_{(x,t)\in R(\lambda,h)} |u(x,t,h)|_1 \le C_0 \|f\|.$$

III. Example

The parabolic boundary criteria appears to be difficult to verify. In reality it is merely a refined RYABENKY and GODUNOV [3] normal mode analysis, completely analogous to the analysis due to KREISS [4], [5], and the author [6] in the hyperbolic case. We consider the following simple example:

$$u_t = u_{xx}, u(x,0) = f(x), 0 \le x, t < \infty, u_x(0,t) = 0. \tag{3.1}$$

Approximate this by the difference scheme:

$$u(x,t+k) = \lambda u(x-h,t) + (1-2\lambda)u(x,t) + \lambda u(x+h,t) \tag{3.2}$$

$x = 0, h, 2h, \dots, 0 < \lambda < \frac{1}{2}$ and $u(-h,t+k) - u(h,t+k) = 0$.

This scheme is clearly consistent, invertible and maximum norm stable (all the coefficients of the amplification matrix are positive and less than one), and obeys all the relevant assumptions.

For this problem eq. (2.5) becomes:

$$\frac{1}{2\pi i} \oint_{|z|=1} r \frac{(z^2-1)}{(\lambda z^2 + (1-2\lambda-r)z + \lambda)} (G_{11}\lambda + ((1-2\lambda-r) + \lambda z)G_{12}) dz = 1.$$

$$\lambda z^2 + (1-2\lambda-r)z + \lambda = \lambda(z-x(r))\left(z - \frac{1}{x(r)}\right),$$

where $|x(r)| < 1$ for $|r| \ge 1$ except at $r = 1$, $x(1) = 1$. Thus $x(r)$ is analytic for $|r| \ge 1, r \ne 1$ and has a branch point at $r = 1$. We integrate and obtain:

$$(x(r)G_{11} - G_{12}) = \frac{1}{r}.$$

We wish the G_{ij} to be analytic, and hence single valued, near $r = 1$. Thus we also need

$$\left(\frac{1}{x(r)}G_{11} - G_{12}\right) = \frac{1}{r}.$$

The unique solution to these two equations:

$$G_{11} = 0, \quad G_{12} = -\frac{1}{r}.$$

This vector function is clearly analytic in the proper region.

IV. Proof of Main Results

It is clear that for any fixed $\xi_2, \ldots, \xi_s$, the operator on the left sides of (2.1)′ and (2.2)′ is merely a finite dimensional perturbation of a semi-infinite Toeplitz Matrix. Hence, by the work of GOHBERG and KREIN [7], the conditions (a), (b) and (c) of Theorem 2.1 are necessary and sufficient for the invertibility of the operator for any fixed $\xi_2, \ldots, \xi_s$; hence necessary for uniform invertibility. Sufficiency follows because the inverse is easily shown to be a uniformly continuous function of all the ξ_i variables, $i = 2, \ldots, s$.

We shall assume in what follows that the operator is indeed uniformly invertible.

We shall now obtain estimates for the fundamental solution to (2.1), (2.2). It is clear that we may define linear operators L and M so that (2.1), (2.2) may be rewritten

$$Lu(x,t + k,h) = Mu(x,t,h). \tag{4.1}$$

Moreover, L and M are clearly independent of h

$$u(x,t + k,h) = L^{-1}Mu(x,t,h).$$

The fundamental solution to (4.1) is a matrix function $G(x,x_0,qk,pk)$, defined for $(x,x_0) \in R(h)$, $0 \le pk \le qk \le t$, having the properties:

(1) $G(x,x_0,qk,pk) = \delta(x-x_0)I, (\delta(x-x_0) = \prod_{j=1}^{s} \delta(x_j-x_{0_j}))$

(2) $G(x,x_0,(q + 1)k,pk) = L^{-1}MG(x,x_0,qk,pk).$

It is thus clear that

$$G(x,x_0,qk,pk) = (L^{-1}M)^{q-p}\delta(x-x_0)I$$

$$= \frac{1}{2\pi i} \oint_{|r|=A} r^{q-p}(r-L^{-1}M)^{-1}\delta(x-x_0)I\,dr$$

for $|A| > \|L^{-1}M\|$.

We are thus interested in obtaining the solution to

$$(r - L^{-1}M)f = \delta(x - x_0)I$$
$$(rL - M)f = L\delta(x - x_0)I.$$

(4.2)

Without loss of generality we may choose

$$x_0 = \{x_{0_1}, 0, \ldots, 0\},$$

since everything is translation invariant with respect to the tangential variables.

We now Fourier transform all the tangential variables and denote $f(jh, \xi_2, \ldots, \xi_s)$ by $\hat{f}_j$, $[L\delta(x - x_0)I](jh, \xi_2, \ldots, \xi_s)$ by $\hat{u}_j$. We may then replace (4.2) by a semi-infinite set of equations:

$$\hat{u}_j = \sum_{t=-a_1}^{b_1} (rB_t(\xi) - C_t(\xi))\hat{f}_{j+t}, \quad j = 0, 1, \ldots,$$

(4.3)

$$u_j = \sum_{t=-a_1}^{l} [rP_t^{(j)}(\xi) - Q_t^{(j)}(\xi)]\hat{f}_t, \quad j = -a_1, \ldots, -1.$$

We now introduce the following sequence of $(a_1 + b_1)$ n-component vectors:

$$F_k = [\hat{f}_{k+b_1-1}, \ldots, \hat{f}_{k-a_1}]', \quad k = 0, 1, \ldots,$$
$$G_k^{(1)} = [(rB_{b_1}(\xi) - C_{b_1}(\xi))^{-1}\hat{u}_k, 0, \ldots, 0]', \quad k = 0, 1, \ldots.$$

Eq. (4.3) becomes:

$$F_{k+1} = M(r)F_k + G_k(r), \quad k = 0, 1, \ldots,$$

(4.4)

$$\begin{bmatrix} u_{-1} \\ \vdots \\ u_{-a_1} \end{bmatrix} = L_0(r)F_0 + L_1(r)F_1 + \cdots + L_{l-b_1+1}F_{l-b_1+1}.$$

We define the matrices $P(r)$, $M(r)$, and $L_i(r)$ as follows:

Let $M(r, \xi)$ for $|r| > 1, \xi_2, \ldots, \xi_s$ real, be an $n(a_1 + b_1)$ order square matrix whose i,j element we take to be an n order square matrix with

$$[(M(r, \xi)]_{1j} = -[(C_{b_1}(\xi) - rB_{b_1}(\xi))^{-1}(C_{b_1-j}(\xi) - rB_{b_1-j}(\xi))]$$
$$[M(r, \xi)]_{ij} = [I]\delta(j - i + 1), \quad i > 1,$$

($\delta(i)$ is the Kronecker delta function).

Then, for any complex number z, it is true that there exists an integer P_0 with

$$\det[M(r, \xi) - zI] = (-1)^{P_0}[\det(C_{b_1}(\xi) - rB_{b_1}(\xi))]^{-1}$$
$$\det\left\{\sum_{j=-a_1}^{b_1} [C_j(\xi) - rB_j(\xi)]z^{j+a_1}\right\}.$$

Moreover, if $\det\left\{\sum\limits_{j=-a_1}^{b_1}[C_j(\xi)-rB_j(\xi)]z^{j+a_1}\right\}\neq 0$, then $[M(r,\xi)-zI]^{-1}$ exists with

$$[M(r,\xi)-zI]_{ij}^{-1} = -\left\{\sum_{p=-a_1}^{b_1}[C_p(\xi)-rB_p(\xi)]z^{p+a_1}\right\}^{-1}$$
$$\sum_{t=b_1+1-j}^{b_1}[C_t(\xi)-rB_t(\xi)]z^{j+t-i+a_1-1} + K_{ij}(z)$$

where $K_{ij}(z)=0$ if $j\leq i$, $K_{ij}(z)=z^{j-i-1}$ if $j>i$.

Thus, in particular, for parabolic difference schemes, if $z=e^{i\xi_1}$, ξ real, and $|r|>1-\delta'|\xi|_2^{2m}$, this inverse exists.

The proof is left to the reader.

Let $P(r)$ be the projection on $n(a_1+b_1)$ dimensional vector space defined by:

$$P(r,\xi) = \frac{1}{2\pi i}\oint_{|z|=1}[zI-M(r,\xi)]^{-1}dz, \quad |r|>1, \quad \text{real}\quad \xi.$$

We also define $a_1n\cdot(a_1+b_1)n$ order matrices:

$$L_0(r,\xi) = \begin{bmatrix} [Q_{b_1-1}^{(-1)}(\xi)-rP_{b_1-1}^{(-1)}(\xi)] & \dots & [Q_{-a_1}^{(-1)}(\xi)-rP_{-a_1}^{(-1)}(\xi)] \\ [Q_{b_1-1}^{(-a_1)}(\xi)-rP_{b_1-1}^{(-a_1)}(\xi)] & \dots & [Q_{-a_1}^{(-a_1)}(\xi)-rP_{-a_1}^{(-a_1)}(\xi)] \end{bmatrix}$$

$$L_i(r,\xi) = \begin{bmatrix} [Q_{b_1-1+i}^{(-1)}(\xi)-rP_{b_1-1+i}^{(-1)}(\xi)] & \dots & [0] & \dots & [0] \\ [Q_{b_1-1+i}^{(-a_1)}(\xi)-rP_{b_1-1+i}^{(-1)}(\xi)] & \dots & [0] & \dots & [0] \end{bmatrix}$$

$i=1,2,\dots,l-a_1+1$. (If $l\leq a_1=1$, then of course $L_i=[0]$ if $i\geq 1$, etc.)

Consider the equation:

$$\frac{1}{2\pi i}\oint_{|z|=1}[L_0(r,\xi)+L_1(r,\xi)z+\cdots+L_{l-b_1+1}(r,\xi)z^{l-a_1+1}]$$
$$[z-M(r)]^{-1}dzF_0 = [\hat{u}_{-1}(\xi),\dots,\hat{u}_{-a_1}(\xi)]' = H^l(\xi)$$

(^{l} denotes transpose), for a_1 arbitrary n vector functions of ξ, the equation to be solved for $P(r)F_0$. We may easily show that the *parabolic boundary criterion* is obeyed if and only if for $|r|\geq 1$ and all $\xi_2,\dots,\xi_s$ real with $|\xi_j|<\pi$ except $r=1$ and $0=\xi_2=\xi_3=\cdots=\xi_s$ the equation has a unique solution which may be expressed as:

$$P(r)F_0 = \frac{1}{2\pi i}\oint_{|z|=1}[z-M(r)]^{-1}dzG(r,\xi)H(\xi)$$

such that the $n(a_1 + b_1) \cdot a_1 n$ matrix, $G(r,\xi)$ is independent of ξ_1 and is analytic in $e^{i\xi_2}, \ldots, e^{i\xi_s}$, and r for $|r| \geq 1 - \delta^{III}$, $\|e^{i\xi_j}| - 1| \leq \delta^{IV}$ for constants $\delta^{III}, \delta^{IV} > 0$.

The solution to this system (4.4) for $|r| \geq 1$ and all real ξ, $|\xi_i| < \pi$, except for $r = 1$ and $\xi = 0$, may be written as: (ignoring the ξ dependence)

$$P(r)F_k = M^k(r)P(r)F_0 + \sum_{j=0}^{k-1} M^{k-j-1}(r)P(r)G_j(r) \qquad (4.5)$$

$$[I - P(r)]F_k = -\sum_{j=k}^{\infty} M^{k-j-1}(r)[I - P(r)]G_j(r),$$

where $P(r)F_0$ is obtained from:

$$[L_0(r) + L_1(r)M(r) + \cdots + L_{l-a_1+1}(r)M^{l-a_1+1}(r)]P(r)F_0 \qquad (4.6)$$

$$= \begin{bmatrix} \hat{u}_{-1} \\ \hat{u}_{-a_1} \end{bmatrix} - (L_0(r)(I - P(r))F_0 + L_1(r)(M(r)(I - P(r))F_0 + G_0(r)) + \cdots$$

$$+ L_{l-a_1+1}(M^{l-a_1+1}(r)(I - P(r))F_0 + M^{l-a_1}(r)G_0(r) + \cdots + G_{l-a_1}(r))).$$

This equation has a unique solution for $P(r)F_0$ for all $|r| \geq 1$ and real ξ_i, $|\xi_i| \leq \pi$ except $r = 1$ and $0 = \xi_2 = \xi_3 = \ldots = \xi_s$, by the parabolic boundary criterion.

The solution (4.5) must converge in this r, ξ region by the following argument: The spectrum of $M(r)$ on $P(r)$ is made up of just those roots of $\det[zI - M(r)] = 0$ which lie inside the unit circle. The spectrum of $M(r)$ consists of the zeros of $\det\left\{ \sum_{j=-a_1}^{b_1} [C_j(\xi) - rB_j(\xi)]z^{j+a_1} \right\}$. By parabolicity, continuity of zeros of a polynomial as a function of its coefficients, Theorem 2,1, the fundamental theorem of algebra, and Assumption 2.1, there are exactly $(a_1 + b_1)n$ of them, $a_1 n$ of which lie within the unit circle, $b_1 n$ of which lie outside the unit circle, thus the spectral radius of $M(r)$ on $P(r)$ is less than one as is the spectral radius of $M^{-1}(r)$ on $I - P(r)$.

Now:

$$(\widehat{\delta(x - x_0)I})(jh, \xi_2, \ldots, \xi_s) = \delta(jh - x_0)I$$

$$(\widehat{L\delta(x - x_0)})(jh, \xi_2, \ldots, \xi_s) = 0 \quad \text{if} \quad j > \frac{x_{0\,1}}{h} + a_1$$

$$\text{or} \quad j < \left(\frac{x_{0\,1}}{h} - b_1 - l \right)$$

otherwise

$$(\overline{L\,\delta(x-x_0))}(jh,\xi_2,\dots,\xi_s) = B_{\left(\frac{x_{01}}{h}-j\right)}(\xi) \quad \text{if} \quad j \geq 0$$

$$\text{or} \quad P^{(j)}_{\left(\frac{x_{0_1}}{h}\right)}(\xi) \quad \text{if} \quad j < 0.$$

Thus, in general, $\hat{u}_j = 0$ except for a set of mesh points of length $\leq a_1 + l + 1 + b_1$ containing $j = x_{0\,1}/h$.

We may now easily estimate the terms in (4.5) except for a slight difficulty involving $M^k(r)\,P(r)\,F_0$. We save this estimate (which uses the parabolic boundary criterion and our symmetry and consistency assumptions) for last. We first estimate terms of the form

$$M^{\frac{x_1}{h}-j-1}\,P(r)\,G_j(r) \quad \text{if} \quad \frac{x_1}{h} \geq j+1.$$

We have:

$$\frac{1}{(2\pi i)^{s+1}}\int\dots\int d\xi_2\dots d\xi_s e^{+i\left(\frac{x_2}{h}\xi_2+\cdots+\frac{x_s}{h}\xi_s\right)}$$

$$\int dr\,r^{q-p}\oint d\xi_1 e^{i\left(\frac{x_1}{h}-j\right)\xi_1}[M(r)-e^{i\xi_1}I]^{-1}G_j.$$

Integrate with respect to r first.

$$\frac{1}{2\pi i}\oint r^{q-p}[M(r)-e^{i\xi_1}I]^{-1}\left[(r\,B_{a_1}(\xi)-C_{a_1}(\xi))^{-1}\hat{u}_j,0,\dots,0\right]dr$$

$$= \frac{1}{2\pi i}\oint r^{q-p}\left[\sum_{t=-a_1}^{b_1}(C_t(\xi)-r\,B_t(\xi))e^{it\xi_1}\right]^{-1}[e^{i(b_1-1)\xi_1},\dots,e^{ia_1\xi_1}]\hat{u}_j\,dr.$$

We are interested in the b_1-st n-vector above. The result is

$$-\left(\sum_{t=-a_1}^{b_1}B_t(\xi)e^{it\xi_1}\right)^{-1}\left(\left(\sum_{t=-a_1}^{b_1}C_t(\xi)e^{it\xi_1}\right)\left(\sum_{t=-a_1}^{b_1}B_t(\xi)e^{it\xi_1}\right)^{-1}\right)^{q-p}\hat{u}_j.$$

The resulting integral may be majorized using Lemma 3.3 of [1] in exactly the same way as in the proof of Theorem 3.1 of that paper.

We next estimate terms of the form

$$-M^{\frac{x}{h}-j-1}(r)[I-P(r)]\,G_j(r) \quad \text{for} \quad \frac{x}{h} \leq j.$$

We then have

$$\frac{-1}{(2\pi i)^{s+1}}\int\dots\int d\xi_2,\dots,d\xi\,e^{i\left(\frac{x_2}{h}\xi_2+\cdots+\frac{x_s}{h}\xi_s\right)}$$

$$\int dr\,r^{q-p}\int_{\Gamma-\Gamma'} z^{\frac{x_1}{h}-j-1}dz[M(r)-zI]^{-1}dz\,G_j$$

the $\int\limits_{\Gamma - \Gamma'}$ indicates integration over the boundary of an annulus whose outer boundary is $\Gamma : |z| = R$, R arbitrarily large, inner boundary is $\Gamma' : |z| = 1$. We perform the r integration first. Again the resulting term is

$$-\left(\sum_{t=-a_1}^{b_1} B_t(\xi) z^t \right)^{-1} \left(\sum_{t=-a_1}^{b_1} C_t(\xi) z^t \right) \left(\sum_{t=-a_1}^{b_1} B_t(\xi) z^t \right)^{-1} \right)^{q-p} \hat{u}_j$$

which is bounded independent of $q - p$ for $|z| = R$ since Assumption 2.1 guarantees that the spectral radius of $(C_{b_1}(\xi))(B_{b_1}^{-1}(\xi))$ is less than one. The integral over Γ is zero since the integrand's behavior is like $|R|^{-b_1 + \frac{x}{h} - j - 1}$. The integral over Γ' may be estimated again as in Theorem 3.1 of Widlund.

We must next estimate $M^k P(r) F_0$. We first take $G_j = 0$ for $j \geq 0$. We must hence estimate

$$\int \dots \int dr \, d\xi_2, \dots, d\xi_s \oint z^k [z - M(r)]^{-1} G(r, \xi)$$

$$[\hat{u}_{-1}, \dots, \hat{u}_{-a}]' dz \, e^{-i\left(\frac{x_2}{h} \xi_2 + \dots + \frac{x_s}{h} \xi_s \right)}$$

We do the integration first again. By the assumption on the analyticity of $G(r, \xi)$, we may expand

$$G(r, \xi) = \sum_{k=0}^{\infty} G^{(k)}(\xi) r^{-k}, \ |r| \geq 1 - \delta'''.$$

Each $G^{(k)}(\xi)$ is analytic if $\left| |e^{i\xi} j| - 1 \right| \leq \delta^{IV}$, and by the Cauchy integral formula, there exists a constant T with $\left| G^{(k)}(\xi) \right| \leq T(1 - \delta''')^k$.

We may carry out the integration as in WIDLUND, [1], obtaining a bound of the form:

$$C_\lambda^1 h^s e^{-\langle x \cdot B \rangle} \sum_{t=0}^{q-p} \frac{e^{3c|B|(q-p+1-t)k}(1 - \delta''')^t}{(q - p + 1 - t)^{s/2m}}$$

but

$$\sum_{t=0}^{(q-p)} \frac{e^{3c|B|\frac{3}{2}m(q-p+1-t)k}(1 - \delta''')^t}{(q - p + 1 - t)^{s/2m}} =$$

$$\frac{e^{3c|B|\frac{3}{2}m(q-p+1)k}}{(q - p + 1)^{s/2m}} \sum_{t=0}^{q-p} \frac{(1 - \delta''')^t e^{-3c|B|\frac{3}{2}m t k}}{\left(1 - \dfrac{t}{q - p + 1} \right)^{s/2m}} \leq$$

$$C^{IV} \frac{e^{3c|B|\frac{3}{2}m(q-p+1)k}}{(q - p + 1)^{s/2m}},$$

C^{IV} some fixed constant. $\quad \dots$
Thus the estimates still hold for this term.

Next we consider the case when all the $G_k = 0$ except G_j for $j \geq l + a_1$. The right side of (4.6) is now

$$[L_0(r) M^{-j-1}(r) + L_1(r) M^{-j}$$
$$+ \cdots + L_{l-a_1+1}(r) M^{-j-l-a_1}(r)][I - P(r)] G_j.$$

The solution for $M^k P(r) F_0$ is

$$M^k P(r) F_0 = \frac{1}{2\pi i} \oint z^k [z - M(r)]^{-1} [-I + G(r)(L_0(r) + L_1(r)$$
$$+ \cdots + L_{l-a_1+1} M^{l-a_1+1})] dz \, M^{-j-1}(r) G_j.$$

The expression

$$[-I + G(r, \xi)(L_0(r, \xi) + \cdots + L_{l-a_1+1}(r, \xi) M^{l-a_1+1}(r, \xi)] M^{-j-1}(r, \xi)$$

is analytic for $|r| \geq 1 - \delta'''$, $\||e^{i\xi j}| - 1| \leq \delta^{IV}$. Moreover, if we multiply by $P(r)$ on the right and $z^k [z - M(r)]^{-1}$ on the left and integrate around the unit circle, we get zero. Thus the term $M^{-j-1}(r, \xi)$ may be thought of as acting on the range of $[I - P(r)]$. We shall show that there exists a circle $|r| = 1 - \delta^{VI}$, $\delta^{III} \geq \delta^{VI} > 0$, on which $M^{-j-1}(r, \xi + ihn)$ is uniformly bounded independently of j when it acts on the range of $I - P(r, \xi + ihn)$ for all real ξ and $|hn| < \delta^{VII}$, $\delta^{IV} \geq \delta^{VII} > 0$.

Let some $\xi_i \neq 0$, $i \geq 2$, each $|\bar{\xi}_j| \leq \pi$. Then the roots k_i of

$$\det \left[\sum_{j=-a_1}^{b_1} (C_j(\xi + ihn) - k_i B_j(\xi + ihn)) e^{j(i\xi_1 - hn_1)} \right] = 0$$

are strictly less than $1 - \delta(\xi)$ for $|hn|_\infty < \delta(\bar{\xi})$ if $|\xi - \bar{\xi}|_\infty < \delta(\bar{\xi})$ by (3.7) of [1]. If each $\bar{\xi}_j = 0$, then:

$$\det \left[\sum_{k=-a_1}^{b_1} (B_k(0) - r C_k(0)) x^{k+a_1} \right] = 0$$

for r not real, has solutions for x of the form:

$$|x_i(r)| < 1, \quad i = 1, 2, \ldots, a_1 n$$
$$|y_i(r)| > 1, \quad i = 1, 2, \ldots, b_1 n$$

by the symmetry assumption. We are interested in the behavior of the roots which approach 1 as r approaches 1. By consistency, for $(x - 1)$ small

$$\left(\sum_{k=-a_1}^{b_1} B_k(0) x^k \right)^{-1} \left(\sum_{j=-a_1}^{b_1} C_j(0) x^j \right)$$
$$= I + \frac{\mu}{2m} A_{(2m,0,\ldots,0)} \left(x - \frac{1}{x} \right)^{2m} + 0 \left(\left(x - \frac{1}{x} \right)^{2m+1} \right).$$

Thus if we let $(x - 1/x) = a_1(r - 1)^{k_0}$ + lower order terms, we may expand the determinant above, set it equal to zero, and equate corresponding lowest order powers of $(r - 1)$. We recall that the eigenvalues of $A_{(2m,0,...,0)}$ all have positive real part if m is odd, negative real part if m is even. We can easily show that the solutions $x_i(r)$ and $y_i(r)$ which approach one as $r \to 1$ are of the form

$$z_j(r) = 1 - \left(\frac{r-1}{\mu d_k}\right)^{\frac{1}{2m}} + \text{lower order terms},$$

d_k is any one of the eigenvalues of $A_{(2m,0,...,0)}$, the $\frac{1}{2m}$ roots above has $2m$ possible definitions. It is thus clear that the relevant part of $\dfrac{(M(r,0) - I)}{(r-1)^{1/2\,m}}$ acting on $[I - P(r,0)]$ is uniformly diagonalizable in some

neighborhood of $r = 1$ by the assumption that the d_i are distinct. Thus, since the spectrum of $M^{-1}(r,0)$ on $[I - P(r,0)]$ lies in the closure of the unit circle, it is indeed power bounded. If we now let $r, \xi_2, ..., \xi_s$, and $hn_2, ..., hn_s$ be such that r is in this neighborhood and $\max\limits_{c,j \geq 2} \dfrac{|\xi_i|, |hn_j|}{(r-1)^{1/2m}}$ $< \delta^{IV}$, then by continuity of the roots of a polynomial as a function of its coefficients, it follows that the relevant part of

$$\left[\frac{M(r,\xi + ihn) - I}{(r-1)^{1/2m}}\right] \quad \text{on} \quad \left[I - P(r,\xi + ihn)\right]$$

is again uniformly diagonalizable. By the symmetry assumption, the roots $Z_j(r,\xi + ihn)$ are not equal to one in absolute value as long as r is not on certain contour lines depending on $\xi_2, ..., \xi_s, hn_2, ..., hn_s$. We may choose $|r| = 1 - \delta^{XI}$ and r in the diagonalizable neighborhood and off these contour lines. As r approaches the contour lines the roots $y_j(r,\xi + ihn)$ may approach one in absolute value, but will not be less than one. This guarantees that $M^{-1}(r,\xi + ihn)$ on $[I - P(r,\xi + ihn)]$ has spectrum contained in the closed unit circle and by its diagonalizability, is power bounded on $|r| = 1 - \delta^{XI}$ in this r, ξ, ihn_s neighborhood. We may cover the unit circle $\sum\limits_{i=2} |\xi_i|^2 = 1$ by a finite number of these ξ neighborhoods, thus piecing together our claim.

Having this result, we may now carry out the integration in exactly the same fashion as in the case when $G_j = 0$ for $j \geq 0$ obtaining the same estimates modulo fixed constants.

Finally, we consider the case for which $G_k = 0$ except G_j for some j, $0 \leq j < l - a_1$. The right side of (4.6) then becomes

$$[L_0(r)M^{-j-1}(r) + \ldots + L_j(r)M^{-1}(r)][I - P(r)]G_j$$
$$-[L_{j+1}(r) + \ldots + L_{l-a_1+1}(r)M^{l-a_1-j}(r)]P(r)G_j$$
$$= -[L_0(r) + \ldots + L_{l-a_1+1}(r)M^{l-a_1+1}(r)]M^{-j-1}(r)P(r)G_j$$
$$+ [L_0(r)M^{-j-1}(r) + \cdots + L_j(r)M^{-1}(r)]G_j.$$

The solution for $M^k(r)P(r)F_0$ is

$$M^k(r)P(r)F_0 = \frac{1}{2\pi i}\oint z^k[z - M(r)]^{-1}[-I + G(r)L_0(r)$$
$$+ L_1(r)M(r) + \ldots + L_j(r)M^{-1}(r)]M^{-j-1}(r)G_j.$$

For these $(l-a_1)G_j$ the proper estimates are easily obtained.

We thus have the following expression for the fundamental solution: (WIDLUND [1], Theorem 3.1).

Lemma 4.1. There exists a constant C^1 independent of (x, x_0) in $(R(h) \cdot R(h))$ and h so that for all $pk, qk, p \le q, h \le h > 0$:

$$|G(x, x_0, qk, pk)| \le C^1 \frac{h^s}{(qk - pk + k)^{s/2m}} \exp(3C|B|_{2m}^{2m}(qk - pk + k)$$
$$- \langle x - x_0, B \rangle$$

for all B such that $|hB_i| \le C_0$, C_0 is defined in (3.5) of [1].

We may now prove our Main Theorem. It is clear that the solution to (4.1) may be written

$$u(x, t, h) = \sum_{x_0 \in R(h)} G(x, x_0 t, 0) f(x_0)$$

or

$$|u(x, t, h)|_1 \le \|u(\ , t, h)\| \frac{C^1}{(t + k)^{s/2m}} h^s \sum_{x_0 \in R(h)} \exp(3C|B|_{2m}^{2m}(t + k)$$
$$- \langle x - x_0, B \rangle).$$

The proof then follows in the same way as that of Theorem 4.2 of [1].

Q.E.D.

V. References

1. WIDLUND, O. B.: Stability of Parabolic Difference Schemes in the Maximum Norm. Numer. Math. **8**, 186 – 202 (1966).
2. – On the Stability of Parabolic Difference Scheme. Math. Comp., 1 – 13 (1965).
3. GODUNOV, S. K., and V. S. RYABENKII: Theory of Difference Schemes, An Introduction. Amsterdam: North-Holland Publishing Company 1964.

4. KREISS, H. O.: Difference Approximations for the Initial-Boundary Value Problem for Hyperbolic Differential Equations. p. 141 – 166. In: Numerical Solutions of Nonlinear Partial Differential Equations. New York: Edited by D. Greenspan, John Wiley and Sons 1966.
5. — Stability Theory for Difference Approximations for Mixed Initial Boundary Value Problems I. Math. Comp. 22 703—714 (1968).
6. OSHER, S.: Systems of Difference Equations with General Homogeneous Boundary Conditions. A. M. S. Transactions 137, 177—201 (1969).
7. GOHBERG, I. C., and M. G. KREIN: Systems of Integral Equations on a Half-Line with Kernels Depending on the Difference of Arguments". A. M. S. Transactions 14, 217 – 280 (1960).

Stability and Well-Posedness for Difference Schemes and Partial Differential Equations for Time Dependent Problems in Half-Space

S. J. OSHER

I. Strict Well Posedness of Hyperbolic Partial Differential Equations in Half-Space

This result is due to H.-O.KREISS and was first presented at a meeting on these problems at the University of Denver. The method of proof involves transform techniques.

We are concerned with the system

$$u_t = A u_{x_1} + \sum_{j=2}^{m} B_j \frac{\partial u}{\partial x_j} \tag{1.1}$$

$$0 \leq x_1$$

$$-\infty < x_j < \infty, \quad j = 2, \ldots, m$$

$$t \geq 0$$

$$u = \{u^{(1)}(x, t), \ldots, u^{(n)}(x, t)\}^T$$

A and all the B_j are constant n by n matrices. A is diagonal

$$A = \begin{bmatrix} a_1 & & & & & \\ & \ddots & & & & \\ & & a_l & & & \\ & & & a_{l+1} & & \\ & & & & \ddots & \\ 0 & & & & & a_n \end{bmatrix}, \qquad \begin{matrix} a_1 < 0, \ldots, a_l < 0 \\ \\ a_{l+1} > 0, \ldots, a_n > 0. \end{matrix}$$

We take initial data:

$$u(x, 0) = 0.$$

Let

$$u^{\mathrm{I}} = \{u^{(1)}(x,t), \dots, u^{(l)}(x,t)\}^T$$
$$u^{\mathrm{II}} = \{u^{(l+1)}(x,t), \dots, u^{(n)}(x,t)\}^T$$

The boundary conditions are

$$u^{\mathrm{I}}(0, x_2, \dots, x_m, t) = R u^{\mathrm{II}}(0, x_2, \dots, x_m, t) + g(x_2, \dots, x_m, t)$$

R is some constant l by $(n-l)$ matrix.

Definition. The differential eq. (1.1) is said to be hyperbolic if there exists a matrix

$$\hat{H}(w) = \sum_{0 \le |r| \le 2p} H_r w^r, p \quad \text{some non-negative integer,} \quad \text{real} \quad w =$$

$(w_1, \dots, w_m)$, $|w| = \sum_{i=1}^{2p} |w_i|$, each H_r is a constant n by n matrix, such that $\hat{H}(w)$ is symmetric and there exists a constant $c > 0$ with

$$c^{-1}|w|^{2p} \le \hat{H}(w) \le c|w|^{2p}$$

for sufficiently large w. Finally,

$$\hat{H}P + P\hat{H} = 0, \quad P = i\left(w_1 A + \sum_{j=2}^{m} B_j w_j\right).$$

We assume in what follows that (1.1) is indeed hyperbolic in this sense. Consider the matrix

$$M(s, w_2, \dots, w_m) = A^{-1}\left(s\mathrm{I} - i \sum_{j=2}^{m} B_j w_j\right), \quad \mathrm{Re}\, s > 0,$$

each w_i is real. It can easily be shown that M has l eigenvalues K_j with $\mathrm{Re}\, K_j < 0$, $n-l$ eigenvalues with $\mathrm{Re}\, K_j > 0$. It thus follows that the system of equations $-\dfrac{d\varphi}{dx} = M\varphi$ has exactly l linearly independent solutions which go to zero as $x \to \infty$. These are of the form:

$$\Psi_i(x, s, w) = \sum_{j=1}^{l} P_{ij}(x, s, w) e^{K_j(s,w)x}, \quad i = 1, 2, \dots, l,$$

each $P_{ij}(x, s, w)$ is a vector polynomial in x. Assume $|\Psi_i(0, s, w)| = 1$, $i = 1, 2, \dots, l$.

Definition
Let

$$[E(s,w)]_{l \times l} = \left[[I]_{l \times l} [-R]_{(l \times n-l)}\right]\left[\begin{bmatrix} \Psi_1(0,s,w) \end{bmatrix} \dots \begin{bmatrix} \Psi_l(0,s,w) \end{bmatrix}\right]_{n \times l}$$

We may now state Kreiss' main result.

Main Theorem. Suppose (1.1) is hyperbolic and

$$\det\left[E(s,w)\right] \neq 0 \quad \text{for} \quad \text{Re } s \geq 0, \quad w_i \text{ real}, \quad i = 2,\dots,m.$$

It then follows that for g having compact support in $x_2,\dots,x_m$ and p continuous derivatives in t and x the solution $u(x,t)$ exists and:

$$(u(\cdot,t), H(D)u(\cdot,t)) \leq K(t)\|g\|_{t,p}^2$$

where $K(t)$ depends only on t,

$$H(D) = \sum_{|r| \leq 2p} H_r \left(\frac{\partial}{\partial x_1}\right)^{r_1} \cdots \left(\frac{\partial}{\partial x_m}\right)^{r_m},$$

$$(u(\cdot,t), v(\cdot,t)) = \int\limits_{-\infty}^{\infty} dx_2 \dots \int\limits_{-\infty}^{\infty} dx_s \int\limits_{0}^{\infty} dx_1 \, u(x,t)v(x,t),$$

$$\|g\|_{t,p}^2 = \sum_{|j| \leq p} \left\| \frac{\partial^{j_1}}{\partial t^{j_1}} \frac{\partial^{j_2}}{\partial x_2^{j_2}} \cdots \frac{\partial^{j_m}}{\partial x_m^{j_m}} g \right\|_t^2,$$

$$\text{and} \quad \|g\|_t^2 = \int\limits_0^t dt \int\limits_{-\infty}^{\infty} dx_2 \dots \int\limits_{-\infty}^{\infty} dx_s |g(x_2,\dots,x_m,t)|^2.$$

Thus the problem is strictly well-posed in the sense that the solution and certain of its derivatives are estimated in terms of the boundary data and the same number of its derivatives.

II. Stability of Hyperbolic Partial Difference Schemes Using Toeplitz Matrices

This result will appear in a paper of mine in the Transactions of the American Mathematical Society.

We wish to approximate the following system:

$$u_t = A u_x \tag{2.1}$$
$$0 \leq x, t$$
$$u = \{u^{(1)}(x,t),\dots,u^{(n)}(x,t)\}^T$$

A is a diagonal constant matrix having the same structure as in the first section.

We take initial data: $u(x,0) = f(x)$, $f(x)$ with compact support and continous, and boundary conditions:

$$u^{\mathrm{I}}(0,t) = R u^{\mathrm{II}}(0,t).$$

(This problem is easily shown to be well-posed, in fact using the above notation $H(w) \equiv I$ and $E(s,w) \equiv I$). We introduce a rectangular lattice of mesh points $R(h) = \{x, x = nh, n = -a, -a + 1, \ldots, 0, \quad\}$, a some non-negative integer.

$$R(\mu, h) = \{(x, t); \quad x \in R(h), \quad t = nk = n\mu h, \quad n = 0, 1, \ldots\}.$$

We always assume that $\mu = k/h$ is a fixed positive constant.

We shall study difference approximations to (1.1) of the following type:

$$\left(I + \sum_{j=-a}^{b} B_j S^j\right) u(x, t + k, h) = \left(I + \sum_{j=-a}^{b} C_j S^j\right) u(x, t, h) \quad (2.2)$$

S is the shift operator

$$S u(x, t, h) = u(x + h, t, h)$$
$$(x, t) \in R(\mu, h), \quad x \geq 0$$

$u(x, 0, h) = f(x)$ (where $f(x)$ is extended continuously to $-ah \leq x \leq 0$), C_j and B_j are diagonal constant matrices depending only on μ.

We must now approximate the boundary conditions. The necessity for this is clear in the explicit case, i.e., when $B_j \equiv 0$. Then we have

$$u(0, t + k, h) = u(0, t, h) + \sum_{j=-a}^{b} C_j u(j h, t, h)$$

and we need some recipe for obtaining $u(x, t, h)$ for $x = -h, -2h, \ldots, -ah$. We do this in the following manner:

$$\sum_{j=-a}^{l} P_j^{(i)} S^j u(0, t + k, h) = \sum_{j=-a}^{l} Q_j^{(i)} S^j u(0, t, h), \quad i = -1, \ldots, -a. \quad (2.3)$$

The $P_j^{(i)}$ and $Q_j^{(i)}$ are constant (but not necessarily diagonal) n by n matrices.

In order that (2.2) and (2.3) be relevant to (1.1) it is necessary that the difference scheme be consistent.

Definition (2.1). The difference scheme is said to be consistent with the differential equation if every sufficiently smooth solution to (1.1) satisfies (2.2) and (2.3) modulo terms of order k^2.

Definition (2.2). For any function $u(x, t, h)$ defined on $R(h)$, we let

$$\|u(0, t, h)\|_h^2 = \sum_{x \in R(h)} h |u(x, t, h)|_2^2.$$

Let $l_{2, h}$ be the obvious Hilbert space with this norm.

Definition (2.3). We say that (2.2) and (2.3) form a stable difference approximation to (2.1) if

    (i) The difference scheme is consistent

    (ii) (2.2) and (2.3) are uniformly invertible for $u(\cdot, t + k, h)$ on $l_{2, h}$.

    (iii) There exists a constant $K > 0$ independent of t, h, and $f(x)$ such that

$$\|u(\cdot,t,h)\|_h^2 \leq K_2\|f\|_h^2$$

for $0 \leq t \leq T$, $T > 0$, $0 \leq h \leq h_0$, $h_0 > 0$.

By the well known Lax-Richtmyer Theorem, stability is equivalent to convergence for consistent difference schemes. Thus, the main result concerns stability. We state our results here and outline the proofs below.

Theorem 2.1. The following conditions are necessary and sufficient for invertibility of the difference scheme on $l_{2,h}$.

(1) $\det\left\{\sum\limits_{j=-a}^{b} B_j e^{ij\theta}\right\} \neq 0$ for any real θ, $-\pi \leq \theta < \pi$.

(2) $\dfrac{1}{2\pi} \Delta \arg \det\left\{\sum\limits_{j=-a}^{b} B_j e^{ij\theta}\right\} = 0$ $-\pi \leq \theta \leq \pi$.

(3) Eqs. (2.2) and (2.3) with right hand sides zero have only the unique solution zero.

We next consider $l_{2,h}$ solutions to the following equation:

$$\left[I(1-r) + \sum_{j=-a}^{b} (C_j - r B_j) S^j\right] \varphi(x,h) = 0, \qquad x \geq 0,$$

for r any constant with $|r| > 1$. It is easily seen that there exists exactly an linearly independent solutions, and in fact they are of the form:

$$\psi_i(x,h) = \sum_{j=1}^{an} \sum_{k=0}^{a-1} P_{ijk}\left(\frac{x}{h}, r\right) (\tau_j(r))^{x/h-a+k}$$

where $\tau_j(r)$ is one of the roots of the equation

$$\det\left\{\sum_{j=-a}^{b} (C_j - r B_j) \tau^{j+a} + I(1-r)\tau^a\right\} = 0$$

which lies within the unit circle, and $P_{ijK}(x/h, r)$ is a polynomial in x/h with vector coefficients. Let us assume that each $\psi_i(jh, h)$ has norm one for some $j \leq -1$. We then define the matrices:

$$A_j = [\psi_1(jh,h), \quad \psi_2(jh,h), \ldots, \psi_{an}(jh,h)]_{n \times an}$$

$$j = -a, \quad -a+1, \ldots, l,$$

$$\text{and} \quad R_j(r) = \begin{bmatrix} [Q_j^{(-1)} - r P_j^{(-1)}] \\ [Q_j^{(-a)} - r P_j^{-a}] \end{bmatrix}_{(an \times n)}$$

$$j = -a, \quad -a+1, \ldots, l,$$

and finally

$$[E(r)]_{(an \times an)} = \sum_{j=-a}^{l} R_j(r) A_j(r).$$

This boundary matrix plays exactly the same role as did $(E(s, w)$ in well-posedness of the equation.

We may now state our main theorem.

Main Theorem. Let (2.2) and (2.3) be consistent with (1.1) and invertible. The following conditions are then sufficient for stability.

(1) Dissipativity, i.e., there exists $\delta > 0$ and p a positive integer, such that the solutions to

$$\det \left[\sum_{j=-a}^{b} (C_j e^{ij\theta} - K B_j e^{ij\theta}) + I(1-K) \right\} = 0$$

have the property that

$$|K_i(\theta)| \leq 1 - \delta \theta^{2p}, \quad i = 1, 2, \ldots, n$$
$$\theta \text{ real}, \quad \pi < \theta \leq \pi$$

(This condition may be weakened slightly, but it is necessary if we wish to extend these results to include variable coefficients).

$$(2) \det [E(r)] \neq 0 \quad \text{for} \quad |r| \geq 1.$$

Proof. We first recognize that eqs. (2.2) and (2.3) may be replaced by an operator equation on $l_{2,h}$

$$(T_0 + S_0) u(\cdot t + h, h) = (T_1 + S_1) u(\cdot t, h)$$

T_0 and T_1 are simple Toeplitz operators, i.e., simple because their kernels are Laurent polynomials, not general l_1 functions. S_0 and S_1 are finite dimensional perturbations involving the boundary conditions. Conditions (1), (2), and (3) of Theorem 2.1 guarantee that (1) T_0 has closed range, (2) T_0 has index zero, (3) $T_0 + S_0$ has an empty null space. These conditions are well known to be necessary and sufficient for invertibility of a completely continuous perturbation of a Toeplitz operator.

Obtaining sufficient conditions for stability involves the following steps:

(1) We derive a general theorem on power-boundedness of operators on a Hilbert space:

Theorem 2.2. Let L be a bounded linear operator on a Hilbert space H. Let P be an orthogonal projection on H. Assume

(1) $\left\| (I-P) L \right\| \leq 1$
(2) $\left\| L(I-P) \right\| \leq 1$
(3) There exists a sequence of non-negative numbers $\{a_n\}$ such that

6 Hyperbolic Equations and Waves

$$\sum_{n=0}^{\infty} a_n = a < \infty$$

and $\|P L^n P\| \leq a_n$ for all $n = 0, 1, 2, \ldots$.

Then $\|L^n\| \leq (1 + a^2)^{\frac{1}{2}}$ for all $n = 0, 1, \ldots$.

(2) We let $L = (T_0 + S_0)^{-1} (T_1 + S_1)$. We can easily obtain a P such that conditions (1) and (2) are valid. This projection is finite dimensional and is chosen so that $I - P$ wipes out the operator near the boundary. Then condition (1) of our main theorem guarantees the validity of (1) and (2) of Theorem 2.2. In order to verify condition (3), we let

$$P[(T_0 + S_0)^{-1} (T_1 + S_1)] P$$
$$= \frac{1}{2 \pi i} \oint_{|r| = m} P r^n (r - (T_0 + S_0)^{-1} (T_1 + S_1))^{-1} dr P,$$

m bigger than the spectral radius of $(T_0 + S_0)^{-1} (T_1 + S_1)$. The expression

$$P(r - (T_0 + S_0)^{-1} (T_1 + S_1))^{-1} P$$
$$= P(r (T_0 + S_0) - (T_1 + S_1))^{-1} (T_0 + S_0) P$$

may be represented as a finite dimensional matrix mapping the range of P into itself, analytic for $|r| \geq 1$ except for poles when $\det \{E(r)\} = 0$. Thus condition (2) guarantees that it is analytic for $|r| \geq 1 - \delta'$, $\delta' > 0$.

Now by the Cauchy integral formula, condition (3) of Theorem 2.2 is valid with $a_n = K(1 - \frac{\delta'}{2})^n$, $K > 0$; the result now follows.

                                                                    q.e.d.

III. Maximum Norm Stability for Parabolic Difference Schemes in Half-Space

This result was first presented at the above mentioned meeting at the University of Denver. What follows is a very brief description of the problem and the result.

Consider a partial differential equation

$$\left[I \frac{\partial}{\partial t} - P(\partial) \right] u(x, t) = 0 \tag{3.1}$$

where $P(\partial) = \sum_{|r| = 2m} A_r \partial^{r_1} \ldots \partial_j^r s$,

$x = (x_1, \ldots, x_s)$

$r = (r_1, \ldots, r_s)$, $r_i > 0$, $|r| = \sum |r_i|$, $\partial_j = \frac{\partial}{\partial x_j}$.

The A_r are complex valued n by n matrices and the differential operator is uniformly parabolic in the sense of PETROVSKY. The equation is to be solved for $0 \leq x_1$, $-\infty < x_i < \infty$, $i = 2, \ldots, s$. We also restrict ourselves to operators $P(\partial)$ such that $A_{(2m,0,0,0)}$ has distinct eigenvalues.

We consider the initial boundary value problem (3.1) with boundary conditions

$$[K u(x_1, \ldots, x_s t)]_{x_1 = 0} = 0 \tag{3.2}$$

for K some appropriate matrix differential operator with constant coefficients, and initial conditions

$$u(x, 0) = f(x). \tag{3.3}$$

We are not here concerned with well-posedness of this problem, rather only with stability conditions for difference approximations.

We introduce rectangular lattices of mesh points

$$R(h) = \{x, x_i = n_1 h, n_1 = -a_1, \ldots, 1, 0, 1, \ldots$$
$$n_1 = 0, \quad \pm 1, \quad \pm 2, \ldots$$
$$i = 2, \ldots, s\}.$$

$R(\lambda, h) = \{(x, t); \ x \in R(h), \ t = nk = n\lambda h^{2m}, \ n$ is an integer with $0 \leq n \leq T/k = $ integer $> 0\}$. We shall always assume that $\lambda = K/h^{2m}$ is a positive constant.

We shall study difference approximations to the equation of the following type.

$$\left(I + \sum_{j=-a}^{b} B_j S^j\right) u(x, t + k, h) = \left(I + \sum_{j=-a}^{b} C_j S^j\right) u(x, t, h); \tag{3.4}$$

$$j = (S_1, \ldots, S_s), \quad S^j = (S_1^{j_1}, \ldots, S_s^{j_s})$$
$$S_1^j u(x, t + k, h) = u(x_1, \ldots, x_i + j_i h, x_{i+1}, \ldots, x_s, t + k, h)$$
$$(x, t) \in R(\lambda, h), \quad 0 \leq t \leq T - K, \quad x_1 \geq 0,$$

$u(x, 0, h) = f(x)$, C_j, B_j constant matrices with boundary conditions of the form:

$$\sum_{j=-a}^{l} P_j^{(i)} S^j u(0, x_2, \ldots, x_s, t + K, h) = \sum_{j=-a}^{l} Q_j^{(i)} S^j u(0, x_2, \ldots, x_s, t, h) \tag{3.5}$$

$$i = 1, \ldots, -a_1.$$

Again the $P_j^{(i)}$ and $Q_j^{(i)}$ are constant matrices. Our results are as follows:

Theorem 3.1. The difference scheme is uniquely invertible if

(a) $\det\left\{I + \sum_{j=-a}^{b} B_j e^{i(j_1 \xi_1 + \cdots + j_s \xi_s)}\right\} \neq 0$

for any real ξ_i real, $|\xi_i| \leq \pi$, $i = 1, 2, \ldots, s$

(b) $\dfrac{1}{2\pi} \underset{-\pi \leq \xi_1 \leq \pi}{\Delta \arg} \det\left\{ I + \sum\limits_{j=-a}^{b} B_j e^{i(j_1 \xi_1 + \cdots + j_s \xi_s)} \right\} = 0$

for real ξ_i, $i = 2, \ldots, s$

(c) Eqs. (3.4) and (3.5) for right hand sides zero have only the unique solution zero.

Definition and Assumption. Let (3.4) be a parabolic difference scheme, i.e.,

$$\left\{ k \sum B_j e^{i(j_1 \xi_1 + \cdots + j_s \xi_s)} - \sum C_j e^{i(j_1 \xi_1 + \cdots + j_s \xi_s)} + k - I \right\}$$

has eigenvalues satisfying

$$|k_i| \leq 1 - \delta'' |\xi|_2^{2m}, \quad i = 1, 2, \ldots, n, \quad \delta'' > 0.$$

and

$$\sum B_j e^{i(j_1 \xi_1 + \cdots + j_s \xi_s)} = 0\left(|\xi|_2^{2m}\right)$$

$$\sum C_j e^{i(j_1 \xi_1 + \cdots + j_s \xi_s)} = 0\left(|\xi|_2^{2m}\right)$$

Moreover let each K_j be real for $\xi_1, \ldots, \xi_2$, in fact require that if we replace ξ_j by $\xi_j + ihn_j$, $j = 2, \ldots, s$ for real ξ_j and hn_j sufficiently small in absolute value, then the imaginary part of each K_j depends only on $\xi_j + ihn_j$, $j = 1, 2, \ldots, s$.
Finally:

$$\left\{ \sum_{j_s} \sum_{j_2} (C_{a_1, j_2 \cdots j_s} - r B_{a_1, j_2 \cdots j_s}) e^{i(j_2 \xi_2 + \cdots + j_s \xi_s)} \right\} \neq 0$$

for real ξ and $|r| \geq 1$.

We next do something completely analogous to the method involving $E(s, w)$ in the first result of Kreiss and involving $E(r)$ in our work above. We obtain a somewhat more involved criterion which we call the parabolic boundary criterion. Essentially it says that under certain conditions the analogous matrix $E(r, \xi_2, \ldots, \xi_s)$ can have a vanishing determinant of first order at $r = 1$, and $0 = \xi_2 = \ldots \xi_s$, but it vanishes nowhere else in the region $|r| \geq 1$, real ξ_i, $|\xi_i| \leq \pi$.

Main Theorem. Under these assumptions, if the difference scheme obeys the parabolic boundary criterion, then there exists a constant C_0 independent of h and $f(x)$ such that:

$$\sup_{(x,t) \in R(\lambda, h)} |u(x, t, h)| \leq C_0 \sup_{x \in R(h)} |f(x)|.$$

On the General Theory of Mixed Problems

R. HERSH

We start with a simple example. Consider the wave equation in a half-plane,

$$u_{tt} = u_{xx} + u_{yy} \quad \text{in} \quad t > 0, x > 0,$$
$$u = u_t = 0 \quad \text{on} \quad t = 0, x > 0, \tag{1}$$
$$B(D_t, D_x, D_y)u = f(t, y) \quad \text{on} \quad t > 0, x = 0.$$

B is a "general" or "arbitrary" polynomial with constant (real or complex) coefficients.

Question One: For which B is the problem solvable? (i.e., solve it if it can be solved.) Question Two: For which B does the solution retain the main feature of the wave equation in unbounded space — finite speed of propagation?

Our open-minded approach to the boundary operator B is in the spirit of modern work in partial differential equations, which looks at *all* operators having some desirable intrinsic property, not just one or two special ones sanctified by tradition. Such breadth is indeed compelled by the variety of problems actually arising in modern engineering and applied physics.

To answer Question One, it is enough to consider the particular case $f = \delta(t)\delta(y)$. This choice avoids the need to check for compatibility conditions at the corner $x = 0$, $t = 0$. If a solution exists in this case, it can be used as a convolution kernel to generate general solutions.

Now if a solution u exists and if

$$\tilde{u} = \int_0^\infty \int_{-\infty}^\infty e^{\tau t + i\eta y} u(t, x, y) \, dy \, dt \tag{2}$$

exists, then $\tilde{u}(\tau, x, \eta)$ satisfies

$$\tau^2 \tilde{u} = \tilde{u}_{xx} - \eta^2 \tilde{u}, \quad B\left(\tau, \frac{\partial}{\partial x}, i\eta\right)\tilde{u} = 1 \quad \text{at} \quad x = 0. \tag{3}$$

Moreover, standard theorems on Fourier and Laplace transforms tell us that $\tilde{u}$ is holomorphic in $\operatorname{Re} \tau > M$ for some M, and is integrable there with respect to η and $\operatorname{Im} \tau$.

A function satisfying these conditions is

$$\tilde{u} = e^{x\xi}/B(\tau, \xi, i\eta) \tag{4}$$

$$\text{where} \quad \xi = \sqrt{\tau^2 + \eta^2}, \quad \operatorname{Re}\xi < 0.$$

Observe that ξ is single-valued and holomorphic in $\operatorname{Re}\tau > 0$ for any fixed real η.

Definition 1. B is stable for the wave equation if, for some M, $B(\tau, \xi, i\eta) \neq 0$ for $\operatorname{Re}\tau > M$ and η real.

From eq. (3) it is evident that if B is stable for the wave equation we can obtain u from $\tilde{u}$ by applying an inverse Fourier transformation and an inverse Laplace transformation to $\tilde{u}$, integrating along the real η-axis and along $\operatorname{Re}\tau = M + \varepsilon$ for some $\varepsilon > 0$. Even if B goes to zero at infinity, $1/B$ can be estimated by a polynomial in τ, η, so the inverse transforms converge at least in the generalized or distribution sense. On the other hand, if B is not stable for the wave equation, (4) does not make sense. Moreover, we can show that in this case the associated problem with homogeneous boundary condition

$$u_{tt} = u_{xx} + u_{yy}, \quad Bu = 0 \quad \text{on} \quad x = 0, \quad t > 0, \tag{1'}$$

$$u, u_t \quad \text{given at} \quad t = 0, \quad x > 0$$

is not well posed. Indeed, to say B is not stable is to say that there exist sequences $\langle \tau_j \rangle$ and $\langle \eta_j \rangle$ such that $\operatorname{Re}\tau_j \to +\infty$, η_j real, $B(\tau_j, \xi(\tau_j, \eta_j), i\eta_j) = 0$. But then $u_j = \delta_j \exp(t\tau_j + x\xi_j + iy\eta_j)$ satisfies (1'), has Cauchy data less than 1, and is arbitrarily large at any point $t > 0$, $x = 0$.

We summarize the discussion up to this point:

Theorem 1. Problem (1) has a solution if B is stable for the wave equation; problem (1) is incorrectly posed with respect to perturbations of the initial data if B is unstable for the wave equation.

Corollary 1. If in particular

$$Bu = b_0 u_0 + b_1 u_x + b_2 u_y + b_4 u_t, \tag{5}$$

with real coefficients b_i, then (1) is solvable if and only if none of the following is true:

$\{b_0 = b_1 = b_4 = 0\}$; $\{0 < b_1/b_4 < 1\}$; $\{b_1 = b_4 \text{ and } b_0/b_1 \geq 0\}$. The proof is an exercise in elementary algebra, checking our definition of stability against these conditions. It is carried out explicitly in [6].

Corollary 2. If (1) is solvable for an operator B_0, then it is solvable for any power of B_0, i.e., if $B = B_0^n$.

Proof. The zeroes of $[B_0(\tau, \xi, i\eta)]^n$ are the same as those of $B_0(\tau, \xi, i\eta)$.

This means, in particular, that we can solve the wave equation with the n'th -order normal derivative prescribed at $x = 0$ (not only the zero'th or the first, as is customary).

It is noteworthy that the situation is radically simpler if we descend to one dimension. Recall that instability of B means in particular that B vanishes at infinitely many points in the τ-plane. Now, if the y-variable and the η-variable are no longer present, $B(\tau, \xi, \eta)$ reduces to $B(\tau, -\tau)$. Since it is an algebraic function of one variable, it can fail to be stable only by vanishing identically. But then $B(\tau, \xi)$ must be a multiple of $\tau + \xi$. Thus we obtain a very simple stability test, not involving any irrational algebraic function.

Theorem 2. For the one-dimensional version of (1), $u_{tt} = u_{xx}$, $B(D_t, D_x)$ is correct for the wave equation (a solution exists) unless $B \equiv 0 \bmod (D_t + D_x)$.

(One-dimensional problems were considered from a general viewpoint by SOBOLEV [16]. His correctness condition is expressed in terms of the roots of the differential operator.)

Now what about Question Two? To see that this is a genuine problem, consider

$$u_{tt} = u_{xx} + u_{yy}, \quad u(0, x, y) = u_t(0, x, y) = 0; \tag{6}$$
$$u_t - u_{yy} = \delta(t)\delta(y) \quad \text{on} \quad x = 0, \quad t \geq 0.$$

Problem (6) describes a motion where the state variable u is vibrating in the interior of the medium and diffusing along the boundary. Notice that this phenomenon depends essentially on having a boundary of dimension ≥ 1. The problem is clearly solvable. We find u by first solving Cauchy's problem for the heat equation and then using the solution as the boundary value for our two-dimensional vibration. It is also clear that the parabolic boundary condition in (6) means that we cannot have any of the three properties — finite speed of propagation, time reversibility, and local uniqueness — which usually hold in hyperbolic problems. In this sense the boundary condition dominates over the differential operator, as Prof. LIGHTHILL remarked here in another connection.

In view of this example, we need a condition on B in addition to stability which will preserve hyperbolicity.

Definition 2. B is hyperbolic for the wave equation if it is stable, and if in addition, constants c, K exist such that the roots $\tau(\eta)$ of the equation $B(\tau, \xi(\tau, \eta), i\eta) = 0$ satisfy $\text{Re}\,\tau < c|\eta| + K$ for all complex η. (Clearly this criterion excludes (6).)

Theorem 3. If B is stable and hyperbolic for the wave equation, the solution of problem (1) is unique. It propagates in the y-direction with speed at most c. If B is stable but not hyperbolic, the solution of (1) is *not* unique.

Proof. This theorem is a special case of Corollaries 1 and 4 of [6] and of the non-uniqueness theorem of [7]. The speed of propagation is obtained from the Paley-Wiener theorem. Uniqueness then follows by HOLMGREN's duality argument. An explicit construction shows non-uniqueness.

(The example (6) was introduced in [6] for purely theoretical purposes. It was a surprise to find subsequently a similar problem derived in [2] as a model for the study of thermal transients in nuclear power reactors.)

Up to now we have focussed on a special differential equation, the wave equation. But the problem we pose and the method we use both have a much wider scope. Consider, for instance, the transverse vibration of a semi-infinite plate. This is governed by a non-hyperbolic, fourth-order equation. (By non-hyperbolic we mean having unbounded domain of dependence.) Two boundary conditions are needed:

$$
\begin{aligned}
u_{tt} + \Delta^2 u &= 0 \quad \text{in} \quad t > 0, \quad x > 0, \\
u &= u_t = 0 \quad \text{on} \quad t = 0, \quad x > 0, \\
B_1(D_t, D_x, D_y)u &= f_1, \quad \text{on} \quad x = 0, \quad t > 0 \\
B_2(D_t, D_x, D_y)u &= f_2, \quad \text{on} \quad x = 0, \quad t > 0.
\end{aligned}
\tag{7}
$$

$\left(\text{Here } \Delta = \dfrac{\partial^2}{\partial x^2} + \dfrac{\partial^2}{\partial y^2}.\right)$ We could replace the f's by delta functions, but this time we leave them general, assuming only that they possess transforms f.

There are three standard problems:

$$
\begin{aligned}
B_1 u &= u, \quad B_2 u = u_x \quad \text{(clamped edge)} \\
B_1 u &= u, \quad B_2 u = u_{xx} \quad \text{(supported edge)} \\
B_1 u &= u_{xx}, \; B_2 u = u_{xxx} \quad \text{(free edge)}
\end{aligned}
$$

If we take transforms as before, we find

$$
\tilde{u} = c_1 e^{\xi_1 x} + c_2 e^{\xi_2 x} = \left[e^{\xi_1 x} \, e^{\xi_2 x} \right] \begin{bmatrix} c_1 \\ c_2 \end{bmatrix}
$$

where ξ_1 and ξ_2 are the two roots of $\tau^2 + (\xi^2 - \eta^2)^2 = 0$ with negative real parts, $\xi_i = \sqrt{\eta^2 \pm i\tau}$. Again ξ is holomorphic in $\operatorname{Re} \tau > 0$. Substituting $\tilde{u}$ into (7), we find

$$
\begin{bmatrix} B_1(\tau, \xi_1, i\eta) & B_1(\tau, \xi_2, i\eta) \\ B_2(\tau, \xi_1, i\eta) & B_2(\tau, \xi_2, i\eta) \end{bmatrix} \begin{bmatrix} c_1 \\ c_2 \end{bmatrix} = \begin{bmatrix} \tilde{f}_1 \\ \tilde{f}_2 \end{bmatrix}
$$

Denoting the matrix on the left by $\tilde{B}$, we make a

Definition 3. $\tilde{B}$ is stable for the vibrating plate if, for some M, $\tilde{B}$ is non-singular for $\operatorname{Re} \tau > M$ and η real.

Theorem 4. If $\tilde{B}$ is stable for the vibrating plate, (7) has a solution. If $\tilde{B}$ is not stable, (7) is incorrect for perturbations of the initial data.

Proof. Almost the same as for theorem 1. The fact that we have to do with a 2-by-2 matrix instead of a single algebraic function $\tilde{B}$ requires some obvious alterations in the previous proof. The details are in [6].

Example. Suppose $\dot{B}_1(u) = \left(\dfrac{\partial}{\partial x}\right)^{a_1} u, \; B_2 u = \left(\dfrac{\partial}{\partial x}\right)^{a_2} u.$ Then it turns out, again as an exercise in elementary algebra, that (7) is solvable if and only if $|a_1 - a_2| = 1$ or 2. Notice how this result includes the three standard problems as special cases.

On the other hand, for the one-dimensional case, the "vibrating beam," (no y- or η-variables, $\Delta^2 = \partial^2/\partial x^2$), (7) is correct unless $|a_1 - a_2| \equiv 0 \bmod 4$.

We have no analog of theorem 2, because there is no limit to the speed of propagation for this equation even in the absence of boundaries.

The example just considered shows that we are not limited to second-order equations or to hyperbolic equations.

Our next example will be a system of equations, Maxwell's equations of electromagnetic theory.

Let $U = [U_1, U_2, U_3, U_4, U_5, U_6]$; in $x > 0, t > 0$,
$$U_t + A_1 U_x + A_2 U_y + A_3 U_z = 0, \quad \text{where}$$

$$A_1 = \begin{pmatrix} 0 & \begin{matrix} 0 & 0 & 0 \\ 0 & 0 & 1 \\ 0 & -1 & 0 \end{matrix} \\ \begin{matrix} 0 & 0 & 0 \\ 0 & 0 & -1 \\ 0 & 1 & 0 \end{matrix} & 0 \end{pmatrix}, \quad A_2 = \begin{pmatrix} 0 & \begin{matrix} 0 & 0 & -1 \\ 0 & 0 & 0 \\ 1 & 0 & 0 \end{matrix} \\ \begin{matrix} 0 & 0 & 1 \\ 0 & 0 & 0 \\ -1 & 0 & 0 \end{matrix} & 0 \end{pmatrix},$$

$$A_3 = \begin{pmatrix} 0 & \begin{matrix} 0 & 1 & 0 \\ -1 & 0 & 0 \\ 0 & 0 & 0 \end{matrix} \\ \begin{matrix} 0 & -1 & 0 \\ 1 & 0 & 0 \\ 0 & 0 & 0 \end{matrix} & 0 \end{pmatrix} \tag{8}$$

At $t = 0$, let $U = 0$.
At $x = 0$, $t > 0$, let BU be given,
$$B = (b_{ij}), \quad i = 1,\ldots 6$$
$$j = 1,2.$$

In this example the boundary $x = 0$ is a characteristic surface. Nevertheless, there is no difficulty in finding a correctness condition analogous to those obtained above, by using Fourier transformation in y and z together with Laplace transformation in t.

We now find that $\tilde{U}$ satisfies

$$\tau \tilde{U} + A_1 \tilde{U}_x + (i\eta A_2 + i\zeta A_3) \tilde{U} = 0.$$

From this it follows that $\tilde{U} = \sum c_j e^{\xi_j x} W_j$ where the $\xi_j(\tau, \eta, \zeta)$ are roots of $\det(\tau I + A_1 \xi + i\eta A_2 + i\zeta A_3) = 0$ having $\operatorname{Re} \xi_j < 0$ and corresponding null vectors W_j. It turns out that $\xi = \sqrt{\tau^2 + \eta^2 + \zeta^2}$ is the only such root. It has multiplicity two, with nullvectors

$$W_1 = (-i\zeta\eta, 0, \tau\xi, -i\eta\zeta, \tau^2 + \eta^2, \zeta\eta) \quad \text{and}$$
$$W_2 = (i\eta\tau, -\tau\xi, 0, -i\zeta\xi, \zeta\eta, \tau^2 + \zeta^2)$$

which are independent for $\operatorname{Re} \tau > 0$. Substituting back, we can solve for c_1 and c_2 if and only if the determinant of a certain 2-by-2 matrix is non-zero; the formula is given in full on p. 253 of [6], where it takes five lines of type to write down. I have never been able to solve this condition in terms of the coefficients b_{ij} in full generality. However, the situation becomes much simpler if we assume, for example, that B is real and $b_{ij} = 0$ for $j = 4, 5, 6$. (I.e., the boundary condition involves only the electric field vector (u_1, u_2, u_3).) The standard condition in electromagnetic theory is just

$$B = \begin{pmatrix} 0 & 1 & 0 & 0 & 0 & 0 \\ 0 & 0 & 1 & 0 & 0 & 0 \end{pmatrix}.$$

Now the stability determinant reduces to three terms, and its real part is just

$$\begin{vmatrix} b_{12} & b_{13} \\ b_{22} & b_{23} \end{vmatrix} \neq 0.$$

In other words, we can prescribe any two components of (U_1, U_2, U_3) as long as they do not determine U_1, the normal component. Classically, one prescribes precisely the tangential components.

For more examples and discussion on MAXWELL's equations, LAMÉ's equations in elasticity, the heat equation and the SCHRÖDINGER equation, the reader is referred to [6].

It is obvious now that there is a general principle here which lies behind all our special cases. We always consider a half-space $x > 0$ and a differential operator $P(D_t, D_x, D_y)$ with constant coefficients. By taking Laplace transform in t and Fourier transform in y (or y and z, or $y_1 \ldots y_n$, depending on the number of space variables) we obtain for $\tilde{u}$ an ordinary differential equation in x with constant coefficients, whose solution is a sum of finitely many exponentials.

We have to use only those exponentials $e^{\xi x}$ with $\operatorname{Re} \xi < 0$; this guarantees that inverse transforms exist. So an essential ingredient of our construction is that for $\operatorname{Re} \tau > M$ the roots ξ of $\det P(\tau, \xi, i\eta) = 0$ should fall into two distinct groups, the "admissible" roots with $\operatorname{Re} \xi < 0$ and "inadmissible" ones with $\operatorname{Re} \xi > 0$. It is shown in [5] and [6] that this is so if P is hyperbolic or parabolic, or more generally, correct in the sense of PETROWSKY.

Then the number of boundary conditions to impose just equals the number of roots ξ with $\operatorname{Re} \xi < 0$.

Those solutions of the transformed equation which correspond to admissible roots ξ constitute a linear space. Let $E^-(\tau, i\eta)$ be a matrix whose columns are a basis for this space. Then the boundary operator $B(D_t, D_x, D_y)$ is a matrix with as many rows as the columns of E^- and as many columns as the rows of E . Let $\tilde{B} = (B(\tau, D_x, i\eta)E^-)|_{x=0}$.

Then [6] Theorem 1 is correct if the wave equation is replaced by an arbitrary constant-coefficient equation or system of equations which is correct in the sense of PETROWSKY, and if in Definition 1, B is replaced by $\det \tilde{B}$.

Corollary 2 is still true where there is only one boundary condition, as in the heat equation for example.

Theorem 2 generalizes as follows.

Theorem 5. If $P(\tau, \xi)$ is a polynomial in two complex variables, and in some right half-plane $\operatorname{Re} \tau > M$ there exists exactly one root $\xi(\tau)$ with $\operatorname{Re} \xi < 0$ satisfying $P(\tau, \xi(\tau)) \equiv 0$, then the problem $\{P(D_t, D_x)u = 0$ in $t > 0$, $x > 0$; $B(D_t, D_x) = \delta(t)$ in $t > 0$, $x = 0$; with zero Cauchy data in $t = 0$, $x > 0\}$ is solvable unless $B \equiv 0$ modulo that irreducible factor of P corresponding to the admissible root.

For example, any boundary operator is stable for the one-dimensional heat equation if it is not a multiple of the heat operator $D_t - D_x^2$.

The proof is a simple algebraic consequence of the fact that in one space variable B can fail to be stable for P only if $B(\tau, \xi(\tau)) \equiv 0$ identically.

If P is hyperbolic (not necessarily strictly hyperbolic), Theorem 3 is still true when the wave equation is replaced by $Pu = 0$ and in Definition 2 B is replaced by $\det \tilde{B}$. (See [6] and [7].)

Thus we have answered Questions One and Two above reasonably well, in the case of constant coefficients and a half-space. We conclude by mentioning some variants and extensions of these ideas.

(1) We can apply the same transform procedures to *difference* equations in a half-space [9], obtaining a representation for the solution in terms of the transform of the boundary data. This leads easily to sharp criteria for stability and convergence of constant-coefficient finite-difference schemes, if one views the solution as being stepped out in the

x- rather than the t-direction. An alternative approach to these problems, using the theory of Toeplitz operators, was introduced by STRANG [17] for a single first-order equation, and has recently been carried through for first-order systems by OSHER [14].

(2) Instead of a boundary problem in a half-space, we can consider a "conjugacy" or "transmission" problem in two half-spaces, $x > 0$ and $x < 0$. The goal then becomes determination of all correct conjugacy conditions for a given pair of differential operators, P_L and P_R, which govern the motion in the left and right half-spaces.

Among the examples treated in [8] are:

$$P_L = D_t - \Delta, \quad P_R = D_t^2 - \Delta$$
$$P_L = D_t - \Delta, \quad P_R = \Delta$$
$$P_L = D_t^2 - \Delta, \quad P_R = \Delta$$

where $\Delta = D_x^2 + D_y^2$.

The last two examples are noteworthy because they are *not* PETROW-SKY-correct. As a system of equations they would not define a well-posed Cauchy problem; nevertheless, for suitable conjugacy or boundary conditions, they are solvable in the mixed problem. The third example above is related to FRIEDRICHS' "hemi-hyperbolic" systems; the equation in the right half-space can be thought of as describing a vibrating medium with infinite wave speed.

(3) Another variant [10] is to consider $0 < x < 1$ instead of $0 < x < \infty$. It turns out, rather surprisingly, that even this slight twist leads to new phenomena. The solution to this problem can be given as an infinite series of reflected solutions, provided the interior differential operator P satisfies a certain new algebraic condition: For all M there exists $K(M)$ such that if $\det P(\tau, \xi, i\eta) = 0$ and $\operatorname{Re} \tau > K$, then $|\operatorname{Re} \xi| > M$. This condition is not satisfied by all PETROWSKY-correct operators; it is satisfied by hyperbolic or PETROWSKY-parabolic operators, among others.

These examples and applications have, I hope, given the main reason for considering mixed problems from a general viewpoint — there is an ample supply of surprising facts and amusing theorems to be gathered up. It is easy to list open questions and unsolved problems.

(1) If P is a symmetric first-order hyperbolic system, the *energy* $E(t)$ is defined as $\int\int u \cdot u\, dx\, dy$. FRIEDRICHS [4] and LAX and PHILLIPS [12] have given a condition on B, called "dissipativity", which is sufficient to imply

$$E(t) \le E(0). \tag{9}$$

For dissipative problems, they prove existence even if P and B have variable coefficients. In the case of two dependent variables, STRANG [18] recently showed that dissipativity is necessary for (9). What about more than two unknowns?

(2) In the theory of Cauchy's problem (pure initial-value problem) the notion of *strict* hyperbolicity plays a prominent role. This is an algebraic condition on the operator which enables one to estimate the solution $u(t)$ in terms of the initial data $u(0)$, using the same norm for $u(t)$ and $u(0)$. Let us use the term strong estimates for estimates of this type. If one has a strong estimate for the solution, an iterative procedure permits one to solve the equation even if it is perturbed by arbitrary terms of lower order.

What condition in the mixed problem would be an appropriate generalization of strict hyperbolicity? This question has been considered by SARASON [15] and more recently by H. O. KREISS [19]. Both authors use the Fourier-Laplace transform, and both consider a first-order system P with zero-order boundary operator B. Both find conditions on P and B that yield strong estimates. The form of the estimates and of the conditions are quite different. With the help of a pseudo-differential "symmetrizing" operator, KREISS is able to find solvability conditions which permit variable coefficients even in the leading terms of the equation.

(3) Instead of adding extra conditions to obtain strong estimates, an alternative approach is to try to use weak estimates to go from constant to variable coefficients. This may seem an unpromising line of attack, but it is the only one that can keep the full generality of the results we have for constant coefficients. In order to keep this generality for the differential and boundary operators, one might have to restrict the data to very smooth functions (members of a Gevrey class). This viewpoint seems to have much in common with that of Leray in his recent work with Ohya on non-strictly hyperbolic Cauchy problems. Perhaps his method of formal norms is the tool now needed in the general theory of mixed problems.

Added in proof. In [21] Da Prato uses the notion of semi-group of growth t^{-n} (as $t \searrow 0$) to solve a variety of variable-coefficient mixed problems with Dirichlet boundary conditions. In [22] Ikawa proves that the constant-coefficient half-space mixed problem for the wave equation is not "well-posed in the L_2 sense" if $\dfrac{\partial}{\partial x} + b\dfrac{\partial}{\partial y}$ is the boundary operator, $b \neq 0$. Of course, our Corollary 1 shows that this problem *has* a solution, but our requirements for well-posedness are less stringent than Ikawa's. Thus Ikawa's result, like Da Prato's, indicates that a satisfactory treatment of general mixed problems calls for a function-space framework broader than those one is accustomed to using in the pure Cauchy problem.

References

1. AGMON, S.: Problèmes mixtes pour les équations hyperboliques d'ordre supérieur, Colloques sur les équations aux dérivées partielles. C. N. R. S 13−18 (1962).

2. BIRKHOFF, GARRET, and ROBERT E. LYNCH: Numerical Solution of The Telegraph and Related Equations, Numerical Solution of Partial Differential Equations. 373 pp. Proceed. Symp. Univ. Maryland 1966.

3. DUFF, G. F. D.: Mixed Problems for linear systems of first order equations. Canada Jour. of Math. **10**, 127−160 (1958).

4. FRIEDRICHS, K. O.: Symmetric Positive Linear Differential Equations. Comm. Pure and Appl. Math., Vol. XI, No. 3, 1958.

5. HERSH, REUBEN: Mixed Problems in Several Variables, Jour. Math. and Mech., Vol. 12, No. 3, 317−334 (1963).

6. − Boundary Conditions for Equations of Evolution. Arch. Rational Mech. and Anal., Vol. 16, No. 4, 243−263 (1964).

7. − On Surface Waves with Finite and Infinite Speed of Propagation. Arch. Rational Mech. and Anal., Vol. 19, No. 4, 308−316 (1965).

8. − On Vibration, Diffusion, Equilibrium Across a Plane Interface. Arch. Rational Mech. and Anal., Vol. 21, No. 5, 368−390 (1966).

9. − On the Theory of Difference Schemes for Mixed Initial Boundary Value Problems. SIAM Jour. Numer. Anal., Vol. 5, No. 2 (1968).

10. − The Method of Reflection for Two-sided Problems of General Type. Denver Univ.-NSF Conference on Stability Theory of Initial-Boundary Value Problems (1968).

11. LAX, P. D., and K. O. FRIEDRICHS: Boundary Value Problems for First Order Operators. Comm. on Pure and Appl. Math., Vol. XVIII, No. 1/2 (1965).

12. −, and R. S. PHILLIPS: Local Boundary Conditions for Dissipative Symmetric Linear Differential Operators. Comm. on Pure and Appl. Math., Vol. XIII, No. 3 (1960).

13. MIZOHATA, S.: Quelques Problèmes Au Bord, Du Type Mixte Pour Des Équations Hyperboliques. College de France 1966−1967.

14. OSHER, STANLEY: Systems of Difference Equations with General Homogeneous Boundary Conditions. Brookhaven National Laboratory, Report No. BNL 11857, 1/16/68.

15. SARASON, LEONARD: On Hyperbolic Mixed Problems. Arch. for Rational Mech. and Anal., Vol. 18, No. 4, 310−334 (1965).

16. SOBOLEV, S. L.: On Mixed Problems for Partial Differential Equations with Two Independent Variables, Doklady. Akad. Nauk USSR 1958.

17. STRANG, G.: Wiener-Hopf Difference Equations. J. Math. Mech. **13**, 85−96 (1964).

18. – Hyperbolic Initial-Boundary Value Problems in Two Unknowns I Diff. Eq. 1969.

19. KREISS, H. O.: Initial boundary value problems for hyperbolic systems, Uppsala University, Department of Computer Sciences, May 1969.

20. LERAY, J., and Y. OHYA: Systèmes Linéaries, Hyperboliques non Stricts. Colloque de Liège, C. N. R. B. 1964.

21. DA PRATO, GIUSEPPE: Problèmes Au Bord de Type Mixte Pour Des Équations Paraboliques ou Hyperboliques, Collège de France (1967—1968).

22. IKAWA, MITSURU: On the Mixed Problem for the Wave Equation with an Oblique Derivative Boundary Condition, Proceed. Japan Acad., Vol. 44, No. 10, 1033—1037 (1968).

J. Inst. Maths Applics **1,** 1–28.

Group Velocity

M. J. Lighthill †

This paper gives a survey of the theory of group velocity for one-dimensional and three-dimensional, isotropic and anisotropic, homogeneous and inhomogeneous, conservative and dissipative, linear and non-linear, classical and relativistic systems exhibiting wave propagation under free and forced-motion conditions.

1. Introduction

IN choosing a subject for this lecture, I had in mind the aim which permeates the policy of the I.M.A. journals and Programme Committee, namely, to choose subject matter that will be of interest to as many as possible of our members, and in particular, to mathematicians working in more than just one branch or field of application of mathematics. It occurred to me that these conditions would be met by a general survey of what is known today about group velocity. The theory of group velocity is an essentially mathematical theory that has been developed over the years with an eye on a great variety of spheres of application. I am going to give an account of what seem to me the most important parts of our knowledge of this subject, largely ignoring history and concentrating rather on mathematics (and, of course, on its applications!).

I should like to apologize to those who attended Professor Synge's three very erudite lectures here at King's College recently on the same subject. I recognized that the subject was indeed the same too late to be able to change significantly the material of this lecture, but I hope that even those who attended both may still find useful an elementary and relatively brief account of the subject, that proceeds by degrees from simpler cases to more and more general ones.

I am sure that the properties of group velocity always surprise somebody who meets them for the first time. The most striking feature of waves is, without doubt, their capability of carrying energy over long distances; sometimes, of course, as well as the energy, they carry also, more or less imperfectly, information. With many waves, furthermore, the velocity of crests and troughs and their regular progress is extremely evident. It is natural to imagine, then, that this "phase velocity" is also the velocity with which energy is propagated by the wave, particularly because in some simple cases, including sound waves and waves on a flexible string, the two velocities are indeed the same. For the vast bulk of wave motions occurring in nature, however, the phase velocity, with which the crests and troughs are propagated, and the group velocity, with which the energy is propagated, have quite different magnitudes; and

† An Inaugural Lecture by the President of the Institute of Mathematics and its Applications delivered in King's College, London, on 25 November, 1964.

where the system possesses any degree of anisotropy they are normally also quite different in direction.

Waves of interest include one-dimensional waves, such as may propagate along a string, or a transmission line; two-dimensional waves, such as may propagate over a water surface, or other surfaces of separation between different phases of matter; and three-dimensional waves, propagating freely in space. One-dimensional systems, and isotropic two- and three-dimensional systems, that is systems without quantitative differences between waves travelling in different directions, have at least the consoling property that energy is propagated in the same direction as that in which the crests move. The magnitudes of the phase and group velocities are, however, not equal for any waves whose phase velocity takes different values for waves of different length. This state of affairs is usually described as dispersion, because it means that if we imagine any general disturbance split up into components of different wavelength, all these components will progress at different speeds, and therefore will tend to get separated out, that is "dispersed", into a long wave train with the wavelength varying rather gradually along it.

In this process of dispersion, the energy associated with waves of a given length is propagated at the group velocity, say u, of those waves. Hence, after a time t has elapsed, waves of that length will be found a distance ut farther on. Anyone who imagined that, because the crests travel with the phase velocity v, those waves should be observed at the quite different distance vt, would fail to find them. There he would find waves, if any, of quite different length, namely those whose group velocity has the value v. If he were mesmerized, in fact, into trying to follow individual crests, he would find that these evolve into crests of waves of continually changing length, or even disappear altogether. Only if he rigorously fixed his gaze on a point moving with the group velocity u would he find that he was continually observing waves of the same length. Thus, although it is variations in the phase velocity which cause the phase and group velocities to be different, it is variations in the group velocity which produce the dispersion.

Waves, as I said earlier, are often used as carriers, not merely of energy, but also of information. Needless to say, the transmission of information is not assisted by dispersion, and the dispersed waveform is hard to unscramble into the original waveform. This is why every effort is made in transmission lines to match constants so that dispersion becomes negligible. When this is successfully done, then the phase and group velocities once more coincide. However, there are relatively few systems where this is possible.

Many different approaches to the mathematics of group velocity have been used. They can be divided into, first, kinematic approaches, which consider a general linear combination of waves of varying length with frequency functionally related to wavelength, and work out how the waves of different lengths become dispersed from one another, and, secondly, dynamic approaches, which are basically quadratic rather than linear. They prove, in particular, that energy flow takes place at the group velocity and so make it possible to predict not only where waves of a given length will be, but also what their amplitudes will be. I propose to describe both kinematic and dynamic methods in this lecture, for one-dimensional and multi-dimensional systems, with and without isotropy, with and without homogeneity and with and without

7   Hyperbolic Equations and Waves

dissipation. The work will be confined to systems satisfying linear equations, except for a brief final description of extensions to non-linear problems, in which Prof. Whitham (1965a,b) and others have been active in recent years.

2. The Kinematics of Wave Crests (One-dimensional)

I shall begin, then, with the simple case of one-dimensional wave motions, to which it is relatively easy to reduce the case of isotropic multi-dimensional wave motions. The simplest method of deriving the formula for group velocity in this case, as indeed in more general cases, is to consider the kinematics of wave crests (Lighthill & Whitham, 1955; see also Havelock, 1914). This procedure is suitable at any rate when dispersion has already caused the lengths of adjoining waves to differ by only a small fraction. Under these circumstances, it is so rare for a crest to cease to be a crest, or to divide into two crests, that we can to a good approximation apply a law of conservation of numbers of crests. This law takes the form of a hydrodynamical equation of continuity:

$$\frac{\partial k}{\partial t}+\frac{\partial \omega}{\partial x} = 0, \tag{1}$$

where ω is the number of wave crests per second, i.e. the frequency, and k is the number of wave crests per cm, i.e. the wavenumber (the reciprocal of the wavelength). Equation (1) means that the change in number of crests in a fixed length equals the difference between inflow of crests at one end and outflow at the other. If in equation (1) we write $\omega = kv$, where v is the phase velocity, the analogy with the equation of continuity becomes particularly apparent.

In a wave motion in which the frequency ω is a function of the wavenumber k, equation (1) becomes

$$\frac{\partial k}{\partial t}+u\frac{\partial k}{\partial x} = 0, \tag{2}$$

where $u = d\omega/dk$ will be called the group velocity. Equation (2) says that, in the (x,t)-plane, along paths $x-ut =$ constant (travelling with speed u), the wavenumber k remains unchanged (Fig. 1). Waves of a given length are found, in fact, at points

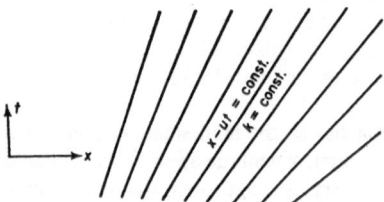

FIG. 1. Homogeneous one-dimensional system. Geometrical significance of equation (2) which governs wave dispersion.

along such paths, and there is a different path for each wavelength. Thus, the basic fact about group velocity for such a system has already been derived.

Here we have taken the system as homogeneous, but an inhomogeneous system is not much more complicated to treat. This is a system where the relationship between frequency and wavenumber changes gradually with position, so that $\omega = \omega(k,x)$. We cannot then deduce equation (2) for k from equation (1). We can, however, multiply equation (1) by

$$u = \left(\frac{\partial \omega}{\partial k}\right)_x \tag{3}$$

(where the subscript x signifies a derivative keeping x constant) and deduce

$$\frac{\partial \omega}{\partial t} + u\frac{\partial \omega}{\partial x} = 0, \tag{4}$$

because

$$\left(\frac{\partial k}{\partial t}\right)_x \left(\frac{\partial \omega}{\partial k}\right)_x = \left(\frac{\partial \omega}{\partial t}\right)_x. \tag{5}$$

Equation (4) says that the frequency ω remains constant along paths (Fig. 2) satisfying

$$\frac{dx}{dt} = u. \tag{6}$$

FIG. 2. Inhomogeneous one-dimensional system. Geometrical significance of equation (4) which governs wave dispersion.

Here, u will vary with position along such a path, but in a known manner because it can be expressed as a function of x and of the frequency ω which is constant along the path. Wave packets of a given frequency propagate, then, with velocity u although their crests travel with the quite different velocity $v = \omega/k$.

All this argument can be put into strictly physical language, in terms of an observer who travels (say) with velocity U, and who therefore passes $Uk - \omega$ crests per second. If another observer follows on behind him, passing every point a time T later, and the two observers follow a path on which the frequency ω remains constant, then the number of wavecrests between the observers is ωT which is constant, so that the value of $Uk - \omega$, the number of wavecrests per second passed by each, must be the same. It follows that

$$U = \frac{\Delta\omega}{\Delta k}, \tag{7}$$

which is the change in frequency during time T at a fixed point divided by the change in wavenumber during the same time interval at the same point—in full agreement with expression (3).

It is, of course, quite possible for waves of two more different frequencies to have the same value of the group velocity u; in this case, all of them may be found travelling together along the same path, where an apparently irregular motion would be observed

7*

and Fourier analysis would be needed to resolve it into a small number of sine wave components. The argument I have given could nevertheless be justified by applying it separately to these different components. Another case where apparent local chaos is intelligible only in the light of Fourier analysis is when one wave packet passes through another wave packet which has lower group velocity.

The solution of equation (6) can be written in various forms, but the most convenient is one in which we begin by writing the wavenumber k as a function of the frequency ω rather than the other way round. This is because ω remains constant along the path. Then the equation of such paths in an (x,t) diagram is

$$
\begin{aligned}
t &= \int \frac{dx}{u} + \text{constant} \\
&= \int \left(\frac{\partial k}{\partial \omega} \right)_x dx + \text{constant} \\
&= \frac{\partial}{\partial \omega} \int k(\omega,x) dx + \text{constant},
\end{aligned}
\tag{8}
$$

a form requiring us merely to integrate a known function with respect to one of its variables and differentiate it with respect to the other.

If, furthermore, we argue that the total energy in each part of the frequency spectrum remains constant, then it follows that the energy between two adjacent paths must remain constant. Obviously this can be used to infer changes in amplitude if the relationship of energy density to amplitude is known. This method can be used just as reliably for two- and three-dimensional propagation in isotropic systems, where the normal law of dependence of energy density on amplitude is modified by an additional factor x^{n-1}, where n is the number of dimensions. However, in the neighbourhood of a caustic, i.e. an envelope of paths, the method gives a locally infinite value of the energy density, which in reality is not found; the true peak in energy density near the caustic cannot be calculated by this simple approximate method, which as we shall see has to be replaced by a more refined one within a few wavelengths of the caustic.

3. The Method of Stationary Phase (One-dimensional)

The method so far described gives results quickly, but is not demonstrably firm on its foundations, which therefore need bolstering up with some rather more rigorous analysis of the result of linearly combining waves of different length when the frequency is a function of the wavelength. The classical argument of Stokes (1876) infers a surprising amount from the simple special case when waves with just two values of wavelength are combined:

$a \cos 2\pi(k_1 x - \omega_1 t) + a \cos 2\pi(k_2 x - \omega_2 t)$

$$
= \{2a \cos 2\pi[(k_2 - k_1)x - (\omega_2 - \omega_1)t]\} \cos 2\pi[(k_1 + k_2)x - (\omega_1 + \omega_2)t].
\tag{9}
$$

The first term on the right is a slowly varying amplitude for the rapidly varying second term, so that the formula can be interpreted (Fig. 3) as a series of packets travelling with the velocity

$$
U = \frac{\omega_2 - \omega_1}{k_2 - k_1}
\tag{10}
$$

and incapable of exchanging energy with one another through the nodal points where this amplitude is zero. It might be expected from this that a general wave packet, for which neighbouring portions of wave have only slightly different wavenumbers, would

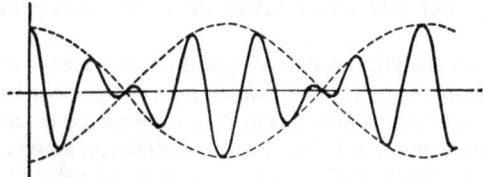

FIG. 3. The linear combination of two sine-waves interpreted as a series of wave packets (equation (9)).

propagate at a speed equal to the limit of (10) for low frequency difference, namely,

$$u = \frac{d\omega}{dk}. \tag{11}$$

Obviously, the argument is still far too special, but it contains one idea that is applicable to much more general cases. A point where the amplitude of (9) takes the full value $2a$, equal to the sum of the amplitudes of the component waves, travels along at the velocity (10) for one very good reason: this point passes wavecrests of the two wave systems at rates, $Uk_1 - \omega_1$ and $Uk_2 - \omega_2$, which are *equal*. It is possible, therefore, for these wavecrests to remain in phase, which for any other value of U could not happen.

We now consider a much more general combination of waves of different lengths in a homogenous system where ω is a function of k. Such a general linear combination is

$$\phi = \int_{-\infty}^{\infty} f(k) \exp\{2\pi i[kx - \omega(k)t]\}\, dk. \tag{12}$$

We consider the problem of estimating this integral when t is large and x may also be large. We assume that ω is an analytic function of k. We also assume that that the form taken initially, when $t = 0$, by the dependent variable ϕ in which we are interested, namely

$$\phi = \int_{-\infty}^{\infty} f(k)\, e^{2\pi i kx} dk, \tag{13}$$

represents a disturbance confined to a limited region, that is, to a limited interval of values of x. This requires that $f(k)$ also is an analytic function of k.

We shall see that the main contribution to the integral (12) from a given small interval of wavenumber is found when x and t have values such that the phase $kx - \omega(k)t$ is stationary, i.e. practically constant throughout the interval, so that once more it is possible for the crests of different wave components in this interval to reinforce one another, instead of tending to cancel out by interference. We can prove this most easily (Jeffreys & Jeffreys, 1950, p. 505) by writing the phase in equation (12) as $t\psi(k)$, so that

$$\psi(k) = k\frac{x}{t} - \omega(k). \tag{14}$$

We then transform the integral into one with respect to ψ, namely,

$$\phi = \int f \frac{dk}{d\psi} e^{2\pi it\psi} d\psi, \tag{15}$$

the integral being over that set of values of ψ that corresponds to the interval $-\infty < k < \infty$.

If $\psi(k)$, which is an analytic function of k, is also one-valued, with $\psi'(k)$ everywhere positive, or everywhere negative, then $f \, dk/d\psi$ is an analytic function of ψ itself, and the theory of Fourier integrals then tells us that, for large t, ϕ is very small, of smaller order than any inverse power of t. The situation is different, however, if $\psi'(k)$ becomes zero anywhere. This usually implies that more than one value of k corresponds to each value of ψ, but, more important for our purpose, it means that $dk/d\psi$ is singular. Where $\psi'(k)$ has a simple zero, $dk/d\psi$ has an inverse-square-root singularity; where $\psi'(k)$ has a double zero, $dk/d\psi$ has an inverse-two-thirds-power singularity, and so on. We can then use the theory of the asymptotic behaviour of Fourier integrals (Lighthill, 1958), which tells us that the asymptotic value of equation (15) for large t is determined by the behaviour of $f \, dk/d\psi$ at its *worst singularities*. These are the zeros of $\psi'(k)$, in other words, the points of stationary phase. In the commonest case, when only simple zeros are present, there is a contribution from the inverse-square-root singularity corresponding to each. If the zeros of $\psi'(k)$ are $k_1, k_2, \ldots, k_m$, the asymptotic form of the integral (15) is

$$\phi \sim \sum_{r=1}^{m} \frac{f(k_r) \exp \left[2\pi it\psi(k_r) + \frac{1}{4}\pi i \, \text{sgn} \, \psi''(k_r) \right]}{[t \, | \, \psi''(k_r) \, | \,]^{\frac{1}{2}}}$$

$$= \sum_{r=1}^{m} \frac{f(k_r) \exp \left\{ 2\pi i [k_r x - \omega(k_r)t] - \frac{1}{4}\pi i \, \text{sgn} \, \omega''(k_r) \right\}}{[t \, | \, \omega''(k_r) \, | \,]^{\frac{1}{2}}}. \tag{16}$$

Here sgn $\psi''(k_r)$ is $+1$ where $\psi''(k_r) > 0$ so that ψ has a minimum and is -1 where $\psi''(k_r) < 0$ so that ψ has a maximum.

Physically, the zeros of $\psi'(k)$ are the points where

$$x = t\omega'(k), \tag{17}$$

that is, the values of the wavenumber k for which the speed of propagation $\omega'(k)$ of the energy associated with that wavenumber carries it a distance x during the (large) time t. There is only one term in the sum (16) when the group velocity $\omega'(k)$ takes each value only once, but in general, as mentioned before, two or more wave packets may be superimposed on one another and travel along together because they happen to have the same value of the group velocity.

The amplitude falls off like the inverse square root of the time because after time t the energy is spread over a distance proportional to t, so that the energy density varies as t^{-1} and the amplitude, therefore, as $t^{-\frac{1}{2}}$. Specifically, the energy between wavenumbers k_r and $k_r + dk$ is initially $Ef^2(k_r) \, dk$, where E is some constant, and later spreads out to fill a distance

$$\left| t\omega'(k_r) - t\omega'(k_r + dk) \right| = t \, | \, \omega''(k_r) \, | \, dk. \tag{18}$$

Then, therefore, the energy density is

$$\frac{Ef^2(k_r)}{t \, | \, \omega''(k_r) \, |}, \tag{19}$$

consistently with the value

$$\frac{f(k_r)}{[t \, | \, \omega''(k_r) \, | \,]^{\frac{1}{2}}} \tag{20}$$

for the amplitude. Thus, in the sum (16) the only extra piece of information that we would not have been able to predict from the kinematic analysis is the phase factor

$$\exp \left[-\tfrac{1}{4}\pi i \, \text{sgn} \, \omega''(k_r) \right] \tag{21}$$

which simply decreases the phase by 45° where ω'' is positive and increases it by the same amount where ω'' is negative.

The special case, when $\omega''(k_r) = 0$ for a particular solution $k = k_r$ of the equation $x/t = \omega'(k)$, needs individual treatment, and corresponds to the caustic that I mentioned earlier (Fig. 4). The simple form (16) of the solution is then not applicable.

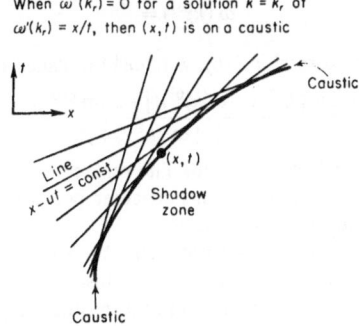

When $\omega''(k_r) = 0$ for a solution $k = k_r$ of
$\omega'(k_r) = x/t$, then (x,t) is on a caustic

FIG. 4. Homogeneous one-dimensional systems; geometry of a caustic.

The integral (15) with respect to ψ has an inverse-two-thirds-power singularity, and the asymptotic contribution from waves of wavenumber k_r is

$$\frac{f(k_r) \exp \{2\pi i[k_r x - \omega(k_r)t]\} (\tfrac{1}{3})! \sqrt{3}}{[\tfrac{1}{3}\pi t \omega'''(k_r)]^{\frac{1}{3}}} , \tag{22}$$

falling off only as the inverse cube root of the time. Sometimes we need to approximate ϕ not actually on the caustic, as here, but rather near it. Then it is best to evaluate the contribution from $k = k_r$, where $\omega''(k_r)$ is nearly but not exactly zero, by replacing $\omega(k)$ in the original Fourier integral (13) by the first four terms of its Taylor expansion about $k = k_r$, and evaluating the remaining integral in terms of the Airy function (Jeffreys & Jeffreys, 1950, p. 508). This function decays exponentially on the shadow side of the caustic, and is oscillatory on the other side. In between it rises to a peak close to the value (22).

The analysis I have just given in the particular case of systems homogenous in x has confirmed the value of the approximation derived from the kinematics of wavecrests. Now, instead of duplicating the extension to the non-homogenous case, which is cumbersome by this method, I shall give rather a brief analysis of the extension to dissipative systems, considering only such small dissipation as reduces the wave amplitude by only a small fraction in one wave period. I suppose in fact that for waves of wavenumber k the amplitude is reduced by a factor $e^{-2\pi\sigma(k)t}$ in time t, where $\sigma(k) \ll \omega(k)$. (We assume further in what follows that $\sigma'(k) \ll \omega'(k)$.) Then $\omega(k)$ must be replaced by

$$\omega(k) - i\sigma(k) \tag{23}$$

in the integral (12). The analysis then goes through as before, but the k_r must satisfy

$$\omega'(k_r) - i\sigma'(k_r) = \frac{x}{t}. \tag{24}$$

An approximate solution for small σ is

$$k_r = k_r^{(0)} + \frac{i\sigma'(k_r^{(0)})}{\omega''(k_r^{(0)})}, \tag{25}$$

where $k_r^{(0)}$ is the solution with no dissipation, satisfying

$$\omega'(k_r^{(0)}) = \frac{x}{t}. \tag{26}$$

When we use the earlier expression (16), with $\omega(k)$ replaced by (23), it becomes

$$\phi \sim \sum_{r=1}^{m} \frac{f(k_r^{(0)}) \exp\{2\pi i[k_r^{(0)}x - \omega(k_r^{(0)})t] - 2\pi\sigma(k_r^{(0)})t - \frac{1}{4}\pi i \, \text{sgn} \, \omega''(k_r^{(0)})\}}{\{t \mid \omega''(k_r^{(0)}) \mid\}^{\frac{1}{2}}} \tag{27}$$

to a first approximation, when σ is small. The replacement here of k_r by $k_r^{(0)}$ can be shown to introduce very small errors, in particular because $k_r^{(0)}$ is a point where the phase term in square brackets is stationary.

The physical meaning of the asymptotic form (27) is simply that every wave packet exhibits a time rate of decay of amplitude to a close approximation the same as for periodic waves of the same wavelength, and that in other respects the theory of group velocity is not affected by the presence of a small amount of dissipation. By contrast, when the dissipation rate $\sigma(k)$ is comparable with the total variation in $\omega(k)$, as is the case in anomalous dispersion of an electromagnetic wave, the theory of group velocity is affected very greatly indeed (Stratton, 1941).

4. The Kinematics of Wave Crests (Homogeneous Anisotropic Systems)

But it is time to consider the propagation of waves in anisotropic systems. These are important, of course, in relation to waves such as electromagnetic or elastic waves in anisotropic media such as crystals; they are also important in relation to waves in the presence of an external field; for example, rotation about an axis makes many types of mechanical waves anisotropic; so does a gravitational field when the medium has non-uniform density, and a magnetic field when it is conductive. Now, it is reasonable, encouraged by the success of the method which used the kinematics of wave crests in homogeneous one-dimensional systems, where the conclusions were fully substantiated by the more rigorous analysis, to apply the same method to anisotropic three-dimensional systems, first homogeneous and then non-homogeneous (Whitham, 1960; see also Hamilton, 1837). In making this jump we are leaving out the quite important two-dimensional systems, but that gap is, obviously, rather easy to fill in.

In this three-dimensional case, the geometry and kinematics of the wavecrests can be specified (Fig. 5) by means of a phase function $U(x_1, x_2, x_3, t)$, which is continuous and changes by unity between one crest and the next; more generally, the phase function U changes by 1 in any distance of one wavelength perpendicular to the surfaces of constant U. Thus, we suppose that each of the local physical quantities takes the form

$$\mathscr{R}\{a(x_1,x_2,x_3,t) \exp [2\pi i U(x_1,x_2,x_3,t)]\},\tag{28}$$

in which a complex amplitude, in general different for each quantity, is multiplied by a phase factor the same for each quantity. Obviously, different physical quantities may

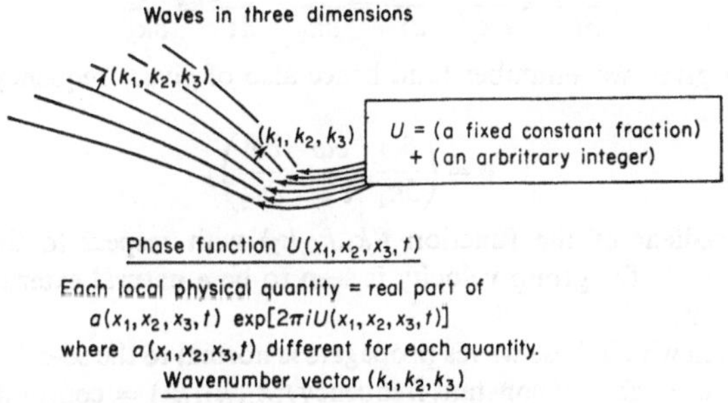

FIG. 5. Phase function and wavenumber vector for waves in three dimensions.

have their crest or peak values at different places, but the function U describes all of them; the crests for every physical quantity lie on curves of the form

$$U = \text{(a fixed constant fraction)} + \text{(an arbitrary integer)}.$$

We suppose, furthermore, that a functional relationship exists between frequency and wavenumber. The wavenumber in three dimensions has direction as well as magnitude (Fig. 5); its magnitude, as usual, is the reciprocal of the wavelength, and its direction is perpendicular to the wavecrests. If the components of the wavenumber are (k_1,k_2,k_3), then locally the phase factor varies in proportion to

$$\exp [2\pi i(k_1x_1 + k_2x_2 + k_3x_3 - \omega t)].\tag{29}$$

This means that

$$k_1 = \frac{\partial U}{\partial x_1}, \; k_2 = \frac{\partial U}{\partial x_2}, \; k_3 = \frac{\partial U}{\partial x_3}, \; \omega = -\frac{\partial U}{\partial t}.\tag{30}$$

I have said that, in a homogeneous anisotropic system, we assume that the frequency is a function of the wavenumber vector:

$$\omega = f(k_1,k_2,k_3).\tag{31}$$

In other words,

$$-\frac{\partial U}{\partial t} = f\left(\frac{\partial U}{\partial x_1}, \frac{\partial U}{\partial x_2}, \frac{\partial U}{\partial x_3}\right).\tag{32}$$

We may then ask how wave packets of given wavenumber vector (k_1,k_2,k_3), and hence also given frequency, propagate. We obtain the answer by differentiating equation (32) with respect to x_α (where α may be 1, 2 or 3). This differentiation gives

$$-\frac{\partial^2 U}{\partial x_\alpha \partial t} = \frac{\partial f}{\partial k_1} \frac{\partial^2 U}{\partial x_\alpha \partial x_1} + \frac{\partial f}{\partial k_2} \frac{\partial^2 U}{\partial x_\alpha \partial x_2} + \frac{\partial f}{\partial k_3} \frac{\partial^2 U}{\partial x_\alpha \partial x_3},\tag{33}$$

which can also be written as

$$\frac{\partial k_\alpha}{\partial t} + \frac{\partial f}{\partial k_1}\frac{\partial k_\alpha}{\partial x_1} + \frac{\partial f}{\partial k_2}\frac{\partial k_\alpha}{\partial x_2} + \frac{\partial f}{\partial k_3}\frac{\partial k_\alpha}{\partial x_3} = 0. \tag{34}$$

This states simply that the wavenumber vector k_α remains constant under changes in the time and the three position co-ordinates which are in these ratios:

$$\frac{dx_1}{dt} = \frac{\partial f}{\partial k_1}, \quad \frac{dx_2}{dt} = \frac{\partial f}{\partial k_2}, \quad \frac{dx_3}{dt} = \frac{\partial f}{\partial k_3}. \tag{35}$$

Thus, waves of given wavenumber (and hence also of given frequency) propagate at a velocity

$$\mathbf{u} = \left(\frac{\partial \omega}{\partial k_1}, \frac{\partial \omega}{\partial k_2}, \frac{\partial \omega}{\partial k_3}\right), \tag{36}$$

which is the gradient of the function $f(k_1, k_2, k_3)$ with respect to the wavenumber vector. This formula for group velocity is seen to be a natural extension of the one-dimensional result.

The direction in which these waves propagate is normal to the so-called wavenumber surface, that is, the surface of constant frequency $f(k_1, k_2, k_3) = $ constant. Furthermore, since they are waves with fixed values of k_1, k_2 and k_3, the direction and the speed of propagation remain constant for the wave packet, which therefore travels in a straight line. On the other hand, both the speed and direction of movement of the wavecrests are quite different, being represented actually by the vector

$$\mathbf{v} = \left(\frac{\omega k_1}{k_1^2 + k_2^2 + k_3^2}, \frac{\omega k_2}{k_1^2 + k_2^2 + k_3^2}, \frac{\omega k_3}{k_1^2 + k_2^2 + k_3^2}\right). \tag{37}$$

To sum up, energy travels in straight lines in a homogeneous anistropic system, with a direction and speed given by the group velocity vector (36), and both are in general different from the phase velocity's direction and magnitude.

We can illustrate this result by considering the particular case when the waves originate from a small region during a small interval of time (Lighthill, 1960). In this case, waves can spread out from this region in straight lines in all directions in which normals to the wavenumber surface ($f = $ constant) lie. Some of these directions may, actually, be normal to that surface at several points. In such a direction, waves with wavenumber corresponding to each one of these points may be found, and will be found if waves with the wavenumber in question were among those originally generated. However, the magnitude of the group velocity will in general be different for each, so that the wave packets with different wavenumbers will at a given time have reached different points along that direction.

Figure 6 illustrates this for waves generated at an approximately constant frequency during a short time interval in a homogeneous plasma in a uniform magnetic field, when the magnetic pressure is small compared with the plasma pressure. If the frequency, divided by the gyrofrequency of the ions, is negligibly small, then the wavenumber surface is a plane (marked "0" in the figure). This means that energy can be propagated only in the direction normal to the plane.

Magnetohydrodynamicists are, in fact, familiar with the proposition that disturbances are, in such a case, transmitted only along magnetic lines of force. However, when the ratio of the frequency to the ion gyrofrequency is not negligibly small, the difficult, in fact, to calculate the shape of one of the crests from this condition. Such a surface of constant phase for the waves generated turns out to be what the geometers have for a long time called the reciprocal polar of the wavenumber surface, that is the locus of the poles of the tangent planes to that surface. This follows from the fact that the point x_α on a surface of constant phase, corresponding to a given point k_α on the wavenumber surface, must lie in the direction perpendicular to the tangent plane at that point, and is furthermore at a distance inversely proportional to the distance from the origin to that plane, because the scalar product of x_α and k_α must be constant on a surface of constant phase. For the problem we have just been discussing some surfaces of constant phase for continuously excited waves are shown in the lower diagram on the right of Fig. 6. The somewhat complicated look of these waves is in only apparent contradiction with the very simple law governing their propagation: namely, that the energy travels in straight lines.

5. The Kinematics of Wavecrests (Inhomogeneous Anisotropic Systems)

It is desirable to turn now to the inhomogeneous case, when the relation between frequency and wavenumber varies with position, although at any fixed point it does not change with time. Thus we assume that

$$\omega = f(k_\alpha, x_\alpha). \tag{38}$$

Naturally, we expect that energy will continue to be propagated locally at the same velocity

$$u_\alpha = \frac{\partial \omega}{\partial k_\alpha}, \tag{39}$$

although this will now vary from point to point. Also, because the equations of motion do not involve the time explicitly, energy in waves of a given frequency should remain always in waves of the same frequency. However, the equations do involve the space co-ordinates explicitly, so that, as these waves propagate into different regions of space, their wavenumber vector k_α can be expected to change. We may hope, therefore, first, to prove that, when changes in x_α and t are in a proportion given by the velocity (39), that is, when

$$\frac{dx_\alpha}{dt} = \frac{\partial f}{\partial k_\alpha}, \tag{40}$$

the frequency ω remains constant, and, secondly, we may hope to find how the wavenumber vector k_α varies.

To do both these things, we write equation (38) as

$$-\frac{\partial U}{\partial t} = f\left(\frac{\partial U}{\partial x_\alpha}, x_\alpha\right), \tag{41}$$

and differentiate with respect to x_α, giving

$$-\frac{\partial^2 U}{\partial x_\alpha \partial t} = \frac{\partial f}{\partial k_1}\frac{\partial^2 U}{\partial x_\alpha \partial x_1} + \frac{\partial f}{\partial k_2}\frac{\partial^2 U}{\partial x_\alpha \partial x_2} + \frac{\partial f}{\partial k_3}\frac{\partial^2 U}{\partial x_\alpha \partial x_3} + \frac{\partial f}{\partial x_\alpha}.$$

This can be rewritten as

$$\frac{\partial k_\alpha}{\partial t} + \frac{\partial f}{\partial k_1}\frac{\partial k_\alpha}{\partial x_1} + \frac{\partial f}{\partial k_2}\frac{\partial k_\alpha}{\partial x_2} + \frac{\partial f}{\partial k_3}\frac{\partial k_\alpha}{\partial x_3} = -\frac{\partial f}{\partial x_\alpha}, \tag{42}$$

and we can interpret this by saying that when x_1, x_2, x_3 and t change according to equation (40), k_α changes according to the equation

$$\frac{dk_\alpha}{dt} = -\frac{\partial f}{\partial x_\alpha}. \tag{43}$$

This very simple rule, governing how the wavenumber associated with a given packet of energy varies as the packet is propagated through space, is very strongly reminiscent of the Hamiltonian form of the equations of motion of a dynamical system, and it comes as no surprise to learn that this rule also was discovered by Hamilton; see for example equation (T^2) on p. 182 of Hamilton (1837). Equations (40) and (43) have to be solved simultaneously to find out how the wave packet is propagated through x_α space and k_α space. With modern computational aids this is no great problem, however, whereas the direct computation of the waves themselves, involving a complicated partial differential equation with four independent variables, is expected to remain out of reach of such aids for some time to come.

The check, that, when f is a function of k_α and x_α without explicit dependence on the time, the two equations (40) and (43) imply that the frequency, i.e. the value of f itself, remains constant for the wave packet, is now very straightforward, being identical with the proof that, for a dynamical system whose Hamiltonian is without explicit dependence on the time, every motion of the system carries a constant value of the Hamiltonian, that is, of the total energy. The fact that frequency behaves like energy in this way and wavenumber like momentum is directly related to the fact that quantum mechanics associates with every particle a wave of frequency proportional to its energy and of wavenumber proportional to its momentum.

As one simple illustration of the practical application of these results, I should like to refer to the propagation of sound through an atmosphere in non-uniform motion. The effects of non-uniform winds influence greatly the propagation of sound over long distances and needed to be studied very seriously when accurate calculations of the possible intensity on the ground of "booms" from supersonic aircraft were being made (Warren, 1964). Now, the relation between frequency and wavenumber at a point where the sound speed is a and the wind speed is (V_1, V_2, V_3) is

$$\omega = V_1 k_1 + V_2 k_2 + V_3 k_3 + a\sqrt{(k_1^2 + k_2^2 + k_3^2)}. \tag{44}$$

This gives, first, that the group velocity, with which energy is propagated, is

$$u_\alpha = \frac{\partial \omega}{\partial k_\alpha} = V_\alpha + a\frac{k_\alpha}{\sqrt{(k_1^2 + k_2^2 + k_3^2)}}, \tag{45}$$

namely, the vector sum of the wind velocity and of a vector whose magnitude is a and whose direction is normal to the wavecrests. This result may appear rather obvious, but one of the very few mistakes in Rayleigh's *Theory of Sound* occurs in Section 289, where the energy is assumed in this problem to be propagated exactly at right angles to the wavecrests. Secondly, it gives that, as energy travels along one of these rays of sound, at the group velocity, the wavenumber vector changes according to the law

$$\frac{dk_\alpha}{dt} = -\frac{\partial V_1}{\partial x_\alpha}k_1 - \frac{\partial V_2}{\partial x_\alpha}k_2 - \frac{\partial V_3}{\partial x_\alpha}k_3 - \frac{\partial a}{\partial x_\alpha}\sqrt{(k_1^2 + k_2^2 + k_3^2)}. \tag{46}$$

In the most general case, the rays must be computed by numerically solving this, with the equation for dx_α/dt, as a system of ordinary differential equations, which is relatively straightforward. One particular type of inhomogeneity, however, is of great practical importance and allows us to reduce the problem to a simple integration. When the relation between ω and k_α depends on only one space co-ordinate, say x_1, which in the atmospheric problem might be the altitude, then the wavenumber components k_2 and k_3 remain constant, and the differential equation for the component k_1 has the simple integral ω = constant. This enables the rays to be calculated directly, by integrating equation (40) with k_1 given by the condition ω = constant.

For example, in the atmospheric case with only the V_2 component of wind non-zero, equation (44) for the frequency ω gives

$$k_1 = \sqrt{[(\omega - V_2 k_2)^2 a^{-2} - (k_2^2 + k_3^2)]}. \tag{47}$$

The equation (40) for the rays can then be written

$$\frac{dx_2}{dx_1} = \frac{V_2\omega + k_2(a^2 - V_2^2)}{a\sqrt{[(\omega - V_2 k_2)^2 - a^2(k_2^2 + k_3^2)]}},$$

$$\frac{dx_3}{dx_1} = \frac{k_3 a}{\sqrt{[(\omega - V_2 k_2)^2 - a^2(k_2^2 + k_3^2)]}}, \tag{48}$$

which can be numerically integrated with exceptional ease since the right-hand sides are just known functions of x_1. The acoustic amplitudes are then determined from the condition that energy propagates along ray tubes in inverse proportion to the cross-sectional area of a ray tube. A convenient form of this condition is that

$$u_1 \mathscr{E} \frac{\partial(x_2, x_3)}{\partial(k_2, k_3)} = \text{constant along a ray}, \tag{49}$$

where $\mathscr{E}$ is the energy density.

Hitherto I have discussed inhomogeneous systems in which the relation between ω and k_α varies with the space co-ordinates only. In the still more general case, which however is less often of practical importance, when it depends explicitly also on the time, the results can be put in a similar form most neatly if we use the idea from relativity of regarding the time simply as a fourth co-ordinate, x_4, and the frequency as minus a wavenumber component,

$$\omega = -\frac{\partial U}{\partial x_4} = -k_4. \tag{50}$$

The most general relation between frequency, wavenumber, time and the space co-ordinates can then be written

$$F(k_\lambda, x_\lambda) = 0, \tag{51}$$

where λ goes from 1 to 4 and k_λ is the derivative $\partial U/\partial x_\lambda$ of the phase function with respect to x_λ.

If now we differentiate (51) with respect to x_λ, we obtain

$$\sum_{\mu=1}^{4} \frac{\partial F}{\partial k_\mu} \frac{\partial^2 U}{\partial x_\lambda \partial x_\mu} + \frac{\partial F}{\partial x_\lambda} = 0, \tag{52}$$

which says that if a ray in space–time is defined by the equations

$$\frac{dx_\mu}{d\tau} = \frac{\partial F}{\partial k_\mu}, \tag{53}$$

where τ is a parameter, then along such a ray

$$\frac{dk_\lambda}{d\tau} = -\frac{\partial F}{\partial x_\lambda}. \tag{54}$$

Thus, the Hamiltonian form of the equations is preserved in this still more general case. Professor Synge in his lectures pointed out that the same equations would describe an even more general situation, namely, the situation defined by specifying that a ray is a path in space–time between two points such that the integral

$$\int (k_1\, dx_1 + k_2\, dx_2 + k_3\, dx_3 + k_4\, dx_4) \tag{55}$$

along the path is stationary, where the k_λ are constrained to satisfy $F(k_\lambda, x_\lambda) = 0$. This integral represents the phase change between the two points, so that stationary phase is being applied even though, in this greatly generalized theory, the quantity in brackets is no longer assumed equal to the exact derivative of a phase function U.

6. The Method of Stationary Phase (Decay of an Initial Disturbance in Three Dimensions)

I shall not pursue this any farther, however, but rather go back to the case of the homogeneous system and check the results inferred from the kinematics of wave-crests against an asymptotic analysis of the result of linearly combining waves of different wavenumber vector when the frequency is a function of the wavenumber vector. It is interesting to sketch briefly the methods involved in this, even though there is not time to describe them in detail. I shall consider separately two types of problem; the first is that of how a disturbance, initially confined to a small region, spreads out in space.

We shall suppose that it can be Fourier analysed as

$$\phi = \int_{-\infty}^{\infty} \int_{-\infty}^{\infty} \int_{-\infty}^{\infty} f(k_\alpha) \exp\{2\pi i[k_1 x_1 + k_2 x_2 + k_3 x_3 - \omega(k_\alpha)t]\}\, dk_1\, dk_2\, dk_3, \tag{56}$$

so that initially (when $t = 0$) we have

$$\phi = \int_{-\infty}^{\infty} \int_{-\infty}^{\infty} \int_{-\infty}^{\infty} f(k_\alpha) \exp[2\pi i(k_1 x_1 + k_2 x_2 + k_3 x_3)]\, dk_1\, dk_2\, dk_3. \tag{57}$$

The same methods as in the one-dimensional case can be used to show (see, for example, Chako, 1965) that the asymptotic behaviour of the integral (56) for large t is dominated by terms associated with the points, if any, where the phase term in square brackets is stationary, that is, where

$$x_\alpha = \frac{\partial \omega}{\partial k_\alpha} t = u_\alpha t. \tag{58}$$

This says that waves of a given wavenumber travel a distance t times the group velocity in time t.

In order to derive the contribution from one such point, say $k_\alpha^{(r)}$, where the phase is stationary, we might hope to reduce the problem to a one-dimensional one, by expanding the phase in a Taylor series up to terms quadratic in $k_\alpha - k_\alpha^{(r)}$ and then wavenumber surface splits into two. When, for example, the ratio is one-quarter, the two sheets are those marked "$\frac{1}{4}$" in the figure. The energy of waves whose wavenumber corresponds to a given point on the surface propagates in the direction of the normal to the surface at that point, and in this particular case the directions normal to the left-hand sheet all lie within a certain cone N_1; while those normal to the right-hand sheet lie within a somewhat larger cone N_2. All the energy that is created remains, therefore, within this larger cone N_2, while the speed with which energy of a given wavenumber is propagated is the gradient of frequency along the normal to the wavenumber surface at the corresponding point. The upper diagram on the right shows the waves, that were created in a certain short time interval, at a given later

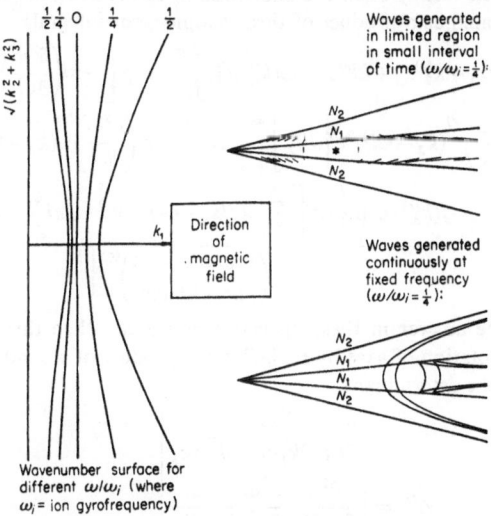

FIG. 6. Diagrams illustrating waves in a homogeneous plasma in a uniform magnetic field, when the magnetic pressure is small compared with the plasma pressure.

time t. Those associated with the left-hand sheet all lie within the cone N_1, and have travelled considerably further than those associated with the right-hand sheet (which lie within N_2). The spacing and angle of the crests is representative of the direction and magnitude of the local wavenumber vector. From a physical point of view the interest lies in the wide dispersion of the waves exhibited already when the ratio of frequency to ion gyrofrequency is one-quarter, a situation completely different from that in which the ratio is very small, when all the energy would be concentrated at the point marked with a star.

This upper diagram illustrates, then, the problems arising when the waves are all generated during a small interval. Quite a different appearance is assumed by the waves when they are, instead, generated continuously, at a fixed frequency, within a small region. The different magnitude of the group velocity in different directions is not then particularly important; time is unlimited, and so waves are found at all distances from the region in every direction in which they propagate. The angle of the crests for waves propagated in a given direction is, however, the same as in Fig. 5. It is not estimating the integral as a product of terms involving separate integrations with respect to k_1, k_2 and k_3. This is impeded in practice by the fact that the quadratic terms include not only the squares of $k_\alpha - k_\alpha^{(r)}$ but also products of two of these. The quickest way to get an answer is temporarily to rotate the axes in (k_1,k_2,k_3) space in such a way that in the new axes

$$\frac{\partial^2 \omega}{\partial k_\alpha \partial k_\beta} = 0 (\alpha \neq \beta) \quad \text{at} \quad k_\alpha = k_\alpha^{(r)}. \tag{59}$$

(See Lighthill (1960, Appendix B), where the analysis given is more complicated, however, because the relationship between ω and k is supposed given in an implicit rather than explicit form.) Then the contribution to the integral (56) from $k_\alpha = k_\alpha^{(r)}$ can be approximated as a product of three simple error integrals

$$f(k_\alpha^{(r)}) \exp\{2\pi i [k_1^{(r)} x_1 + k_2^{(r)} x_2 + k_3^{(r)} x_3 - \omega(k_\alpha^{(r)}) t]\} \int_{-\infty}^{\infty} \exp\left[-\pi i \left(\frac{\partial^2 \omega}{\partial k_1^2}\right)^{(r)} (k_1 - k_1^{(r)})^2 t\right] dk_1$$

$$\int_{-\infty}^{\infty} \exp\left[-\pi i \left(\frac{\partial^2 \omega}{\partial k_2^2}\right)^{(r)} (k_2 - k_2^{(r)})^2 t\right] dk_2 \int_{-\infty}^{\infty} \exp\left[-\pi i \left(\frac{\partial^2 \omega}{\partial k_3^2}\right)^{(r)} (k_3 - k_3^{(r)})^2 t\right] dk_3$$

$$= \frac{f(k_\alpha^{(r)}) \exp\left\{2\pi i \left[\sum_{\alpha=1}^{3} k_\alpha^{(r)} x_\alpha - \omega(k_\alpha^{(r)}) t\right] - \frac{1}{4}\pi i \sum_{\alpha=1}^{3} \text{sgn} \frac{\partial^2 \omega}{\partial k_\alpha^2}\right\}}{t^{\frac{3}{2}} \left|\left(\frac{\partial^2 \omega}{\partial k_1^2} \frac{\partial^2 \omega}{\partial k_2^2} \frac{\partial^2 \omega}{\partial k_3^2}\right)^{(r)}\right|^{\frac{1}{2}}}. \tag{60}$$

Having found the answer in these special axes we can then throw it into a form invariant under rotation of axes by replacing the product of second derivatives in the denominator by the determinant

$$\Delta^{(r)} = \begin{vmatrix} \dfrac{\partial^2 \omega}{\partial k_1^2} & \dfrac{\partial^2 \omega}{\partial k_1 \partial k_2} & \dfrac{\partial^2 \omega}{\partial k_1 \partial k_3} \\[2mm] \dfrac{\partial^2 \omega}{\partial k_2 \partial k_1} & \dfrac{\partial^2 \omega}{\partial k_2^2} & \dfrac{\partial^2 \omega}{\partial k_2 \partial k_3} \\[2mm] \dfrac{\partial^2 \omega}{\partial k_3 \partial k_1} & \dfrac{\partial^2 \omega}{\partial k_3 \partial k_2} & \dfrac{\partial^2 \omega}{\partial k_3^2} \end{vmatrix}_{(k_\alpha = k_\alpha^{(r)})} \tag{61}$$

which is invariant under rotation of axes and in the special axes where the cross-derivatives vanish takes that value. In other words, if we write the asymptotic form of ϕ as the sum

$$\phi \sim \sum_{r} \frac{f(k_\alpha^{(r)}) \exp\left\{2\pi i \left[\sum_{\alpha=1}^{3} k_\alpha^{(r)} x_\alpha - \omega(k_\alpha^{(r)}) t\right] + i\theta^{(r)}\right\}}{t^{\frac{3}{2}} |\Delta^{(r)}|^{\frac{1}{2}}} \tag{62}$$

over all $k_\alpha^{(r)}$ where $\sum_{\alpha=1}^{3} k_\alpha x_\alpha - \omega(k_\alpha) t$ is stationary, the phase addition term $\theta^{(r)}$ being $\frac{3}{4}\pi$ at a minimum (where the signs $\partial^2 \omega/\partial x_\alpha^2$ in (60) are all -1), $-\frac{3}{4}\pi$ at a maximum, and $\frac{1}{4}\pi$ or $-\frac{1}{4}\pi$ at a saddle-point according as the function increases along two or along only one of the three principal directions, then each term is invariant under change of axes and therefore is valid in any system of axes, not just the one in which it was derived.

We can interpret this result physically, by using (58) to note that the volume of space occupied at time t by energy whose wavenumber vector lies in an elementary volume $dk_1\, dk_2\, dk_3$, centred on the value $k_z^{(r)}$, is

$$
dx_1\, dx_2\, dx_3 = \left[\left|\frac{\partial(x_1,x_2,x_3)}{\partial(k_1,k_2,k_3)}\right|\right]_{k_z=k_z^{(r)}} dk_1\, dk_2\, dk_3
$$

$$
= t^3 \left[\left|\frac{\partial(u_1,u_2,u_3)}{\partial(k_1,k_2,k_3)}\right|\right]_{k_z=k_z^{(r)}} dk_1\, dk_2\, dk_3 = t^3\, |\,\Delta^{(r)}\,|\, dk_1\, dk_2\, dk_3. \qquad (63)
$$

Accordingly, as this energy spreads out to fill greater and greater volume, the energy density must vary inversely as $t^3|\,\Delta^{(r)}\,|$ and therefore the amplitude would be expected to vary as the inverse square root of this as we found above.

7. The Method of Stationary Phase (Forced Motions in Three Dimensions)

After this brief treatment of the propagation and decay of an initially limited disturbance, I shall describe, though again only briefly, a second problem treated by the method of stationary phase in three dimensions, but intrinsically different in that the disturbance is of fixed frequency and is maintained by means of a steady source operating at that frequency within a limited region. This, then, is a problem of forced motion. Specifically, I suppose that the quantity to be determined, ϕ, satisfies the equation

$$
P\left(\frac{1}{2\pi i}\frac{\partial}{\partial x_1},\ \frac{1}{2\pi i}\frac{\partial}{\partial x_2},\ \frac{1}{2\pi i}\frac{\partial}{\partial x_3},\ -\frac{1}{2\pi i}\frac{\partial}{\partial t}\right)\phi = e^{-2\pi i \omega_0 t} f(x_1,x_2,x_3), \qquad (64)
$$

where P is some polynomial in the partial differential operators shown, and the source term of fixed frequency ω_0 on the right vanishes outside a limited region. The problem is to determine the form of ϕ at distances from that region large compared with its size.

In cases relevant to this lecture, that is, when the homogeneous equation (without a forcing term) $P\phi = 0$ possesses solutions in the form of waves, and the problem in the forced motion is to determine the wavenumbers and amplitudes of the waves found at large distances along each direction out from the source region, there is a well-known mathematical difficulty, namely, that equation (64) does not have a unique solution tending to zero at infinity. On the contrary, it has a large multiplicity of such solutions. Out of all these, however, only one is of any physical interest. This is the one commonly described as satisfying the "radiation condition". I should like to say something about how to obtain this unique solution of physical interest in rather general wave problems.

Out of many ways of deriving it, I am inclined to think that the most convenient and, physically, the most logical is to require that the steady-state wave motion must be "arrivable at by switching on (the source) and waiting" (see Lighthill, 1960, pp. 412–414 and also p. 430). Different methods of switching on can be considered, varying from the most abrupt, in which the right-hand side of equation (64) takes the form shown for $t > 0$ but is zero for $t < 0$, to far more gradual methods. The most gradual method is that in which the right-hand side is replaced by $\exp[2\pi(\varepsilon - i\omega_0)t] f(x_1,x_2,x_3)$ where ε is very small, so that the forcing term has grown to its present strength from zero during all the time from $t = -\infty$; we then find a solution ϕ proportional to $\exp[2\pi(\varepsilon - i\omega_0)t]$. This is, evidently, equivalent to allowing the frequency ω to have a small positive imaginary part ε (which is later allowed to tend to zero). All these methods of switching on produce identical steady-state solutions, except in the not very physically important condition when the system is unstable to disturbances of certain wavenumbers and the forcing term f includes components with those wavenumbers, for which wave solutions increasing in amplitude exponentially with time are possible. In this condition the more abrupt methods of switching on can trigger off the instability while the very gradual method would not do so. In the other cases (which are of far greater practical importance) the different ways of switching on all give identical results. I shall not refer again to any except the very gradual method.

In this, we obtain a solution tending to zero at large distances from the source region by supposing that f and ϕ have three-dimensional Fourier transforms, in terms of which they can be written as

$$f = \int_{-\infty}^{\infty} \int_{-\infty}^{\infty} \int_{-\infty}^{\infty} F(k_1,k_2,k_3) \exp[2\pi i(k_1 x_1 + k_2 x_2 + k_3 x_3)] \, dk_1 \, dk_2 \, dk_3 \qquad (65)$$

and as $\phi = \exp[2\pi(\varepsilon - i\omega_0)t]\phi_\varepsilon$, where

$$\phi_\varepsilon = \int_{-\infty}^{\infty} \int_{-\infty}^{\infty} \int_{-\infty}^{\infty} \Phi_\varepsilon(k_1,k_2,k_3) \exp[2\pi i(k_1 x_1 + k_2 x_2 + k_3 x_3)] \, dk_1 \, dk_2 \, dk_3. \qquad (66)$$

Then equation (64) tells us

$$P(k_1,k_2,k_3,\omega_0 + i\varepsilon)\Phi_\varepsilon = F. \qquad (67)$$

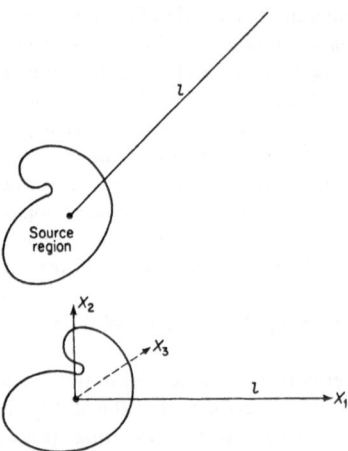

FIG. 7. To make estimates at large distances along a particular straight line l stretching away from the source region, we temporarily use the special axes shown.

The problem is to find ϕ_0, the limit of ϕ_ε as $\varepsilon \to 0$ from above, at large distances from the source region along any straight line l stretching away from it. As in the previous problem, this estimation is carried out most easily if we first effect a rotation of axes. In order to estimate ϕ_0 on a particular line l, we temporarily choose axes (Fig. 7) such that l is the positive x_1-axis. On l, therefore, x_2 and x_3 are zero, while x_1 is positive, and

$$\phi_\varepsilon = \int_{-\infty}^{\infty} \int_{-\infty}^{\infty} dk_2 \, dk_3 \int_{-\infty}^{\infty} \frac{F(k_1,k_2,k_3) \, e^{2\pi i k_1 x_1} dk_1}{P(k_1,k_2,k_3,\omega_0 + i\varepsilon)}. \tag{68}$$

We estimate the inner integral in equation (68) by moving the path of integration to one (Fig. 8) on which the imaginary part of k_1 takes a suitably chosen positive (constant) value h. The integral over the new path is small of order $e^{-2\pi h x_1}$ for large x_1, and we shall regard quantities of this order of magnitude as negligible. Between the two paths the only singularities of the integrand are poles at zeros of the denominator, since the vanishing of f outside a finite region implies that its Fourier transform F is regular. We must therefore calculate the sum of the residues at those poles.

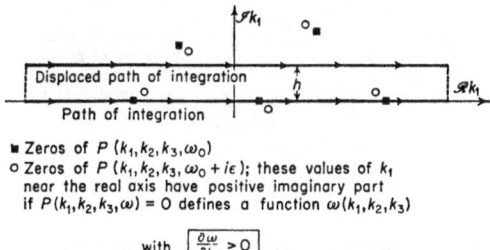

FIG. 8. Illustrating how the radiation condition is derived by displacement of the path of integration in equation (68).

Now, for the systems capable of wave propagation in which we are chiefly interested, there are zeros of the denominator for *real* k_1 when $\varepsilon = 0$ (at least in some ranges of k_2, k_3 and ω; these represent plane-wave solutions). For small positive ε these zeros may be expected to be displaced from the real axis, and will then contribute to the sum of residues only if they are displaced into the region where the imaginary part of k_1 takes *positive* values, between 0 and h. Evidently, giving ω a small positive imaginary part will give the k_1 for which $P = 0$ a positive imaginary part if the equation $P(k_1,k_2,k_3,\omega) = 0$ specifies the frequency ω as a function of the wavenumber (k_1,k_2,k_3) such that

$$\frac{\partial \omega}{\partial k_1} > 0. \tag{69}$$

We have already obtained the result, therefore, that wave energy will be found along the line l only if the component of the group velocity along l is positive. This, then, is the physical significance of the radiation condition. The contribution to the inner integral in equation (68) from each pole will be

$$2\pi i \frac{F(k_1,k_2,k_3) \, e^{2\pi i k_1 x_1}}{\partial P(k_1,k_2,k_3,\omega_0)/\partial k_1} \tag{70}$$

in the limit as $\varepsilon \to 0$ when this inequality is satisfied, and will be zero otherwise. Obviously P may have other zeros, with imaginary parts positive even for $\varepsilon = 0$, but if h is taken less then all those imaginary parts then these singularities do not lie between the two paths for small enough ε. The asymptotic form of ϕ_0 is then given by substituting for the inner integral in (68) a sum of terms (70) from all real k_1 satisfying $P(k_1,k_2,k_3,\omega_0) = 0$ together with the inequality (69). In other words,

$$\phi_0 \sim 2\pi i \int\int_S \frac{F(k_1,k_2,k_3)\, e^{2\pi i k_1 x_1} dk_2\, dk_3}{\partial P(k_1,k_2,k_3,\omega_0)/\partial k_1}, \tag{71}$$

where the surface S of integration (Fig. 9) is the part of what I earlier called the wave-number surface (defining waves with $\omega = \omega_0$) on which $\partial\omega/\partial k_1 > 0$. This then is the answer for equations which possess plane-wave solutions; for those which do not, the limiting process involving ε does not have to be introduced, and the answer comes out even more simply in terms of those zeros of P which have *smallest* positive imaginary part.

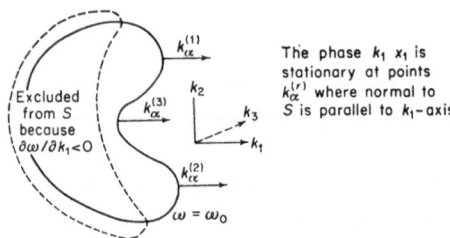

FIG. 9. The surface of integration S, and the points of stationary phase, in the integral (71).

I must return to this answer (71) now and simplify it still further. This is done by the method of stationary phase. The phase $k_1 x_1$ is stationary on the wave-number surface S at those points $k_\alpha^{(r)}$ where the normal to S is parallel to the k_1-axis (Fig. 9). Having already fixed the k_1-axis, along the direction l in which we are estimating ϕ_0, we now find it easiest to calculate the contribution to (71) from each such position of stationary phase by a temporary choice of the k_2- and k_3-axes along the principal directions of curvature of the surface. If we take the associated curvatures κ_2 and κ_3 positive where concave to the positive k_1-direction and negative where convex, then locally the surface S has the approximate equation

$$k_1 = k_1^{(r)} + \tfrac{1}{2}\kappa_2(k_2 - k_2^{(r)})^2 + \tfrac{1}{2}\kappa_3(k_3 - k_3^{(r)})^2, \tag{72}$$

which, substituted in (71), makes it easy to calculate the contribution ϕ_r to the asymptotic form of ϕ_0 from the point $k_\alpha^{(r)}$ where the normal to S is in the k_1-direction as

$$\phi_r = 2\pi i \frac{F(k_\alpha^{(r)})\exp\left[2\pi i k_1^{(r)}x_1 + \tfrac{1}{4}\pi i(\operatorname{sgn}\kappa_2 + \operatorname{sgn}\kappa_3)\right]}{[\partial P(k_1,k_2,k_3,\omega_0)/\partial k_1]_{k_\alpha = k_\alpha^{(r)}} x_1 \sqrt{|\kappa_2\kappa_3|}}. \tag{73}$$

These contributions from different points $k_\alpha^{(r)}$, which must be added up, are not in general in identical axes, so that they must first be put into forms invariant under rotation of axes as

$$\phi_r = \frac{2\pi F(k_\alpha^{(r)})\exp\left[2\pi i(k_1^{(r)}x_1 + k_2^{(r)}x_2 + k_3^{(r)}x_3) + i\theta^{(r)}\right]}{(x_1^2 + x_2^2 + x_3^2)^{\frac{1}{2}}\left[\left(\dfrac{\partial P}{\partial k_1}\right)^2 + \left(\dfrac{\partial P}{\partial k_2}\right)^2 + \left(\dfrac{\partial P}{\partial k_3}\right)^2\right]_{k_\alpha = k_\alpha^{(r)}}^{\frac{1}{2}} |K^{(r)}|^{\frac{1}{2}}} \tag{74}$$

where $K^{(r)} = \kappa_2\kappa_3$ is the Gaussian curvature (product of the two principal curvatures) at $k_\alpha = k_\alpha^{(r)}$, and the phase addition term $\theta^{(r)}$ takes, when $K^{(r)} > 0$, the value 0 where the surface is convex to the direction P increasing and π where it is concave to that direction, and, when $K^{(r)} < 0$, the value $\frac{1}{2}\pi$ or $-\frac{1}{2}\pi$ according as that direction is parallel or antiparallel to l. The asymptotic solution along l,

$$\phi = \phi_0 \, e^{-2\pi i \omega_0 t} \sim e^{-2\pi i \omega_0 t} \Sigma \phi_r, \tag{75}$$

where the summation is taken over points $k_\alpha^{(r)}$ of the surface $\omega = \omega_0$ where the normal to the surface is parallel to l and ω increases in the direction of l, is a solution with amplitude decreasing as the inverse first power of the distance $R = \sqrt{(x_1^2 + x_2^2 + x_3^2)}$.

The proportionality to the inverse square root of the modulus of the Gaussian curvature $K^{(r)}$ at the point $k_\alpha^{(r)}$ can be understood physically as follows. The normals from a small area dS around the point $k_\alpha^{(r)}$ fill a cone (Fig. 10) whose cross-sectional

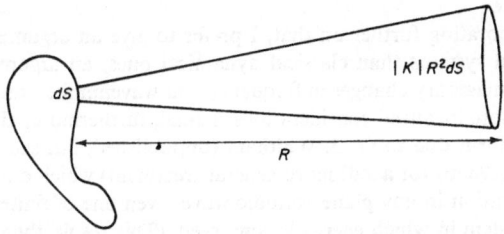

FIG. 10. Physical intepretation of the expression for waves produced in forced motion (equations (74) and (75)). The energy is diminished by a factor $|K|^{-1}R^{-2}$ in distance R, and the amplitude by $|K|^{-\frac{1}{2}}R^{-1}$.

area increases with distance R like $|K^{(r)}| R^2 dS$. The energy created in the source region with wavenumbers lying in this elementary area dS is therefore diminished by a factor $|K^{(r)}|^{-1}R^{-2}$ at distance R, which makes the factor on amplitude $|K^{(r)}|^{-\frac{1}{2}}R^{-1}$.

There is not time to discuss further properties of the asymptotic solution (75). As in the one-dimensional case, the method is easily modified to give the proper asymptotic form near a caustic (i.e. a locus of cusps of wave crests) and in other singular cases (Lighthill, 1960, pp. 408–411).

8. Energy Propagation Velocity

I began this lecture by saying that I would describe both the kinematic and the dynamic approaches to the mathematics of group velocity. Hitherto I have described basically kinematic approaches, which, for a general linear combination of waves with frequency and wavenumber functionally related, evaluate how the waves of different wavenumbers become dispersed from one another. From time to time I have shown the results on amplitude variation to be consistent with the assumption that the energy in waves of each wavenumber is propagated at the group velocity, but I have still not proved explicitly that this is so.

This result is in many ways the most surprising of all, for whereas in a motion with varying wavenumber a derivative, like $\partial \omega / \partial k_\alpha$, may be expected to be important, the fact that, *in a perfectly periodic motion of fixed wavenumber*, energy is propagated at a velocity which can be expressed as a ratio of changes of frequency and wavenumber in going to a neighbouring wave solution, appears distinctly odd. To prove this result for a general periodic wave motion, how are we to introduce the concept of the changes in frequency and wavenumber on passing to a neighbouring solution?

One very amusing answer to this question was supplied by Rayleigh (1877) in the one-dimensional case, and can easily be extended to the three-dimensional case (Lighthill, 1960, Appendix A). This answer supposes that, in a general dynamical system, a pure imaginary change in wavenumber is made, and shows that the corresponding imaginary change in the frequency with which the system oscillates in small disturbances would be replaced by zero frequency change if the motion of every particle in the system were resisted by an additional small force proportional to its momentum. The energy flow across a plane in this steady state is then calculated by balancing it against the dissipative action of those forces throughout the region beyond that plane.

Instead of elaborating further on that, I prefer to give an argument applicable to still more general systems than classical dynamical ones, an argument that utilizes real rather than imaginary changes in frequency and wavenumber, and that calculates energy flow by more localized considerations. I shall, furthermore, give the argument in a generalized form due to G. B. Whitham (unpublished, but see the forthcoming paper Whitham (1965b) for a still more general treatment) which obtains the velocity of energy propagation in any plane periodic wave, even one of finite amplitude, in a homogeneous system in which energy is conserved. (This result, then, unlike those in the rest of the paper, is valid even for non-linear equations of motion.)

Any such system homogeneous in space can be specified (see, for example, Goldstein (1950), p. 350) by a single function: a Lagrangian density $\mathscr{L}$ (or Lagrangian per unit volume), which is a function of (say) n local variables (such as displacements, field strengths, variables of state, etc.) $\eta_1, \eta_2, \ldots, \eta_n$, together with their first derivatives

$$\dot{\eta}_i = \frac{\partial \eta_i}{\partial t} \quad \text{and} \quad \overset{\alpha}{\eta}_i = \frac{\partial \eta_i}{\partial x_\alpha}. \tag{76}$$

There is no direct dependence of $\mathscr{L}$ on x_1, x_2, x_3 and t, only an indirect dependence, due to the dependence of η_i, $\dot{\eta}_i$ and $\overset{\alpha}{\eta}_i$ on those variables.

Possible developments of the system in time are specified by Hamilton's principle, that the time integral of the Lagrangian is stationary. The usual form of the principle states that

$$\delta \int_{t_1}^{t_2} dt \int_V \mathscr{L} d\tau = 0 \tag{77}$$

for any changes $\delta \eta_i$ in the function $\eta_i(x_1, x_2, x_3, t)$ which vanish at the beginning and end $t = t_1$ and $t = t_2$ of the arbitrary time interval of integration and on the boundary of the arbitrary volume of integration V. This implies the equations of motion

$$\frac{\partial}{\partial t}\left(\frac{\partial \mathscr{L}}{\partial \dot{\eta}_i}\right) + \sum_{\alpha=1}^{3} \frac{\partial}{\partial x_\alpha}\left(\frac{\partial \mathscr{L}}{\partial \overset{\alpha}{\eta}_i}\right) - \frac{\partial \mathscr{L}}{\partial \eta_i} = 0. \tag{78}$$

This formulation of the equations of a homogeneous system is of great generality. For a classical dynamical system $\mathscr{L}$ is the kinetic energy minus the potential energy, per unit volume. For a relativistic dynamical system the Lagrangian for each particle can be written in terms of its rest mass m, potential energy V and velocity $c\beta$ (where c is the speed of light) as $-mc^2\sqrt{(1-\beta^2)} - V$. From this Lagrangian for each particle a

Lagrangian density can be calculated. Electromagnetic fields E and H contribute (in rationalized m.k.s. units) $\frac{1}{2}\varepsilon E^2 - \frac{1}{2}\mu H^2$ to the Lagrangian density $\mathcal{L}$, and, when charge and current densities ρ and j are also present, additional terms $-\rho\phi + j \cdot A$, where ϕ and A are the scalar and vector potentials. Other types of field can be taken into account similarly in the Lagrangian formulation. In all cases the density of total energy is

$$\mathscr{E} = \sum_{i=1}^{n} \dot{\eta}_i \frac{\partial \mathcal{L}}{\partial \dot{\eta}_i} - \mathcal{L}, \tag{79}$$

and it follows from the equation of motion (78) that the rate of change of the energy density $\mathscr{E}$ with time can be written as

$$\frac{\partial \mathscr{E}}{\partial t} = -\sum_{\alpha=1}^{3} \frac{\partial I_\alpha}{\partial x_\alpha} \tag{80}$$

where I_α is an "energy flux vector"

$$I_\alpha = \sum_{i=1}^{n} \dot{\eta}_i \frac{\partial \mathcal{L}}{\partial \dot{\eta}_i^\alpha}. \tag{81}$$

Equation (80) says that the energy in a rectangular element changes at a rate equal to the differences between the energy flux across opposite faces of the element. For periodic plane waves, we may define an energy propagation velocity u_α as

$$u_\alpha = \frac{\langle I_\alpha \rangle}{\langle \mathscr{E} \rangle} = \frac{\left\langle \sum_{i=1}^{n} \dot{\eta}_i \partial \mathcal{L}/\partial \dot{\eta}_i^\alpha \right\rangle}{\left\langle \sum_{i=1}^{n} \dot{\eta}_i \partial \mathcal{L}/\partial \dot{\eta}_i - \mathcal{L} \right\rangle} \tag{82}$$

where for example $\langle \mathscr{E} \rangle$ signifies the mean energy density $\mathscr{E}$, that is, $\mathscr{E}$ averaged over an integral number of wavelengths or periods. In terms of u_α, the averaged form of equation (80) can be written

$$\frac{\partial \langle \mathscr{E} \rangle}{\partial t} + \sum_{\alpha=1}^{3} \frac{\partial (\langle \mathscr{E} \rangle u_\alpha)}{\partial x_\alpha} = 0, \tag{83}$$

showing that mean energy is convected by the group velocity vector field u_α in exactly the same way as the mass of a compressible fluid is by the hydrodynamic velocity field.

For periodic plane waves, Hamilton's principle can be used in a slightly special form to obtain an expression for the energy propagation velocity u_α. This special form states that such a wave satisfies (77) for all changes $\delta\eta_i$ that are periodic with the same frequency and wavenumber as the η_i themselves, provided that $t_2 - t_1$ is an integral multiple of the (time) period, and V is a rectangular box with four of its sides perpendicular to the wave fronts, and the other two parallel to wave fronts and an integral number of wavelengths apart. For then we have

$$\delta \int_{t_1}^{t_2} dt \int_V \mathcal{L} d\tau = \int_{t_1}^{t_2} dt \int_V \sum_{i=1}^{n} \left(\frac{\partial \mathcal{L}}{\partial \eta_i} \delta\eta_i + \frac{\partial \mathcal{L}}{\partial \dot{\eta}_i} \delta\dot{\eta}_i + \sum_{\alpha=1}^{3} \frac{\partial \mathcal{L}}{\partial \dot{\eta}_i^\alpha} \delta\dot{\eta}_i^\alpha \right) d\tau$$

$$= \int_{t_1}^{t_2} dt \int_V \sum_{i=1}^{n} \left[\frac{\partial \mathcal{L}}{\partial \eta_i} - \frac{\partial}{\partial t}\left(\frac{\partial \mathcal{L}}{\partial \dot{\eta}_i}\right) - \sum_{\alpha=1}^{3} \frac{\partial}{\partial x_\alpha}\left(\frac{\partial \mathcal{L}}{\partial \dot{\eta}_i^\alpha}\right) \right] \delta\eta_i d\tau +$$

$$\int_V d\tau \int_{t_1}^{t_2} \frac{\partial}{\partial t}\left(\sum_{i=1}^{n} \frac{\partial \mathcal{L}}{\partial \dot{\eta}_i} \delta\eta_i \right) dt + \int_{t_1}^{t_2} dt \int_V \sum_{\alpha=1}^{3} \frac{\partial}{\partial x_\alpha}\left(\sum_{i=1}^{n} \frac{\partial \mathcal{L}}{\partial \dot{\eta}_i^\alpha} \delta\eta_i \right) d\tau = 0, \tag{84}$$

because the first integral vanishes by (78) while, in the second, the inner integral vanishes because $t_2 - t_1$ is an integral multiple of a period, and, in the third, the inner integral vanishes because it equals a surface integral with the contributions from each pair of faces of the rectangular box V cancelling out.

Before using this result to calculate u_α, I will remark that, in the denominator of the expression (82) for u_α, the mean Lagrangian density $\langle \mathcal{L} \rangle$ vanishes in the special case of waves of infinitesimal amplitude. For classical dynamical systems, this is the familiar result (Rayleigh, 1877) that mean kinetic energy equals mean potential energy in waves of infinitesimal amplitude. For more general systems subject to Hamilton's principle, it follows most easily by changing each η_i to $(1+\varepsilon)\eta_i$. This changes $\mathcal{L}$ to $(1+\varepsilon)^2\mathcal{L}$, because the hypothesis of infinitesimal amplitude is equivalent to $\mathcal{L}$ being homogeneous of the second degree in all its variables. The variational result (84) can therefore be true only if

$$\int_{t_1}^{t_2} dt \int_V \mathcal{L} d\tau \qquad (85)$$

is itself zero (otherwise, multiplying it by $(1+\varepsilon)^2$ would not give it zero variation), and this means that $\langle \mathcal{L} \rangle = 0$.

However, for waves of finite amplitude, this argument fails since $\mathcal{L}$ is not necessarily a homogeneous function of its variables. Simple examples show, in fact, that $\langle \mathcal{L} \rangle$ is, in general, not zero for periodic plane waves of finite amplitude. The corresponding term in the denominator of equation (82) for u_α cannot then be omitted.

One inclines, naturally, to ask whether

$$u_\alpha = \frac{\partial \omega}{\partial k_\alpha} \qquad (86)$$

for plane waves of finite amplitude. However, a serious obstacle to the possible truth of such an equation is that for these the frequency ω is in general a function not only of k_1, k_2 and k_3 but also of some quantity representing the amplitude of the wave. (It can in general vary also with other parameters, called "pseudo-frequencies" by Whitham (1965b); but this possibility is not allowed for below.) The derivative in (86) can therefore have meaning only if it is understood in the sense "keeping constant both the wavenumber components other than k_α and also some measure of the amplitude". Now, we have already noted that the mean Lagrangian density $\langle \mathcal{L} \rangle$, although zero for infinitesimal amplitude, has in general different values for finite values of the amplitude. Whitham's result is that equation (86) remains true if the measure of amplitude which is kept constant in calculating the derivative is $\langle \mathcal{L} \rangle / \omega$. That is,

$$u_\alpha = \left(\frac{\partial \omega}{\partial k_\alpha} \right)_{\langle \mathcal{L} \rangle / \omega}. \qquad (87)$$

The classical result for infinitesimal amplitude is evidently a special case of this, since $\langle \mathcal{L} \rangle / \omega$ remains constant and equal to zero for all infinitesimal-amplitude waves.

To prove (87), let plane periodic waves have the form

$$\eta_i = f_i\left(\omega t - \sum_{\alpha=1}^{3} k_\alpha x_\alpha \right), \qquad (88)$$

where the functions $f_1(z), \ldots, f_n(z)$ are all periodic functions of z with period 1. Then equation (84) holds for any infinitesimal changes whatever in the f_i which leave them still with period 1.

Consider now a more drastic perturbation, in which also the frequency and wavenumber change. If

$$\eta_i + \delta\eta_i = F_i\left(\Omega t - \sum_{\alpha=1}^{3} K_\alpha x_\alpha\right), \tag{89}$$

where

$$\Omega = \omega + \delta\omega, \quad K_\alpha = k_\alpha + \delta k_\alpha, \quad F_i(z) = f_i(z) + \delta f_i(z), \tag{90}$$

and the $\delta f_i(z)$ like the $f_i(z)$ have period 1, then

$$\delta\langle\mathscr{L}\rangle = \delta\int_0^1 \mathscr{L}[f_i(z), \omega f_i'(z), -k_\alpha f_i'(z)] \, dz$$

$$= \int_0^1 \sum_{i=1}^{n} \left(\frac{\partial\mathscr{L}}{\partial\eta_i}\delta f_i + \frac{\partial\mathscr{L}}{\partial\dot\eta_i}\omega\delta f_i' - \sum_{\alpha=1}^{3} \frac{\partial\mathscr{L}}{\partial\dot\eta_i}k_\alpha\delta f_i'\right) dz +$$

$$\left(\int_0^1 \sum_{i=1}^{n} \frac{\partial\mathscr{L}}{\partial\dot\eta_i}f_i' dz\right)\delta\omega - \sum_{\alpha=1}^{3}\left(\int_0^1 \sum_{i=1}^{n} \frac{\partial\mathscr{L}}{\partial\dot\eta_i}f_i' dz\right)\delta k_\alpha. \tag{91}$$

Of the three terms on the right-hand side, the first represents the changes resulting from changes in the f_i without changes in ω and k_α and must vanish by equation (84) (Hamilton's principle for changes in the η_i which maintain frequency and wavenumber), while the second and third parts, on multiplication by ω, can be expressed in terms of the mean values that appear in the expression for u_α; that is,

$$\omega\delta\langle\mathscr{L}\rangle = \left\langle\sum_{i=1}^{n} \dot\eta_i\frac{\partial\mathscr{L}}{\partial\dot\eta_i}\right\rangle\delta\omega - \sum_{\alpha=1}^{3}\left\langle\sum_{i=1}^{n} \dot\eta_i\frac{\partial\mathscr{L}}{\partial\dot\eta_i}\right\rangle\delta k_\alpha. \tag{92}$$

We can now obtain the relationship between $\delta\omega$ and δk_α, which makes $\langle\mathscr{L}\rangle/\omega$ remain constant, by substituting

$$\omega\delta\langle\mathscr{L}\rangle = \langle\mathscr{L}\rangle\delta\omega \tag{93}$$

in equation (92). If we use expression (82), this gives

$$\delta\omega = \sum_{\alpha=1}^{3} u_\alpha\delta k_\alpha, \tag{94}$$

which finally proves (87).

The quantity $\langle\mathscr{L}\rangle/\omega$ which is kept constant in (87) is the integral of the Lagrangian density with respect to time over a single period. It is the quantity which remains stationary, when we go from a periodic solution η_i to neighbouring values $\eta_i + \delta\eta_i$ which are periodic with the same frequency and wavenumber but are not, in general, solutions. This explains what may appear surprising in the proof I have given, namely, that I did not use the fact that the functions (89), with their perturbed frequency and wavenumber, are solutions of the equations of motion. It was not necessary, because the value of $\langle\mathscr{L}\rangle/\omega$ would be the same for them as for neighbouring functions which are not solutions!

It seems likely that this result of Whitham's, that the velocity of energy propagation in plane periodic waves of finite amplitude, in a general three-dimensional homogeneous conservative system, is equal to the gradient of the frequency with respect to the wavenumber vector in changes, from one plane periodic wave to another of neighbouring wavenumber vector, in which the measure of amplitude represented by the integral of the Lagrangian with respect to time over a single period remains constant, is the first step in a major process of extension of group velocity and ray theory to non-linear dispersive systems just as far-reaching as was Whitham's work (1956) on non-dispersive systems of intermediate amplitude, based on the principle that the ray geometry will be approximately as for an infinitesimal-amplitude system but that the law of propagation of waveforms along ray tubes takes a new form.

9. Concluding Remarks

I believe I have now said as much as could well be said at one stretch about my subject. Two points about its relationship with modern physics still, perhaps, need to be made. First, that relativity forbids energy to be propagated at a speed exceeding the velocity of light c. It follows that no wave motion of a conservative system can have a group velocity greater than c, although phase velocities greater than c are rather common. (The restriction of this conclusion to conservative systems must be emphasized; as noted at the end of section 3, in the presence of large dissipation the theory of group velocity needs major modifications.)

For example, the relationship between frequency and wavenumber in an electromagnetic wave propagating in an ionized gas is

$$\omega = \left(\frac{4\pi n_e e^2}{m_e} + c^2 k^2\right)^{\frac{1}{2}}$$

where n_e is the number of electrons per cm³ and m_e and e are the mass and charge (in e.s.u.) of the electron. It follows that the phase velocity $v = \omega/k$ exceeds the speed of light c; however, the group velocity is

$$u = \frac{d\omega}{dk} = \frac{c^2 k}{\left(\dfrac{4\pi n_e e^2}{m_e} + c^2 k^2\right)^{\frac{1}{2}}} < c.$$

The second point, a related one, concerns the waves that Schrödinger's wave mechanics associates with a particle. They have frequency and wavenumber proportional to the particle's energy and momentum, as I mentioned earlier. In relativistic mechanics, this gives

$$\omega = \frac{E}{h} = \frac{mc^2}{h\sqrt{(1-\beta^2)}}, \quad k = \frac{p}{h} = \frac{mc\beta}{h\sqrt{(1-\beta^2)}}$$

for a particle of rest mass m and velocity $c\beta$. The phase velocity of the associated waves is therefore

$$\frac{\omega}{k} = \frac{c}{\beta} > c,$$

but the group velocity, with which they carry energy, is

$$\frac{d\omega}{dk} = \frac{d\omega/d\beta}{dk/d\beta} = c\beta < c.$$

The waves carry energy, in fact, simply at the velocity of the particle itself, and indeed are equivalent to the particle as energy carriers.

To conclude, then, the mathematical theory I have described in this lecture touches upon several interesting and important parts of physical and engineering science as well as upon more than one major branch of mathematics. Much of the theory is old, and during its long life has found very many applications. This part will find even more applications now that computational aids make the problem of ray tracing by Hamilton's equations (40) and (43) so simple a matter. This can be done easily, now, in the case of complicated internal wave motions in stratified media, rotating media or plasma in magnetic fields, and will be done increasingly. At the same time there are signs that further extensions to this powerful theory are being, and will continue to be made.

REFERENCES

CHAKO, N. 1965 *Phil. Trans. Roy. Soc. A.* In the press.

GOLDSTEIN, H. 1950 *Classical Mechanics.* Addison Wesley, Reading, Massachusetts.

HAMILTON, W. R. 1837 Third Supplement to *An Essay on the Theory of Systems of Rays*; reprinted in *Mathematical Papers* (1931), Vol. 1, pp. 164–293. Cambridge University Press.

HAVELOCK, T. H. 1914 *The Propagation of Disturbances in Dispersive Media*, p. 4. Cambridge University Press.

JEFFREYS, N. & JEFFREYS, B. S. 1950 *Methods of Mathematical Physics*, 2nd ed. Cambridge University Press.

LIGHTHILL, M. J. 1958 *An Introduction to Fourier Analysis and Generalised Functions*, p. 55. Cambridge University Press.

LIGHTHILL, M. J. 1960 *Phil. Trans. Roy. Soc. A*, **252**, 397–430.

LIGHTHILL, M. J. & WHITHAM, G. B. 1955 On kinematic waves. *Proc. Roy. Soc. A*, **229**, 281–345.

RAYLEIGH, LORD, 1877 *On progressive waves*; reprinted in *Theory of Sound* (1894), 2nd ed. Vol. 1, pp. 475–480. Cambridge University Press.

STOKES, G. G. 1876 *Smith's Prize examination question no.* 11; reprinted in *Mathematics and Physics Papers* (1905), Vol. 5, p. 362. Cambridge University Press.

STRATTON, J. A. 1941 *Electromagnetic Theory*, p. 333. McGraw-Hill, New York.

WARREN, C. H. E. 1964 *J. Sound Vib.* **1**, 175–178.

WHITHAM, G. B. 1956 *J. Fluid Mech.* **1**, 290–318.

WHITHAM, G. B. 1960 *J. Fluid Mech.* **9**, 347–352.

WHITHAM, G. B. 1965a *Proc. Roy. Soc. A*, **283**, 238–261.

WHITHAM, G. B. 1965b *J. Fluid Mech.* In the press.

J. Fluid Mech. (1967), *vol.* 27, *part* 4, *pp.* 725–752

On Waves Generated in Dispersive Systems to Travelling Forcing Effects, with Applications to the Dynamics of Rotating Fluids

M. J. LIGHTHILL

A theory of the generation of dispersive waves by travelling forcing effects, that may be steady, oscillatory or transient in character, is given for a general homogeneous system. Small disturbances to the system are supposed stable, and governed by a linear partial differential equation with constant coefficients which admits solutions in the form of plane waves satisfying an, in general, anisotropic dispersion relation $P(\sigma, \mathbf{k}) = 0$ between frequency σ and wave-number vector $\mathbf{k}$.

If the forcing region, supposed of limited extent, travels with constant velocity $\mathbf{U}$, then oscillatory forcing terms of frequency σ_0 (which would be replaced by 0 in the limiting case of a steady forcing effect, while taking, for a typical transient one of duration T, values from 0 to about $10/T$) produce waves of frequency $\sigma_0 + \mathbf{U} . \mathbf{k}$ (the Doppler effect). For any such waves, the wave-number $\mathbf{k}$ satisfies the equation $P(\sigma_0 + \mathbf{U} . \mathbf{k}, \mathbf{k}) = 0$, representing a surface in wave-number space here called $S(\sigma_0)$, and their position relative to that of the forcing region is determined by having been generated when that region was in an earlier position, and having subsequently progressed with the group velocity. This implies the rule, also derived analytically in §§2 and 3, that waves with a particular value of $\mathbf{k}$ on $S(\sigma_0)$ are found in a direction, stretching out from the forcing region, which is one of the directions normal to $S(\sigma_0)$ at $\mathbf{k}$, namely the one pointing towards $S(\sigma_0 + \delta)$. This rule is supplemented by results on wave amplitudes and shapes of crests.

The theory is applied (§§4 and 5) to Rossby waves excited in a beta-plane ocean by travelling patterns of wind stress. If a steady wind-stress pattern moves westward, semicircular waves of length $2\pi \sqrt{(U/\beta)}$ trail behind it, but signals are found also directly ahead, consisting of the disturbance integrated in the west–east direction and subjected to a 'low-pass filter' with respect to its north–south components of wave-number. An eastward-travelling pattern, by contrast, produces only a wake-like disturbance, calculated in detail in §4. The waves generated for intermediate directions of travel are identified, and the strong tendencies in all cases for westward intensification of transient currents are noted.

For example, a wind-stress pattern travelling 30° N. of E. leaves a trailing wedge of currents from W. to 30° S. of W. in the steady case. The influence on this conclusion of a finite duration T of such a pattern is investigated in §5 by Fourier analysis in time. The fate of Fourier components of frequency σ_0 depends on the ratio $L = \sigma_0/\sqrt{(U\beta)}$. If this is less than 1 for all σ_0 up to about $10/T$, then the disturbance retains its trailing character; on the other hand, any components with $L > 1$ have a much greater directional spread. Tidal terms make fairly small changes to the results, except that (in an ocean of depth H) they make the directional spread disappear for L greater than about $\sqrt{(\beta g H/4f^2 U)}$.

Excitation of gravity waves in non-rotating fluid is briefly considered, including generation on deep water by a travelling oscillating disturbance (§6), and generation in a uniformly stratified fluid by a vertically moving obstacle (§7). The predicted wave shapes in the latter case, with cusps at a finite distance behind the obstacle, agree excellently (figures 7 and 8) with experiments by Mowbray (1966).

An exceptional case, in that part of $S(\sigma_0)$ is doubly covered, is generation by steady ($\sigma_0 = 0$) motion of an obstacle along the axis of uniformly rotating homogeneous fluid, the surface $S(0)$ being a sphere and two coincident planes. Whereas waves corresponding to points on the sphere trail behind the obstacle, the appropriate normals on the two planes point in opposite directions inside the sphere (§8), permitting the well-known formation of the 'Taylor column' ahead of the obstacle at low Rossby numbers.

Still more complicated, because fully three-dimensional, is the case when the obstacle moves at right angles to the axis of rotation (§9). At finite though small Rossby number it is impossible for the Taylor column formed near the body to extend to large distances from it, where on the contrary the disturbance is shown to take the form of slightly trailing cones, shown in cross-section in figure 12, containing waves whose crests have cusps on the boundaries of the cones. An estimate of the length of the Taylor column, as body dimension divided by Rossby number (for small enough kinematic viscosity), is made by considering the fit between the Taylor-column and wave-cone regions.

1. Introduction

This paper is concerned with a homogeneous system whose undisturbed condition is stable, and in which small disturbances are possible, taking the form of plane waves satisfying an, in general, anisotropic dispersion relation. Within a region which is moving at constant velocity U through the system, a forcing process acts. The forcing function may be steady (independent of time), or oscillate with a fixed frequency, or be zero for times $t < 0$ and a prescribed function of time for $t > 0$. In all these cases the complex wave pattern generated by the forcing process is studied.

The general theory is given in §§2 and 3. The examples of its use which follow (§§4 to 9) are derived mainly from rotating fluid dynamics, forming a kind of continuation of the author's recent survey of that subject (Lighthill 1966, hereafter referred to as S). In rotating fluids, including the atmosphere and the oceans, many types of dispersive wave system are possible, and it is desirable to know how they can be excited by different forcing processes. Several possible forcing effects move relative to the fluid; for example, when an atmospheric disturbance travels over an ocean which it is perturbing, or when a disturbance that is fixed relative to the earth perturbs a wind blowing over it (in this case the disturbance is travelling relative to the air itself).

Previous discussions of currents generated either by steady (S, §5) or transient (Longuet-Higgins 1965b) wind-stress distributions have taken their speed of travel over the ocean to be either zero or large compared with a typical group velocity. The discussion in §§4 and 5 below shows, however, that the intermediate case is of great importance. In fact the velocity of travel of the forcing effect in relation to its characteristic frequencies and wave-numbers appears to be of dominant significance in this problem, just as in the problem of the sound radiated by travelling eddies in a turbulent jet (Lighthill 1963).

Many problems involving forcing effects that are steady or of fixed frequency have been treated in the literature; for example the problem of §8 (motion of an obstacle along the axis of rotating fluid). Difficulties have often been experienced, however, because the ensemble of solutions vanishing at infinity is a vast one, and *ad hoc* methods of selecting the solution satisfying the 'radiation condition' at infinity have been of very variable simplicity and effectiveness. This paper points out (§2) the extremely simple general rule, easy to infer from published papers but not so well known as it ought to be, for specifying the wave pattern that this solution involves.

A steady forcing effect may be so strong as to generate in its neighbourhood large disturbances not governed even approximately by the linear equations appropriate to small disturbances. The present theory can nevertheless be used to infer characteristics of the wave pattern set up far from the forcing region, where the disturbances are small enough for linear equations to apply. This is because the (admittedly unknown) non-linear terms in the equations which operate in the near field can be simply regarded as an additional forcing term whose region of application travels at the same speed.

For example, a large steady disturbance moving slowly through an extended body of uniformly rotating fluid at right angles to the axis can generate locally a 'Taylor column', but for non-zero Rossby number this cannot extend to infinity even in an inviscid fluid. In fact, the far disturbances must be small, and so take the form of inertial waves stationary with respect to the forcing disturbance. Their nature is worked out in §9, where some suggestions are made also about how they match with the near-field 'Taylor-column' solution.

In addition, two problems in non-rotating fluids are considered. First, in §6, as a link with the classical Kelvin ship-wave problem, the waves generated on deep water by an oscillatory disturbance travelling at speed U are studied. Results given by Eggers (1957) and others are confirmed, in opposition to incorrect predictions by Sretensky (1954), and subsumed within the general theory. It is shown that only oscillations of radian frequency less than $1\cdot62g/U$ can produce waves outside the normal ship-wave wedge. Secondly, in §7, the shape of gravity waves excited in a uniformly stratified fluid by a vertically moving steady disturbance are calculated, in good agreement (figure 8) with experiments by Mowbray (1966).

2. General theory for steady or periodic forcing terms

Consider a system such that small disturbances to the undisturbed state are governed by a linear partial differential equation with independent variables x, y, z and t and constant coefficients, which we write

$$P\left(i\frac{\partial}{\partial t}, -i\frac{\partial}{\partial x}, -i\frac{\partial}{\partial y}, -i\frac{\partial}{\partial z}\right)\phi = 0, \tag{1}$$

where P is a polynomial and ϕ is some variable specifying the disturbance. Then a plane wave

$$\phi = \phi_0 \exp\{i(-\sigma t + lx + my + nz)\} = \phi_0 \exp\{i(-\sigma t + \mathbf{k}.\mathbf{r})\} \tag{2}$$

can exist if the dispersion relation

$$P(\sigma, l, m, n) = 0 \tag{3}$$

is satisfied. On the other hand, because the undisturbed state is assumed stable, no solution of (3) exists with l, m, n real and the imaginary part of σ positive.

A forcing region is one where the right-hand side of (1) is replaced by a non-zero 'forcing term', which may represent the action of external forces on the system, and may also include substantial terms non-linear in the solution ϕ wherever disturbances are not small. In the forcing regions moving with uniform velocity $\mathbf{U}$ which this paper considers, steady forcing terms

$$f(\mathbf{r} - \mathbf{U}t), \tag{4}$$

where $\mathbf{r} = (x, y, z)$, are of particular interest. Sinusoidally varying forcing terms

$$e^{-i\sigma_0 t}f(\mathbf{r} - \mathbf{U}t) \tag{5}$$

are also of interest, and to save writing the theory will be given in this section in the more general case (5). The reader is asked, however, to bear in mind continually the specially important case $\sigma_0 = 0$.

We suppose that $f(\mathbf{r}) = f(x, y, z)$ vanishes outside a limited forcing region around the origin, and therefore can be written as a Fourier integral

$$f(\mathbf{r}) = \int_{-\infty}^{\infty}\int_{-\infty}^{\infty}\int_{-\infty}^{\infty} F(\mathbf{k})e^{i\mathbf{k}.\mathbf{r}}\,dl\,dm\,dn, \tag{6}$$

where $F(\mathbf{k}) = F(l, m, n)$ is a regular function for all l, m, n. The equation

$$P\left(i\frac{\partial}{\partial t}, -i\frac{\partial}{\partial x}, -i\frac{\partial}{\partial y}, -i\frac{\partial}{\partial z}\right)\phi = e^{-i\sigma_0 t}f(\mathbf{r} - \mathbf{U}t), \tag{7}$$

with (6) used to rewrite the right-hand side, then has the formal solution

$$\phi = \int_{-\infty}^{\infty}\int_{-\infty}^{\infty}\int_{-\infty}^{\infty} \frac{F(\mathbf{k})\exp\{i[-\sigma_0 t + \mathbf{k}.(\mathbf{r} - \mathbf{U}t)]\}}{P(\sigma_0 + \mathbf{U}.\mathbf{k}, l, m, n)}\,dl\,dm\,dn. \tag{8}$$

This solution is not unique, however, in any system for which plane wave solutions satisfying (3) exist, because the denominator can then vanish for real l, m, n, in which case many determinations of the integral (8) are possible. However, only one is of physical significance, namely that obtained when the source strength has been built up to its present strength from zero, and the system has reached its steady state. This is obtained, for example, by replacing σ_0 by $\sigma_0 + i\epsilon$ (which gives an extra factor $e^{\epsilon t}$ in the forcing term (5) and also, because no σ with positive imaginary part satisfies (3), gives the integral (8) a determinate value) and then letting ϵ tend to zero. We shall see in §3 that the same answer is obtained also if other modes of build-up of the forcing effect to its steady-state value are employed.

Lighthill (1965, §7) has given the method for evaluating integrals such as (8) at distances from the forcing region large compared with its dimensions, using methods described earlier by Lighthill (1960). The asymptotic form of ϕ, as defined by replacing σ_0 by $\sigma_0 + i\epsilon$ in (8) and letting $\epsilon \to 0$, can be described as follows.

In wave-number $(l, m, n) = \mathbf{k}$ space, at each point of the surface

$$P(\sigma_0 + \mathbf{U} \cdot \mathbf{k}, l, m, n) = 0 \tag{9}$$

on which the denominator of (8) vanishes, we draw an arrow normal to the surface, choosing from the two normal directions the one pointing towards the surface

$$P(\sigma_0 + \mathbf{U} \cdot \mathbf{k} + \delta, l, m, n) = 0 \quad \text{with } \delta \text{ small and positive.} \tag{10}$$

In other words, the arrow is in the direction σ increasing. Then the waves (if any) found *in some particular direction* stretching out from the forcing region are those with $\mathbf{k} = (l, m, n)$ given by a *point* (if any) on the wave-number surface (9) *where the arrow is in that particular direction*. Their amplitude is asymptotically

$$\frac{4\pi^2}{|K|^{\frac{1}{2}}R} \frac{F(\mathbf{k})}{|\nabla P(\sigma_0 + \mathbf{U} \cdot \mathbf{k}, l, m, n)|}, \tag{11}$$

where $R = |\mathbf{r} - \mathbf{U}t|$ means distance from the forcing region, ∇ is the operator grad with respect to $\mathbf{k} = (l, m, n)$ and K is the Gaussian curvature (product of principal curvatures) of the surface (9).

More strictly, ϕ is asymptotic to (11) provided $K \neq 0$, and falls off less rapidly than R^{-1} if $K = 0$. For example, we shall be concerned in what follows with cases of purely two-dimensional propagation, where there is no dependence on z at all. The wave-number surface is then cylindrical (so that $K = 0$); and its intersection with the plane $n = 0$, which we shall call the wave-number curve, alone determines the form of the waves generated. Equation (11) remains true with the first factor replaced by $(2\pi)^{\frac{3}{2}}/|\kappa|^{\frac{1}{2}}R^{\frac{1}{2}}$, where κ is the curvature of the wave-number curve. As a second example, a plane portion of the wave-number surface generates waves without attenuation, the first factor in (11) being replaced simply by 2π. Other examples are given by Lighthill (1960).

In a direction such that more than one point of the wave-number surface (9) has the arrow pointing in that direction, waves corresponding to each such point can be found superimposed on one another. The amplitude of each separately is determined by the above rules.

To explain physically the basic rule concerning the surface (9) and the arrows thereon, we note first that waves whose frequency is σ_0 *relative to* a forcing region moving at velocity $\mathbf{U}$ must have absolute frequency $\sigma_0 + \mathbf{U} \cdot \mathbf{k}$ (the Doppler effect), and so their wave-number vector *must* lie on the surface (9). Furthermore, the group velocity for waves satisfying (3) is

$$\left(\frac{\partial\sigma}{\partial l}, \frac{\partial\sigma}{\partial m}, \frac{\partial\sigma}{\partial n}\right) = -\frac{\nabla P}{\partial P/\partial\sigma}. \tag{12}$$

Now, at time $t = 0$ when the forcing region is around the origin, the position of a wave group created earlier, at time $t = -T$ when it was around the point $-\mathbf{U}T$, and propagating since then at the group velocity, must be

$$-\mathbf{U}T - \frac{\nabla P}{\partial P/\partial\sigma}T, \tag{13}$$

which does indeed lie in the direction of the arrow defined above.

Lighthill (1965), following Whitham (1960), gives also an interpretation of the amplitude variation (11). However, the only feature of this used below is the fairly obvious one, that waves will be generated corresponding only to those parts of the wave-number surface (9) for which the Fourier transform $F(\mathbf{k})$ of the forcing term is not negligibly small.

The shapes of wave-crests and other surfaces of constant phase can be deduced from the above rules (Lighthill 1960). Each is in fact the 'reciprocal polar' of the wave-number surface (9), that is, the locus of the poles of its tangent planes with respect to the origin. Analytically, it is the locus of the points

$$A\frac{\nabla P(\sigma_0 + \mathbf{U}.\mathbf{k}, \mathbf{k})}{\mathbf{k}.\nabla P(\sigma_0 + \mathbf{U}.\mathbf{k}, \mathbf{k})}, \tag{14}$$

where A is a constant.

In the special case $\sigma_0 = 0$ the surface (9) becomes

$$P(\mathbf{U}.\mathbf{k}, l, m, n) = 0, \tag{15}$$

which may be interpreted as a statement that a steady forcing effect can generate only waves whose crests are stationary relative to the velocity of the forcing region. This physically plausible idea (which can also be expressed by saying that the component of the forcing region's velocity in the direction of the phase velocity is exactly equal to the phase velocity) is proved to be of general validity by the present mathematical arguments.

3. General theory for transient forcing terms

Before giving examples of the results for steady or periodic forcing terms, we shall briefly obtain the corresponding results for a transient type of forcing term $f(\mathbf{r} - \mathbf{U}t, t)$, where $f(\mathbf{r}, t)$ is zero for $t < 0$, and also, as in §2, is assumed zero outside a limited region of space around the origin. Under these circumstances it can be written as a Fourier integral

$$f(\mathbf{r}, t) = \int_{i\epsilon-\infty}^{i\epsilon+\infty} e^{-i\sigma t}d\sigma \int_{-\infty}^{\infty}\int_{-\infty}^{\infty}\int_{-\infty}^{\infty} F(\sigma, \mathbf{k})e^{i\mathbf{k}.\mathbf{r}}dl\,dm\,dn, \tag{16}$$

where $\epsilon > 0$ and $F(\sigma, \mathbf{k})$ has no singularities where the imaginary part of σ is positive.

9 Hyperbolic Equations and Waves

The formal solution corresponding to (8) when the right-hand side of (7) is replaced by $f(\mathbf{r} - \mathbf{U}t, t)$ is

$$\phi = \int_{i\epsilon-\infty}^{i\epsilon+\infty} e^{-i\sigma t} d\sigma \int_{-\infty}^{\infty} \int_{-\infty}^{\infty} \int_{-\infty}^{\infty} \frac{F(\sigma, \mathbf{k}) \exp[i\mathbf{k}.(\mathbf{r} - \mathbf{U}t)] dl \, dm \, dn}{P(\sigma + \mathbf{U}.\mathbf{k}, l, m, n)}. \quad (17)$$

The asymptotic behaviour of (17) at large distances from the source will here be considered, first for a forcing term of finite duration and then in the case of one which becomes purely oscillatory after a finite time.

Lighthill (1965, §6) gave the solution for a source of finite duration, using the ideas of Lighthill (1960, appendix B) in a simplified form. The results will here be quoted in terms of the forms of the surfaces (9) for different values of the frequency σ_0.

The waves (if any) of frequency σ_0 found in some particular direction stretching out from the forcing region are those with $\mathbf{k} = (l, m, n)$ given by a point (if any) on the wave-number surface (9) for which the arrow is in that direction. Their amplitude is proportional to

$$\frac{F(\sigma_0, l, m, n)}{R^{\frac{3}{2}}}, \quad (18)$$

where $R = |\mathbf{r} - \mathbf{U}t|$ as before, and where the explicit form of the factor of proportionality depends like that in (11) on the geometry of the surfaces (9) but will not here be required. The energy density, which is proportional to the square of (18), falls off like R^{-3} instead of like R^{-2} because dispersion makes a transient disturbance grow outwards as a wave group filling a region whose volume expands in all three of its dimensions. The position at which the waves are found at time t is given by t times the appropriate group velocity (12).

This asymptotic result for the transient disturbance is derived from the singularities of the integrand in (17), which for $\sigma = \sigma_0$ are on the surface (9). On the other hand, for a source whose action continues indefinitely, additional singularities can arise in the integrand, due to singularities in F itself. For example, if

$$f(\mathbf{r}, t) = e^{-i\sigma_0 t} f(\mathbf{r}) \quad (19)$$

meaning that the simple harmonic forcing term (5) is 'switched on' at time $t = 0$, then

$$F(\sigma, \mathbf{k}) = \frac{F(\mathbf{k})}{2\pi i(\sigma - \sigma_0)}, \quad (20)$$

and this possesses a singularity at $\sigma = \sigma_0$. The asymptotic behaviour of (17) is then best obtained by first using the method of §2 for the inner integral (in which, as in §2, the imaginary part of σ is $+\epsilon$), after which integration with respect to σ gives exactly the result of §2, because division by $2\pi i(\sigma - \sigma_0)$ followed by integration replaces σ by σ_0.

This means that we obtain asymptotically the same steady-state solution, by suddenly switching on the source term (5) and waiting, as we did in §2 by allowing the strength to grow gradually from zero like $e^{\epsilon t}$. A much more general forcing term is the sum of (19) and an arbitrary source term of finite duration. This represents, indeed, a completely general forcing term starting from zero at time

$t = 0$ and becoming sinusoidal after a finite time. The solution in this case is the sum of the R^{-1} term of §2 and the $R^{-\frac{3}{2}}$ term of (18). Ultimately the former dominates, and we once again receive the solution which satisfies the radiation condition as the limiting result after the source has for a long time assumed the sinusoidal form.

Summary of sections 2 and 3

Sections 2 and 3 can be summarized by saying that the waves generated by forcing effects moving at velocity **U** are determined above all by the shapes of the surfaces $S(\sigma_0)$ in wave-number space given by equation (9). Arrows normal to $S(\sigma_0)$ pointing in the direction of $S(\sigma_0 + \delta)$ indicate in what direction stretching out from the source region waves of given frequency σ_0 and wave-number (l, m, n) will be found. But only those parts of $S(\sigma_0)$ where the forcing term's Fourier transform ($F(\mathbf{k})$ for a steady disturbance or $F(\sigma, \mathbf{k})$ for a transient disturbance) takes significant values will produce significant waves. In directions corresponding to those parts, the surfaces of constant phase for an oscillating disturbance of frequency σ_0 have the shape of the reciprocal polar of the surface $S(\sigma_0)$.

4. Rossby waves excited by a travelling steady disturbance

The method of this paper will first be applied to a travelling steady forcing effect generating Rossby waves in a 'beta-plane ocean'. Studies by Longuet-Higgins (1964, 1965 a) appear to indicate that waves in an ocean of uniform depth at frequencies large compared with the Coriolis parameter can be approximated reasonably well by divergenceless Rossby waves on a beta-plane, although a still better approximation for the lower wave-numbers is obtained by including a tidal term in the dispersion equation (see also S, (46)); the effect of this term on the results will be noted at the end of the present section.

Divergenceless Rossby waves on a beta-plane (S, (36)) satisfy

$$\frac{\partial}{\partial t}\left(\frac{\partial^2 \psi}{\partial x^2} + \frac{\partial^2 \psi}{\partial y^2}\right) + \beta \frac{\partial \psi}{\partial x} = 0, \tag{21}$$

where the x-direction is eastward and β is the gradient of Coriolis parameter in the northward y-direction. Multiplying (21) by $-i$ (for convenience) before comparison with (1), we obtain for this two-dimensional system

$$P(\sigma, l, m) = \sigma(l^2 + m^2) + \beta l. \tag{22}$$

A classic problem is the generation of Rossby waves by a steady westward-moving forcing effect. Such a means of excitement, relatively rare in the ocean, corresponds in the atmosphere to a commoner situation, generation by a steady eastward-moving wind blowing past a topographical feature.

If the forcing effect moves with velocity $(-U, 0)$, then the wave-number curve $S(0)$, given by equation (9) with $\sigma_0 = 0$, is

$$-Ul(l^2 + m^2) + \beta l = 0, \tag{23}$$

which consists of the straight line $l = 0$ and the circle

$$l^2 + m^2 = \beta/U. \tag{24}$$

9*

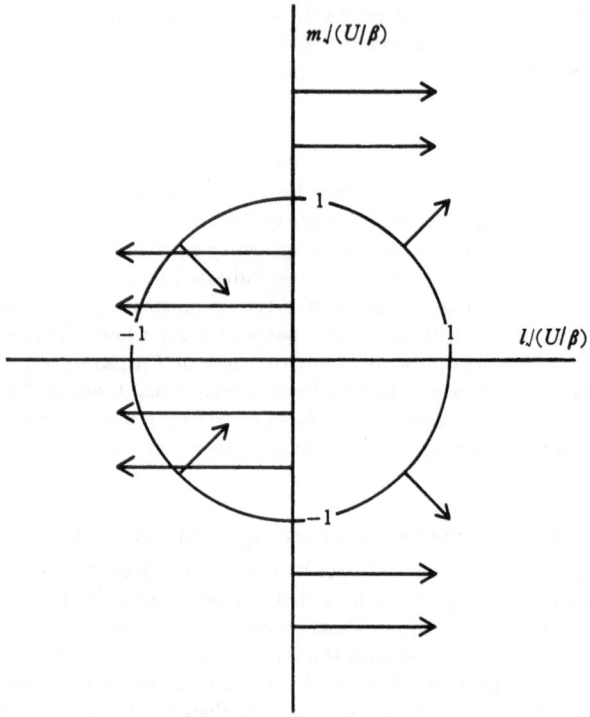

FIGURE 1. Wave-number curve for Rossby waves generated on a beta-plane ocean by a steady forcing effect travelling westward, with velocity $(-U, 0)$.

The literature has laid particular emphasis on the waves (24), of uniform length $2\pi \sqrt{(U/\beta)}$ and arbitrary direction.

But according to §2 it is essential to study not only the wave-number curve $S(0)$, here given by (23), but also the arrows normal to it pointing towards $S(+\delta)$, the curve defined by $P(-Ul+\delta, l, m) = 0$. Now it is easy to show that the change in l (say) for fixed m in going from $S(0)$ to $S(\delta)$ for small δ is asymptotically $(l^2 + m^2)\delta/[U(3l^2 + m^2) - \beta]$, and hence that the required arrows are as in figure 1. This means that the waves satisfying (24), with circular wave-crests, trail to the east of the westward-moving disturbance, filling the eastward-facing hemisphere behind it. The physical explanation of this was given in S, §7.

In addition, disturbances with l and σ zero, that is, independent of x and t, are possible, with those whose meridional wave-number $|m|$ exceeds $\sqrt{(\beta/U)}$ appearing behind the forcing region (that is, to the east), and those for which it is less appearing in front (to the west). Physically, this results from the rule due to Longuet-Higgins (1964) that the group velocity of Rossby waves makes an angle with the eastward direction twice that which the wave-number vector **k** makes, and is of magnitude β/k^2. For waves with $l = 0$, that is, with east–west crests, the group velocity β/m^2 is westward (although the phase velocity is zero), and exceeds U if and only if

$$|m| < (\beta/U)^{\frac{1}{2}}. \qquad (25)$$

These waves propagate (on the dissipationless model here used) without attenuation, because the associated part of the wave-number curve is a straight line. The disturbance that extends ahead (westward) of the obstacle is then the transverse disturbance created at the obstacle† modified by a 'low-pass filter' passing

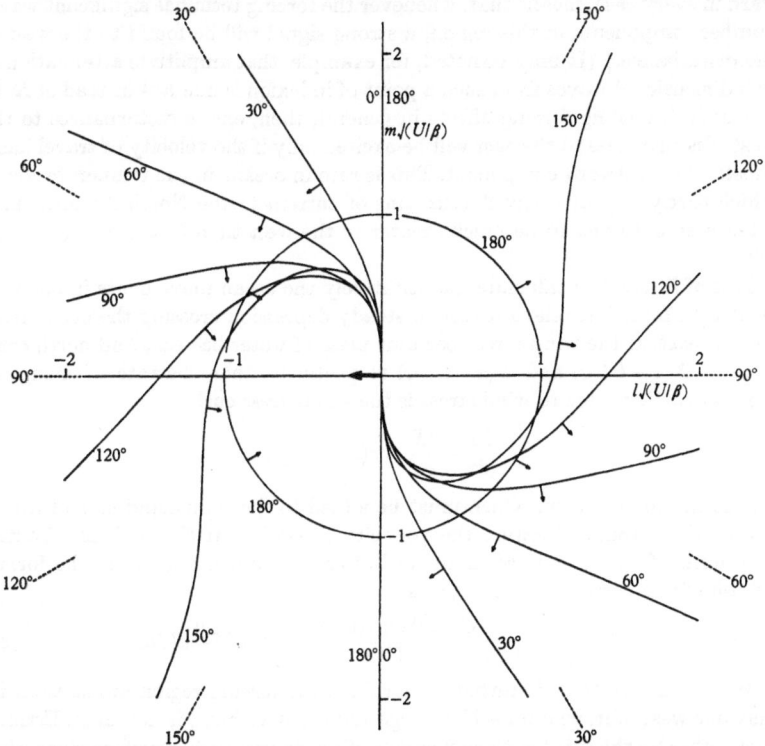

Figure 2. Wave-number curves for Rossby waves generated on a beta-plane ocean by a steady forcing effect travelling with uniform velocity U in directions making positive angles $\alpha = 0°, 30°, 60°, 90°, 120°, 150°$ and $180°$ (marked on the curves) with the eastward direction. Arrows are omitted on the m-axis, which both is the whole curve $\alpha = 0°$ (arrows westward only) and also is part of the curve $\alpha = 180°$ (arrows as in figure 1). ・・・・・, asymptotes.

only wave-numbers below $\sqrt{(\beta/U)}$. The disturbance extending to the east of the obstacle has been subjected to the complementary high-pass filter.

The situation with an eastward-moving steady forcing effect is much simpler. The sign of the first term in (23) being changed, the wave-number curve is merely the axis $l = 0$, and all arrows point to the west. There is therefore merely a long

† Strictly speaking, after integration in the east–west direction, because the $F(\mathbf{k})$ term in (11) for $l = 0$ represents the integral of $f(\mathbf{r})$ with respect to x from $-\infty$ to ∞.

straight unattenuated disturbance trailing behind the forcing region to the west of it, exactly as found in the experiments of Fultz & Long (1951).

Wave-number curves for steady forcing effects moving at an angle α measured in the positive sense from the eastward direction are given in figure 2. They satisfy

$$U(l\cos\alpha + m\sin\alpha)(l^2 + m^2) + \beta l = 0. \tag{26}$$

The point of inflexion of each curve at the origin, where the arrow points westward in every case, means that, whenever the forcing term has significant wavenumber components in this region, a strong signal will be found to the west of the disturbance.† (It may be noted, for example, that amplitude attenuation of two-dimensional waves from such a point of inflexion is like $R^{-\frac{1}{3}}$ instead of $R^{-\frac{1}{2}}$.)

Steady travelling forcing effects in general, then, excite disturbances to the west. Disturbances to the east will be excited only if the velocity of travel has a substantial westward component. This is rare in oceans in the temperate zones, which partly explains why fluctuations of current in the North Atlantic have often been observed to be much greater to the west than to the east (Swallow 1961).

It is of interest to calculate quantitatively the ocean movements in one particularly simple but relevant case, a steady depression crossing the ocean from west to east. If the wind-stress per unit mass of water has east and north components $X(x - Ut, y)$ and $Y(x - Ut, y)$ respectively, then the rate of change of vertical vorticity due to wind stress is the wind-stress curl

$$\frac{\partial Y}{\partial x} - \frac{\partial X}{\partial y} = c(x - Ut, y), \tag{27}$$

and so the forcing term which must be added to the right-hand side of (21) is $-c(x - Ut, y)$ (minus because the vorticity is $-\nabla^2\psi$). If $C(l, m)$ is the Fourier transform of $c(x, y)$, defined as in (6) but in two dimensions, then the formal solution for ψ corresponding to (8) is

$$\psi = i\int_{-\infty}^{\infty}\int_{-\infty}^{\infty} \frac{C(l, m)\exp\{i[l(x - Ut) + my]\}}{Ul(l^2 + m^2) + \beta l}\, dl\, dm. \tag{28}$$

We saw above that disturbances far from the forcing region are substantial only due west of it, where $x - Ut$ is large and negative but y is not large. Estimation, either by the rule for plane portions of the wave-number surface given after equation (11), or by direct asymptotic calculation of (28) (using the determination obtained by replacing Ul by $Ul + i\epsilon$ and letting $\epsilon \to 0$) gives that, as $x - Ut \to -\infty$,

$$\psi \sim 2\pi \int_{-\infty}^{\infty} \frac{C(0, m)e^{imy}}{Um^2 + \beta}\, dm \tag{29}$$

$$= \frac{1}{2\sqrt{(\beta U)}}\int_{-\infty}^{\infty} dx_1 \int_{-\infty}^{\infty} c(x_1, y_1)\exp\left[-|y_1 - y|\sqrt{(\beta/U)}\right]dy_1, \tag{30}$$

where to obtain (30) from (29) Parseval's theorem has been used.

---

† Figure 2 shows how, as the wave-number increases relative to $\sqrt{(\beta/U)}$, the direction in which the waves are found, measured in the positive sense from the westward direction, increases from 0 to a maximum just greater than α before falling to its final asymptotic value α.

Two limiting forms of (30) are of interest. First, when U, the velocity of convection of the forcing region, supposed of dimension L, is small compared with βL^2, then the form (30) of ψ as $x - Ut \to -\infty$ becomes approximately

$$\psi = \frac{1}{\beta} \int_{-\infty}^{\infty} c(x_1, y) \, dx_1. \tag{31}$$

This agrees with the solution of Sverdrup's classical steady-flow problem, that is,

$$\beta(\partial\psi/\partial x) = -c(x, y), \tag{32}$$

provided that ψ be taken zero to the east of the disturbance. Often elaborate boundary-layer arguments have been used to justify this boundary condition on ψ (S, §5), but the present work shows it as an immediate consequence of proper application of the radiation condition.

Secondly, when U is large compared with βL^2, equation (30) differentiated with respect to y gives approximately

$$u = \frac{\partial\psi}{\partial y} \doteq \frac{1}{2U} \int_{-\infty}^{\infty} dx_1 \left(\int_{y}^{\infty} c(x_1, y_1) \, dy_1 - \int_{-\infty}^{y} c(x_1, y_1) \, dy_1 \right), \tag{33}$$

which with expression (27) for c means simply

$$u = \frac{1}{U} \int_{-\infty}^{\infty} X(x_1, y) \, dx_1, \tag{34}$$

stating that water, uninfluenced by the beta-effect, has been accelerated directly by the force $X(x - Ut, y)$ per unit mass as the forcing region passes it. Water south of the centre of the depression is dragged eastward behind it, and water north of the centre is pushed westward. In this case, the group velocity β/k^2 of disturbance is small compared with U, so that only the water which the disturbance has actually passed over can be affected.

By contrast, in the case $U \ll \beta L^2$, the result (31), which in terms of u and X can be written

$$u = -\frac{1}{\beta} \frac{d^2}{dy^2} \int_{-\infty}^{\infty} X(x_1, y) \, dx_1, \tag{35}$$

will hold even far to the west of where the depression may have originated, since the group velocity is much greater than U. When expression (35) is valid, it is smaller than (34), and vice versa; whereas when U and βL^2 are of the same order, the expression

$$u = \frac{1}{U} \int_{-\infty}^{\infty} dx_1 \left[X(x_1, y) - \frac{1}{2} \left(\frac{\beta}{U} \right)^{\frac{1}{2}} \int_{-\infty}^{\infty} X(x_1, y_1) \exp\{ -|y_1 - y| \sqrt{(\beta/U)} \} \, dy_1 \right], \tag{36}$$

obtainable from (30) by differentiating with respect to y and then integrating by parts, shows that u again normally falls below the limiting value (34).

Probably the most interesting conclusion from this section is that, if the problem of steady wind-driven ocean currents is regarded as a limiting case of currents driven by a travelling forcing effect as the speed of travel tends to zero, then Sverdrup's solution with no disturbance to the east is obtained. (The conclusion is unaltered for a westward-moving forcing region, because the limitation (25) to wave-numbers less than $\sqrt{(\beta/U)}$ ceases to be restrictive as $U \to 0$.) This boundary condition is appropriate, therefore, for more fundamental reasons than have usually been given; physically, because group velocity is westward for north-south wave-numbers.

The results of this section are not much changed when the tidal term f^2/gH for an ocean of constant depth H is added (S, (46)) to $l^2 + m^2$ in (22), (23) and (24). The critical wave-number for a westward-moving forcing effect becomes

$$\left(\frac{\beta}{U} - \frac{f^2}{gH}\right)^{\frac{1}{2}}. \tag{37}$$

The curves in figure 2 are slightly modified, to pass through the origin at a small positive angle

$$\tan^{-1}\left[\frac{\sin \alpha}{(\beta gH/f^2 U) + \cos \alpha}\right] \tag{38}$$

to the m-axis. Accordingly, the low wave-number disturbances are to be found at this small positive angle to the westward direction. These changes are not really important provided that U is small compared with

$$\frac{\beta gH}{f^2} = (11 \, \text{m/s}) \left(\frac{\cos \theta}{\sin^2 \theta}\right) (H \, \text{in km}), \tag{39}$$

which is likely to be the case except perhaps at rather high latitudes θ, or small depths H.

5. Rossby waves excited by a travelling transient disturbance

In this section the generation of Rossby waves in a beta-plane ocean by travelling forcing effects of transient character is studied. As in §4, the effect of a tidal term is considered only at the end of the section.

Longuet-Higgins (1965b) gave an excellent account of transient currents generated (i) by a stationary transient forcing effect, and (ii) by a transient forcing effect that moves 'very rapidly', that is, much faster than the group velocity of the waves produced. The present solution is valid not only in these two extreme cases, but also for those intermediate speeds of travel which are often important in practice. It is complementary also in another way to the work of Longuet-Higgins (1965b), which is concerned with instantaneous impulse-type (delta-function) forcing effects, so that waves of all frequencies, however high, can be produced. Here we consider disturbances of non-zero duration, which normally will not excite waves of very high frequency. We shall see that their exclusion makes significant qualitative differences to the conclusions.

The wind-stress curl, then, has the form $f(\mathbf{r} - \mathbf{U}t, t)$ as in §3, where the function $f(\mathbf{r}, t)$ vanishes except in a finite region of $\mathbf{r}$ and within a finite time interval, and can be expressed as a Fourier integral as in (16). We shall assume such smooth variation of $f(\mathbf{r}, t)$ that its Fourier transform $F(\sigma, k)$ is small for frequencies σ exceeding a frequency σ_1 characteristic of the disturbance, or for wave-numbers k exceeding a characteristic wave-number k_1. Here σ_1 might be about 10 divided by the 'duration' T of the disturbance, since, for example, a Gaussian transient proportional to $\exp[-10(t/T)^2]$, with amplitude 8 % of its maximum at $t = \pm \frac{1}{2}T$, has Fourier transform less than 8 % of *its* maximum for $\sigma > 10/T$.

For each frequency σ_0 less than σ_1, the waves generated are those specified by the curve $S(\sigma_0)$ in the wave-number plane. This curve, which if the direction of travel makes a positive angle α with the eastward direction has the equation

$$[\sigma_0 + U(l\cos\alpha + m\sin\alpha)](l^2 + m^2) + \beta l = 0 \qquad (40)$$

by (9) and (22), is drawn in figure 3 for various σ_0 when $\alpha = 30°$. This particular case was chosen because it appeared in §4 that steady forcing effects when α is relatively small can generate Rossby waves only in a limited sector trailing

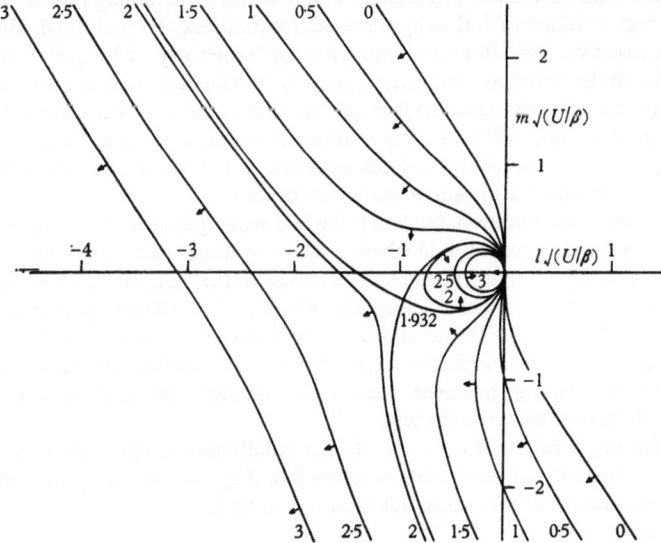

FIGURE 3. Wave-number curves for Rossby waves generated on a beta-plane ocean by an oscillatory forcing effect travelling with uniform velocity U in a direction making a positive angle $30°$ with the eastward direction. The number marked on each curve is the value of $L = \sigma_0/\sqrt{(U\beta)}$, where σ_0 is the frequency.

behind the disturbance, and it is desirable to find out (without limitation to the over-special case $\alpha = 0$) whether this conclusion remains valid for transient forcing effects travelling in directions typical of temperate-zone conditions.

It is seen that the shape of $S(\sigma_0)$ depends critically on the value of a frequency parameter

$$\sigma_0/\sqrt{(U\beta)} = L, \qquad (41)$$

say. It is close to the form $S(0)$ used in §4 to investigate steady forcing effects only when L is small. Big changes of form occur for values of L around unity, and for

$$L > 2\cos\tfrac{1}{2}\alpha \qquad (42)$$

(here $L > 1\cdot932$) the curve splits into two. For larger values of L, the two parts approximate closer and closer to a straight line and a circle. For a non-travelling forcing effect ($U \to 0$, $L \to \infty$) the circle alone is left (as would be expected from the work of Longuet-Higgins 1965b).

When $\sigma_0 = \sigma_1$, the highest frequency characteristic of the forcing effect, values of L satisfying (42) may be found for disturbances of relatively short duration. For example, a forcing effect of duration 4 days travelling at 10 m/s at latitude 45° has $L = 2.5$ for $\sigma_0 = \sigma_1$. However, all frequencies below σ_1 will in general be significant, and an important zero-frequency component is in particular present if the time integral of the disturbance is non-zero. All the values of L in figure 3 are likely to be found together, therefore.

Figure 3 shows that for the lower values of L, around 0 to 1·0, the currents generated trail once more in a narrow wedge behind the forcing region, but that for the higher values of L Rossby waves all round it may be generated. However, only waves of very small wave-number $k = \sqrt{(l^2 + m^2)}$, say with $k \sqrt{(U/\beta)}$ less than about 1, will be found all round it, physically for the same reason that led to the condition (25). For the example just quoted, this would limit such waves to those of length exceeding 5000 km. For a different example, with a forcing effect of duration 14 days travelling at 4 m/s at latitude 45°, the maximum value of L would be 1 and all disturbances would be trailing.

A still more pronounced tendency for the wave pattern to trail exclusively behind the disturbance is found when gravity effects are taken into account, as at the end of §3, by adding a term f^2/gH in equation (22). This term must be added to $l^2 + m^2$ also in (40), which modifies the curves in figure 3 mainly *near the origin*, where $l^2 + m^2$ is small. The modification to the curve $L = 0$, already noted in §4, is that it passes through the origin at the small positive angle (38) to the m-axis. The other curves, however, cease to pass through the origin, where they are displaced, in fact, towards the left.

For the larger values of L, the new term actually reduces the size of the nearby circular branch of the curve, somewhat as found by Longuet-Higgins (1965a, b) with U neglected. It approximates then to the circle

$$\sigma_0 \left(l^2 + m^2 + \frac{f^2}{gH} \right) + \beta l = 0, \tag{43}$$

with radius

$$\left(\frac{\beta^2}{4\sigma_0^2} - \frac{f^2}{gH} \right)^{\frac{1}{2}}, \tag{44}$$

and vanishes altogether when σ_0 exceeds

$$\frac{\beta \sqrt{(gH)}}{2f} = (0.7 \cot \theta)(\text{depth in km})^{\frac{1}{2}}(\text{days})^{-1}, \tag{45}$$

or when L exceeds $(\beta gH/4f^2U)^{\frac{1}{2}}$. In the first example quoted above, the values of L for which the waves were not trailing, namely 1 to 2·5, *all* correspond to values of σ_0 which already exceed the limit (45) for depths less than 2 km; and, even for a depth of 4 km, only a minimal amount of wave energy, with L between 1 and 1·5, could be found ahead of the forcing region.

6. Surface gravity waves generated by a travelling oscillating disturbance

Among the best-known wave combinations due to travelling forcing effects is Kelvin's pattern of surface gravity waves, set up by a ship in steady motion, and shown by him to be confined within a wedge of semi-angle $19\frac{1}{2}°$. The method of this paper is now used to study surface gravity waves generated by a travelling disturbance that is not steady but oscillatory with frequency σ_0, so that Kelvin's ship waves are the special case $\sigma_0 = 0$.

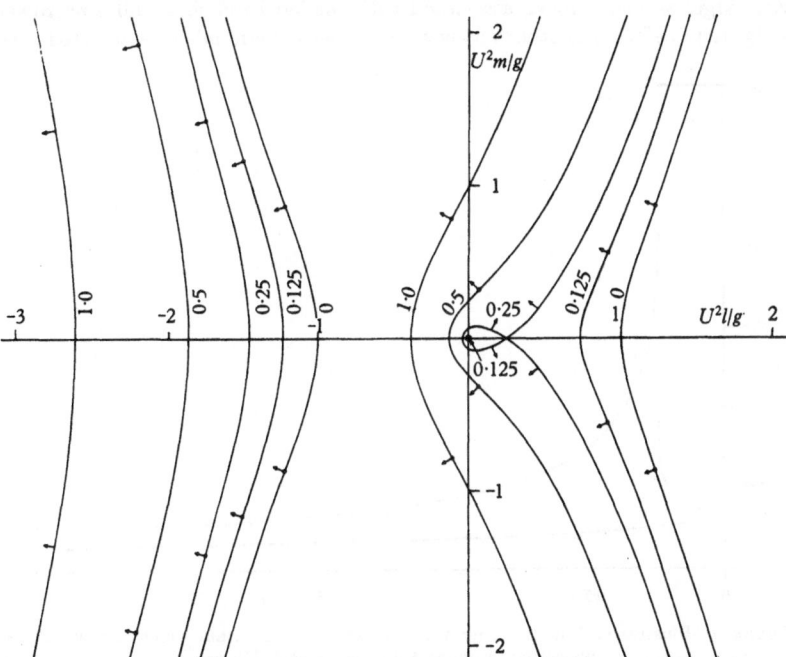

FIGURE 4. Wave-number curves for surface gravity waves, generated by an oscillatory forcing effect with various frequencies σ_0, travelling with velocity $(U, 0)$ over deep water. The numbers 0, 0·125, 0·25, 0·5 and 1·0 on the curves give the values of $U\sigma_0/g$ in each case. On certain branches of the curves, which correspond to a wedge of waves, points of inflexion (corresponding to waves on the boundary of the wedge) are marked by a spot.

This problem, like some others studied in this paper, has been treated by various writers, with very different results. Our object here is to show that the correct placing of the waves follows immediately from the general theory of §2, which, in fact, supports the work of Eggers (1957) and Newman (1959) and others against that of Sretensky (1954).

For this two-dimensional system the dispersion relation takes the form (3), where

$$P(\sigma, l, m) = \sigma^4 - g^2(l^2 + m^2). \tag{46}$$

Hence equation (9) for the surface $S(\sigma_0)$ becomes

$$(\sigma_0 + Ul)^4 = g^2(l^2 + m^2),\qquad(47)$$

where the forcing effect has frequency σ_0 and travels with velocity $(U, 0)$.

The surface $S(\sigma_0)$ is shown in figure 4 for various values of the ratio $U\sigma_0/g$. Kelvin's ship waves have wave-numbers on $S(0)$, and fill a backward-trailing wedge of semi-angle $19\frac{1}{2}°$ because the arrows on $S(0)$ (pointing towards $S(+\delta)$) do so. When $U\sigma_0/g$ takes small positive values (for example, $0\cdot125$) the two sheets of which $S(0)$ is composed are in $S(\sigma_0)$ both displaced to the left. Certain waves with larger wave-numbers, associated with the left-hand sheet, fill a narrower wedge than before, and other waves with smaller wave-numbers, associated with

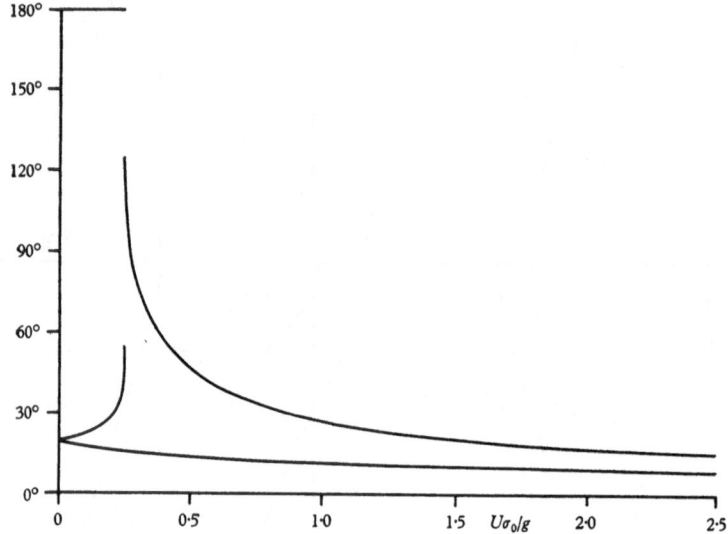

FIGURE 5. Illustrating how the semi-angles of the wedges within which the waves re-presented in figure 4 lie change with $U\sigma_0/g$.

the right-hand sheet, fill a wider wedge. (The point of inflexion, marked on each curve, corresponds to waves on the boundary of such a wedge.) $S(\sigma_0)$ includes, in addition, a small oval region involving very small wave-numbers (of the order of σ_0^2/g), and, if the forcing effect has components with such wave-numbers, then the associated waves are disposed in all directions around it.

Transition to a new régime occurs at $U\sigma_0/g = 0\cdot25$ (where $S(\sigma_0)$ crosses itself), beyond which there are only two branches of $S(\sigma_0)$, and the waves associated with each lie within a certain wedge. Figure 5 plots the semi-angles of the wedges within which the waves associated with the left-hand and right-hand branches lie as a function of $U\sigma_0/g$. (For $U\sigma_0/g < 0\cdot25$, a value of $180°$ is also included, to represent the fact that waves associated with the small oval branch are found in all directions around the forcing region.)

One interesting conclusion is that waves are found outside the Kelvin wedge, associated with the steady part of any composite travelling disturbance, only for frequencies σ_0 satisfying $U\sigma_0/g < 1\cdot63$. Another is that waves are found in front of the obstacle (that is, in the forward-facing semicircle) only if $U\sigma_0/g < 0\cdot27$.

7. Internal gravity waves generated by a vertically moving steady disturbance

Gravity waves in a uniformly stratified medium are next discussed. Generation by a vertically moving steady disturbance is of particular interest owing to the importance in certain stably stratified regions of the atmosphere of the phenomenon known as a 'thermal', that is, a rising localized region of hot air. The question of what gravity waves, if any, a 'thermal' generates may be treated

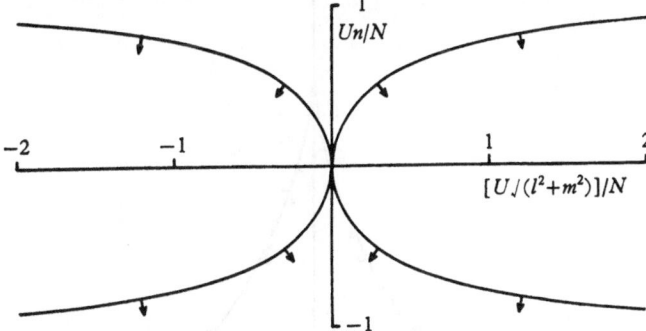

FIGURE 6. Wave-number surface for internal gravity waves generated by steady vertical motion of an obstacle with velocity $(0, 0, U)$ through a uniformly stratified medium with Väisälä–Brunt frequency N.

approximately by regarding it as a travelling steady disturbance (see Warren (1960), who gave an excellent analysis of wave-making resistance on this assumption). Another reason for interest in this case is that experiments suitable for detailed comparison with theory have been made by Mowbray (1966), using a tank of uniformly stratified salt solution.

With the z-axis vertical, the vertical component of velocity w in gravity waves satisfies

$$\frac{\partial^2}{\partial t^2}\,\nabla^2 w = N^2\left(\frac{\partial^2 w}{\partial x^2} + \frac{\partial^2 w}{\partial y^2}\right), \tag{48}$$

where N is the Väisälä–Brunt frequency. Additional terms significant only if the wavelength is comparable with the scale height have been omitted from (48), and in what follows a uniformly stratified medium with constant N will alone be considered. Accordingly, in this three-dimensional problem,

$$P(\sigma, l, m, n) = \sigma^2(l^2 + m^2 + n^2) - N^2(l^2 + m^2), \tag{49}$$

and for steady forcing effects travelling with velocity $(0, 0, U)$ equation (9) with $\sigma_0 = 0$ for the surface of revolution $S(0)$ becomes

$$U^2 n^2(l^2 + m^2 + n^2) = N^2(l^2 + m^2). \tag{50}$$

Figure 6 gives the surface (50) in meridian section, with the arrows aimed towards $S(+\delta)$. This is a problem where all waves trail behind the forcing region. They are found, in fact, only *below* a rising obstacle, or *above* a falling one. Shorter waves, with wave-number comparable with N/U or greater, are found close to the path of the obstacle, but longer waves, with wave-number small compared with N/U, are found (if the forcing effect includes components with those wave-numbers) at places whose radius vector from the forcing region makes all angles up to 90° with the path.

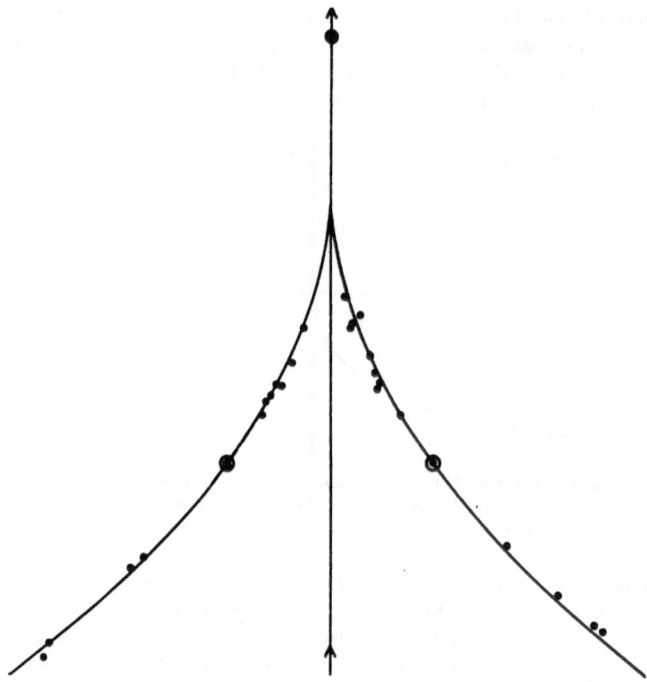

FIGURE 8. Shape of a surface of constant phase for internal gravity waves generated by steady vertical motion of a sphere (shown at top of diagram) through a uniformly stratified medium. Shape is normalized so that the points on the lines through the obstacle making an angle $\tan^{-1}(\frac{1}{4})$ with the vertical are in the ringed positions. Curve: theoretical shape. Points: experimental results from figure 7 and from another similar photograph (Mowbray 1966).

This is a case where the shapes of wave-crests (surfaces of constant phase) are well worth calculating, for comparison with Mowbray's experimental observations. Figure 7 (see p. 152) is a schlieren photograph of a sphere being raised at uniform velocity through uniformly stratified salt solution. The loci of maximum darkness are surfaces of constant phase (in fact, nodal surfaces with respect to density). The shape of several of these was measured and plotted on a single diagram (figure 8) after being scaled so that the points at an angle $\tan^{-1}(\frac{1}{4})$ behind the centre of the sphere (ringed in figure 8) are the same for each.

The curve in figure 8 is the theoretical surface of constant phase, given by equation (15) as

$$A\left(\frac{(U^2n^2 - N^2)\,l,\,(U^2n^2 - N^2)\,m,\,U^2(l^2 + m^2 + 2n^2)\,n}{N^2(l^2 + m^2)}\right),\qquad(51)$$

where l, m, n satisfy (50). It was plotted in terms of the parameter Un/N as

$$\frac{N\,\sqrt{(x^2 + y^2)}}{AU} = \frac{(1 - Un/N)^{\frac{3}{2}}}{(Un/N)^2},\quad \frac{Nz}{AU} = \frac{2}{Un/N} - (Un/N).\qquad(52)$$

The agreement with the experimental points is seen to be good.

8. Motion of obstacle along axis of rotating fluid

Various special geometrical features of the surface $S(\sigma_0)$, as discussed in §2 above, and in more detail by Lighthill (1960), can call for special treatment. One not there mentioned, and yet arising in a problem that has been studied at great length in the literature of rotating fluids, as well as being of independent interest, is the case when part of $S(\sigma_0)$ is twice covered; that is, when the surface contains two coincident portions. The difficulties experienced by many writers on the subject now to be described can be partly related to this rather unusual circumstance.

Through a large body of homogeneous fluid, in uniform rotation with angular velocity Ω, an obstacle moves with constant velocity U along the axis of rotation, which is the z-axis. As explained in §2, the motion at large distances from the obstacle would be expected to constitute a small perturbation of the state of uniform rotation, and therefore (S, §6) to take the form of inertial waves, in which all components of the fluid velocity $\mathbf{v}$ (in a rotating frame of reference) satisfy

$$\frac{\partial^2}{\partial t^2}\nabla^2\mathbf{v} + 4\Omega^2\frac{\partial^2\mathbf{v}}{\partial z^2} = 0.\qquad(53)$$

Comparison with (1) shows that

$$P(\sigma, l, m, n) = \sigma^2(l^2 + m^2 + n^2) - 4\Omega^2 n^2\qquad(54)$$

in this problem. For a steady disturbance ($\sigma_0 = 0$), equation (9) for the wave-number surface becomes

$$U^2n^2(l^2 + m^2 + n^2) - 4\Omega^2 n^2 = 0.\qquad(55)$$

The zero-frequency wave-number surface $S(0)$ given by (55) is shown in figure 9, and has evidently a strong resemblance to the Rossby-wave case illustrated in figure 1. It includes the sphere

$$l^2 + m^2 + n^2 = (2\Omega/U)^2,\qquad(56)$$

just as the circle (24) was included in the Rossby-wave case. Wave-numbers on this sphere correspond to waves of uniform length $\pi U/\Omega$ and arbitrary direction. The fact, at first sight surprising, that any such wave must remain stationary with respect to the moving disturbance, is well known to students of the literature, as also is their combination in various standard normal modes, e.g. of Bessel-function type. As in figure 1, the directions of the arrows on the sphere are such that these waves are found only behind the forcing region.

In a manner equally reminiscent of figure 1, the surface $S(0)$ in figure 9 includes a straight portion, here the plane $n = 0$. An essential point of difference, however, is that the surface $S(0)$ defined by the quartic equation (55) consists of the sphere

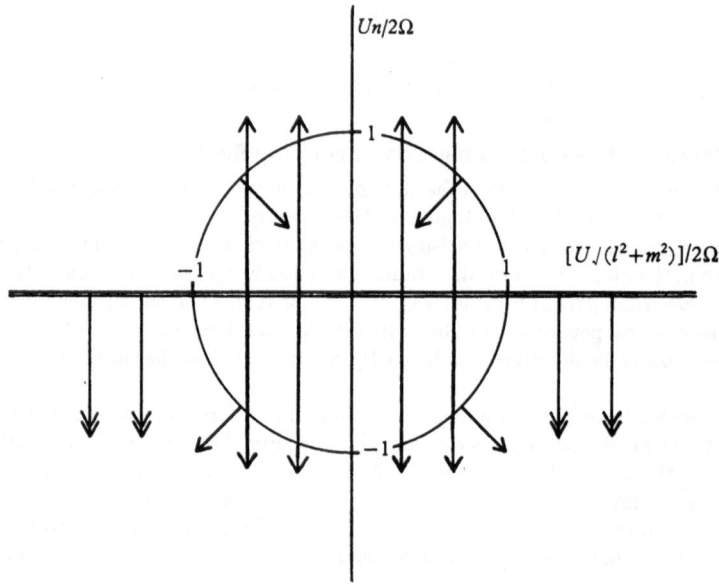

FIGURE 9. Wave-number surface for inertial waves generated by steady axial motion o an obstacle with velocity $(0, 0, U)$ through fluid rotating at angular velocity $(0, 0, \Omega)$. It consists of a sphere and two coincident planes.

(56) and the plane *taken twice*: it is a sphere and *two* (coincident) planes. It may be expected, therefore, that arrows along the appropriate normal must be drawn on both planes, and that the normal directions appropriate to each plane may or may not coincide.

The actual directions for waves with n (and σ) zero, that is, for disturbances independent of z (and t), are shown in figure 9. Disturbances whose transverse wave-number $\sqrt{(l^2 + m^2)}$ exceeds $2\Omega/U$ trail behind the obstacle, because the arrows on *both* planes point in the negative z-direction, and those for which it is less than $2\Omega/U$ are found partly behind and partly in front, because the arrows on each plane point in opposite directions. This fact can be deduced in various ways, of which perhaps the easiest is actually to draw $S(\sigma)$ for various σ as in figure 10 below, and to observe that for small positive σ the plane $n = 0$ splits into two sheets, which lie on different sides of it where

$$\sqrt{(l^2 + m^2)} < 2\Omega/U, \qquad (57)$$

and otherwise lie both below it. For a more analytical deduction, see below.

The physical explanation of the result is that these waves with zero phase velocity, whose stationary crests are parallel to the axis of rotation, have a group velocity $2\Omega/\sqrt{(l^2+m^2)}$ directed along the axis of rotation (either up or down it). This exceeds U (so that forward influence becomes possible) if and only if (57) is satisfied. The waves propagate (for the inviscid fluid here treated) without attenuation, because the associated part of the wave-number surface is plane. After a long enough time, those for which (57) is satisfied extend arbitrarily far, both ahead of and behind the obstacle, in a 'Taylor column'.

Admittedly obstacles whose transverse dimension, say a, is small (so that the Rossby number

$$U/2\Omega a \tag{58}$$

is large) cannot significantly excite waves satisfying (57). But, as the ratio (58) decreases, transverse disturbances satisfying (57) can increasingly be excited by the obstacle. The disturbance that extends ahead of the obstacle is then the transverse disturbance created by the obstacle, modified as in §4 by a 'low-pass filter' passing only wave-numbers below $2\Omega/U$.

In contrast with §4, however, the disturbance that extends behind the obstacle is not subjected merely to the complementary high-pass filter; it includes, in fact, also some low-wave-number terms. To obtain an estimate of their magnitude, the method leading to equation (11) cannot be used without change because the integral to be estimated has a double-pole singularity on doubly covered portions of the wave-number surface. The modifications to the method that are needed are as follows.

With P as in (54) and σ_0 replaced by $i\epsilon$, equation (8) becomes

$$\phi = \int_{-\infty}^{\infty}\int_{-\infty}^{\infty} \exp\{i(lx+my)\}dl\,dn \int_{-\infty}^{\infty} \frac{F(l,m,n)\exp\{in(z-Ut)\}dn}{(Un+i\epsilon)^2(l^2+m^2+n^2)-4\Omega^2n^2}, \tag{59}$$

and the problem is to estimate the inner integral when $|z-Ut|$ is large. When ϵ is positive but very small, the double pole at $n=0$ is split into two simple poles at

$$n_1 = \frac{i\epsilon}{\dfrac{2\Omega}{\sqrt{(l^2+m^2)}}-U} \quad \text{and} \quad n_2 = \frac{i\epsilon}{\dfrac{2\Omega}{\sqrt{(l^2+m^2)}}+U}. \tag{60}$$

When (57) is satisfied these are on opposite sides of the real axis, so that by Jordan's lemma there is a contribution to the inner integral from the pole $n=n_1$ when $z-Ut$ is positive and from $n=n_2$ when it is negative; but, when (57) is not satisfied, both n_1 and n_2 have negative imaginary parts and so there is no contribution at all for $z-Ut > 0$; this agrees with the direction of the arrows in figure 9.

More precisely, when (57) is satisfied, calculation of the residues at the poles gives a contribution to the inner integral in each case of

$$-\frac{\pi}{2\Omega\epsilon\sqrt{(l^2+m^2)}} F(l,m,n)\exp\{in(z-Ut)\}, \tag{61}$$

where $n=n_1$ for $z-Ut>0$ and $n=n_2$ for $z-Ut<0$. When (57) is not satisfied, both contributions appear for $z-Ut<0$ (and none for $z-Ut>0$) but that from $n=n_1$ has its sign changed.

Particularly when (57) *is* satisfied, there is apparently a difficulty in taking the limit of expressions (61) as $\epsilon \to 0$. The difficulty disappears, however, when we realize that in this problem, unlike the Rossby-wave case (see footnote in §5), F is necessarily zero for $n = 0$. To see this, we note that the basic equation (53) was obtained (S, §6) by applying the operation $(\partial/\partial t)$ curl to Helmholtz's equation for the vorticity in its linearized form S, (23). Hence, if Helmholtz's equation is transformed by a travelling forcing effect into a form with a forcing term,

$$\frac{\partial}{\partial t}\,\mathrm{curl}\,\mathbf{v} = 2\Omega\,\frac{\partial \mathbf{v}}{\partial z} + \mathbf{g}(x, y, z - Ut), \tag{62}$$

equation (53) becomes

$$\frac{\partial^2}{\partial t^2}\,\nabla^2 \mathbf{v} + 4\Omega^2\,\frac{\partial^2 \mathbf{v}}{\partial z^2} = -\left(\frac{\partial}{\partial t}\,\mathrm{curl} + 2\Omega\,\frac{\partial}{\partial z}\right)\mathbf{g}. \tag{63}$$

The right-hand side of (63) can be written $\mathbf{f}(x, y, z - Ut)$, if

$$\mathbf{f} = (\partial/\partial z)(U\,\mathrm{curl}\,\mathbf{g} - 2\Omega\mathbf{g}), \tag{64}$$

whose Fourier transform $\mathbf{F}(l, m, n)$ evidently contains a factor n and so vanishes at $n = 0$.

The limit of (61) (which represents, as we saw, the inner integral in the expression (59) for ϕ) is found simply, therefore, by de l'Hôpital's rule, and is

$$-\frac{\pi i(\partial F/\partial n)_{n=0}}{2\Omega[2\Omega \mp U\,\sqrt{(l^2 + m^2)}]}, \tag{65}$$

with the upper sign in the limit for $n = n_1$ and the lower in the limit for $n = n_2$. Disturbances are found ahead of the obstacle only when (57) is satisfied, and furthermore are only the n_1 disturbances, for which the negative sign taken in (65). We see that such disturbances are progressively amplified as $\sqrt{(l^2 + m^2)}$ increases towards the limit (57), even though no disturbance at all appears ahead of the obstacle beyond that limit.

The physical reason why a forcing effect of given axial extent (represented by its first moment $(\partial F/\partial n)_{n=0}$) can excite a forward-moving wave component most powerfully when its group velocity only slightly exceeds the speed of travel of the forcing effect is that the time available before the wave component escapes from the forcing region is then greatest. Increase of (65) to extremely large values would be restricted, however, by dissipation, by non-linearity, or by finiteness of duration of the forcing effect.

For low values of the Rossby number (58), the obstacle can generate substantial disturbances satisfying (57). In the steady state these will be found, as (65) indicates, with a greater amplitude ahead of the obstacle than behind it, in agreement with the experiments of Taylor (1922, 1923) and Long (1953). Although theoretical work on this problem took time to catch up with experiment, the need for such a disturbance ahead of the obstacle was clearly argued by Stewartson (1958).

When the fluid, instead of being unbounded, is contained within a circular cylinder of radius b, whose axis is the axis of rotation along which the obstacle

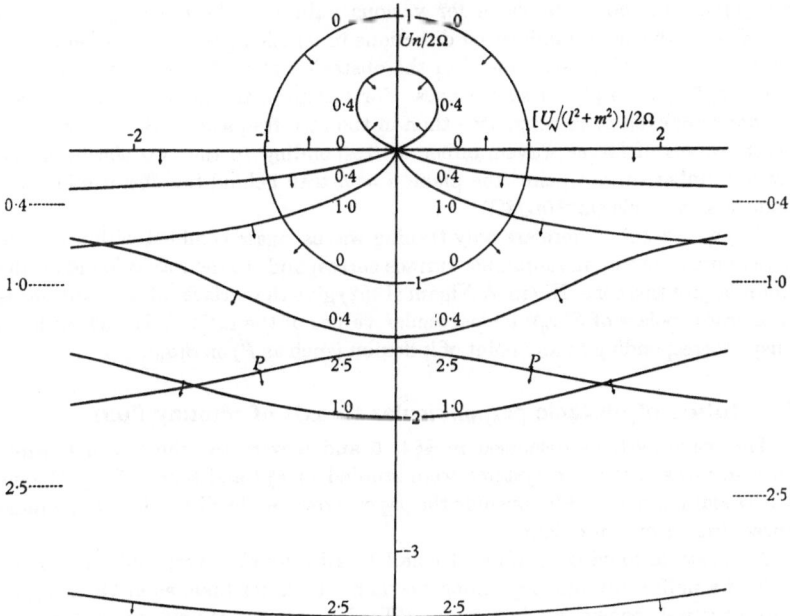

FIGURE 10. Wave-number surfaces for inertial waves generated by axial motion, with velocity $(0, 0, U)$, of an oscillatory forcing effect with various frequencies σ_0, through fluid rotating at angular velocity $(0, 0, \Omega)$. The numbers 0, 0·4, 1·0 and 2·5 on the curves give the value of $\sigma_0/2\Omega$ in each case. - - - - - -, asymptotes.

moves, the spherical waves satisfying (56) have to be combined in axisymmetrical normal modes (wave-guide modes of Bessel-function type) satisfying the boundary condition on the cylindrical surface. The theory states, in agreement with the experiments of Long (1953), that these waves appear only behind the obstacle. However, a disturbance independent of z is found ahead of the obstacle provided that disturbances with wave-numbers satisfying (57) can be combined into a solution satisfying the boundary condition. This requires that

$$U/2\Omega b < j_1^{-1} = 0\cdot 261,$$

where j_1 is the least positive zero of the Bessel function J_1, in agreement with arguments of Trustrum (1964).

Nigam & Nigam (1962) applied the methods of Lighthill (1960) (essentially those of this paper) to the more general case of waves made by a periodic forcing effect with frequency σ_0 moving along the axis of rotation in unbounded fluid. For example, an oscillating obstacle would make such waves, normally in addition to those that would be generated by its steady motion.

The surface $S(\sigma_0)$, which is now singly covered, has the equation

$$(\sigma_0 + Un)^2(l^2 + m^2 + n^2) - 4\Omega^2 n^2 = 0, \tag{66}$$

and figure 10 shows its shape for various values of the ratio $\sigma_0/2\Omega$. When $\sigma_0/2\Omega < 1$, the arrows indicating directions in which waves will be found show that they are still present ahead of the obstacle but that a cone of semi-angle $\sin^{-1}(\sigma_0/2\Omega)$ is empty of such waves. Furthermore, the limits on their wave-number are even more restrictive than in the zero-frequency case (57). By contrast, waves of larger wave-number, corresponding to the two sheets of the wave-number surface below the plane $n = 0$, trail behind the obstacle inside a cone, also of angle $\sin^{-1}(\sigma_0/2\Omega)$.

When $\sigma_0/2\Omega > 1$ there are only trailing waves, again confined within a cone. The point P on the wave-number surface corresponds to the waves found on the boundary of this cone. Nigam & Nigam (1962) give the surfaces of constant phase (reciprocal polars of $S(\sigma_0)$) for particular values of the ratio $\sigma_0/2\Omega$. These have cusps corresponding to any point of inflexion (such as P) on $S(\sigma_0)$.

9. Motion of obstacle perpendicular to axis of rotating fluid

The wave systems discussed in §§4, 5 and 6 were two-dimensional, while three-dimensional wave systems were studied in §§7 and 8 only for problems with axial symmetry. We conclude the paper, however, by discussion of a genuine three-dimensional problem.

An obstacle moving steadily at small Rossby number perpendicular to the axis of a uniformly rotating homogeneous fluid (rather than, as in §8, along it) can, as is well known, set in motion a 'Taylor column' of fluid, approximately cylindrical in shape with generators parallel to the axis, and moving with the obstacle at right angles to the axis. Experimental work on this subject has used, on the whole, somewhat limited volumes of fluid, and it remains uncertain how far along the axis the Taylor column extends in practice.

From the theoretical point of view, two limitations on its extent would be expected, viscous (non-zero Ekman number) and inertial (non-zero Rossby number). In different situations either of these may dominate. Morton (1966) has studied the limitation due to viscosity for zero Rossby number. Here the limitation due to non-zero Rossby number will be studied for an inviscid fluid.

For non-zero Rossby number it cannot be supposed that the Taylor column extends all the way to infinity, and indeed at very large distances from the body disturbances must be supposed to become small, and therefore subject to linear analysis. Moreover, even if this assumption were false, the linear analysis of the far field which follows would, according to the precedents of §§4 and 8, be expected to show up any possible propagation without amplitude reduction.

Accordingly, the near field is regarded as a travelling steady forcing effect which, in the far field, generates small disturbances. These must take the form of inertial waves, for which $P(\sigma, l, m, n)$ is given by (54). If the velocity of travel is $(U, 0, 0)$ then equation (9) for the wave-number surface $S(0)$ appropriate to a steady disturbance takes the form

$$U^2 l^2 (l^2 + m^2 + n^2) - 4\Omega^2 n^2 = 0. \tag{67}$$

Figure 11 indicates the shape of the surface (67) by plotting the contours $m = $ constant in the (l, n) plane. The arrows normal to the surface pointing towards $S(+\delta)$ are easily shown to trail (as in §7) behind the direction of motion of the forcing region; their projections on to the (l, n) plane are shown in figure 11.

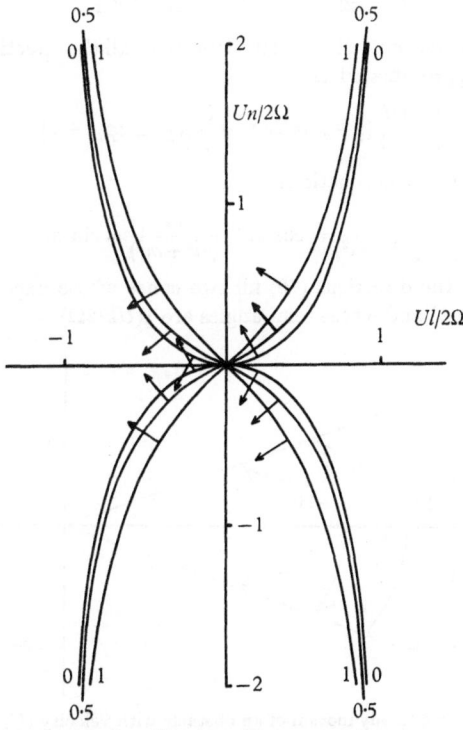

FIGURE 11. Wave-number surface (illustrated by means of contours of constant m in the (l, n) plane) for steady transverse motion of an obstacle with velocity $(U, 0, 0)$ through fluid rotating at angular velocity $(0, 0, \Omega)$. Contours for constant values 0, 0·5 and 1 (marked on the curves) of the ratio $Um/2\Omega$ are shown.

An obstacle of dimension a whose Rossby number, given by (58), is small generates waves whose characteristic wave-number k satisfies

$$0 \leqslant Uk/2\Omega \leqslant \epsilon, \tag{68}$$

where ϵ is a typical maximum Rossby number of the waves generated and would be expected to be proportional to (58). Accordingly, only the part of figure 11 which lies within a sphere of radius ϵ and centre the origin corresponds to waves which that obstacle can generate.

Within that part of the surface, the arrows all make a small angle with the z-direction, in agreement with the idea that to a first approximation the disturbance does not vary with z (as in the Taylor column). However, by studying their departure from the z-direction, we can investigate quantitatively how the disturbance trails behind the obstacle.

It follows from (61) and (62) that

$$\left|\frac{n}{l}\right| = \frac{Uk}{2\Omega} \leqslant \epsilon \quad \text{and} \quad \left|\frac{Ul}{2\Omega}\right| \leqslant \frac{Uk}{2\Omega} \leqslant \epsilon, \tag{69}$$

and using the fact that both these quantities are small the direction of the normal to (67) may be approximated as

$$\left(-\frac{Uk}{2\Omega}(\tfrac{3}{2}+\tfrac{1}{2}\cos 2\phi), \ -\frac{Uk}{2\Omega}(\tfrac{1}{2}\sin 2\phi), \ \pm 1\right), \tag{70}$$

where ϕ is defined by the equations

$$\frac{l}{\sqrt{(l^2+m^2)}} = \cos\phi, \quad \frac{m}{\sqrt{(l^2+m^2)}} = \sin\phi. \tag{71}$$

For fixed $Uk/2\Omega$ the directions (70) fill two cones whose axes have directions $(-\tfrac{3}{2}(Uk/2\Omega), 0, \pm 1)$ and whose semi-angles are $\tfrac{1}{2}(Uk/2\Omega)$.

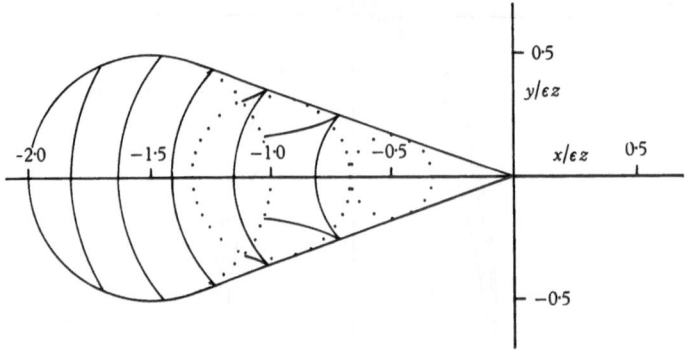

FIGURE 12. Transverse steady motion of an obstacle with velocity $(U, 0, 0)$ through fluid rotating at angular velocity $(0, 0, \Omega)$ produces at large axial distances z from the obstacle a region of waves (plain curves) whose cross-section is as shown. Dotted lines: loci of constant wavelength.

Waves are found, then, only in such cones trailing behind the obstacle, where $Uk/2\Omega$ takes all values from 0 to ϵ. The region filled by such cones, where waves are found, is shown (in cross-section by a plane $z = $ constant) in figure 12. The dotted circles (cross-sections of the above-mentioned cones) are loci of fixed wave-number k.

Within the region in figure 12, the shape of the surfaces of constant phase is given parametrically by equation (15), with $\sigma_0 = 0$. This gives

$$\frac{x}{\sqrt{|AUz/2\Omega|}} = -\frac{\tfrac{3}{2}+\tfrac{1}{2}\cos 2\phi}{\sqrt{|\cos\phi|}}, \quad \frac{y}{\sqrt{|AUz/2\Omega|}} = -\frac{\tfrac{1}{2}\sin 2\phi}{\sqrt{|\cos\phi|}}, \quad |\cos\phi| \geqslant \frac{AU}{2\Omega z\epsilon^2}, \tag{72}$$

and a few such surfaces, with equal phase difference between each, are shown (again in cross-section by a plane $z = $ constant) in figure 12. Those that extend to the straight lines $|y/x| = 2^{-\frac{3}{2}}$, which form part of the boundary of the region within which waves are found, have cusps thereon.

Only order of magnitude estimates can be given concerning the matching of the far-field behaviour depicted in figure 12 with the near-field 'Taylor-column' behaviour. The cross-section in figure 12 has dimension of order $\epsilon|z|$ and can be expected to match with a Taylor column whose cross-section has dimension a in some transition region situated around $|z| = a/\epsilon$. This indicates that Taylor columns extend for a distance of order the dimension of the obstacle divided by the Rossby number (except in circumstances when viscosity is large enough to limit them to a smaller length).

The author gratefully acknowledges Mrs N. A. Lighthill's help with the computations.

REFERENCES

EGGERS, K. 1957 *Schiff Hafen* **9**, 874.

FULTZ, D. & LONG, R. R. 1951 *Tellus* **3**, 61.

LIGHTHILL, M. J. 1960 *Phil. Trans.* A **252**, 397.

LIGHTHILL, M. J. 1963 *Am. Inst. Aeron. Astron. J.* **1**, 1507.

LIGHTHILL, M. J. 1965 *J. Inst. Math. Appl.* **1**, 1.

LIGHTHILL, M. J. 1966 *J. Fluid Mech.* **26**, 411 (referred to throughout this paper as **S**).

LONG, R. R. 1953 *J. Met.* **10**, 197.

LONGUET-HIGGINS, M. S. 1964 *Proc. Roy. Soc.* A **279**, 446.

LONGUET-HIGGINS, M. S. 1965a *Proc. Roy. Soc.* A **284**, 40.

LONGUET-HIGGINS, M. S. 1965b *Deep Sea Res.* **12**, 923.

MORTON, B. R. 1966 Private communication.

MOWBRAY, D. E. 1966 Private communication.

NEWMAN, J. N. 1959 *J. Ship Res.* **3**, 1.

NIGAM, S. D. & NIGAM, P. D. 1962 *Proc. Roy. Soc.* A **266**, 247.

SRETENSKY, L. N. 1954 *Trudy mosk. mat. Obshch.* **3**, 3.

STEWARTSON, K. 1958 *Q.J.M.A.M.* **11**, 39.

SWALLOW, J. 1961 *Sci. Progr.* **49**, 281.

TAYLOR, G. I. 1922 *Proc. Roy. Soc.* A **102**, 180.

TAYLOR, G. I. 1923 *Proc. Roy. Soc.* A **104**, 213.

TRUSTRUM, K. 1964 *J. Fluid Mech.* **19**, 415.

WARREN, F. W. G. 1960 *J. Fluid Mech.* **7**, 209.

WHITHAM, G. B. 1960 *J. Fluid Mech.* **9**, 347.

Journal of Fluid Mechanics, Vol. 27, part 4   *Plate* 1

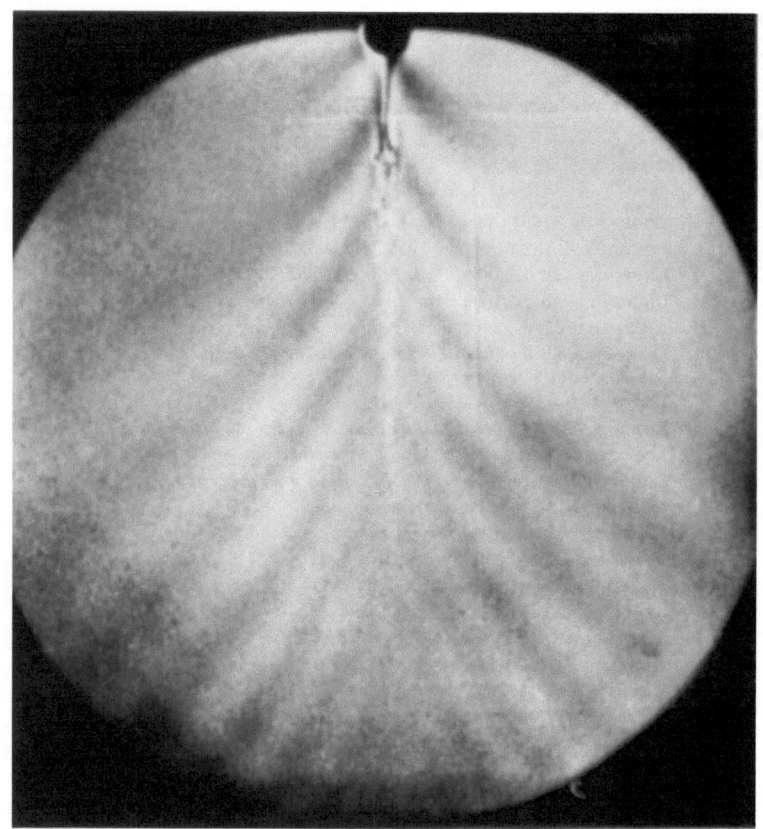

FIGURE 7. (see p. 142) Schlieren photograph (Mowbray 1966) of waves generated by a sphere of diameter 2·54 cm rising vertically at a speed of 1·02 cm/s in a solution of sodium chloride whose density falls with height at a rate of 0·0020 gm/cm³ per cm.

Proceedings of the Royal Society of London, **A 299**, 6 (1967)

Variational Methods and Applications to Water Waves

G. B. Whitham

This paper reviews various uses of variational methods in the theory of nonlinear dispersive waves, with details presented for water waves. The appropriate variational principle for water waves is discussed first, and used to derive the long-wave approximations of Boussinesq and Korteweg & de Vries. The resonant near-linear interaction theory is presented briefly in terms of the Lagrangian function of the variational principle. Then the author's theory of slowly varying wavetrains and its application to Stokes's waves are reviewed. Luke's perturbation theory for slowly varying wavetrains is also given. Finally, it is shown how more general dispersive relations can be formulated by means of integro-differential equations; an important application of this, developed with some success, is towards resolving long-standing difficulties in understanding the breaking of water waves.

1. Variational principle for water waves

Certain investigations in nonlinear wave theory can be given a general form if the basic equations are governed by a variational principle

$$\delta \int \int L \, d\mathbf{x} \, dt = 0. \tag{1}$$

At the same time, the mathematical manipulations, which may be formidable otherwise, become simple in terms of the 'Lagrangian function' L. There seems to be no general method, other than experienced guesswork, for finding variational principles for given systems of equations. However, they are known for many important cases. Strangely enough, a suitable variational formulation for water waves does not seem to be given in the literature and certainly is not widely known. Water waves are the prime example considered in this paper, as being typical for dispersive waves, so the first two sections present the appropriate variational principle and the approximations for long waves.

In fluid dynamics, it is known that Hamilton's principle with L equal to kinetic energy minus potential energy must apply since, as a last resort, the fluid may be treated as a system of particles. However, the direct formulation of Hamilton's principle gives difficulty in the Eulerian description and various side conditions have to be introduced by means of Lagrange multipliers (see, for example, Serrin 1959).

For irrotational water waves, at least, a more convenient variational principle, free of side conditions, is (1) with

$$L = \int_0^{h(\mathbf{x},\, t)} \{\phi_t + \tfrac{1}{2}(\nabla\phi)^2 + gy\} \, dy, \tag{2}$$

where y is the vertical coordinate, $\mathbf{x} = (x_1, x_2)$ is the horizontal coordinate, $\phi(\mathbf{x}, y, t)$ is the velocity potential, $y = h(\mathbf{x}, t)$ is the equation of the free surface and g is the acceleration of gravity. Variations of ϕ within the flow region lead to

$$\nabla^2\phi = 0, \tag{3}$$

the variation of h gives the pressure condition

$$\phi_t + \tfrac{1}{2}(\nabla\phi)^2 + gy = 0 \quad \text{on} \quad y = h(\mathbf{x}, t) \tag{4}$$

and, with some integration by parts, the variation of ϕ at the upper surface leads to the 'natural' boundary condition

$$h_t + \phi_{x_i} h_{x_i} - \phi_y = 0 \quad \text{on} \quad y = h(\mathbf{x}, t). \tag{5}$$

This formulation was pointed out explicitly by Luke (1967). Bateman (1944) writes down a form of which (2) is a special case, but he does not note that the free surface conditions (the main difficulty in water waves) also follow from (2).

It should be noted that Hamilton's principle would have

$$L_1 = \int_0^{h(\mathbf{x}, t)} \{\tfrac{1}{2}(\nabla\phi)^2 - gy\}\, dy; \tag{6}$$

variations of this function would give Laplace's equation for the flow, but incorrect boundary conditions at the surface. It is easily shown that

$$L_1 = -L - [\phi\phi_y]_{y=0} - [\phi(h_t + \phi_{x_i} h_{x_i} - \phi_y)]_{y=h} - \int_0^h \phi\, \nabla^2\phi\, dy + \frac{\partial}{\partial t}\int_0^h \phi\, dy + \frac{\partial}{\partial x_i}\int_0^h \phi\phi_{x_i}\, dy.$$

Apart from the divergence expression, the extra terms all concern conservation of mass. When ϕ is a *solution* of the water wave equations, these extra terms vanish and L_1 differs from $-L$ by the divergence expression. For the theory described in §4, the average value of L is used; since the divergence would average to zero, the average values of L and $-L_1$ are the same.

2. LONG WAVES

Boussinesq equations

Approximations for long waves may be derived by expanding ϕ in a power series in y. The solution of Laplace's equation subject to $\partial\phi/\partial y = 0$ on the bottom $y = 0$ is

$$\phi = f(\mathbf{x}, t) - \tfrac{1}{2}y^2\nabla^2 f(\mathbf{x}, t) + O(h^4/\lambda^4), \tag{7}$$

where λ is a typical wavelength. Then the Lagrangian in (2) becomes

$$L = h(f_t + \tfrac{1}{2}f_{x_i}^2) + \tfrac{1}{2}gh^2 - \tfrac{1}{6}h^3\{\nabla^2 f_t + f_{x_i}\nabla^2 f_{x_i} - (\nabla^2 f)^2\} + O(h^5/\lambda^5). \tag{8}$$

The term in h^3 is the dispersive correction to the usual shallow water theory. The variational equations for (8) give two differential equations for the functions $f(\mathbf{x}, t)$ and $h(\mathbf{x}, t)$. In this form the equations are complicated and it is simpler to work with the value of the potential at the surface, i.e.

$$F(\mathbf{x}, t) = f - \tfrac{1}{2}h^2\nabla^2 f + O(h^4/\lambda^4),$$

instead of f, the value on the bottom.

For, apart from a term

$$\frac{\partial}{\partial t}(\tfrac{1}{3}h^3\nabla^2 f) + \frac{\partial}{\partial x}(\tfrac{1}{3}h^3 f_{x_i}\nabla^2 f),$$

which does not contribute in the variational principle since it can be integrated out,

$$L = h(F_t + \tfrac{1}{2}F_{x_i}^2) + \tfrac{1}{2}gh^2 - \tfrac{1}{6}h^3(\nabla^2 F)^2 + O(h^5/\lambda^5). \tag{9}$$

The variational equations from this Lagrangian are

$$\delta F: \quad h_t + (hF_{x_i})_{x_i} + \nabla^2(\tfrac{1}{3}h^3\nabla^2 F) = 0, \tag{10}$$

$$\delta h: \quad F_t + \tfrac{1}{2}F_{x_i}^2 + gh - \tfrac{1}{2}h^2(\nabla^2 F)^2 = 0. \tag{11}$$

The highest order derivatives give the dispersive correction to shallow water theory. It is usually considered sufficient to have the linearized form for these correction terms; that is

$$\left.\begin{aligned}
h_t + (hF_{x_i})_{x_i} + \tfrac{1}{3}h_0^3\nabla^4 F &= 0, \\
F_t + \tfrac{1}{2}F_{x_i}^2 + gh &= 0.
\end{aligned}\right\} \tag{12}$$

This additional approximation assumes that the amplitude parameter a/h_0 is small in addition to h_0^2/λ^2. The fully linearized shallow water equations correspond to $a/h_0 \to 0$, $h_0^2/\lambda^2 \to 0$ and may be written

$$h_t + h_0 F_{x_i x_i} = 0, \quad F_t + gh = 0.$$

Equations (12) include the next order corrections in a/h_0 and h_0^2/λ^2.

An alternative form is obtained when the mean value of ϕ over the depth is introduced in place of F. The mean value $\mathscr{F}$ is given by

$$\begin{aligned}
\mathscr{F} &= f - \tfrac{1}{3}h^2\nabla^2 f + O(h^4/\lambda^4) \\
&= F + \tfrac{1}{6}h^2\nabla^2 F + O(h^4/\lambda^4),
\end{aligned}$$

and, in place of (12), we have

$$\left.\begin{aligned}
h_t + (h\mathscr{F}_{x_i})_{x_i} &= 0, \\
\mathscr{F}_t + \tfrac{1}{2}\mathscr{F}_{x_i}^2 + gh - \tfrac{1}{3}h_0^2\nabla^2\mathscr{F}_t &= 0.
\end{aligned}\right\} \tag{13}$$

From the first of these, $h_t = -h_0\nabla^2\mathscr{F}$ plus smaller terms, so (13) can be written in an equivalent form

$$\left.\begin{aligned}
h_t + (h\mathscr{F}_{x_i})_{x_i} &= 0, \\
\mathscr{F}_t + \tfrac{1}{2}\mathscr{F}_{x_i}^2 + gh + \tfrac{1}{3}h_0 h_{tt} &= 0.
\end{aligned}\right\} \tag{14}$$

Finally, if the mean horizontal velocity $U_i = \mathscr{F}_{x_i}$ is introduced, we may write

$$\left.\begin{aligned}
\frac{\partial h}{\partial t} + \frac{\partial(hU_j)}{\partial x_j} &= 0, \\
\frac{\partial U_i}{\partial t} + U_j\frac{\partial U_i}{\partial x_j} + g\frac{\partial h}{\partial x_i} + \tfrac{1}{3}h_0\frac{\partial^3 h}{\partial x_i\,\partial t^2} &= 0,
\end{aligned}\right\} \tag{15}$$

and this seems to be the form usually quoted for Boussinesq's equations. The Lagrangian corresponding to (14) is

$$L = h(\mathscr{F}_t + \tfrac{1}{2}\mathscr{F}_{x_i}^2) + \tfrac{1}{2}gh^2 - \tfrac{1}{6}h_0 h_t^2.$$

11*

Korteweg & de Vries equation

Korteweg & de Vries (1895) obtained an equation for waves propagating in one direction only. It may be obtained as the 'simple wave' solution of the shallow water equations corrected for the third-order dispersion term in (15). It may be verified that

$$
\left.
\begin{aligned}
U &= c_0\left(\frac{\eta}{h_0} - \frac{1}{4}\frac{\eta^2}{h_0^2}\right) + \tfrac{1}{6}c_0 h_0 \eta_{xx}, \\
\eta_t + c_0\left(1 + \frac{3}{2}\frac{\eta}{h_0}\right)\eta_x + \tfrac{1}{6}c_0 h_0^2 \eta_{xxx} &= 0, \\
\eta &= h - h_0,
\end{aligned}
\right\}
\tag{16}
$$

is a solution of (15) with errors of second order in a/h_0 and h_0^2/λ^2.

It is not clear how (16) could be obtained from the variational principle (9). However, a variational principle can be found directly when (16) is written in the form

$$
\left.
\begin{aligned}
\psi_{xt} + c_0(1 + \psi_x)\psi_{xx} + \tfrac{1}{6}c_0 h_0^2 \chi_{xx} &= 0, \\
\psi_{xx} - \chi &= 0,
\end{aligned}
\right\}
\tag{17}
$$

where $3\eta/2h_0 = \psi_x$. Equations (17) follow from

$$
\delta \iint \{\tfrac{1}{2}\psi_x \psi_t + \tfrac{1}{2}c_0\psi_x^2 + \tfrac{1}{6}c_0\psi_x^3 + \tfrac{1}{12}c_0 h_0^2(\chi^2 + 2\chi_x \psi_x)\}\, dx\, dt = 0.
\tag{18}
$$

3. Resonant interactions

One way to tackle nonlinear waves is to use a perturbation theory for small amplitude, based on the linearized theory as the lowest order approximation. The naïve expansion gives rise to secular terms growing linearly in t, owing to resonance of higher order products of linear terms with the original terms in the linear theory. This topic has been studied extensively by the contributors to this discussion and most of the following papers will centre around this approach.

It seems worthwhile to note briefly how variational methods are used for resonant interactions in order to contrast this approach with the one of slowly varying waves described in the next section. A simple example will suffice for this purpose, but it should be stressed from the start that the algebraic calculations increase considerably in other examples. This one is unusually easy.

The simple example[†] is

$$
u_t + 3u^2 u_x + u_{xxx} = 0.
$$

To obtain a variational principle, we introduce $u = \phi_x$, $\chi = \phi_{xx}$, and write the equivalent pair

$$
\left.
\begin{aligned}
\phi_{xt} + 3\phi_x^2 \phi_{xx} + \chi_{xx} &= 0, \\
\phi_{xx} - \chi &= 0.
\end{aligned}
\right\}
\tag{19}
$$

[†] This was proposed by Professor D. J. Benney as a simple model for discussions of resonant interactions.

The Lagrangian is
$$L = \tfrac{1}{2}\psi_x\phi_t + \phi_x\chi_x + \tfrac{1}{2}\chi^2 + \tfrac{1}{4}\phi_x^4 \tag{20}$$

Consider now a superposition of waves expressed as

$$\phi = \Sigma \frac{1}{ik_\alpha} A_\alpha(t)\, e^{ik_\alpha x} \quad (\alpha = \pm 1, \ldots, \pm N),$$

where $k_{-\alpha} = -k_\alpha$ and $A_{-\alpha} = A_\alpha^*$ in order that ϕ be real. (The asterisk denotes complex conjugates.) The Lagrangian L becomes, apart from a divergence term,

$$L = \Sigma\Sigma \left\{ \frac{1}{2ik_\alpha} A_\beta \frac{dA_\alpha}{dt} + \tfrac{1}{2}k_\alpha k_\beta A_\alpha A_\beta \right\} \exp\{i(k_\alpha + k_\beta)x\}$$
$$+ \tfrac{1}{4}\Sigma\Sigma\Sigma\Sigma A_\alpha A_\beta A_\gamma A_\delta \exp\{i(k_\alpha + k_\beta + k_\gamma + k_\delta)x\}. \tag{21}$$

In the variational principle L is integrated over an arbitrary rectangle in the (x,t) plane. Take a rectangle with $-l \leqslant x \leqslant l$ and consider

$$\hat{L} = \lim_{l \to \infty} \frac{1}{2l} \int_{-l}^{l} L\, dx. \tag{22}$$

Only the 'resonant terms' with

$$k_\alpha + k_\beta = 0, \quad k_\alpha + k_\beta + k_\gamma + k_\delta = 0,$$

contribute in the limit. The resonant duets are $(k_\alpha, k_\beta) = (k_n, -k_n)$ in either order, where $n = 1, 2, \ldots, N$. In the quartets there are various possibilities. There will always be
$$(k_n, k_n, -k_n, -k_n) \text{ in some order,}$$
and
$$(k_m, k_n, -k_m, -k_n), m \neq n, \text{ in some order.}$$

But special cases may be posed in the given initial modes; for example, wavenumbers $k_0, k_+ = k_0 + \mu, k_- = k_0 - \mu$ are supposed to be among the initial set in the resonance problem considered by Dr Benjamin. Then the quartets

$$(k_0, k_0, -k_+, -k_-) \quad \text{and} \quad (-k_0, -k_0, k_+, k_-) \text{ resonate.}$$

The Lagrangian $\hat{L}$ becomes

$$\hat{L} = \Sigma \frac{i}{2k_n}\left(A_n \frac{dA_n^*}{dt} - A_n^* \frac{dA_n}{dt}\right) - \Sigma k_n^2 A_n A_n^*$$
$$+ \tfrac{3}{2}\Sigma(A_n A_n^*)^2 + 6\sum\sum_{m \neq n} A_m A_m^* A_n A_n^* + 3\{A_0^2 A_+^* A_-^* + A_0^{*2} A_+ A_-\}. \tag{23}$$

The variational principle is reduced to

$$\delta \int_{t_1}^{t_2} \hat{L}\, dt = 0.$$

In the linear theory, only the quadratic terms are retained in $\hat{L}$ and variation with respect to A_n^* gives

$$\frac{i}{k_n}\frac{dA_n}{dt} = -k_n^2 A_n. \tag{24}$$

The solution is $A_n = a_n e^{-i\omega_n t}$ with ω_n satisfying the linear dispersion relation

$$\omega_n = -k_n^3. \tag{25}$$

With the full expression (23), variation with respect to A_+^*, for example, gives

$$\frac{i}{k_+}\frac{dA_+}{dt} = -\{k_+^2 - 3A_+A_+^* - 6\Sigma' A_n A_n^*\}A_+ + 3A_0^2 A_-^*,\qquad(26)$$

where Σ' denotes summation over all modes except the one with $k = k_+$. The second term in the coefficient of A_+ is the change in frequency due to the non-linear effects of a single mode; the third term is the change in frequency due to nonlinear coupling with the other modes. The term $3A_0^2 A_-^*$ gives the change in A_+ due to the special resonance between k_0, k_+ and k_-. If A_+ and A_- are small compared with A_0, A_0 can be assumed to be unaffected by A_+ and A_- to first order, so that

$$A_0 = a_0 e^{-i\omega_0 t}, \quad \omega_0 = -k_0^2 + 3k_0 a_0^2$$

and a_0 is taken to be real without loss of generality.

Then (26), and the same equation with k_+, A_+ replaced by k_-, A_-, have approximate solutions

$$\left.\begin{array}{c} A_+ = a_+ e^{-i\omega_+ t}, \quad A_- = a_- e^{-i\omega_- t}, \\ \omega_\pm = -k_\pm^3 + 6k_\pm a_0^2 + 3k_0(a_0^2 - \mu^2) + i\sqrt{\{9k_0^2\mu^2(2a_0^2 - \mu^2)\}}. \end{array}\right\}\qquad(27)$$

This is the kind of 'instability' discovered by Dr Benjamin for deep water waves. The 'resonance' of the frequencies, $\omega_+ + \omega_-^* = 2\omega_0$, which can be seen directly from the exponents of the exponentials in (26), should be specially noted.

In more general examples, the complications mentioned earlier arise because the resonance may not appear until two orders beyond linear theory. This whole question, and the use of diagram techniques to manage these complications, is discussed in the paper by Dr Hasselmann.

4. AVERAGING FOR SLOWLY VARYING NONLINEAR WAVETRAINS

The interaction approach is limited to nearly linear problems. With the idea that true nonlinear concepts may be missed in this approach, an attempt was made to find some fully nonlinear solutions in addition to the periodic uniform wavetrains which were known to exist in typical problems. Some simplifying feature other than linearization was sought and the obvious possibility seemed to be slowly varying wavetrains, with solutions close to the exact solutions for the uniform wavetrains. A general theory was developed (see Whitham 1965a, b) and the main steps will be reviewed in this section. A simple way of comparing the 'interaction approach' and the 'slowly varying approach' is to think of the elementary discussion of beats. For a linear dispersive problem, a solution with two neighbouring modes may be written either as

$$a_1 \cos(k_1 x - \omega_1 t) + a_2 \cos(k_2 x - \omega_2 t)$$

or as

$$a\cos(kx - \omega t - \epsilon),$$

where

$$\left.\begin{array}{l} a^2 = a_1^2 + a_2^2 + 2a_1 a_2 \cos\{\tfrac{1}{2}(k_2 - k_1)x - \tfrac{1}{2}(\omega_2 - \omega_1)t\}, \\ \tan\epsilon = \dfrac{a_1 - a_2}{a_1 + a_2}\tan\{\tfrac{1}{2}(k_2 - k_1)x - \tfrac{1}{2}(\omega_2 - \omega_1)t\}, \\ k = \tfrac{1}{2}(k_1 + k_2), \quad \omega = \tfrac{1}{2}(\omega_1 + \omega_2). \end{array}\right\}\qquad(28)$$

In a nonlinear analogue we may discuss either changes in a_1, k_1, a_2, k_2 corresponding to the interaction of modes or the changes in the slowly varying functions a, k, ω.

In developing the theory of a slowly varying nonlinear wavetrain, the mathematical manipulations became impossible except for the easiest cases, until it was realized that all relevant expressions could be determined in terms of a Lagrangian function. But then the whole derivation could be given from the variational principle.

Take the case of water waves with

$$L = \int_0^{h(\mathbf{x}, t)} \{\phi_t + \tfrac{1}{2}(\nabla^2\phi) + gy\}\, dy.$$

There exist uniform wavetrains in which

$$\phi = \beta_i x_i - \gamma t + \Phi(\theta, y), \quad \theta = \kappa_i x_i - \omega t, \quad h = H(\theta), \tag{29}$$

where $\kappa_i, \omega, \beta_i, \gamma$ are constant parameters. The terms proportional to x_i and t must be included in ϕ for complete generality. The solution is normalized so that the change of the phase θ in one period is 2π and so that Φ is the periodic part of ϕ; these conditions may be written

$$[\theta] = 2\pi, \quad [\Phi] = 0, \tag{30}$$

where $[\ ]$ denotes the change in one period. Then, κ_i is the wavenumber, ω is the frequency, β_i is the mean horizontal velocity. There are two further parameters, which may be taken to be the mean height b and the amplitude of the waves a. Thus the solution depends on two triads (κ, ω, a) and (β, γ, b). Two relations between these parameters are provided by the normalization conditions (30), but it is crucial to leave them independent at this stage.

For the full theory of water waves, this uniform solution is not known explicitly. For the long wave approximations in §2, the uniform wavetrain solution is known in terms of elliptic functions. It is also known more generally in the Stokes approximation for small amplitude waves; this would be going back to the near-linear case, but the general point of view is valuable.

A slowly varying wavetrain is close to this uniform solution and may be approximated by the same expressions with (κ, ω, a), (β, γ, b) slowly varying functions of space and time in the sense that the relative change of each of them in one wavelength or one period is small. Then, an *average* Lagrangian $\mathscr{L}$ is derived as

$$
\begin{aligned}
\mathscr{L}(\kappa, \omega, a; \beta, \gamma, b) &= \frac{1}{2\pi}\int_0^{2\pi} L\, d\theta \\
&= \frac{1}{2\pi}\int_0^{2\pi}\int_0^H \{-(\gamma + \omega\Phi_\theta) + \tfrac{1}{2}(\beta_i + \kappa_i\Phi_\theta)^2 + \tfrac{1}{2}\Phi_y^2 + gy\}\, dy\, d\theta. \tag{31}
\end{aligned}
$$

It is then argued that the 'averaged equations' for the slowly varying functions (κ, ω, a), (β, γ, b) can be obtained from the 'averaged variational principle',

$$\delta\iint\mathscr{L}(\kappa, \omega, a; \beta, \gamma, b)\, dx\, dt = 0, \tag{32}$$

with

$$
\left.
\begin{aligned}
\kappa_i &= \frac{\partial\theta}{\partial x_i}, \quad \omega = -\frac{\partial\theta}{\partial t}, \\
\beta_i &= \frac{\partial\psi}{\partial x_i}, \quad \gamma = -\frac{\partial\psi}{\partial t},
\end{aligned}
\right\} \tag{33}
$$

as the appropriate generalization of $\theta = \kappa_i x_i - \omega t$, $\psi = \beta_i x_i - \gamma t$ in the uniform solution. This variational principle is intuitively correct but is not expressed as a formal perturbation procedure. It was not clear how to apply a formal procedure to higher orders in a variational approach. But Luke (1966) has established how a detailed formal procedure on the differential equation leads to the same results in a special case; the main steps are reviewed in §5.

The variational principle (32), with restraints (33), yields

$$\mathcal{L}_a = 0, \quad \mathcal{L}_b = 0, \tag{34}$$

$$\frac{\partial}{\partial t}\mathcal{L}_\omega - \frac{\partial}{\partial x_i}\mathcal{L}_{\kappa_i} = 0, \quad \frac{\partial}{\partial t}\mathcal{L}_\gamma - \frac{\partial}{\partial x_i}\mathcal{L}_{\beta_i} = 0. \tag{35}$$

Since only the derivatives of θ and ψ are involved, it is more convenient to add the consistency relations

$$\left.\begin{aligned}\frac{\partial \kappa_i}{\partial t} - \frac{\partial \omega}{\partial x_i} &= 0, \quad \text{curl } \boldsymbol{\kappa} = 0,\\[2mm]\frac{\partial \beta_i}{\partial t} - \frac{\partial \gamma}{\partial x_i} &= 0, \quad \text{curl } \boldsymbol{\beta} = 0,\end{aligned}\right\} \tag{36}$$

rather than substitute (33). The functional relations (34) give exactly the normalization conditions (30). The dispersion relation $[\theta] = 2\pi$ is $\mathcal{L}_a = 0$.

4·1. *Adiabatic invariants*

The problem of slowly varying wavetrains is analogous to that of slowly varying oscillations in classical mechanics. The elementary problem usually quoted is to find the variations in the amplitude of a simple pendulum when the string is pulled slowly over the support. In general it concerns the behaviour of a Hamiltonian system when an external parameter varies slowly with time. The theory is usually developed from Hamilton's equations with much use of canonical transformations. The corresponding transformations do not exist in the case of more independent variables (Rüssman 1961), so similar methods cannot be found in the waves problem. On the other hand, the averaged Lagrangian method can be applied to the mechanics problem, at least in simple cases.

Consider a mechanical system with Lagrangian $L(q, \dot{q}, \lambda)$ where $\lambda(t)$ denotes an external parameter. Suppose there are periodic solutions $q = q(\theta)$, $\dot{\theta} = \omega$, for $\lambda = \text{constant}$, with an energy integral

$$\dot{q}(\partial L/\partial \dot{q}) - L = E.$$

These correspond to the uniform wavetrains; here (E, ω) are parameters corresponding to (a, κ, ω), etc. Now calculate the average Lagrangian

$$\begin{aligned}\mathcal{L}(\omega, E, \lambda) &= \frac{1}{2\pi}\int_0^{2\pi} L \, d\theta\\[2mm]&= \frac{1}{2\pi}\int_0^{2\pi} \frac{\partial L}{\partial \dot{q}} \dot{q} \, d\theta - E\\[2mm]&= \frac{\omega}{2\pi}\oint p \, dq - E, \tag{37}\end{aligned}$$

where $p = \partial L/\partial \dot{q}$ is calculated as a function $p(q, \lambda, E)$ from the energy equation.

The variational equations are

$$\mathcal{L}_E = 0, \quad \mathrm{d}\mathcal{L}_\omega/\mathrm{d}t = 0.$$

If the action integral,
$$\frac{1}{2\pi} \oint p \, \mathrm{d}q,$$

is denoted by $I(\lambda, E)$, they reduce to

$$\frac{1}{\omega} = \frac{\partial I}{\partial E}, \quad I = \mathrm{const}.$$

These are the classical results and

$$I = \frac{1}{2\pi} \oint p \, \mathrm{d}q = \mathcal{L}_\omega$$

is the 'adiabatic invariant'.

In the waves problem, then, the first equation in (35) may be interpreted as the balance between the changes in a timelike adiabatic variable $\mathcal{L}_\omega$ and the changes in spacelike adiabatic variables $\mathcal{L}_{\kappa_i}$, as the energy is transferred slowly to different parts of the wavetrain. In simple cases, the averaged Lagrangian follows closely the form in (37) (see (57) below). Also the function W in an early version (Whitham 1965a) is an analogue of I.

4.2. *Linear theory*

While developed specifically for nonlinear waves, this theory provides a new general treatment of linear dispersive waves. First, the water waves example is considered, then the general situation is discussed.

In the linearized theory of one-dimensional water waves, (29) becomes

$$H = h_0 + b + a \cos \theta,$$

$$\Phi = \frac{a\omega}{\kappa} \frac{\cosh \kappa y}{\sinh \kappa h_0} \sin \theta,$$

where h_0 is the undisturbed depth. It is expected that (β, γ, b) will not be required in a linear theory, but they will be carried along in calculating $\mathcal{L}$ to keep the comparison with the nonlinear case. The averaged Lagrangian $\mathcal{L}$ is calculated by substitution in (31) and we have

$$\mathcal{L} = (\tfrac{1}{2}\beta^2 - \gamma)(h_0 + b) + \tfrac{1}{2}gb^2 + \tfrac{1}{4}ga^2 \left(1 - \frac{\omega^2}{g\kappa \tanh \kappa h_0}\right).$$

As expected, changes in mean velocity β and mean height b uncouple from the wave motion and we may take $b = \beta = \gamma = 0$ to obtain the usual linear theory. Then

$$\mathcal{L} = \tfrac{1}{4}ga^2 \left(1 - \frac{\omega^2}{g\kappa \tanh \kappa h_0}\right).$$

Quite generally, for a linear system the Lagrangian L is quadratic in the perturbations so that it will always turn out that

$$\mathcal{L} = G(\omega, \kappa) a^2. \tag{38}$$

Since $\mathscr{L}_a = 0$, we have $\qquad\qquad\qquad G(\omega, \kappa) = 0,$ $\qquad\qquad\qquad$ (39)

and it is clear that the function G will always give the dispersion relation.

The other variational equation is

$$\frac{\partial \mathscr{L}_\omega}{\partial t} - \frac{\partial \mathscr{L}_{\kappa_i}}{\partial x_i} = 0,$$
(40)

which becomes $\qquad\qquad \dfrac{\partial (G_\omega a^2)}{\partial t} - \dfrac{\partial (G_{\kappa_i} a^2)}{\partial x_i} = 0.$ $\qquad\qquad$ (41)

While this is like an energy equation, it corresponds rather to an equation for the adiabatic invariants discussed above. The companion equations to be solved with (41) are

$$\frac{\partial \kappa_i}{\partial t} + \frac{\partial \omega}{\partial x_i} = 0, \quad \text{curl } \kappa = 0, \quad G(\omega, \kappa) = 0;$$
(42)

ω can be treated as a function of κ and we have

$$\frac{\partial \kappa_i}{\partial t} + C_j(\kappa) \frac{\partial \kappa_i}{\partial x_j} = 0,$$
(43)

where $\qquad\qquad\qquad C_j(\kappa) = -G_{\kappa_j}/G_\omega$ $\qquad\qquad\qquad$ (44)

is the linear group velocity. From (43) and (44), it is clear that (41) can be transformed into

$$\frac{\partial}{\partial t}(F(\kappa) a^2) + \frac{\partial}{\partial x_i}(C_i F(\kappa) a^2) = 0$$
(45)

for any function $F(\kappa)$. In particular $F(\kappa) = 1$ transforms (41) to

$$\frac{\partial a^2}{\partial t} + \frac{\partial (C_i a^2)}{\partial x_i} = 0.$$
(46)

This may be treated as the 'energy equation', but the physical energy is, in general, another choice of $F(\kappa)$, see §4·5.

Equations (43) and (46) can be solved by integration along characteristics

$$d\mathbf{x}/dt = \mathbf{C};$$
(47)

κ remains constant along characteristics and, once κ is found, the changes of a are determined from

$$\frac{da}{dt} = -\frac{1}{2}\frac{\partial C_i}{\partial x_i} a.$$
(48)

Equation (48) shows the decrease in amplitude due to divergence of the group. Thus the group velocity is a double characteristic velocity for the system and determines the propagation of changes in κ and a. It should also be noted that an 'energy velocity' may be defined as energy flux divided by energy density. According to (45) this is also $\mathbf{C}$ for a linear system. Conceptually, however, the characteristic velocity is different from the energy velocity, and it turns out that the two are not the same for a nonlinear system.

4·3. *Type of the equations for a nonlinear system*

In the nonlinear theory $\mathscr{L}_a = 0$ does not give a relation independent of a, and

$$\omega = \omega(\kappa, a),$$

even in a simple case in which the other variables β, γ, b do not arise. Thus, the equations for κ, a are no longer uncoupled and constitute a system of differential equations to be studied. The first important question is whether the equations are elliptic or hyperbolic. This can be decided by standard methods. As a simple first step beyond linear theory one can suppose, for a one-dimensional case, that

$$\omega = \omega_0(\kappa) + \omega_1(\kappa)\, a^2$$

to bring in the nonlinear effects, but assume that the 'energy equation'

$$\frac{\partial a^2}{\partial t} + \frac{\partial}{\partial x}\,(C_0(\kappa)\, a^2) = 0$$

still holds with $\qquad\qquad C_0(\kappa) = \omega_0'(\kappa).$

When this is coupled with $\quad \dfrac{\partial \kappa}{\partial t} + \dfrac{\partial}{\partial x}\,(\omega_0 + \omega_1 a^2) = 0,$

the characteristic velocities are easily found to be

$$C = C_0(\kappa) \pm a\sqrt{(\omega_1 C_0')} + O(a^2). \tag{49}$$

The equations are hyperbolic when $\omega_1 C_0' > 0$ and elliptic when $\omega_1 C_0' < 0$.

In the hyperbolic case, the double characteristic velocity of linear theory splits into two separate velocities and provides a generalization of the group velocity to nonlinear problems.

When the elliptic case is found the indication is that the original uniform wave train is unstable in a certain sense. For, small sinusoidal disturbances in κ and a will be given by solutions of the form

$$e^{i\mu(x - Ct)}, \tag{50}$$

where C is the value calculated from (49) for the unperturbed values of κ and a. When C is complex, corresponding to the elliptic case, the modulations given by (50) grow exponentially; in this sense the wave train is unstable. If this simple argument is applied to Stokes waves in deep water, the elliptic case is found since

$$\omega = \sqrt{(g\kappa)}\,(1 + \tfrac{1}{2}\kappa^2 a^2) + O(a^4). \tag{51}$$

Hence, $\omega_1 C_0' < 0$, the velocities in (49) are imaginary, and Stokes waves in deep water are unstable. At first this result seemed surprising and probably wrong! And it was put aside until a complete discussion of water waves could be given. However, when Lighthill (1965) looked at this whole theory, he came across this result for the Stokes waves and immediately saw it must be correct! He then studied elliptic cases in detail since they had largely been ignored up to that point.

The complete investigation for the Stokes waves in water of arbitrary depth has now been carried through (Whitham 1966). In addition to the nonlinearity introduced in the dispersion relation, there is a coupling of the wave motion with changes in the mean height b and velocity β; for deep water, this can be ignored and the above result is correct. For finite depth this coupling produces effects to counteract the growth of modulations, and for shallow water the equations change type and the wavetrains are stable.

For arbitrary depth, the average Lagrangian is found to be

$$\mathscr{L} = (\tfrac{1}{2}\beta^2 - \gamma)(h_0 + b) + \tfrac{1}{2}gb^2 + \frac{1}{2}\left\{1 - \frac{(\omega - \beta\kappa)^2}{g\kappa\tanh\kappa(h_0 + b)}\right\}E + \frac{1}{2}\frac{\kappa^2 D_0}{g\tanh\kappa h_0}E^2 + \dots,$$

where $E = \tfrac{1}{2}ga^2$,

$$D_0 = \frac{9\tanh^4\kappa h_0 - 10\tanh^2\kappa h_0 + 9}{8\tanh^3\kappa h_0}.$$

The averaged equations are

$$\frac{\partial}{\partial t}\left(\frac{E}{\omega_0}\right) + \frac{\partial}{\partial x}\left(\frac{C_0 E}{\omega_0}\right) = 0,$$

$$\frac{\partial b}{\partial t} + \frac{\partial}{\partial x}\left(\beta h_0 + \frac{E}{c_0}\right) = 0,$$

$$\frac{\partial\kappa}{\partial t} + \frac{\partial}{\partial x}\left(\omega_0 + \frac{\kappa^2 D_0}{c_0}E + \frac{\kappa B_0}{h_0}b + \kappa\beta\right) = 0,$$

$$\frac{\partial\beta}{\partial t} + \frac{\partial}{\partial x}\left(gb + \frac{B_0}{c_0 h_0}E\right) = 0,$$

where
$$\omega_0^2 = g\kappa\tanh\kappa h_0, \quad C_0 = \omega_0'(\kappa), \quad c_0 = \omega_0/\kappa_0,$$

$$B_0 = C_0 - \tfrac{1}{2}c_0.$$

It is found that the equations are elliptic if $\kappa h_0 > 1\cdot36$ and hyperbolic if $\kappa h_0 < 1\cdot36$. Further details may be seen in the paper cited.

The theory has been applied to the Korteweg–de Vries and Boussinesq equations (without any approximation to small amplitude) and to a similar problem in plasma waves. For these and for further general discussion, reference may be made to the earlier papers (Whitham 1965 a, b).

4·4. *Relation with Benjamin's theory of instability*

A slowly varying wavetrain which is nearly linear can be written in the form

$$\phi = \tfrac{1}{2}a\,\mathrm{e}^{\mathrm{i}\theta} + \tfrac{1}{2}a^*\,\mathrm{e}^{-\mathrm{i}\theta},$$

where a, θ_x, θ_t are slowly varying functions. It is assumed that the amplitude is small enough for the wavetrain to keep the sinusoidal form, but the phase function θ will still have the nonlinear dependence on the amplitude given by the nonlinear dispersion relation. If we consider the special case in which these slowly varying functions are close to constant values, we may write

$$a = a_0 + a_1, \quad \theta = \theta_0 + \theta_1, \quad \theta_0 = \kappa_0 x - \omega(\kappa_0, a_0)t,$$

where a_0, κ_0 are constants and a_1, θ_1 represent small perturbations in the amplitude and the phase. Assuming that θ_1 is bounded (unlike θ_0), we may expand the solution to first order in a_1 and θ_1 as

$$\phi = \tfrac{1}{2}a_0 e^{i\theta_0} + \tfrac{1}{2}a_1 e^{i\theta_0} + \tfrac{1}{2}i\theta_1 a_0 e^{i\theta_0} + \text{conjugate}.$$

The small perturbations a_1 and θ_1 will satisfy the linearized approximation of the averaged equations. These have constant coefficients depending upon (κ_0, a_0) and admit solutions in the form

$$a_1 = B_+(t)\, e^{i\mu x} + B_-(t)\, e^{-i\mu x},$$
$$\theta_1 = \Theta_+(t)\, e^{i\mu x} + \Theta_-(t)\, e^{-i\mu x};$$

the functions B_+, etc., are exponentially growing or oscillatory depending upon whether the equations are elliptic or hyperbolic. Finally, then, ϕ may be expressed as

$$\phi = A_0 e^{i\kappa_0 x} + A_+ e^{i(\kappa_0+\mu)x} + A_- e^{i(\kappa_0-\mu)x} + \text{conjugate}.$$

This approach can now be identified with the resonant interaction approach of Benjamin (see §3). Notice that the *two* side band modulations introduced in Benjamin's analysis are equivalent to the coupled modulations of amplitude and phase in the averaging approach. The perturbations a_1 and θ_1 are slowly varying functions provided $\mu \ll \kappa_0$. Benjamin's approach is free of this restriction, but, on the other hand, applies only to near-linear problems.

4·5. *Conservation equations and shocks*

When the averaged equations are hyperbolic, certain solutions will 'break' in the sense that an initially continuous solution becomes multivalued. This is analogous to the appearance of shock waves in gas dynamics. However, in this treatment of water waves, it could correspond simply to the superposition of two parts of the wavetrain and not require discontinuities. The prediction of this occurrence from an initial form close to a single periodic wavetrain is itself interesting. Of course, after this occurrence the averaged equations developed here no longer apply. An extended theory with the possibility of more than one principal mode would be necessary. It is possible that such a theory could be developed and the interaction treatment for nearly linear modes would help in this connexion.

Another intriguing possibility is that a discontinuity in the averaged equations— a 'shock'—is the required solution in some cases. The argument would be that the averaged equations break down, because the assumption of a slowly varying wavetrain is no longer valid. However, just as in gas dynamics, the solution may be saved without appeal to the original detailed equations by fitting in discontinuities which satisfy the appropriate conservation equations. Mathematically this is the appeal to 'weak solutions'.

The jump conditions are taken from conservation equations. But in these non-linear problems, it seems to be an essential feature that there are always more conservation equations than the number of required shock conditions. For example, from the usual differential equations of inviscid gas dynamics one can obtain the equation for conservation of entropy

$$(\rho S)_t + \nabla \cdot (\rho \mathbf{u} S) = 0,$$

in addition to conservation of mass, momentum and energy. But the corresponding jump condition should not be applied across a discontinuity, since it is known by physical arguments that the entropy is not conserved across a shock. A similar situation arises here.

The equations in (35) and (36) are already in conservation form. Other important ones may be obtained from the variational principle (32) by use of Noether's theorem. Since $\mathscr{L}$ is invariant with respect to an arbitrary translation in time, it follows that

$$\frac{\partial}{\partial t}(\omega\mathscr{L}_\omega+\gamma\mathscr{L}_\gamma-\mathscr{L})-\frac{\partial}{\partial x_i}(\omega\mathscr{L}_{\kappa_i}+\gamma\mathscr{L}_{\beta_i}) = 0 \qquad (52)$$

this is the energy equation. Similarly, since $\mathscr{L}$ is invariant with respect to a translation in space, it follows that

$$-\frac{\partial}{\partial t}(\kappa_j\mathscr{L}_\omega+\beta_j\mathscr{L}_\gamma)+\frac{\partial}{\partial x_i}(\kappa_j\mathscr{L}_{\kappa_i}+\beta_j\mathscr{L}_{\beta_i}-\mathscr{L}\delta_{ij}) = 0; \qquad (53)$$

this is the momentum equation. The invariance of $\mathscr{L}$ with respect to arbitrary constant changes in θ and ψ reproduces the variational equations (35). For water waves it turns out that the second equation in (35) corresponds to conservation of mass.

For any conservation equation in the form

$$\frac{\partial P}{\partial t}+\frac{\partial Q_i}{\partial x_i} = 0$$

the corresponding jump condition across a moving discontinuity surface with unit normal n_i and normal speed V is

$$n_i[Q_i] = V[P],$$

where [ ] denotes the magnitude of the discontinuity. However, a choice of the conservation equations has to be made and only the corresponding jump conditions are applicable. This choice must be made from additional information which is not contained in the averaged equations. Since the original differential equations still apply, we choose those conservation equations which are the averaged form of corresponding conservation equations in the original equations. For water waves the choice is then: (i) conservation of energy (52), (ii) conservation of momentum (53), (iii) conservation of mass which is the second one in (35), and (iv) the second set in (36), which result from elimination of ψ and seem to have no general physical significance. (The details of this choice can be found in Whitham (1965b).)

The conservation equations which must be omitted, because they can only be found in the averaged form for slowly varying wavetrains, are

$$\frac{\partial\kappa_i}{\partial t}+\frac{\partial\omega}{\partial x_i} = 0, \quad \operatorname{curl}\boldsymbol{\kappa} = 0, \qquad (54)$$

and

$$\frac{\partial\mathscr{L}_\omega}{\partial t}-\frac{\partial\mathscr{L}_{\kappa_i}}{\partial x_i} = 0. \qquad (55)$$

The set (54) comes from the existence of a phase function θ, but it may be interpreted more forcefully as the conservation of waves both in space and time. When curl $\kappa = 0$, the line integral around a closed circuit is zero, i.e.

$$\oint \kappa.ds = 0;$$

in the wave pattern at any instant there are the same number of waves (e.g. wave crests) entering the contour as leaving. Similarly, from the first equation in (54),

$$\frac{d}{dt}\int_{x_1^{(1)}}^{x_1^{(2)}} \kappa_1 dx_1 = \omega^{(2)} - \omega^{(1)},$$

so that the number of waves in the interval changes at a rate given by the net flux of waves into the interval. But these integrated forms are precisely the ones that cannot be used across a discontinuity; they cannot be established directly but only from (54) which is valid for continuous parts of the solution.

The quantities $\mathcal{L}_\omega$, $\mathcal{L}_{\kappa_i}$ in (55) are similar to adiabatic invariants in classical mechanics (see §4·1). Equation (55) represents a balance between the changes of spacelike adiabatic invariants $\mathcal{L}_{\kappa_i}$ and a timelike adiabatic invariant $\mathcal{L}_\omega$. But this refers to slow changes and is not valid across a discontinuity in the system (a quantum jump!).

Equations (54) and (55) are on a similar footing from this point of view to the entropy in gas shocks. One naturally asks about the sign of the jump at a discontinuity. In a simple case (discussed in Whitham 1965a), it could be shown that as the waves cross a discontinuity the frequency, relative to the moving surface, is always increased. This seems to be the right way round for 'irreversibility' and might be expected to hold generally; a general proof has not yet been found. The result acquires special interest when it is remembered that the original equations are reversible! It may have relevance to the theory of smooth bores with waves behind them, and to the possibility of collisionless shocks in plasmas.

This idea was raised tentatively in earlier papers. Since then, Zabusky & Kruskal (1965) have performed numerical computations on the Korteweg-de Vries equation in which, effectively, two groups of waves follow one another. In that case the waves are found to pass through each other when the second group overtakes the first. It is clear from the outset, of course, that if it can exist at all, this kind of shock requires special conditions. In linear theory two wavetrains can always be superposed to give a new solution; so highly nonlinear waves are needed. Again it seems clear that two groups with very different velocities will go straight through each other with complicated interactions; so very small relative velocity is needed. Some relevant evidence bearing on this last point is the work of Benney & Luke (1964). They study the collision of two cnoidal waves ('cnoidal waves' are the uniform wavetrain solutions of the Korteweg-de Vries equation). Their theory determines the nonlinear interaction between the two, and the interaction terms tend to infinity as the strengths and directions of the two waves approach each other.

5. FULL PERTURBATION EXPANSION

Luke (1966) considers a nonlinear Klein–Gordon equation

$$u_{tt} - u_{xx} + V'(u) = 0, \tag{56}$$

where $V(u)$ is any nonlinear potential energy which yields oscillatory solutions. The Lagrangian is
$$L = \tfrac{1}{2}u_t^2 - \tfrac{1}{2}u_x^2 - V(u),$$
and the average Lagrangian is easily found to be

$$\mathcal{L}(\omega, \kappa, E) = \{2(\omega^2 - \kappa^2)\}^{\frac{1}{2}} \frac{1}{2\pi} \oint \sqrt{\{E - V(u)\}} \, du - E \tag{57}$$

(which is an interesting comparison with (37)).

Luke introduces a perturbation expansion

$$u(x,t) = U(\theta, X, T) + \epsilon U_1(\theta, X, T) + \dots,$$
$$X = \epsilon x, \quad T = \epsilon t, \quad \theta = \epsilon^{-1} \Theta(X, T).$$

This is a generalization of the geometrical optics expansion for linear problems which would have all the $U_n(\theta, X, T) \propto e^{i\theta}$. It is also a generalization of various methods of Krylov–Bogoliubov, and particularly the work of Kuzmak (1959), from ordinary differential equations to partial differential equations. For $\epsilon = 0$, $u(x,t) = U(\theta)$ is the uniform wavetrain. This expansion is substituted in (56) and the coefficients of powers of ϵ are equated to zero. The first two orders give

$$(\omega^2 - \kappa^2) U_{\theta\theta} + V'(U) = 0, \tag{58}$$

$$(\omega^2 - \kappa^2) U_{1\theta\theta} + V''(U) U_1 = 2\omega U_{\theta T} + 2\kappa U_{\theta X} + \omega_T U_\theta + \kappa_X U_\theta, \tag{59}$$

where $\omega = -\theta_t = -\Theta_T, \kappa = \theta_x = \Theta_X$. Equation (58) is the equation for a uniform wavetrain, and may be considered as an ordinary differential equation in the variable θ even though U, ω, κ are also functions of (X, T); the dependence on (X, T) gives the slow variation. A first integral of (58) is

$$\tfrac{1}{2}(\omega^2 - \kappa^2) U_\theta^2 + V(U) = E(X, T). \tag{60}$$

At this stage $\omega(X, T), \kappa(X, T), E(X, T)$ are undetermined.

Turn attention now to (59). As an ordinary differential equation in θ, the right-hand side is known and the left-hand side is linear in U_1. Moreover, a solution of the homogeneous equation is $U_1 = U_\theta$, since the left-hand side is then the derivative of (58). In principle, (59) can be solved by the substitution $U_1 = fU_\theta$. Here it is sufficient to note that (59) can be recast as

$$(\omega^2 - \kappa^2) \frac{\partial}{\partial \theta} (U_{1\theta} U_\theta - U_1 U_{\theta\theta}) = (\omega U_\theta^2)_T + (\kappa U_\theta^2)_X.$$

The right-hand side is periodic in θ. Hence U_1 is bounded in θ only if the integral of the right-hand side over one period vanishes. The condition is

$$\frac{\partial}{\partial T} \int_0^{2\pi} \omega U_\theta^2 \, d\theta + \frac{\partial}{\partial X} \int_0^{2\pi} \kappa U_\theta^2 \, d\theta = 0.$$

It may be verified that this may be rewritten from (57) and (60) as

$$\frac{\partial \mathcal{L}_\omega}{\partial T} - \frac{\partial \mathcal{L}_\kappa}{\partial X} = 0. \tag{61}$$

Thus, the 'orthogonality condition', to avoid secular terms in solving (59), gives the result of the averaged variational principle.

The discussion of higher order terms is an intricate one with introduction of a further orthogonality condition. The details are given by Luke (1966).

6. INTEGRAL EQUATIONS FOR MORE GENERAL DISPERSION

Linear partial differential equations in $(\mathbf{x}, t)$, with constant coefficients, can only give polynomial dispersion functions; the correspondence is

$$\frac{\partial}{\partial t} \leftrightarrow -\mathrm{i}\omega, \quad \frac{\partial}{\partial x} \leftrightarrow \mathrm{i}\kappa.$$

Water waves have
$$\omega^2 = g\kappa \tanh \kappa h_0, \tag{62}$$

but this is through the dependence on an extra coordinate y, which in this sense is not part of the (x, t) space in which the wave propagation occurs. One method of obtaining more general dispersion in an (x, t) problem is to consider, for example,

$$\frac{\partial \eta}{\partial t} + \int_{-\infty}^{\infty} K(x - \xi)\, \eta_\xi(\xi, t)\, \mathrm{d}\xi = 0, \tag{63}$$

where $K(x)$ is a suitably chosen kernel. Solutions

$$\eta = \mathrm{e}^{\mathrm{i}(\kappa x - \omega t)}$$

satisfy (63) provided that

$$c = \frac{\omega}{\kappa} = \int_{-\infty}^{\infty} K(x - \xi)\, \mathrm{e}^{-\mathrm{i}\kappa(x - \xi)}\, \mathrm{d}\xi. \tag{64}$$

This means that any phase velocity $c(\kappa)$ can be obtained by choosing $K(x)$ to be the Fourier transform of $c(\kappa)$;

$$K(x) = \frac{1}{2\pi} \int_{-\infty}^{\infty} c(\kappa)\, \mathrm{e}^{\mathrm{i}\kappa x}\, \mathrm{d}\kappa. \tag{65}$$

For the polynomial cases, $K(x)$ is a sum of δ functions. In the example

$$c(\kappa) = c_0 + c_2 \kappa^2, \quad K(x) = c_0 \delta(x) - c_2 \delta''(x), \tag{66}$$

(63) becomes the linearized Korteweg–de Vries equation

$$\eta_t + c_0 \eta_x - c_2 \eta_{xxx} = 0.$$

An equation combining the general dispersion of the integral with typical non-linearity would be

$$\eta_t + \alpha \eta \eta_x + \int_{-\infty}^{\infty} K(x - \xi)\, \eta_\xi(\xi, t)\, \mathrm{d}\xi = 0. \tag{67}$$

The Korteweg–de Vries equation follows when (66) is used. An interesting extension is to consider other kernels and, in particular, for water waves to take

$$c(\kappa) = \left(\frac{g}{\kappa}\tanh\kappa h_0\right)^{\frac{1}{2}}, \quad K_g = \frac{1}{2\pi}\int_{-\infty}^{\infty} c(\kappa)\, e^{i\kappa x}\mathrm{d}\kappa. \tag{68}$$

The Korteweg–de Vries case takes the first two terms in the long wave expansion $\kappa h_0 \ll 1$.

While the Korteweg–de Vries equation gives solitary waves and cnoidal wave-trains, it is inadequate to give the waves of greatest height with the Stokes 120° angle at the crest. Moreover, the alternative breaking into bores, described by the simpler shallow water equations, is lost because it seems certain that the η_{xxx} term will always prevent breaking (although a proof does not seem to have been given). Both are high frequency effects which are lost by the long wave expansion $\kappa h_0 \ll 1$. Equation (67) is not limited in this way.

Uniform wavetrains are obtained with $\eta = \eta(X)$, $X = x - Ut$, and (67) is

$$(U - \alpha\eta)\eta' = \int_{-\infty}^{\infty} K(X - \zeta)\eta'(\zeta)\,\mathrm{d}\zeta. \tag{69}$$

This can be integrated once to the form

$$A + U\eta - \tfrac{1}{2}\alpha\eta^2 = \int_{-\infty}^{\infty} K(X - \zeta)\eta(\zeta)\,\mathrm{d}\zeta, \tag{70}$$

where A is an integration constant. For the choice K_g in (68), the explicit solution of (70) cannot be given, but it can be shown that a limiting form occurs when $U = \alpha\eta$ and the crest becomes cusped with a vertical tangent. The cusp instead of the Stokes 120° angle is attributed to the remaining inadequacies of the non-linear terms. However, a wave of greatest height is predicted so the desired qualitative effect is in.

The kernel in (68) normalized to $g = 1$, $h_0 = 1$, has the properties

$$K_g(x) = K_g(-x),$$
$$K_g(x) \sim (2\pi x)^{-\frac{1}{2}} \quad \text{as} \quad x \to 0,$$
$$K_g(x) \sim (\tfrac{1}{2}\pi^2 x)^{-\frac{1}{4}} e^{-\frac{1}{2}\pi x} \quad \text{as} \quad x \to \infty,$$
$$\int_{-\infty}^{\infty} K_g(x)\,\mathrm{d}x = 1.$$

If a model kernel $\quad K_0(x) = \tfrac{1}{4}\pi\, e^{-\frac{1}{2}\pi|x|}, \quad c(\kappa) = \dfrac{1}{1 + (2\kappa/\pi)^2} \tag{71}$

is tried, the integrals in (67) and (70) can be eliminated by applying the operator

$$\left(\frac{\partial^2}{\partial x^2} - \tfrac{1}{4}\pi^2\right),$$

since (71) is the Green function for this operator. Further details can then be worked out. With $\alpha = 3c_0/2h_0$, corresponding to the Korteweg–de Vries equation (see (16)), the solitary wave of maximum height has

$$\frac{\eta_{\text{max.}}}{h_0} = \frac{8}{9},$$

and, amazingly (since it is fortuitous) this case has a finite angle of $110°$ at the crest! The finite angle instead of a cusp is because $K_0(x)$ is regular at $x = 0$, whereas $K_o(x)$ has a singularity there. Thus the angle result should not be taken seriously. However, the result for the maximum height may be taken more seriously since it depends upon the whole profile. The result compares reasonably well with McCowan's value of $0\cdot78$ obtained by a sort of Pohlhausen method.

As regards breaking into a bore, (67) also looks hopeful. There is no longer a higher derivative to prevent breaking. The integral could be more analogous to the example

$$\eta_t + (c_0 + \alpha\eta)\,\eta_x + \beta\eta = 0$$

which breaks if the initial slope η_x is ever negative and

$$|\eta_x| > \beta/\alpha.$$

Some first results in this direction have been found by Mr R. L. Seliger and will be published later.

If this type of integro-differential equation proves worthy of much further study, it can be derived from a variational principle and all the theory of slowly varying wavetrains, etc., would follow. The equation (67) is first written with $\alpha\eta = \psi_x$ as

$$\psi_{xt} + \psi_x \psi_{xx} + \int_{-\infty}^{\infty} K(x-\xi)\,\psi_{\xi\xi}(\xi, t)\,\mathrm{d}\xi = 0.$$

This equation follows from the variational principle

$$\delta \iint L\,\mathrm{d}x\,\mathrm{d}t = 0$$

with
$$L = \tfrac{1}{2}\psi_x\psi_t + \tfrac{1}{6}\psi_x^3 + \tfrac{1}{2}\psi_x\int_{-\infty}^{\infty} K(x-\xi)\,\psi_\xi(\xi, t)\,\mathrm{d}\xi.$$

This research was supported by the National Science Foundation and the Office of Naval Research of the U.S. Navy.

APPENDIX: VARIATIONAL PRINCIPLE FOR ROSSBY WAVES

The question of a suitable variational principle for Rossby waves was raised in the discussion. This question was investigated after the meeting by Mr R. L. Seliger and the author with results as follows. The simple formulation for Rossby waves in the 'β-plane' approximation is

$$\left.\begin{aligned}
Du/Dt - f(y)\,v &= -p_x, \\
Dv/Dt + f(y)\,u &= -p_y, \\
u_x + v_y &= 0.
\end{aligned}\right\} \tag{A 1}$$

Variational principles are most conveniently obtained in terms of potentials. In this case we introduce generalized potentials ϕ, α, β with the representation

$$u = \phi_x + \alpha\beta_x - \alpha, \tag{A 2}$$

$$v = \phi_y + \alpha\beta_y - f\beta, \tag{A 3}$$

$$-p = \phi_t + \alpha\beta_t + \tfrac{1}{2}\{(\phi_x + \alpha\beta_x - \alpha)^2 + (\phi_y + \alpha\beta_y - f\beta)^2\}. \tag{A 4}$$

This is an extension of the Clebsch transformation (Lamb's *Hydrodynamics*, p. 248) to include the Coriolis terms. From (A 1) it is easily shown that the equations to be satisfied by ϕ, α, β are

$$\left.\begin{array}{r}D\alpha/Dt + fv = 0, \\ D\beta/Dt - u = 0, \\ u_x + v_y = 0,\end{array}\right\} \tag{A5}$$

where u, v stand for the expressions given in (A 2), (A 3). The variational principle is again in terms of the expression for the pressure. The set (A 5) follows from the variational principle

$$\delta \iiint \{\phi_t + \alpha\beta_t + \tfrac{1}{2}(u^2 + v^2)\}\,dx\,dy\,dt = 0, \tag{A6}$$

where u and v are again expressed by (A 2) and (A 3). The general theory can now be applied to Rossby waves using the Lagrangian in (A 6). Applications and a more detailed discussion of the use of the 'Clebsch potentials' for rotational flows will appear later.

REFERENCES (Whitham)

Bateman, H. 1944 *Partial differential equations.* Cambridge University Press.
Benney, D. J. & Luke, J. C. 1964 *J. Math. Phys.* **43**, 309.
Korteweg, D. J. & de Vries, G. 1895 *Phil. Mag.* (5), **39**, 422.
Kuzmak, G. E. 1959 *Prikl. Mat. Mekh. Akad. Nauk SSSR* **23**, 515. (Translated in *Appl. Math. Mech.* **23**, 730.)
Lighthill, M. J. 1965 *J. Inst. Math. Appl.* **1**, 269.
Luke, J. C. 1966 *Proc. Roy. Soc.* A **292**, 403.
Luke, J. C. 1967 *J. Fluid Mech.* **27**, 395.
Rüssman, H. 1961 *Arch. Rat. Mech. Anal.* **8**, 353.
Serrin, J. 1959 *Handbook of Physics*, vol. VIII, part I, 144. (Edited by S. Flugge.) Springer.
Whitham, G. B. 1965a *Proc. Roy. Soc.* A **283**, 238.
Whitham, G. B. 1965b *J. Fluid Mech.* **22**, 273.
Whitham, G. B. 1967 *J. Fluid Mech.* **27**, 399.
Zabusky N. & Kruskal, M. 1965 *Bell. Telephone Labs. Tech. Rep.*

J. Inst. Maths Applics **1,** 269–306

Contributions to the Theory of Waves in Non-linear Dispersive Systems

M. J. LIGHTHILL

This paper makes contributions to the general theory of wave propagation in conservative systems under conditions when the proportional change in amplitude or wavenumber over a distance of one wavelength is very small. For linear systems, such propagation is governed by the well-known theory of group velocity; there is "frequency dispersion", in the sense that energy in components of different frequency is propagated at different group velocities. For non-linear systems without frequency dispersion, e.g. acoustic systems, a different, but also well-known, modification of the waveform occurs. It may be called "amplitude dispersion", in that different values of an amplitude variable like the pressure are propagated at different speeds.

A much more general theory of non-linear systems, where frequency dispersion and amplitude dispersion would be expected to be in competition, has been given by Whitham (1965b). Energy does not play a key role in the theory, because it is easily transferred between components of different frequencies. The fundamental equation follows from Hamilton's principle in an averaged form.

In examples given by Whitham, changes in, for example, wavenumber (or amplitude) are propagated at two different velocities, because the fundamental equation is hyperbolic. However, in the limiting case of infinitesimal amplitude, the equation is parabolic and only one velocity of propagation (the group velocity) occurs. Thus, Whitham showed that non-linearity can "split" the group velocity.

This paper is concerned with the inference of detailed conclusions from Whitham's theory, to enable comparisons with experiment that will show the range of applicability of the theory. It attempts to obtain these in the simplest case, namely, that of one-dimensional propagation when Whitham's "pseudo-frequencies" are absent.

If the relationship between frequency ω and wavenumber k for infinitesimal amplitude is $\omega = f(k)$, then for finite amplitude the equation is shown to be hyperbolic or elliptic respectively, according as $[\omega - f(k)]f''(k)$ takes positive or negative values. For gravity waves on deep water this product is negative and these, it is inferred, may be good for comparison of theory with experiment in the elliptic case. A new non-linear non-perturbational theory of waves under the combined action of gravity and surface tension is used to indicate that waves at 9·6 c/s on mercury may be suitable for comparison with experiment in the hyperbolic case.

When non-linear effects are only moderate, approximate transformations of Whitham's equation to the axisymmetric Laplace and wave equations respectively, in the elliptic and hyperbolic cases, are used to obtain particular solutions for comparison with experiment. A feature of these solutions is the appearance of discontinuities in wavelength.

For example, when a wavemaker creates gravity waves of fixed frequency whose amplitude first increases and then decreases, the theory predicts that the length of waves in the group decreases ahead of the point of maximum amplitude and increases behind it. This produces in turn a concentration of energy towards the centre of the group, which continues during the whole period before a discontinuity in wavelength actually forms. This solution in the elliptic case is obtained with the aid of the theory of imaginary characteristics.

1. Introduction

WHITHAM (1965a) analysed the effects of non-linearity on dispersion and group velocity, for one-dimensional propagation of waves satisfying equations in "conservational" form. His theory would be expected to apply under conditions when waves of different wavelengths and amplitudes are (or have become) sufficiently dispersed from one another that neither the wavelength nor the amplitude changes by more than a small fraction of itself in a single wavelength. Under these circumstances the dispersion process can be studied by replacing the quantities appearing in the equations (that is, the conserved quantities and their fluxes) by mean values, those values being calculated, locally, by approximating the local waveform by a perfectly periodic waveform with the same wavelength and amplitude (and, where appropriate, other defining parameters), and taking averages over a wavelength.

The equations so obtained, for all the particular problems analysed by Whitham (1965a), were of hyperbolic type, with at least two systems of characteristics in the (x,t)-plane. In the limiting case of infinitesimal amplitude (when the original equations of motion become linear), however, the equations become parabolic and the systems of characteristics degenerate into a single system, whose slope at any point of the (x,t)-plane is the classical group velocity (for a general account of the classical theory see Lighthill, 1965). Thus, Whitham's analysis shows that in these cases non-linearity "splits" the group velocity into two, or sometimes more, different values. Small changes in, for example, wavelength are, in general, propagated through the wave train at all these velocities, so that the region influenced by the changes increases linearly with the time, instead of as the square root of the time as in the classical theory. Whitham's (1965a) paper seems to give the first systematic theory of what Lighthill (1955) called the "competition between frequency dispersion and amplitude dispersion".

In a second paper, Whitham (1965b) makes a most impressive extension of the theory to waves in two or more dimensions in a general conservative system. From Hamilton's principle (that the time integral of the Lagrangian is stationary) in an averaged form, he derives the equations governing wave dispersion in terms of a function $\mathscr{L}$, which is the averaged Lagrangian per unit volume for periodic waves of given frequency and wavenumber vector (the average being taken over an integral number of wavelengths or periods). In certain problems, however, $\mathscr{L}$ is a function of other variables besides the frequency and the wavenumber vector; these are problems analogous to those with cyclic co-ordinates in general dynamics, and the variables in question are analogous to the cyclic constants; Whitham (1956b) calls them "pseudo-frequencies".

In this paper we study the implications of this general theory of Whitham's in a restricted class of homogeneous conservative systems. These are one-dimensional systems without "pseudo-frequencies", in other words, systems where the wavelength and amplitude of a periodic wave define its waveform uniquely. An example of such a system, studied in detail below (Sections 10 and 11), is given by unidirectional propagation of surface waves on deep water.

For any system within this restricted class, a periodic wave of given wavenumber k and amplitude a has a given frequency ω, which is a function of k and a. Furthermore, the average Lagrangian per unit length $\mathscr{L}$ is also determined when k and a are known. Whitham's theory then leads to the same equations whether it is applied before or after the amplitude a has been eliminated from these relationships; in this paper we assume that the elimination of a has been carried out, so that $\mathscr{L}$ is written as a function of ω and k:

$$\mathscr{L} = \mathscr{L}(\omega,k). \tag{1}$$

The dispersion of a general wave train, satisfying the condition that neither the wavelength nor the amplitude changes by more than a small fraction of itself in a single wavelength, can be described by use of a phase function $\theta(x,t)$ (Whitham, 1965b) such that the local frequency and wavenumber are defined as

$$\omega = -\frac{\partial\theta}{\partial t}, \quad k = \frac{\partial\theta}{\partial x}. \tag{2}$$

It is assumed that, to a good approximation, the local averaged value of the Lagrangian per unit length at a point of such a wave train can be taken as the same function (1) of ω and k as in a perfectly periodic wave. Then the averaged form of Hamilton's principle (that the time integral of the Lagrangian of the whole system is stationary) yields at once, by calculus of variations, the differential equation

$$\frac{\partial}{\partial t}\left(\frac{\partial\mathscr{L}}{\partial\omega}\right) = \frac{\partial}{\partial x}\left(\frac{\partial\mathscr{L}}{\partial k}\right) \tag{3}$$

which governs the dispersion of the wave train.

Methods for determining just how gradual the changes in amplitude and wavelength must be for solutions of this equation to represent the dispersion to good approximation are still not known. One of the objects of this paper is to help answer this question in particular cases by deriving particular solutions of equation (3) in forms suitable for comparison with experiment.

The paper also discusses the general implications of equation (3). First, it establishes that the equation is not necessarily hyperbolic; indeed, it is so only in "anticlastic" regions of the surface defined by equation (1); that is, regions where the principal curvatures are of opposite sign. In "synclastic" regions (where they are of the same sign) the equation is elliptic. Where it is hyperbolic, Whitham's splitting of the group velocity into two velocities occurs; but, where it is elliptic, no real characteristics exist and the group velocity is, as it were, annulled.

A very simple criterion exists, which distinguishes the hyperbolic and elliptic cases whenever the amplitude, though not infinitesimal, remains relatively small (so that non-linear terms in the equations are significant, but only those of lowest order). Suppose that for infinitesimal amplitude (that is, in the limit as $a \to 0$) the relation between ω and k is

$$\omega = f(k). \tag{4}$$

Then, if for small but finite amplitude $\omega - f(k)$ has the same sign as $f''(k)$, the dispersion equation is hyperbolic, and, if they have opposite signs, it is elliptic. For a wide class of problems pursued to one order of approximation beyond the linear, a relatively simple form of equation (3) is found below (equation (24)) which both exhibits this fact and permits calculations to be made rather conveniently.

For example, long-crested gravity waves on deep water satisfy

$$\omega = f(k) = (gk)^{\frac{1}{2}} \tag{5}$$

for infinitesimal amplitude, while for amplitude a we have (Lamb, 1932)

$$\omega = (gk)^{\frac{1}{2}}(1 + \tfrac{1}{2}k^2a^2 + \tfrac{1}{8}k^4a^4 + \ldots). \qquad (6)$$

Thus, in this case $\omega - f(k)$ is positive for finite amplitude, although $f''(k)$ is negative. The dispersion is therefore governed by elliptic equations.

Since one aim of this paper is to suggest experiments to test the accuracy of predictions given by the Whitham dispersion equation, both in the hyperbolic and elliptic cases (for which the predictions are qualitatively rather different), a suitable system is needed not only in the elliptic case (when deep-water gravity waves are convenient) but also in the hyperbolic case. It is shown below that deep-water waves under the combined action of gravity and surface tension have a dispersion equation which is hyperbolic within a particular band of frequencies, and a new non-linear theory of waves in this band is developed, from which it appears that for experimental purposes waves of frequency 9·6 c/s in mercury might be convenient.

Whitham (1965a) showed that in dispersion problems governed by hyperbolic equations, it was possible, as in other non-linear hyperbolic problems, for characteristics to "overtake" other characteristics so that, if the equations continued to be valid under these circumstances, the local quantities including ω and k would become many-valued. Whitham conjectured that under these circumstances an almost discontinuous jump in ω and k would appear, permitting the continuance of one-valued solutions which, except at this jump (which he called a "shock"), had only gradually changing values of wavelength and amplitude.

Experimental investigations are suggested below (Sections 9 and 11) to show whether this is in fact possible; the theory clearly leaves room for uncertainty because the fundamental assumption leading to equation (3) becomes invalid before the overtaking happens. In the meantime it is interesting to note that, for dispersion problems governed by elliptic equations, although no characteristics exist that can overtake one another, and although it can be shown (Section 6) that no singularities of the "limiting line" type referred to by Whitham (1965a) can occur, nevertheless a certain type of isolated singularity can occur, beyond which conjectural discontinuous solutions similar to those proposed by Whitham can be constructed (Section 16).

A number of initial-value problems are considered (Sections 5 to 16), including problems regarding the development of trains of deep-water waves. All these problems lead to boundary conditions for equation (3) that are essentially of hyperbolic type. If the equation is elliptic, therefore, they are not problems of the kind usually described as "well-set". Nevertheless, Garabedian's (1958) technique using imaginary characteristics shows that solutions can be obtained provided that the initial conditions satisfy certain "smoothness" requirements.

In one type of initial-value problem (Section 5), a wave train is studied that initially has a considerable spread of frequency, so that dispersion would be substantial even in the absence of non-linear effects, and the aim is to determine their modifying influence. A general solution, valid so long as amplitude remains moderately small, is obtained, and given a physical interpretation.

In the rest of the paper a hypothetical experiment is considered which constitutes a more crucial test of the Whitham theory because there would be no dispersion at all on a linear theory. Specifically, a wavemaker is supposed to create waves of fixed frequency, whose amplitude varies gradually with time.

The development of the resulting wave group takes two quite different forms, according as the dispersion equation is hyperbolic (Sections 7 to 9) or elliptic (Sections 12 to 16). Whitham (1965a) has described its general character in the former case, and the present paper adds only detailed calculations for comparison with experiment.

In the elliptic case, it is shown that the effect of non-linearity is to concentrate the group, at least for a time, and, hence, to cause the amplitude at the centre to increase (which it does not do in the hyperbolic case). The wavenumber increases in the front of the group and decreases at the back, until a discontinuity forms at the centre; the subsequent development, at any rate near the discontinuity, is most uncertain and its determination must await experiment.

These problems were solved by a convenient transformation, valid for moderately small amplitudes, of the dispersion equation into the axisymmetric wave or Laplace equation, in the hyperbolic or elliptic case respectively. It seems likely that substantially more information than is described in this paper remains to be obtained from this transformation.

2. General Condition Governing whether the Dispersion Equation is Elliptic or Hyperbolic

The dispersion equation (3), expanded, becomes

$$\frac{\partial^2 \mathscr{L}}{\partial \omega^2} \frac{\partial \omega}{\partial t} + \frac{\partial^2 \mathscr{L}}{\partial \omega \partial k} \frac{\partial k}{\partial t} = \frac{\partial^2 \mathscr{L}}{\partial \omega \partial k} \frac{\partial \omega}{\partial x} + \frac{\partial^2 \mathscr{L}}{\partial k^2} \frac{\partial k}{\partial x} , \tag{7}$$

which with (2) gives

$$\frac{\partial^2 \mathscr{L}}{\partial \omega^2} \frac{\partial^2 \theta}{\partial t^2} - 2\frac{\partial^2 \mathscr{L}}{\partial \omega \partial k} \frac{\partial^2 \theta}{\partial t \partial x} + \frac{\partial^2 \mathscr{L}}{\partial k^2} \frac{\partial^2 \theta}{\partial x^2} = 0. \tag{8}$$

This is a quasi-linear partial differential equation of the second order, since equations (1) and (2) define its coefficients as functions of $\partial\theta/\partial t$ and $\partial\theta/\partial x$.

It is an equation that can be transformed into a linear one by a Legendre transformation, namely

$$\phi(\omega,k) = kx - \omega t - \theta, \qquad \frac{\partial \phi}{\partial \omega} = -t, \qquad \frac{\partial \phi}{\partial k} = x, \tag{9}$$

giving

$$\frac{\partial^2 \mathscr{L}}{\partial \omega^2} \frac{\partial^2 \phi}{\partial k^2} - 2\frac{\partial^2 \mathscr{L}}{\partial \omega \partial k} \frac{\partial^2 \phi}{\partial \omega \partial k} + \frac{\partial^2 \mathscr{L}}{\partial k^2} \frac{\partial^2 \phi}{\partial \omega^2} = 0. \tag{10}$$

The only disadvantage of this is that one class of solutions is lost. These are the solutions for which $\theta = \theta(x,t)$ is a developable surface (Courant & Hilbert, 1962, p. 35), which, more precisely, are those for which ω is a fixed function of k. However, this is not a great loss as these are (i) the infinitesimal-amplitude solutions and (ii) the simple wave solutions of Whitham (1965a), both of which we already know in great detail. (The analogous "hodograph transformation" in gas dynamics similarly excludes from consideration the well-known Prandtl-Meyer flows.) By contrast, a great advantage of the transformation is that the characteristics of equation (10) are a fixed system of curves, satisfying

$$\frac{\partial^2 \mathscr{L}}{\partial \omega^2} d\omega^2 + 2\frac{\partial^2 \mathscr{L}}{\partial \omega \partial k} d\omega \, dk + \frac{\partial^2 \mathscr{L}}{\partial k^2} dk^2 = 0. \tag{11}$$

Now at each point (ω, k) the equation (11) defines two directions $d\omega/dk$: real, imaginary or coincident. It is easy to show that these are the directions in which those curves point which are the intersections of the surface $\mathscr{L} = \mathscr{L}(\omega, k)$ with the tangent plane to the surface at that point. They are imaginary at a synclastic point of the surface (where the principal curvatures have the same sign); there, because the characteristic directions are imaginary, the equation must be of elliptic type. They are real and distinct, however, at an anticlastic point of the surface, where therefore the equation is of hyperbolic type.

In fact, the differential equation (11) defines the system of curves on the surface $\mathscr{L} = \mathscr{L}(\omega, k)$ which differential geometers call the "asymptotic lines"; lines on the surface which at every point are in one of the directions just defined. These directions are also those of the asymptotes of the hyperbolae to which the curves, in which planes parallel and very close to the tangent plane cut the surface, approximate— hence the name. The asymptotic lines are real only where the surface is anticlastic.

To sum up, the dispersion equation is elliptic at points (ω, k) where the surface $\mathscr{L} = \mathscr{L}(\omega, k)$ is synclastic. At points where the surface is anticlastic, the dispersion equation is hyperbolic, and its characteristics are the "asymptotic lines" of the surface. A line dividing a synclastic from an anticlastic part of the surface is a transition line on which the equation is locally parabolic.

3. The Case of Moderately Small Amplitude (Geometrical Discussion)

This relatively simple conclusion can be reduced to an even simpler one in the case of moderately small amplitude, where the only non-linear effects which are significant are those of lowest order.

For infinitesimal amplitude, ω is a fixed function of k, say,

$$\omega = f(k). \tag{12}$$

Under this condition we also have $\mathscr{L} = 0$, as Whitham (1956b) shows (see also Lighthill, 1965); this includes the classical result that mean kinetic energy equals mean potential energy in waves of infinitesimal amplitude, and in an almost immediate consequence of the averaged form of Hamilton's principle.

Now consider a periodic wave of finite amplitude a; the meaning of a can be made precise by defining it, for example, as the amplitude of the first harmonic of the periodic waveform when Fourier-analysed. In general, ω will be a function of a as well as of k. However, it will be an even function of a, since a change of sign of a is equivalent merely to a phase change of half a period. Thus,

$$\omega = f(k) + a^2 \omega_1(k) + \ldots, \tag{13}$$

and in this Section we shall consider the case when only the terms shown explicitly in (13) are significant.

The average Lagrangian per unit length, $\mathscr{L}$, is also an even function of a. However, the first term in its expansion in powers of a is in general of order a^4. This is because the leading terms in individual components of energy density are of order a^2, but the contribution of these leading terms to the Lagrangian itself vanishes. Accordingly,

$$\mathscr{L} = a^4 \mathscr{L}_1(k) + \ldots, \tag{14}$$

and, when only the terms shown explicitly in (13) and (14) are included, the relationship between $\mathscr{L}$, ω and k becomes

$$\mathscr{L} = \tfrac{1}{2}g(k)[\omega - f(k)]^2, \tag{15}$$

where

$$g(k) = 2\mathscr{L}_1(k)/\omega_1^2(k). \tag{16}$$

Equation (15) represents the general form of the surface $\mathscr{L} = \mathscr{L}(\omega,k)$ for waves of small but finite amplitude; or, what is the same thing, the general form near the curve $\omega = f(k)$, or, what is again the same thing, the general form near the plane $\mathscr{L} = 0$. In particular, the surface is seen to be tangent to that plane at all points of the curve $\omega = f(k)$.

Now, when a single plane is tangent to a surface at all points of a certain curve lying in the plane, then evidently the part of the surface near the curve is synclastic on the side towards which the curve is convex and is anticlastic on the side towards which the curve is concave. This statement will be taken as geometrically obvious in what immediately follows, though its truth will also be deduced from later analytical work (Section 4). The statement evidently assumes that $\omega = f(k)$, besides having continuous second derivative, is actually a curve, and not a straight line, but the latter case is not a very interesting one in the present connection, since it would provide no "frequency dispersion", to "compete" with the "amplitude dispersion".

Now, near each point of the curve $\omega = f(k)$, equation (13) limits (ω,k) to the part of the surface (15) on just one side of the curve; above it if $\omega_1(k) > 0$ and below if $\omega_1(k) < 0$. The dispersion equation therefore will be of elliptic type if, and only if, the surface is synclastic on that side of the curve, that is, if, and only if, the curve is convex towards that side of the curve. In other words, it will be elliptic when, and only when, finite amplitude pushes the point (ω,k) off the curve on to the side towards which the curve is convex, so that $\omega - f(k)$ has the opposite sign to $f''(k)$. Conversely, it will be hyperbolic when finite amplitude pushes (ω,k) off the curve $\omega = f(k)$ on to the side towards which the curve is concave, so that $\omega = f(k)$ and $f''(k)$ are of the same sign.

It was shown in Section 1 how this very simple criterion, that the dispersion equation becomes hyperbolic or elliptic according as $\omega - f(k)$ and $f''(k)$ have the same or opposite signs, demonstrates at once that the equation is elliptic for long-crested gravity waves on deep water, for which $\omega - f(k) > 0$ and $f''(k) < 0$. It is elliptic also for pure capillary waves (Section 10), for which $\omega - f(k) < 0$ and $f''(k) > 0$. However, for waves governed by a combination of gravity and surface tension, the wavenumbers at which $\omega - f(k)$ and $f''(k)$ change sign are different, so that between these two values of the wavenumber (Section 10) $\omega - f(k)$ and $f''(k)$ have the same sign and the dispersion equation is hyperbolic.

4. Approximate Forms of the Dispersion Equation for Moderately Small Amplitude

We now obtain the form of equation (10) under conditions of moderately small amplitude, retaining terms of order $\omega - f(k)$ in the coefficients but rejecting terms of order $[\omega-f(k)]^2$. To demonstrate the form of this, we must include the next term in the expansion of $\mathscr{L}$ besides that given in (15), putting

$$\mathscr{L} = \tfrac{1}{2}g(k)[\omega-f(k)]^2 + \tfrac{1}{3}h(k)[\omega-f(k)]^3. \tag{17}$$

But, although the second term in (17) contributes terms of order $\omega-f(k)$ to the co-efficients in (10), we shall see that they cancel out.

It is convenient to use

$$\tau = \omega-f(k) \tag{18}$$

as a new variable, so that the line $\tau = 0$ corresponds to infinitesimal amplitude and the characteristic directions both coincide with it thereon. When we use τ and k as independent variables, we must replace $\partial/\partial\omega$ in (10) by $\partial/\partial\tau$ and $\partial/\partial k$ by $\partial/\partial k - f'(k)\partial/\partial\tau$. Then (10) becomes

$$\frac{\partial^2\mathscr{L}}{\partial\tau^2}\frac{\partial^2\phi}{\partial k^2} - 2\frac{\partial^2\mathscr{L}}{\partial\tau\partial k}\frac{\partial^2\phi}{\partial\tau\partial k} + \frac{\partial^2\mathscr{L}}{\partial k^2}\frac{\partial^2\phi}{\partial\tau^2} - f''(k)\left(\frac{\partial\mathscr{L}}{\partial\tau}\frac{\partial^2\phi}{\partial\tau^2} + \frac{\partial^2\mathscr{L}}{\partial\tau^2}\frac{\partial\phi}{\partial\tau}\right) = 0. \tag{19}$$

If in (19) we put $\mathscr{L} = \tfrac{1}{2}g(k)\tau^2 + \tfrac{1}{3}h(k)\tau^3$, and then neglect terms of order τ^2, we obtain

$$[g(k)+2h(k)\tau]\left[\frac{\partial^2\phi}{\partial k^2} - f''(k)\frac{\partial\phi}{\partial\tau}\right] - 2g'(k)\tau\frac{\partial^2\phi}{\partial\tau\partial k} - f''(k)g(k)\tau\frac{\partial^2\phi}{\partial\tau^2} = 0. \tag{20}$$

This can be divided through by the first factor in square brackets, and if τ^2 is still neglected we obtain

$$\frac{\partial^2\phi}{\partial k^2} - 2\frac{g'(k)}{g(k)}\tau\frac{\partial^2\phi}{\partial\tau\partial k} - f''(k)\left(\tau\frac{\partial^2\phi}{\partial\tau^2} + \frac{\partial\phi}{\partial\tau}\right) = 0. \tag{21}$$

Thus, the function $h(k)$ does not affect the answer to this order of approximation.

Of the two terms in (21) which carry the factor τ, the second, $f''(k)\tau\partial^2\phi/\partial\tau^2$, has a far greater effect than the first. In particular, and in agreement with the earlier geometrical discussion, this second term changes the type of the equation from elliptic when $\tau f''(k) < 0$ to hyperbolic when $\tau f''(k) > 0$. The characteristics of equation (21) are in these cases imaginary or real, respectively, being given by

$$\frac{d\tau}{dk} = -\frac{g'(k)}{g(k)}\tau \pm \left[f''(k)\tau + \frac{g'^2(k)}{g^2(k)}\tau^2\right]^{\frac{1}{2}}. \tag{22}$$

For small τ the right-hand side can be approximated as

$$\frac{d\tau}{dk} = \pm[f''(k)\tau]^{\frac{1}{2}}, \tag{23}$$

which is the equation for the characteristics of equation (21) with the second term omitted. Thus, among the second-derivative terms in (21) this term produces effects of smaller order than the other two, while the first-derivative term is also more important since it lacks the τ factor.

We conclude that the next approximation after an infinitesimal-amplitude one gives the equation

$$\frac{\partial^2\phi}{\partial k^2} - f''(k)\left(\tau\frac{\partial^2\phi}{\partial\tau^2} + \frac{\partial\phi}{\partial\tau}\right) = 0. \tag{24}$$

This is the equation analogous to the Tricomi equation in the hodograph equations for compressible plane flow of a gas, and the arguments for dropping the cross-derivative terms are the same in the two problems. The form of the equation is different in two respects, however. One is the presence of the first derivative. The other is that the τ-factor accompanies the second derivative with respect to τ itself, rather than with respect to the other variable.

Accordingly, whereas Tricomi's equation has characteristics with cusps on the transition line between the elliptic and hyperbolic regimes, equation (24) has all its characteristics tangent to the transition line $\tau = 0$, which is an envelope of characteristics and itself a characteristic. Specifically, each characteristic, by equation (23), takes, when $f''(k) > 0$, the form

$$\tau = \tfrac{1}{4}\{\int_{k_0}^{k} [f''(k)]^{\frac{1}{2}}\, dk\}^2, \tag{25}$$

where $k = k_0$ is the point where the characteristic touches the line $\tau = 0$. When $f''(k) < 0$, it takes the form

$$\tau = -\tfrac{1}{4}\{\int_{k_0}^{k} [-f''(k)]^{\frac{1}{2}}\, dk\}^2. \tag{26}$$

In neither (25) nor (26) will much error normally arise if $f''(k)$ is replaced by a constant value $f''(k_0)$, since the equation is in any case valid only for small values of τ and therefore small values of $|k - k_0|$.

Calculations with (24) can frequently be assisted if we use the fact that

$$\tau\frac{\partial^2\phi}{\partial\tau^2} + \frac{\partial\phi}{\partial\tau} = \frac{\partial^2\phi}{\partial\sigma^2} + \frac{1}{\sigma}\frac{\partial\phi}{\partial\sigma} \tag{27}$$

with $\sigma = 2\tau^{\frac{1}{2}}$. In the hyperbolic case we put

$$r = 2\left[\frac{\tau}{f''(k)}\right]^{\frac{1}{2}} = 2\left[\frac{\omega - f(k)}{f''(k)}\right]^{\frac{1}{2}}, \tag{28}$$

and then equation (24) becomes

$$\frac{\partial^2\phi}{\partial k^2} = \frac{\partial^2\phi}{\partial r^2} + \frac{1}{r}\frac{\partial\phi}{\partial r}, \tag{29}$$

if terms proportional to r obtained in transforming $\partial^2\phi/\partial k^2$ are neglected. The neglect of these terms is justified for small r, and is equivalent to neglecting the variation of $f''(k)$ along the part of any characteristic that lies within the region of small τ, as suggested after equation (26).

We conclude that the transformation (28), following on the Legendre transformation (9), reduces the dispersion equation to the axisymmetric wave equation (29) if we retain only the first-order effect of such non-linear terms as are present. The space and time co-ordinates x and t are related to the new variables k and r, to the approximation here considered, by the equations

$$x = \frac{\partial\phi}{\partial k} - \frac{2f'(k)}{f''(k)}\frac{1}{r}\frac{\partial\phi}{\partial r}, \tag{30a}$$

$$t = -\frac{2}{f''(k)}\frac{1}{r}\frac{\partial\phi}{\partial r}. \tag{30b}$$

In the elliptic case, the transformation

$$s = 2\left[-\frac{\tau}{f''(k)}\right]^{\frac{1}{2}} = 2\left[-\frac{\omega - f(k)}{f''(k)}\right]^{\frac{1}{2}}, \tag{31a}$$

$$x = \frac{\partial \phi}{\partial k} + \frac{2f'(k)}{f''(k)}\frac{1}{s}\frac{\partial \phi}{\partial s}, \qquad t = \frac{2}{f''(k)}\frac{1}{s}\frac{\partial \phi}{\partial s} \tag{31b}$$

similarly reduces the dispersion equation to the axisymmetric Laplace equation

$$\frac{\partial^2 \phi}{\partial k^2} + \frac{\partial^2 \phi}{\partial s^2} + \frac{1}{s}\frac{\partial \phi}{\partial s} = 0. \tag{32}$$

5. Influence of Finite Amplitude on Dispersion of Wave Trains with a Substantial Frequency Spread

The interesting dispersion problems are initial-value problems of various kinds, of which in this paper we investigate the influence of finite amplitude on two:

(i) problems where the spread of frequency in the wave train is initially considerable;
(ii) problems where it is small or zero, and amplitude variations constitute the main cause of dispersion.

The transformation of the dispersion equation to the forms (29) and (32) will prove to be of crucial importance in the second class of problem, and of suggestive value also in the first.

In this first problem, we assume that initially, at time $t = 0$, the wave train has a known distribution of wavenumber k and amplitude a with distance x. We assume, as always in the Whitham theory, that neither changes by more than a small fraction of itself in one wavelength. We shall assume also that their derivatives with respect to x have the orders of magnitude expected from "smooth" functions satisfying this condition.

The initial conditions, by equation (13), specify ω as well as k as a function of x. Therefore, in the (ω, k)-plane in which the dispersion equation becomes linear, they specify an initial curve. On this curve $t = 0$ and x is a known function. Hence, by (9), both the first derivatives of ϕ are prescribed. This, as foreshadowed in Section 1, is a boundary condition of the kind normally expected in hyperbolic problems. Nevertheless, we shall find that, owing to the smoothness of the boundary conditions, the method used gives solutions even in the elliptic case. The investigation of how this fact is related to general theory is postponed to Section 12.

We suppose that the initial curve in the (ω, k)-plane, on which both first derivatives of ϕ are specified, has the equation

$$\tau = \omega - f(k) = \tau_0(k), \tag{33}$$

where $\tau_0(k)$ is small enough for the assumption of "moderately small" amplitude to be applicable. Similarly, if the variable r were used, both first derivatives of ϕ would be specified on a curve $r = r_0(k)$, where $r_0(k)$ remains small. The latter fact, in relation to the axisymmetric wave equation (29), immediately suggests the use of an approximation like that in "slender-body theory" (Ward, 1955), where all variations with respect to the radial variable r, which remains small, are calculated easily by approximating the effect of axial variations. But, actually, the idea can be applied just as readily to the earlier equation (24) which uses τ, and this is preferred in what follows.

In terms of τ and k, the initial conditions are

$$\frac{\partial \phi}{\partial \tau} = -t = 0, \quad \frac{\partial \phi}{\partial k} = f'(k)\frac{\partial \phi}{\partial \tau} + x = x = x_0(k), \tag{34}$$

say, on $\tau = \tau_0(k)$; whence we can deduce that on this curve

$$\phi = \phi_0(k), \quad \text{where} \quad \phi_0'(k) = x_0(k). \tag{35}$$

We solve (24), subject to the conditions $\phi = \phi_0(k)$ and $\partial\phi/\partial\tau = 0$ on $\tau = \tau_0(k)$, using the approximation that the first term on the left-hand side of (24) is replaced by

$$\phi_0''(k) = x_0'(k). \tag{36}$$

The solution is easily seen to be

$$\phi = \phi_0(k) + \frac{\phi_0''(k)}{f''(k)}\left[\tau - \tau_0(k) - \tau_0(k) \log \frac{\tau}{\tau_0(k)} \right]. \tag{37}$$

The approximation can now be justified *a posteriori* provided that the relative magnitudes of different derivatives are of similar order in ϕ_0 and in f, and provided τ remains small compared with f. For then the contents of the square brackets in (37) are small compared with f, while ϕ_0''/ϕ_0 and f''/f are of the same magnitude; hence the second term in (37) is a small addition to the first; and the same condition on their derivatives can also be shown to follow from the proviso stated, indicating that the replacement of $\partial^2\phi/\partial k^2$ by $\phi_0''(k)$ in (24) was justified as a first approximation.

In interpreting the solution (37), it is convenient to use the notation

$$u = u(k) = f'(k) \tag{38}$$

for the infinitesimal-amplitude group velocity. The quantity which in the infinitesimal-amplitude case would determine the rate of dispersion is

$$\alpha(k) = \left(\frac{\partial u}{\partial x}\right)_{t=0} = \frac{u'(k)}{x_0'(k)} = \frac{f''(k)}{\phi_0''(k)}. \tag{39}$$

(This is a rate of dispersion in the sense that, in time t, waves in a small wavenumber band centred on k would come to occupy an interval extended by a factor $1 + \alpha(k)t$.)

For waves of moderately small amplitude, equations (9) and (37) give firstly

$$t = -\frac{\partial \phi}{\partial \tau} = \frac{\phi_0''(k)}{f''(k)}\left[\frac{\tau_0(k)}{\tau} - 1 \right], \tag{40}$$

which determines τ in terms of k and t as

$$\tau = \frac{\tau_0(k)}{1 + \alpha(k)t}, \tag{41}$$

and secondly
$$x - x_0(k) - u(k)t = \partial\phi/\partial k - \phi_0'(k)$$
$$= \frac{\tau_0'(k)}{\alpha(k)} \log \frac{\tau_0(k)}{\tau} + \frac{\alpha'(k)\tau_0(k)}{\alpha^2(k)}\left[1 - \frac{\tau}{\tau_0(k)} + \log \frac{\tau}{\tau_0(k)} \right]$$
$$= \frac{\tau_0'(k)}{\alpha(k)} \log\left[1 + \alpha(k)t \right] + \frac{\alpha'(k)\tau_0(k)}{\alpha^2(k)}\left[\frac{\alpha(k)t}{1 + \alpha(k)t} - \log\{1 + \alpha(k)t\} \right]. \tag{42}$$

The right-hand side of (42) represents the effect of finite amplitude on dispersion, since the left-hand side is zero in the classical infinitesimal-amplitude theory.

A physical interpretation of the results (41) and (42) is possible. In waves of moderately small amplitude we expect that, for given wavenumber k, the departure τ of the frequency from its infinitesimal-amplitude value is proportional to local energy density. Equation (41) therefore states that, to the present approximation, energy density varies in inverse proportion to the dimensions of the region that, on infinitesimal-amplitude theory, is occupied by waves of the given wavenumber. Equation (42), differentiated with respect to t, then identifies the velocity with which waves of that wavenumber are propagated as

$$\left(\frac{\partial x}{\partial t}\right)_k = u(k) + \frac{\tau_0'(k)}{1+\alpha(k)t} - \frac{\tau_0'(k)\alpha'(k)t}{[1+\alpha(k)t]^2}, \tag{43}$$

the second and third terms indicating the effect of finite amplitude.

The right-hand side of (43) is

$$u(k) + \left(\frac{\partial \tau}{\partial k}\right)_t = \left(\frac{\partial \omega}{\partial k}\right)_t. \tag{44}$$

It is not surprising that this should be the velocity with which waves of wavenumber k are propagated, since the basic equations (2) imply that

$$\frac{\partial k}{\partial t} + \frac{\partial \omega}{\partial x} = 0, \tag{45}$$

which can be written

$$\frac{\partial k}{\partial t} + \left(\frac{\partial \omega}{\partial k}\right)_t \frac{\partial k}{\partial x} = 0. \tag{46}$$

The present approximation is equivalent, then, to requiring that, for each wavenumber k, the excess frequency $\tau = \omega - f(k)$ shall vary with time in proportion to the variation of energy density indicated by infinitesimal-amplitude theory (equation (41)), while the position x at which the waves are found shall vary according to the equation

$$\frac{dx}{dt} = \left(\frac{\partial \omega}{\partial k}\right)_t. \tag{47}$$

It might, possibly, have been inferred independently that this would be the first approximation to what happens when the perturbing affects of finite amplitude are taken into account.

Evidently, the solution takes the same form for both the hyperbolic and elliptic cases. Even in the hyperbolic case, Whitham's characteristic velocities are not in evidence; there is no reason why they should be, as discontinuities in derivatives (namely, those quantities which are propagated along characteristics) have been excluded in the assumptions defining this problem. In both cases, a single velocity (47) (for alternative forms of it see (43) and (44)), namely, the velocity with which waves of wavenumber k propagate, plays the crucial role; on the other hand, the energy propagation velocity, whose calculation by means of Whitham's theory was given at length by Lighthill (1965), does not seem to have particular importance in this problem, or in the other problems discussed in this paper.

6. Notes on Whitham's Conjecture Regarding Discontinuities

In both cases, again, when $\alpha(k)$ is negative (so that initially waves of higher group velocity follow waves of lower group velocity), the approximate solution has a singularity when t reaches the positive value $-1/\alpha(k)$. This is the time when the waves of higher group velocity would, on infinitesimal-amplitude theory, catch up the waves of lower group velocity.

Such catching up represents the kind of eventuality for which Whitham suggests, as one conceivable outcome, the formation of an almost discontinuous jump in k. His theory does not positively predict this, as the conditions under which it holds cease to be valid before the catching up actually takes place. Analogies with classical cases of shock-wave formation may be misleading, since the effects of departures from the simple theory which are important in those cases (e.g. viscosity) are known to be such as will resist overtaking, and resist it effectively, neither of which properties has been demonstrated in the present case. Nevertheless, the conjecture may well prove valid under certain conditions.

Unfortunately, the approximation in Section 5, leading to (41) and (42), throws no light whatever on this question since the assumption of small τ ceases to be valid as soon as expression (41) ceases to be small. In the elliptic case we can revert to Whitham's exact equation, however, and show that the particular kinds of singularity in the solution from which Whitham infers the possibility of a discontinuity of shock-wave type, namely, an interval of three-valuedness which grows in length with time from zero at an initial instant, cannot occur.

Obviously, the absence of real characteristics in the elliptic problem already removes one means of formation of such a three-valued region. But to prove that it cannot occur, we suppose first that ω and k take values which along some curve L in the (x,t)-plane turn back upon themselves. (For the general theory of such "limiting lines" see Howarth (1953, p. 253).) Then L constitutes a singularity in the transformation between the (x,t)- and (k,ω)-planes. At this singularity, where the transformation becomes many-valued, the Jacobian

$$\frac{\partial(x,t)}{\partial(k,\omega)} = \frac{\partial(\partial\phi/\partial k, -\partial\phi/\partial\omega)}{\partial(k,\omega)} = -\frac{\partial^2\phi}{\partial k^2}\frac{\partial^2\phi}{\partial\omega^2} + \left(\frac{\partial^2\phi}{\partial k\partial\omega}\right)^2 \tag{48}$$

must vanish, whence in general it follows that equation (10) can be written

$$\frac{\partial^2\mathscr{L}}{\partial\omega^2}z^2 + 2\frac{\partial^2\mathscr{L}}{\partial\omega\partial k}z + \frac{\partial^2\mathscr{L}}{\partial k^2} = 0, \tag{49}$$

where

$$z = -\frac{\partial^2\phi/\partial\omega\partial k}{\partial^2\phi/\partial\omega^2}. \tag{50}$$

The existence of a real root of (49) means by (11) that real characteristic directions exist, which is impossible for an elliptic equation.

The only circumstance in which this argument breaks down is when the quantity (50) is indeterminate; in other words, when both $\partial^2\phi/\partial\omega^2$ and $\partial^2\phi/\partial\omega\partial k$ (and so also, by (10), $\partial^2\phi/\partial k^2$) are zero. These two conditions cannot in general be satisfied along a curve L, but merely at an isolated point. At such a point all the first derivatives of x and t with respect to k and ω (not merely the Jacobian (48)) are zero. It will be shown, however, in Section 16, that this alternative type of singularity is just as capable as the "limiting line" type of introducing a possible discontinuous solution. Accordingly, the question of what happens in the problem of Section 5 in the elliptic case remains unresolved.

7. Dispersion Caused Principally by Amplitude Variations: Geometrical Considerations in the Hyperbolic Case

A second type of initial-value problem is now considered, namely, one in which the effect of finite amplitude is not just a small correction to a dispersion process that would already be proceeding actively in its absence. On the contrary, the infinitesimal-amplitude approximation would predict no dispersion at all. Such a problem, in many ways, is the most stringent test of the finite-amplitude theory, and would be particularly suitable for comparisons with experiment.

Problems of this type can be constructed in more than one way. A wave train could, for example, be considered, whose wavelength (or frequency) was uniform at $t = 0$, but whose amplitude was variable, and whose subsequent development it was desired to determine. However, a problem more satisfactory, because related to a more readily realized experimental situation, is as follows.

At a point $x = 0$ a fixed wavemaker is operating at a fixed frequency ω_0 to create waves propagating into the region $x > 0$. The amplitude of the waves it makes varies gradually with time t. The effects of finite amplitude on the dispersion of the resulting wave group in the region $x > 0$ are to be determined.

When the dispersion equation is hyperbolic, there is one form of this problem that is especially suitable for testing Whitham's conjecture regarding the appearance of discontinuities. This is the case when the amplitude at the wavemaker changes (and to fix the ideas we shall suppose that it *increases*) only during a finite time interval, and is constant outside it. Whitham (1965a) pointed out in relation to a similar problem that the (x,t)-plane would include "simple-wave" regions, within which limiting lines were likely to appear.

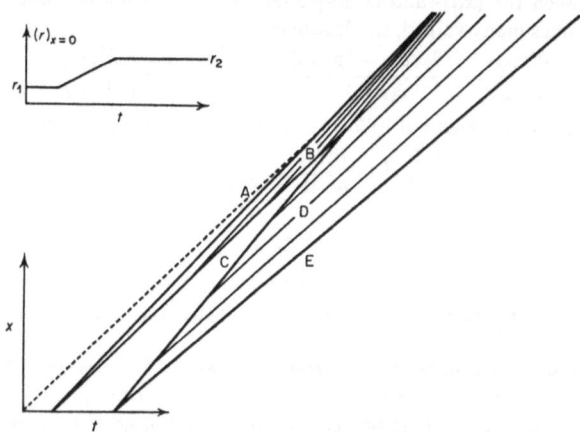

Fig. 1. Characteristics in the (x, t)-plane, due to the action of a wavemaker at $x = 0$, where the measure of amplitude r is caused to increase with time as in the inset graph.

Figure 1 (calculated by a method to be described in Section 9, for a particular law of increase of amplitude at the wavemaker) shows some of the characteristics in the problem just described, when $f''(k) > 0$. The two characteristic directions at any point have the same slope in the (x,t)-plane as in the (k,ω)-plane, as a comparison of (8) and (10) shows, and so, for "moderately small" amplitude, satisfy

$$\frac{dx}{dt} = f'(k) \pm [f''(k)\tau]^{\frac{1}{2}}. \tag{51}$$

Along any one of them, by (23), the quantity

$$2\tau^{\frac{1}{2}} \pm \int [f''(k)]^{\frac{1}{2}} dk \tag{52}$$

remains constant, and it may be checked that the upper or lower sign should be selected according to the selection in (51).

In region B in Fig. 1, the characteristics shown (plain lines) have the upper sign and carry values of (52) determined in the region C, which is influenced by operation of the wavemaker at variable amplitude. The other system of characteristics in region B, those with the lower sign (one of which is shown, dotted, about to enter the region) all come from the region of uniform wave amplitude A (where say $k = k_1$ and $\tau = \tau_1$), and therefore expression (52) is not only constant along each but has the same value throughout this region B, giving

$$2(\tau^{\frac{1}{2}} - \tau_1^{\frac{1}{2}}) = \int_{k_1}^{k} [f''(k)]^{\frac{1}{2}} dk. \tag{53}$$

Region B, in other words, is a "simple-wave" region, the whole of which maps on to a single characteristic curve in the plane.

Similar remarks apply to region D, where

$$2(\tau_2^{\frac{1}{2}} - \tau^{\frac{1}{2}}) = \int_{k_2}^{k} [f''(k)]^{\frac{1}{2}} dk, \tag{54}$$

if τ_2 and k_2 are the uniform values of τ and k in region E. In both regions the characteristics are straight lines (since along one of them in region B, for example, the constancy of (52) with the upper sign, coupled with equation (53), requires both k and τ, and so also the right-hand side of (51), to be constant). However, their points of origin can be determined only by solving the dispersion equation in the complex region C. Only then is it possible to predict, for example, in region B how far beyond the top of the figure a limiting line will be found owing to the running together of characteristics.

In order to solve the problem in region C, the equation in the (k,ω)-plane, that is, equation (10) or its transformed, approximate version (29), may be used. In the (k,ω)-plane the wavemaker occupies part of the line $\omega = \omega_0$, namely, the segment S in which k rises from k_1 to k_2. The boundary conditions are that

$$\frac{\partial \phi}{\partial k} = x = 0, \qquad \frac{\partial \phi}{\partial \omega} = -t = -t_0(k) \quad \text{on} \quad \omega = \omega_0, \tag{55}$$

where $t_0(k)$ is a decreasing function of k.

The region corresponding to C in the (k,ω)-plane consists of the "domain to dependence", say $\overline{C}$, of S; this is illustrated in Fig. 2. It is bounded by S and by two

13*

segments of characteristics. These latter are shown plain, while their continuations to where they touch the curve $\omega = f(k)$ are shown as broken lines. The domain $\bar{C}$ lies above the segment S because

$$\frac{\partial x}{\partial \omega} = \frac{\partial^2 \phi}{\partial k \partial \omega} = -t_0'(k) \tag{56}$$

is positive on S, and x must increase as we move into the region $\bar{C}$.

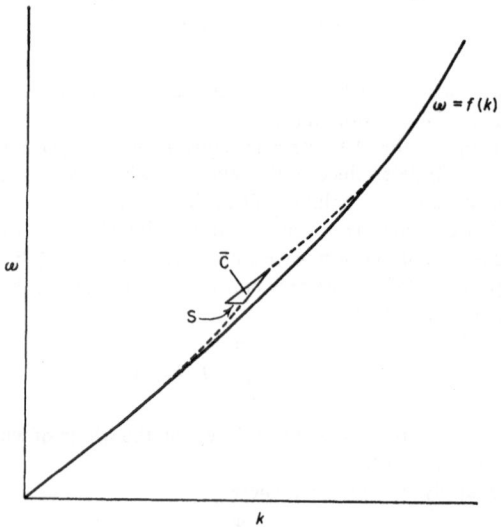

Fig. 2. Domain of dependence $\bar{C}$ of the initial curve S in the $(k\omega)$-plane; here, $\bar{C}$ corresponds to region C in the (x,t)-plane.

The problem of determining ϕ in $\bar{C}$ given the boundary conditions (55) is a well-set hyperbolic problem. Exactly the same problem would have to be solved in the (k,ω)-plane if there were a continuous monotonic variation of the amplitude produced at the wavemaker, without periods of constancy; then the solution of the dispersion equation in $\bar{C}$ would give the solution immediately at all points of the (x,t)-plane.

8. General Analytical Solution Using the Riemann Function

The boundary conditions (55) show that ϕ is a constant, whose value without loss of generality can be taken as zero, on $\omega = \omega_0$. In terms of the variables r and k, the initial conditions are

$$\phi = 0, \quad \frac{\partial \phi}{\partial r} = -\tfrac{1}{2} r f''(k) t_0(k), \quad \frac{\partial \phi}{\partial k} = -f'(k) t_0(k) \quad \text{on} \quad r = 2\left[\frac{\omega_0 - f(k)}{f''(k)}\right]^{\frac{1}{2}}. \tag{57}$$

If we define k_0, u_0 and μ by the equations

$$\omega_0 = f(k_0), \quad u_0 = f'(k_0) \quad \text{and} \quad \mu = f''(k_0), \tag{58}$$

then, to within the approximation for small r made in Section 4, (57) becomes

$$\phi = 0, \quad \frac{\partial \phi}{\partial r} = -\tfrac{1}{2} \mu r t_0(k), \quad \frac{\partial \phi}{\partial k} = -u_0 t_0(k) \quad \text{on} \quad k = k_0 - \frac{\mu r^2}{4u_0}. \tag{59}$$

Now the solution of (29), in the domain of dependence of any curve S, on which $\phi = 0$ and $\partial\phi/\partial r$ and $\partial\phi/\partial k$ are known, is given (Courant & Hilbert, 1962, p. 453, with x and y replaced by $r \pm k$) as

$$\phi_p = \phi(r_p, k_p) = \frac{1}{2}\int_{S^*} R(r,k; r_p, k_p)\left(\frac{\partial\phi}{\partial r}\, dk + \frac{\partial\phi}{\partial k}\, dr\right). \tag{60}$$

Here, R is the Riemann function of equation (29), that is,

$$R(r,k; r_p, k_p) = \left(\frac{r}{r_p}\right)^{\frac{1}{2}} F(z), \tag{61}$$

where F is the hypergeometric function

$$F(z) = F(\tfrac{1}{2}, \tfrac{1}{2}; 1; z), \quad z = \frac{(k_p - k)^2 - (r_p - r)^2}{4 r_p r}, \tag{62}$$

and S^* is the part of S with $z > 0$.

Now,

$$\frac{\partial\phi}{\partial k} = -\frac{\partial\phi}{\partial r}\frac{dr}{dk}, \tag{63}$$

because $\phi = 0$ on S (or, indeed, by direct inspection of (59)), and therefore the first term in round brackets in (60) is smaller by a factor $(dk/dr)^2$ than the second term and can be neglected.

Actually, equation (59) shows, and Fig. 3 illustrates, that k on S varies so little in comparison with r that, to within the approximations so far made, (60) can be regarded as an integral for constant $k = k_0$. This means that the dispersion would be the same whether the wavemaker kept ω accurately constant and varied the amplitude (so allowing k to vary), or cunningly varied the frequency ω, simultaneously with the amplitude, so that it produced waves of constant wavenumber k. In other words, only the variation in $\omega - f(k)$, or (what is the same thing) in r, matters.

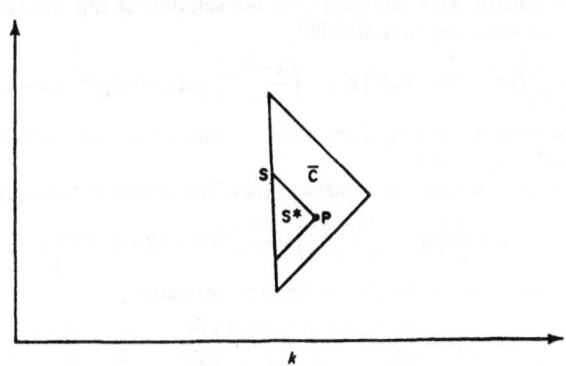

FIG. 3. Domain of dependence $\bar{C}$ in the (k,r)-plane. In this plane the initial curve S is nearly straight.

Equation (60) therefore becomes

$$\phi_p = -\tfrac{1}{2}u_0 \int_{r_p - (k_p - k_0)}^{r_p + (k_p - k_0)} t_0(r)\left(\frac{r}{r_p}\right)^{\frac{1}{2}} F\left[\frac{(k_p - k_0)^2 - (r_p - r_0)^2}{4 r_p r}\right] dr, \tag{64}$$

where $t_0(k)$ has been rewritten as $t_0(r)$ (the time when the value r characterizes the waves produced at the wavemaker). In (64), the argument z of the hypergeometric function F cannot exceed $\frac{1}{4}$ or be less than 0. The function $F(z)$ is tabulated ($\frac{1}{2}\pi F(z)$ is the complete elliptic integral K of the first kind for constant $k = \sqrt{z}$). It varies only between 1 and 1·18 as z goes from 0 to $\frac{1}{2}$, and the first three terms in its series form

$$F(z) = \sum_{n=0}^{\infty} \frac{(\frac{1}{2})^2(\frac{3}{2})^2 \ldots (n-\frac{1}{2})^2}{(n!)^2} z^n \qquad (65)$$

give it to within 0·02; however, $f'(z)$ (which is also needed) varies between 0·25 and 0·54, and six terms in (65) are needed to give it to the same accuracy. On the other hand, the cubic approximation

$$F(z) \doteqdot 1+0·25z+0·09z^2+0·27z^3 \qquad (66)$$

gives both $F(z)$ and $F'(z)$ to within 0·003 throughout the interval, which is good enough for all purposes.

Equation (64) and its two first derivatives are readily enough amenable to calculation. An example to illustrate the results will now be given; however, the most obvious example has a simple analytic solution, obtainable even more easily than through (64); this will be given in the next section.

9. Solution for Waves Created with Linearly Increasing Amplitude

In the special case of waves created at the wavemaker with r increasing linearly with time, say as $r = \alpha t$ from $t = r_1/\alpha$ to $t = r_2/\alpha$, so that the amplitude also increases linearly with time, the boundary conditions (59), as simplified by the discovery that with sufficient accuracy we can merely apply the first and third on $k = k_0$, become

$$\phi = 0, \quad \frac{\partial\phi}{\partial k} = -\frac{u_0}{\alpha}r \quad \text{on} \quad k = k_0. \qquad (67)$$

For the axisymmetric wave equation (29), the solution of this initial-value problem is a simple "second-order conical field"

$$\phi = -\frac{u_0}{4\alpha}\left\{[r^2+2(k-k_0)^2]\sin^{-1}\left(\frac{k-k_0}{r}\right)+3(k-k_0)[r^2-(k-k_0)^2]^{\frac{1}{2}}\right\}, \qquad (68)$$

which can be obtained, e.g. by integrating the elementary solution $[r^2-(k-k_0)^2]^{-\frac{1}{2}}$ three times with respect to k.

To obtain x and t we now use equations (30). The second of these gives

$$t = \frac{u_0}{\alpha\mu}\left\{\sin^{-1}\left(\frac{k-k_0}{r}\right)+\frac{k-k_0}{r^2}[r^2-(k-k_0)^2]^{\frac{1}{2}}\right\}, \qquad (69)$$

while the first gives, to within the present approximation,

$$x - u_0 t = \frac{\partial\phi}{\partial k} - 2\frac{f'(\ )-f'(k_0)}{f''(k)}\frac{1}{r}\frac{\partial\phi}{\partial r}$$

$$= \frac{\partial\phi}{\partial k} - 2\frac{k-k_0}{r}\frac{\partial\phi}{\partial r} = -\frac{u_0}{\alpha r^2}[r^2-(k-k_0)^2]^{\frac{3}{2}}. \qquad (70)$$

The latter result (70) satisfies accurately the boundary condition that $r = \alpha t$ on the wavemaker $x = 0$. It is an equation measuring the small departures in region C from the line $x = u_0 t$. These departures, in for example t, are everywhere less than r_2/α. The former equation (69), by contrast, gives the absolute value of t on the much coarser time-scale $u_0/\alpha\mu$, and to a correspondingly coarser approximation. On this coarser time-scale t is effectively zero when $k = k_0$.

Equations (69) and (70) may be used to compute the shape of region C, bounded as it is by the characteristics

$$k = k_0 + r - r_1 \quad \text{and} \quad k = k_0 + r_2 - r, \tag{71}$$

and to determine the variation in the wave amplitude (as measured by r) throughout C. The results are best exhibited in terms of the non-dimensional variables

$$t' = \frac{\alpha\mu t}{u_0}, \quad x' = \frac{\alpha\mu x}{u_0^2} \quad \text{and} \quad r' = \frac{r\mu}{u_0}. \tag{72}$$

Equations (72) mean that in the (x',t')-plane the characteristics approximate to the straight line $x' = t'$ (corresponding to $x = u_0 t$), while

$$r' = 2\{f''(k)[\omega - f(k)]\}^{\frac{1}{2}}/f'(k) \tag{73}$$

is a non-dimensional measure of amplitude in terms of the frequency change it produces for given wavenumber. (The linear law of variation of amplitude at the wavemaker can be written $r' = t'$.)

In terms of the variables (72), the characteristics (71) can be written

$$x' = t' - r_1' G(t'), \qquad x' = t' - r_2' H(t'), \tag{74}$$

while a line of constant r' takes the form

$$x' = t' - r' F(t'). \tag{75}$$

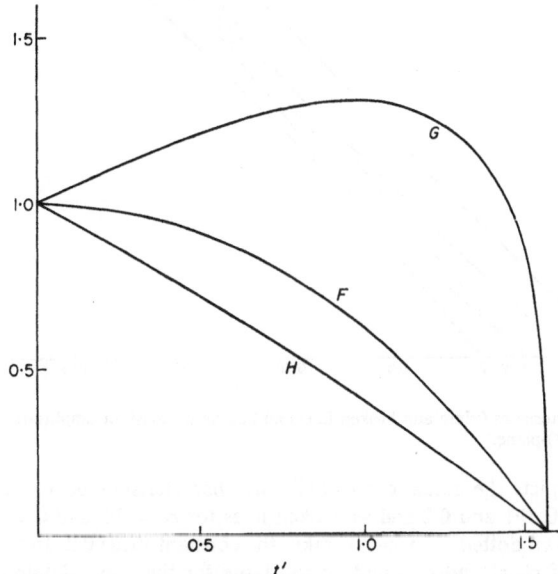

FIG. 4. Functions F, G and H, defining the shapes of characteristics and lines of constant amplitude in region C.

Here, the functions $F(t')$, $G(t')$, $H(t')$ are defined by means of a parameter θ through the equations $t' = \theta + \frac{1}{2}\sin 2\theta$, $F = \cos^3 \theta$, $G = \cos \theta + \frac{1}{2}\sin 2\theta$, $H = \cos \theta - \frac{1}{2}\sin 2\theta$, and Fig. 4 exhibits them in graphical form.

On the other hand, to the approximation used, equations (74) and (75) could perfectly well be replaced by

$$x' = t' - r_1'G(x'), \quad x' = t' - r_2'H(x'), \quad x' = t' - r'F(x'), \tag{76}$$

and this form of the approximation has the advantage that the conditions at the wavemaker $x' = 0$ are satisfied exactly (because $F = G = H = 1$). It has, therefore, been used in the numerical work.

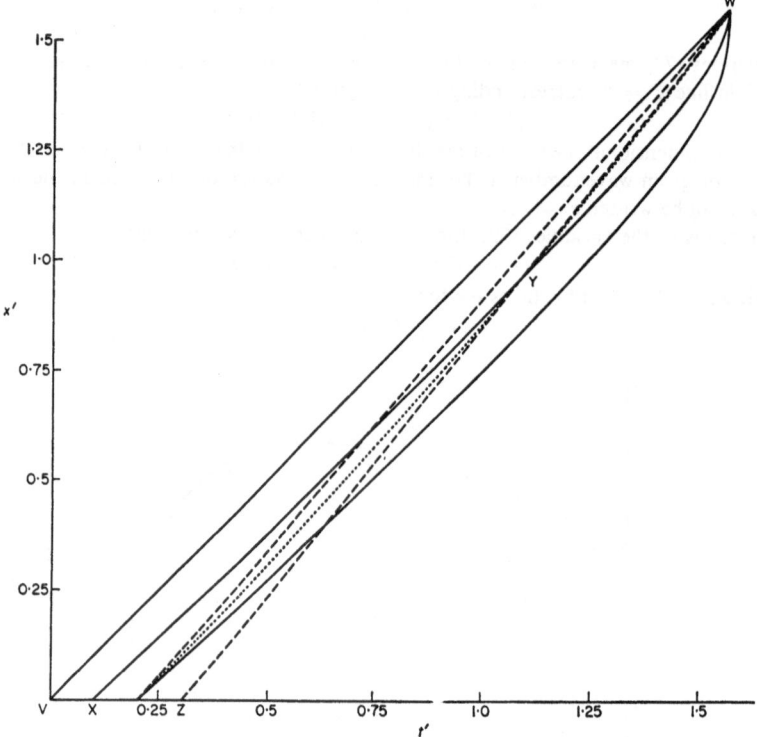

Fig. 5. Characteristics (plain and broken lines) and curve of constant amplitude $r' = 0.2$ (dotted line) in the (x', t')-plane.

Figure 5 depicts the actual curves (76), the characteristics being shown as plain lines for $r_1' = 0, 0.1$ and 0.2 and as broken lines for $r_2' = 0.2$ and 0.3, together with one curve, shown dotted, on which r' takes the constant value 0.2. In many problems these values of r_1', r_2' and r' would be too large for the approximate equation (29) which has been used, to be a good approximation. For illustration on a reasonably sized piece of paper, on the other hand, it was found essential to take them as large as this. The effect of reduction in the values of r_1', r_2' and r' by any given factor is, however, easily imagined, since it is equivalent to reducing the scale of the whole diagram, in the direction perpendicular to the straight line $x' = t'$, by just that factor.

The region C which has been calculated above, and whose relation to other regions of wave propagation is illustrated in Fig. 1, will not consist of the whole of Fig. 5, but only of that part which lies between the two characteristics (74). Thus, for $r_1' = 0.1$ and $r_2' = 0.3$, the region C will be the curvilinear triangle XYZ in Fig. 5. Within XYZ, the equations of all the characteristics take the same form (74) but with r_1' and r_2' replaced by general values lying in the interval between r_1' and r_2'. At the point Y the value of r' is $\frac{1}{2}(r_1' + r_2')$, as Fig. 3 shows, explaining why the curve on which r' takes the constant value 0.2 goes through Y. This form of C for $r_1' = 0.1$ and $r_2' = 0.3$ was used to calculate the complete wave propagation pattern given in Fig. 1; in this process, the simplification $dx'/dt' = 1 + k' \pm \frac{1}{2}r'$ of equation (51) was found useful.

For general r_1' and r_2' the value of t for which the characteristics (74) cross satisfies

$$G(t')/H(t') = r_2'/r_1', \qquad (77)$$

and, since the left-hand side varies from 1 to ∞ as t' goes from 0 to $\frac{1}{2}\pi$, the solution of (77) is always in this interval, while by (74) x' also must be less than $\frac{1}{2}\pi$. Region C cannot therefore extend to a distance of more than

$$x_{max} = \frac{1}{2}\pi\frac{u_0^2}{\alpha\mu} \qquad (78)$$

from the wavemaker. In other words, the extreme position of a point such as Y in Fig. 5 is represented by the point W, where $x' = \frac{1}{2}\pi$.

In the extreme, but particularly interesting, special case when $r_1' = 0$, so that the wavemaker creates no waves before time $t = 0$, after which their amplitude increases linearly with time, region C corresponds to a larger part of Fig. 5, namely, the whole curvilinear triangle VWZ (assuming that r_2' again takes the value 0.3). In this case all the characteristics belonging to the "faster" family (indicated in Fig. 5 by broken lines) go through the point W. So also do the lines of constant r'. The point W is therefore one where lines carrying, not just a small interval of values of the amplitude, but rather a large range of them, all run together. Equation (78) gives its distance from the wavemaker.

This prediction suggests a good test of Whitham's theory of discontinuities. If they exist, then a discontinuity of substantial strength should in this special case come into being instantaneously at the point W. Its distance (78) from the wavemaker can be written

$$x_{max} = \frac{\pi f'^2(k_0)}{4[f''(k_0)]^{\frac{1}{2}}} \bigg/ \left\{\frac{d}{dt}[\omega - f(k)]^{\frac{1}{2}}\right\}_{x=0} \qquad (79)$$

Whether or not the discontinuity described by Whitham occurs, some fairly drastic phenomenon must be expected at this point, owing to the confluence of wave motions at a whole range of different levels of amplitude.

10. The Dispersion Problem for Unidirectional Surface Waves of Finite Amplitude on Deep Water

After reaching this conclusion for general dispersive systems satisfying a hyperbolic form of Whitham's dispersion equation, it is natural at first to think of unidirectional surface waves on deep water as a convenient system on which to try the experiment. However, it was already shown in Section 1 that the dispersion equation in this case, at any rate when gravity is the predominant restoring force, satisfies the condition for ellipticity, namely, that $\omega - f(k)$ and $f''(k)$ have opposite signs. We shall, naturally, wish to consider what will happen also in this case, and for this reason an approximate form of $\mathscr{L}$ is calculated in this section, in preparation for the study of elliptic problems which follows. In the meantime, we consider the position also at the shorter wavelengths, when surface tension is important as an additional restoring force.

In the extreme case of wavelengths so small that gravity g is negligible as a restoring force compared with surface tension γ, an exact solution for unidirectional waves on deep water is known (Crapper, 1957). This shows that $\omega - f(k)$ is negative for waves of finite amplitude; but in this problem

$$f(k) = (\gamma k^3/\rho)^{\frac{1}{2}},$$

so that $f''(k) > 0$. The dispersion equation is therefore still elliptic. Actually, it can be written down exactly, and relatively simply, since $\mathscr{L}$ is given by

$$\mathscr{L} = 2\gamma - \frac{\rho\omega^2}{k^3} - \frac{\gamma^2 k^3}{\rho\omega^2}, \qquad (80)$$

so that (10) becomes

$$\left(1 + \frac{3\gamma^2 k^6}{\rho^2\omega^4}\right)k^2\frac{\partial^2\phi}{\partial k^2} + \left(3 + \frac{3\gamma^2 k^6}{\rho^2\omega^4}\right)2k\omega\frac{\partial^2\phi}{\partial k\partial\omega} + \left(6 + \frac{3\gamma^2 k^6}{\rho^2\omega^4}\right)\omega^2\frac{\partial^2\phi}{\partial\omega^2} = 0, \qquad (81)$$

which is capable of exact reduction to a form somewhat similar to (21) by putting

$$\eta = \frac{\gamma^2 k^6}{\rho^2\omega^4}, \qquad (82)$$

when with k and η as independent variables it becomes

$$(1 + 3\eta)k^2\frac{\partial^2\phi}{\partial k^2} + 12(\eta - 1)\left(k\eta\frac{\partial^2\phi}{\partial k\partial\eta} + \eta^2\frac{\partial^2\phi}{\partial\eta^2}\right) + 6(1 + \eta)\eta\frac{\partial\phi}{\partial\eta} = 0. \qquad (83)$$

This is an equation in which $(\eta - 1)$ plays a role similar to that of τ in (29).

To obtain expressions like (80) for the mean Lagrangian density, it is sometimes easiest to calculate first the mean potential energy density V. In Crapper's solution this is γ times the mean extension of the surface, and is easily obtained as

$$V = 2\gamma\left(\frac{\gamma k^3}{\rho\omega^2} - 1\right). \qquad (84)$$

One can then obtain $\mathscr{L}$ from the general formula

$$\mathscr{L} = 2\omega^2\int_{f(k)}^{\omega} (V/\omega^2)\, d\omega, \qquad (85)$$

which follows immediately from Whitham's expression $\omega\partial\mathscr{L}/\partial\omega - \mathscr{L}$ for the total energy density in problems without "pseudo-frequencies". The author in fact obtained (80) in this way, and then calculated the mean kinetic energy density in Crapper's problem, checking that it indeed was equal to $\mathscr{L} + V$.

The fact that the dispersion equation is elliptic for very small wavenumbers (pure gravity waves) and for very large wavenumbers k (pure capillary waves) does not rule out the possibility that in some intermediate range of wavenumbers the dispersion equation is hyperbolic. Indeed, for these waves which satisfy

$$f(k) = \left(gk + \frac{\gamma}{\rho}k^3\right)^{\frac{1}{2}} \tag{86}$$

$f''(k)$ changes sign where $\gamma k^2/\rho g = \kappa$ takes the value

$$\kappa = (2/\sqrt{3}) - 1 = 0{\cdot}1547. \tag{87}$$

Wilton (1915) found, however, that $\omega - f(k)$ changes sign where $\kappa = 0{\cdot}5$. Physically, these two values of κ correspond to (i) minimum group velocity (ii) a wavenumber carrying the same phase velocity as its first harmonic. Between them, that is for $0{\cdot}1547 < \kappa < 0{\cdot}5$, the dispersion equation is hyperbolic, because $\omega - f(k)$ and $f''(k)$ are both positive.

This is a problem where it is easiest to calculate the mean kinetic energy density T. From its surface-integral expression in general irrotational motions it can be shown that, in terms of coefficients A_n defined by Wilton (1915),

$$\frac{Tk^3}{\omega^2} = \frac{1}{4}\sum_{n=1}^{\infty} nA_n^2 = \frac{1}{4}A_1^2 + \frac{1}{8}\left(\frac{2-\kappa}{1-2\kappa}\right)^2 A_1^4 + O(A_1^6). \tag{88}$$

Wilton shows also that

$$\frac{\omega^2}{gk} - (1+\kappa) = \frac{2\kappa^2 + \kappa + 8}{8(1-2\kappa)}A_1^2 + \frac{24\kappa^5 - 164\kappa^4 - 566\kappa^3 + 1821\kappa^2 - 1322\kappa + 448}{(1-2\kappa)^3(1-3\kappa)}A_1^4 + O(A_1^6), \tag{89}$$

and it follows that

$$\frac{Tk^3}{\omega^2} = \frac{2(1-2\kappa)}{2\kappa^2 + \kappa + 8}\left(\frac{\omega^2}{gk} - 1 - \kappa\right) +$$
$$\frac{3(24\kappa^5 - 116\kappa^4 - 74\kappa^3 + 351\kappa^2 - 110\kappa + 64)}{(2\kappa^2 + \kappa + 8)^3(3\kappa - 1)}\left(\frac{\omega^2}{gk} - 1 - \kappa\right)^2 + \dots \tag{90}$$

We can now use the equation $\partial \mathcal{L}/\partial(\omega^2) = T/\omega^2$ to deduce that

$$\mathcal{L} = \frac{g}{k^2}\left[\frac{1-2\kappa}{2\kappa^2 + \kappa + 8}\left(\frac{\omega^2}{gk} - 1 - \kappa\right)^2 + \right.$$
$$\left. \frac{24\kappa^5 - 116\kappa^4 - 74\kappa^3 + 351\kappa^2 - 110\kappa + 64}{(2k^2 + \kappa + 8)^3(3\kappa - 1)}\left(\frac{\omega^2}{gk} - 1 - \kappa\right)^3 + \dots \right]. \tag{91}$$

In the limiting case of pure gravity waves ($\kappa = 0$) this becomes

$$\mathcal{L} = \frac{g}{8k^2}\left[\left(\frac{\omega^2}{gk} - 1\right)^2 - \left(\frac{\omega^2}{gk} - 1\right)^3 + \dots\right]. \tag{92}$$

Equation (89) confirms that for small but finite amplitude ω^2/gk exceeds its infinitesimal-amplitude value $(1+\kappa)$ provided that $\kappa < 0{\cdot}5$. We may now ask whether the limited range of κ between 0·1547 and 0·5 yields a convenient wave system on which to attempt the experimental test of Whitham's dispersion theory in the hyperbolic case suggested in Section 9. It would require that the fixed frequency at which the wavemaker operates be chosen within the interval

$$0\cdot674\left(\frac{\rho g^3}{\gamma}\right)^{\frac{1}{4}} < \omega < 1\cdot030\left(\frac{\rho g^3}{\gamma}\right)^{\frac{1}{4}}. \tag{93}$$

In water ($\gamma/\rho = 74$ cm³ sec⁻²), for example, it would have to operate at a frequency in cycles per second, $\omega/2\pi$, between 6·4 c/s and 9·8 c/s; in mercury ($\gamma/\rho = 36$ cm³ sec⁻²), between 7·6 c/s and 11·7 c/s.

11. A Non-linear Non-perturbation Theory of Waves under Combined Action of Gravity and Surface Tension

Two main difficulties arise. One is that $0\cdot1547 < \kappa < 0\cdot5$ is an interval of κ where the perturbation procedure leading to the series (89), (90) and (91) has far too limited a region of convergence to be applicable at amplitudes large enough for the experimental test described in Section 9 to be effective. This will be overcome in the present section by a means of a theory not based upon such a perturbation procedure. The second difficulty is that viscosity significantly attenuates these waves of fairly short length. The use of mercury as the working fluid will be seen to give reasonable hope of overcoming this difficulty, while facilitating observation of the waves by optical means.

Wilton (1915) obtained the equation

$$\tfrac{1}{2}\mu + \beta(C + \sum_{n=1}^{\infty} A_n \cos n\xi) + \kappa\beta^{-\frac{1}{2}}[\sum_{n=1}^{\infty} n^2 A_n \cos n\xi + \sum_{n=1}^{\infty} n^3 A_n^2 +$$

$$\sum_{m=1}^{\infty}\sum_{n=m+1}^{\infty} mn(m+n)A_m A_n \cos(m-n)\xi] = 0, \tag{94a}$$

where

$$\beta = (1 + \sum_{n=1}^{\infty} nA_n \cos n\xi)^2 + (\sum_{n=1}^{\infty} nA_n \sin n\xi)^2, \tag{94b}$$

for his coefficients A_n, where C is a constant, ξ is a variable and $\mu = \omega^2/gk$. He wrote down equations obtained by putting the Fourier constants of (94a) equal to zero and solved them by expanding each A_n as a power series in A_1. In A_n, the leading term was proportional to A_1^n. For values of κ in the general neighbourhood of the points $\tfrac{1}{2}, \tfrac{1}{3}, \tfrac{1}{4}, \ldots$, however, the convergence is very bad, and the non-linear effects are greatly enhanced. This is exhibited by the presence of the small divisors $1 - 2\kappa$, $1 - 3\kappa$, etc. in his series (89). Physically, the value $\kappa = n^{-1}$ is significant because it permits the wavenumbers k and nk (its nth harmonic) to have the same phase velocity, and so to be in mutual resonance.

At $\kappa = \tfrac{1}{2}$ itself, where the perturbation scheme has zero radius of convergence, Wilton shows that different power series (the leading terms in A_2 and A_3 being proportional to A_1 and A_1^2 respectively) can be used, with finite radius of convergence; this is possible at $\kappa = \tfrac{1}{3}$ also (but the leading terms in A_2 and A_3 are then proportional to A_1^2 and A_1 respectively), but neither form is possible for any other value of κ. For

$\kappa = n^{-1}$ with $n > 3$, special forms of power series are also possible, but these do not have a term proportional to A_1 itself in any A_n.

The whole region appears, then, to be one in which A_1, A_2 and A_3 can be of similar order of magnitude but other A_n are consistently smaller. A non-perturbational non-linear theory, based on neglect of A_n for $n > 3$, and for simplicity allowing only quadratic interactions between the first, second and third harmonies A_1, A_2 and A_3, can be expected to give more uniformly approximate results in this region of κ. The coefficients of $\cos \xi$, $\cos 2\xi$ and $\cos 3\xi$ in equation (94a) give the simultaneous equations

$$A_0 A_1 + (1+\kappa)A_1 A_2 + (9\kappa - 1)A_2 A_3 = 0, \tag{95a}$$

$$(2\kappa - 1)A_2 + 2A_0 A_2 + (1 - \tfrac{1}{2}\kappa)A_1^2 + (1+3\kappa)A_1 A_3 = 0, \tag{95b}$$

$$(3\kappa - 1)A_3 + \tfrac{3}{2}A_0 A_3 + \tfrac{3}{2}(1-\kappa)A_1 A_2 = 0, \tag{95c}$$

where only the squares of small quantities, including the quantity $A_0 = 2C + 1 + \kappa$, have been included; since $(2\kappa - 1)$ and $(3\kappa - 1)$ are themselves small in the region of interest, equations (95a–c) can be regarded as homogeneously quadratic in small quantities. They are easily solved exactly, without any perturbation procedure, in terms of a single parameter (a convenient one is the ratio A_3/A_1). The term independent of ξ in equation (94) can then be used, with equation (88), to calculate the quantities

$$\frac{\omega^2}{gk} - 1 - \kappa = -A_0 - A_1^2 - 4\kappa A_2^2 + (3 - 18\kappa)A_3^2, \tag{96a}$$

$$\frac{4Tk^3}{\omega^2} = A_1^2 + 2A_2^2 + 3A_3^2. \tag{96b}$$

Examination of the cubic terms neglected in equations (95a–c) indicates that for values of the amplitude such as are discussed below, their effect remains considerably smaller than that of the terms retained. The solutions are not expected to have striking numerical accuracy, but the scheme should give consistent adequate first approximations throughout the neighbourhood of $\kappa = \tfrac{1}{3}$ and $\kappa = \tfrac{1}{2}$.

To seek a value of κ suitable for the experimental test, it is advisable to look in the middle of the permitted interval $0.1547 < \kappa < 0.5$. Near the lower end, where $f''(k)$ vanishes, frequency dispersion is almost absent and the "competition between frequency dispersion and amplitude dispersion" cannot therefore be investigated; it is also a practical difficulty that x_{max}, proportional to the inverse square root of $f''(k)$, becomes inconveniently large. Near the upper end, where the finite-amplitude-correction to the frequency ω vanishes, substantial increases of ω above $f(k)$ are hardly to be expected, and we shall see that they do not occur.

By contrast, the value $\kappa = 0.3$ is found to be quite suitable. The quantities (96), obtained by this approximate theory for $\kappa = 0.3$, are plotted against one another in Fig. 6, in which also the broken line indicates also the exact slope at the origin given by the perturbation-theory expression (90). Note that, very near the origin, the second and third harmonics are small, so that the relative importance of the A_1^2 terms neglected in (95) becomes greater; elsewhere, the relative errors should be substantially less. The figure indicates an approximately linear relationship up to the point marked X. At this point, the frequency ω is a full 2% above its infinitesimal-amplitude value;

that is, $\omega = 1\cdot020f(k)$. (By contrast, for $\kappa = 0\cdot4$ the diagram inset in Fig. 6 shows that ω never rises above $1\cdot004f(k)$.)

The fact that these two quantities vary approximately linearly with one another in no way implies that "linear" approximations in the ordinary sense are valid; the variation of A_1, A_2 and A_3 with one another is, on the contrary, highly non-linear. At X, $A_1 = 0\cdot14$, $A_2 = 0\cdot05$ and $A_3 = 0\cdot03$, and the maximum slope of the surface is $0\cdot25$.

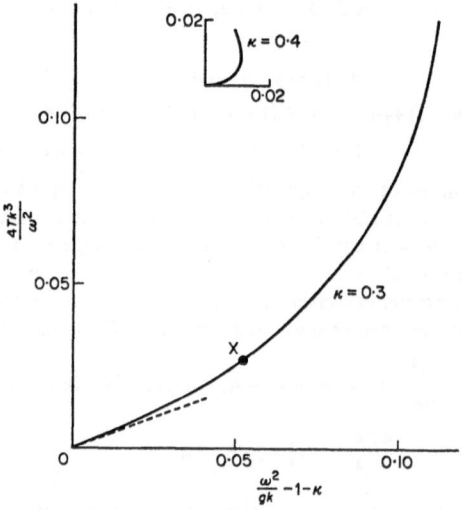

FIG. 6. Non-linear waves under the combined action of gravity and surface tension.

The approximately linear dependence of $\partial\mathscr{L}/\partial(\omega^2) = T/\omega^2$ on $\omega^2 - f^2(k)$ for $\kappa = 0\cdot3$ implies that approximately

$$\mathscr{L} \propto [\omega^2 - f^2(k)]^2 \propto [\omega - f(k)]^2 \quad \text{for} \quad f(k) < \omega < 1\cdot02f(k). \tag{97}$$

The theory of Sections 3 to 9 was based on the assumption of such quadratic dependence of $\mathscr{L}$ on ω, and should therefore be applicable.

We can now investigate whether values of $x_{\max}$ given by equations (78) and (79), and of $t_{\max}$, the time for disturbances to reach $x = x_{\max}$, are experimentally convenient, and whether substantial viscous dissipation of the waves will occur during this time. For $\kappa = 0\cdot3$,

$$t_{\max} = \frac{\pi}{4} \frac{f'(k_0)}{[f(k_0)f''(k_0)]^{\frac{1}{2}}} \bigg/ \frac{d}{dt}\bigg\{\bigg[\frac{\omega - f(k)}{f(k_0)}\bigg]^{\frac{1}{2}}\bigg\}_{x=0} = 1\cdot44\bigg/\frac{d}{dt}\bigg\{\bigg[\frac{\omega - f(k)}{f(k_0)}\bigg]^{\frac{1}{2}}\bigg\}_{x=0} \tag{98}$$

This means that, if the wavemaker takes a time τ to reach $\omega = 1\cdot02f(k_0)$, then $t_{\max} = 10\tau$. To avoid $t_{\max}$ being too large, it would seem reasonable to take $\tau = 1$ sec, so that the build-up of amplitude extends over $9\cdot6$ cycles in mercury (or $8\cdot0$ cycles in water), but it would hardly be a fair test of Whitham's theory to vary the amplitude over its full range in a substantially smaller number of cycles than this.

It is necessary, then, that $t_{\max}$ be approximately 10 sec, giving

$$x_{\max} = t_{\max}f'(k_0) = 154 \text{ cm}$$

in mercury and 184 cm in water, reasonably convenient distances. But the normal viscous dissipation of amplitude in time t is by a factor

$$e^{-2(\mu/\rho)k^2t} = e^{-2(\mu/\rho)(\rho g/\gamma)\kappa t} = e^{-6\mu g/\gamma}, \qquad (99)$$

where μ is the viscosity, and the values $\kappa = 0.3$ and $t = 10$ sec have been inserted. It is important therefore, that $\mu g/\gamma$ be small compared with $\frac{1}{6}$ sec^{-1}. At room temperature the liquid with the lowest value of $\mu g/\gamma$ seems to be mercury, with a value of $\frac{1}{30}$ sec^{-1}. It is uncertain whether the viscous dissipation will be given accurately by (99) for the type of wave motion here considered, and whether or not a dissipation by a factor $e^{-1/5}$ on amplitude would spoil the theory. If its effect were seriously harmful one would have to consider further reduction of τ.

To sum up, a suggested experiment to test Whitham's theory is to make unidirectional deep-water waves in a long tank of mercury with a wavemaker operating at 9·6 c/s, increasing its amplitude to a maximum in, at most, 1 sec. The maximum should correspond to waves with a greatest surface slope of 0·25. The mode of build-up of the waves created should be such that the measure of amplitude given by the variable r increases in proportion to the time; approximately, r is proportional to the square root of the energy of the waves created.

The theory indicates that, at a point about 1·5 m from the wavemaker, different values of the amplitude (initially separated by 15 cm) will have run together. What next happens (according to Whitham's suggestion, the propagation of a well-formed discontinuity of amplitude) should then be carefully observed.

12. General Problems of Elliptic Dispersion Equations

The discussion in this paper has brought out the importance not only of hyperbolic but also of elliptic dispersion equations, and therefore it is desirable to study methods of solving these. For the problem treated in Section 5 the approximate method there developed is applicable whether the equation is of hyperbolic or elliptic type. Equations of the two types require different methods, however, in problems where the dispersion is caused principally by amplitude variation, such as were discussed in the hyperbolic case in Sections 7 to 9.

It has already been pointed out that, even though the equation is elliptic, the boundary conditions are of hyperbolic type. Such problems were rarely studied and frequently dismissed as having no possible physical relevance before Garabedian put the theory and practical solution of them on a sound analytical footing with a series of papers (see for example Garabedian, 1958) making use of imaginary characteristics. These showed that, when the boundary conditions are smooth in a certain sense, uniquely defined solutions exist.

It is purely the *boundary condition* of hyperbolic type which calls for the use of the method of characteristics. The fact that the equation is elliptic merely causes these characteristics to be imaginary. However, explicit solutions such as (64) to the hyperbolic dispersion equation (29), in terms of data on a fixed line $k = k_0$, can be applied to the elliptic dispersion equation (32) after a simple imaginary transformation.

The limits of integration in (64) are points where the characteristics

$$r - r_p = \pm(k_p - k) \qquad (100)$$

through the point (k_p, r_p) intersect the line $k = k_0$ on which the data are given. For equation (32) the characteristics through a point (k_p, s_p) where ϕ is to be determined are $s - s_p = \pm i(k_p - k)$, and these intersect the line $k = k_0$ at the points

$$s = s_p \pm i(k_p - k_0), \tag{101}$$

which can therefore be expected to be the limits of integration in a revised form of (64).

As in Section 8, we may use as approximate boundary conditions

$$\phi = 0, \quad \frac{\partial \phi}{\partial k} = -u_0 t_0(s) \quad \text{on} \quad k = k_0, \tag{102}$$

where $u_0 = f'(k_0)$ and $t_0(s)$ is the time at which the amplitude variable s takes a given value for the waves produced at the wavemaker. When ϕ has been obtained, the equations (31) for x and t can be used, as in Section 9, in the approximate form

$$t = -\frac{2}{\mu} \frac{1}{s} \frac{\partial \phi}{\partial s}, \quad x - u_0 t = \frac{\partial \phi}{\partial k} + 2 \frac{k - k_0}{s} \frac{\partial \phi}{\partial s}, \tag{103}$$

where $f''(k_0)$ has now been replaced by $-\mu$ (since in the example to be studied it will be assumed negative), and $f'(k) - u_0$ has been replaced by $\mu(k - k_0)$. With the conditions (102) on ϕ, the revised form of (64) referred to above can be obtained most simply by transforming (32) into (29) by the transformation $s = ir$, deriving (64) and then replacing r by $-is$ again. Actually, we replace r_p by $-is$ and the variable of integration r by $-iS$, giving $\phi(k, s)$ as

$$\phi = \tfrac{1}{2} u_0 i \int_{s - i(k - k_0)}^{s + i(k - k_0)} t_0(S) \left(\frac{S}{s}\right)^{\frac{1}{2}} F\left(\frac{(k - k_0)^2 + (s - S)^2}{-4sS}\right) dS, \tag{104}$$

with limits of integration as predicted in (102).

The solution exists and is unique, therefore, in any range of (k, s) such that the function $t_0(S)$, describing the variation of the time t with the "amplitude" variable $s = S$ at the wavemaker, can be analytically continued, uniquely, into the area of the complex S-plane filled by the points $s \pm i(k - k_0)$. This implies a smoothness condition on the function $t_0(S)$.

For example, discontinuities in the first derivative of $t_0(S)$, or even in a higher derivative, would make such analytical continuation impossible. Solutions of Whitham's equation do not then exist in the elliptic case, although in the hyperbolic case such discontinuities create no difficulties. Presumably, the approximation involved in Whitham's assumption of gradual variation of amplitude with time at a fixed point makes greater demands (not only "gradualness" but also "smoothness") in the elliptic case.

Even if the function $t_0(S)$ is analytic in a neighbourhood of the (real) interval on which $t_0(S)$ is prescribed, it may have singularities at complex points $S = S_1 \pm i S_2$ (such singularities must occur in conjugate pairs, by Schwarz's Reflexion principle, since $t_0(S)$ is real for real S). Singularities in ϕ can then be expected at $s = S_1$, $k - k_0 = \pm S_2$.

These limitations on the solution (104) do not prevent it having very considerable value in the wide range of cases when $t_0(S)$ can be analytically continued into substantial regions of the complex plane. A simple example, where the only singularities of $t_0(S)$ are on the real axis (corresponding to the greatest and least values of s produced by the wavemaker), is given in the next Section.

In the meantime, the relevant properties of the function $F(z)$, which may be needed by those using (104), are noted. In (104), z can take any value whose real part is negative. When $|z| < 1$ equation (65) for $F(z)$ may be used and, when $|z| > 1$, the equation

$$F(z) = \frac{1}{\pi}(-z)^{-\frac{1}{2}}\left[F\left(\frac{1}{z}\right)\log(-z) + 4\sum_{n=0}^{\infty}\frac{(\frac{1}{2})^2(\frac{3}{2})^2\ldots(n-\frac{1}{2})^2}{(n!)^2 z^n}\right.$$
$$\left.\left(\log 2 - 1 + \frac{1}{2} - \frac{1}{3} + \ldots + \frac{1}{2n}\right)\right]. \tag{105}$$

This region is also one where the series form

$$F(z) = \frac{1}{\pi}(1-z)^{-\frac{1}{2}}\left[F\left(\frac{1}{1-z}\right)\log(1-z) + 4\sum_{n=0}^{\infty}\frac{(\frac{1}{2})^2(\frac{3}{2})^2\ldots(n-\frac{1}{2})^2}{(n!)^2(1-z)^n}\right.$$
$$\left.\left(\log 2 - 1 + \frac{1}{2} - \frac{1}{3} + \ldots + \frac{1}{2n}\right)\right] \tag{106}$$

may be useful, since $|1-z|$ is commonly large enough to render the right-hand side rapidly convergent.

As a last general point on the use of (104), it may be noted that we need solutions with $(k - k_0)$ the same sign as

$$\left(\frac{\partial t}{\partial k}\right)_{k=k_0} = -\frac{2}{\mu}\frac{1}{s}\left(\frac{\partial^2 \phi}{\partial k \partial s}\right)_{k=k_0} = \frac{2}{\mu}\frac{u_0}{s}t_0'(s), \tag{107}$$

in order that the departure of k from k_0 shall correspond to increase in t. In the following Sections we shall consider the case $\mu > 0$, that is, $f''(k) < 0$, so that (for the equation to be elliptic) $\omega - f(k) > 0$. In this case, typified by gravity waves on deep water, $k - k_0$ must have the same sign as $t_0'(s)$. In a problem with $f''(k) > 0$ and $\omega - f(k) < 0$ (such as that of pure capillary waves), some changes of sign would be needed in Section 14.

13. Illustrative Example in the Elliptic Case (Qualitative Discussion)

A particular case is now worked out, with $t_0(S)$ smooth in the sense defined in Section 12. We suppose that the amplitude of the waves created at the wavemaker increases continuously from zero at $t = -\infty$ to a maximum at $t = 0$, and falls continuously back to zero again at $t = +\infty$, according to the law

$$s = \frac{s_1}{1 + (t/T)^2}. \tag{108}$$

Then

$$t_0(S) = \pm T\left(\frac{s_1 - S}{S}\right)^{\frac{1}{2}}. \tag{109}$$

The problems must be worked out with both signs adopted in (109). When the negative sign is adopted, the amplitude is increasing, $t_0'(S)$ is positive, and therefore positive values of $(k - k_0)$ must be chosen. When the positive sign is adopted, so that the amplitude is decreasing again to zero, $t_0'(S)$ is negative, and so negative values of $(k - k_0)$ must be chosen.

As mentioned in Section 12, this example is one where the only singularities of $t_0(S)$ (whichever sign is chosen) are for real S. If, in fact, we define the domain D as the half-plane $\mathscr{R}S > 0$ cut along the real axis from $S = s_1$ to $S = \infty$, then $t_0(S)$ possesses a unique analytic continuation into D, which, inserted into the integral (104), makes the integrand analytic in D. Provided that the path of integration in (104) is forced to remain within D (that is, not to cross the cut), then ϕ can be calculated, for all $s > 0$ and for all k. (When $s > s_1$, the path of integration simply has to be bent to avoid the cut.)

However, although ϕ can be calculated for all $s > 0$ and for all k, only a certain range of values of k and s has physical relevance. This is because, as in the last problem discussed in Section 9, there is a certain time t_{max}, and hence also a certain distance x_{max} from the wavemaker, beyond which the solution breaks down; accordingly, values of s and k for which (103) and (104) give $t > t_{max}$ can be disregarded. The type of breakdown is different, however, from that described in general terms by Whitham and illustrated in Section 9.

We shall see, in fact, that, as the group of waves propagates away from the wavemaker, the total energy of the group gradually becomes concentrated within a narrower distance. At the same time, the peak amplitude, that is, the amplitude at the centre of the group (whose amplitude variation remains symmetrical about this point), increases. (By contrast, it is easy to prove that no increase in peak amplitude occurs in this problem in the hyperbolic case.) This increase becomes more and more rapid, having reached 39% when, as t reaches t_{max}, the rate of increase of peak amplitude becomes infinite. The tendency for the wavenumber to rise above k_0 ahead of the centre of the group and to fall below k_0 behind the centre has at that moment become so enhanced as to produce an infinite gradient, $\partial k/\partial x$, of wavenumber with distance, at the centre.

Evidently the theory breaks down before $t = t_{max}$, since infinite values of $\partial s/\partial t$ and $\partial k/\partial x$ are in direct contradiction with the assumption that amplitude and wavenumber change by only small fractions of themselves in a single period or wavelength. Comments on the possible subsequent development of the wave motion, which can at present only be highly tentative, are made in Section 16.

Whatever that development may be, the theory for $t < t_{max}$ is making a very precise prediction; namely, that non-linear dispersion in a wave group of initially uniform frequency will cause the group to become concentrated, with the local energy density reaching almost twice its initial peak value (corresponding to a 39% increase in amplitude) in a time t_{max} whose value is calculated in Section 14. It will be interesting to test this precise prediction in the case of gravity waves on deep water, and, in the event of approximate agreement, to use experiment to probe into the question of what happens for $t > t_{max}$.

We have seen that only part of the solution for ϕ as a function of k and s, namely the part for which $t < t_{max}$, can be relied on as relevant to what happens to the waves. We shall see in Section 14 that the whole region $t < t_{max}$ is one where the argument of the hypergeometric function F in (104) remains small in absolute value, a fact which greatly facilitates the calculation.

The point $t = t_{max}$, $x = u_0 t$ (the centre of the wave group at the instant of breakdown) is an example of the phenomenon referred to in Section 6, namely, an isolated point where all the first derivatives of x and t vanish. For elliptic equations, this is the only possible case when a solution regular in the (k,s)-plane can have a singularity in the (x,t)-plane. By contrast, we shall see in Section 14 that no singularity in th (x,t)-plane results from the cut in the complex S-plane from $S = s_1$ to $S = \infty$.

14. Calculations of Changes at the Centre of the Group

With $t_0(S)$ as in (106), equation (104) becomes

$$\phi = \pm\tfrac{1}{2}u_0 Ti \int_{s-i(k-k_0)}^{s+i(k-k_0)} \left(\frac{s_1 - S}{s}\right)^{\tfrac{1}{2}} F\left(\frac{(k-k_0)^2 + (s-S)^2}{-4sS}\right) dS. \qquad (110)$$

If $s > s_1$, then when $(k-k_0)$ changes sign the limits of integration in (110) cross the cut, while also the sign at the front changes, so it is a matter for investigation whether the solution does or does not remain analytic. For $(k-k_0)$ very small and positive the lower sign is needed in (110) and the path of integration may be taken, from a starting point just below the cut, along the lower edge to $s = s_1$, and then along the upper edge to a point, just above the cut, opposite to the starting point. On the lower edge, $i(s_1 - S)^{\tfrac{1}{2}}$ is negative and on the upper edge it is positive, while dS is also respectively negative and positive in the two cases, so that (110), with the lower sign taken, becomes the negative expression

$$\phi = -u_0 T \int_{s_1}^{s} \left(\frac{S-s_1}{s}\right)^{\tfrac{1}{2}} F\left(\frac{(s-S)^2}{-4sS}\right) dS. \qquad (111)$$

For $(k-k_0)$ very small and negative, on the other hand, the integral (110) may be taken along the same path in just the opposite sense, while however, the upper sign must be adopted. These two effects just cancel out, and so ϕ takes the same negative value (111). Thus, ϕ is a continuous function of k at $k = k_0$ for $s > s_1$, but takes a negative value (111) instead of the zero value taken for $s < s_1$.

We see also that, when $s > s_1$, $\partial\phi/\partial k$ takes the value zero as k tends to k_0 from either above or below. For when (110) is differentiated with respect to k, the dependence of the limits of integration on k yields terms equal in magnitude but opposite in sign from opposite edges of the cut, while the dependence of the integrand on $(k-k_0)^2$ gives a term which clearly vanishes at $k = k_0$.

Since ϕ, a solution of (32), is continuous and $\partial\phi/\partial k = 0$ on $k = k_0$ for $s > s_1$, it must be an even analytic function of $(k-k_0)$. Thus ϕ, as a function of the *real* variables k and s, is an analytic function, even in $(k-k_0)$, in the half-plane $s > 0$ cut merely from $s = 0$ to $s = s_1$; across this cut, on which $\phi = 0$, $\partial\phi/\partial k$ changes from

$$u_0 T\left(\frac{s_1-s}{s}\right)^{\tfrac{1}{2}} (k > k_0) \quad \text{to} \quad -u_0 T\left(\frac{s_1-s}{s}\right)^{\tfrac{1}{2}} (k < k_0). \qquad (112)$$

From the facts already exposed we can determine the variation of peak amplitude with time. For the evenness of ϕ with respect to $(k-k_0)$ implies, by (103), that the variable $(x-\mu_0 t)$ is an odd function, vanishing when $(k-k_0)$ is zero and $s > s_1$, while t is an even function. It follows that, for given s, the time t is an even function of $(x-u_0 t)$, whence it follows that, for given t, the amplitude is also an even function of $(x-u_0 t)$.

The value of the amplitude s for $x-u_0 t = 0$ is thus a stationary value, which initially is a maximum, and we shall see that in this example it remains a maximum, whose value as a function of time is obtained quite easily be calculating t as a function of s for $k = k_0$, $s > s_1$, from equations (111) and (103), as

$$t = \frac{2u_0 T}{\mu s}\left(y - \frac{y^2}{3} - \frac{y^5}{30} - \frac{y^7}{210} + \frac{y^9}{2520} + \frac{y^{11}}{792} + \cdots\right), \tag{113}$$

where $y^2 = (s - s_1)/s$. (Only the first two terms in (113) are present if the hypergeometric function in (111) is replaced by 1; only the next two terms in the hypergeometric function are needed to determine the remaining four terms in (113).) This gives a variation of the amplitude s at the centre $x = u_0 t$ of the group with time t as in Fig. 7.

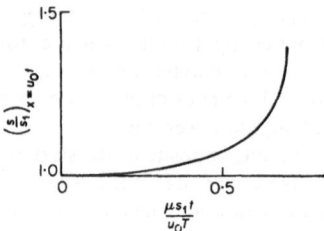

Fig. 7. Variation with time t in the value at the centre of the group of the measure of amplitude s.

The maximum value of (113), where $\partial t / \partial s = 0$, is

$$t_{max} = 0.689\frac{u_0 T}{\mu s_1}, \tag{114}$$

attained where $y^2 = 0.2807$, that is, where

$$s = 1.390 s_1. \tag{115}$$

At this time $t = t_{max}$ the rate of increase of amplitude at the centre of the wave, $x = u_0 t$, becomes infinite. The assumptions of the theory are therefore valid only for $t < t_{max}$. The amplitude has increased by 39%, as (115) shows, when t reaches t_{max}.

Convergence of the series (113) is good for $t \leqslant t_{max}$, since $y^2 < 0.2807$. The first two terms alone (obtained by replacing the hypergeometric function by 1) give quite close values for (114) and (115), the coefficients being replaced by 0.691 and 1.396, and no visible changes to the curve in Fig. 7.

The infinite rate of increase of amplitude with time at the centre when $t = t_{max}$ is accompanied by an infinite spatial gradient of wavenumber, as mentioned in Section 13. For, by (103) and (32), the value at the centre of $\partial(x - u_0 t)/\partial k$ is

$$\left(\frac{\partial^2 \phi}{\partial k^2} + \frac{2}{s}\frac{\partial \phi}{\partial s}\right)_{k=k_0} = \left(\frac{1}{s}\frac{\partial \phi}{\partial s} - \frac{\partial^2 \phi}{\partial s^2}\right)_{k=k_0}, \tag{116}$$

which is zero by (103) where $\partial t / \partial s = 0$. It follows also from this that all the first derivatives of x and t with respect to k and s vanish at $s = 1.390 s_1$, $k = k_0$, which is therefore an isolated singularity of the transformation from the (x,t)-plane to the (k,s)-plane.

15. Changes in the Distribution of Amplitude and Wavelength

We have already seen that, in the calculation of peak amplitude for $t < t_{max}$, good accuracy (within $\frac{1}{2}\%$) is obtained by replacing the hypergeometric function F by 1 in (111). We assume that this may be so also for the calculation of the distribution of amplitude and wavenumber for $t < t_{max}$ and may check *a posteriori* that the argument of F, so calculated, indeed remains small; everywhere, in fact, less than 0·07 in modulus. The approximation gives

$$\phi = \frac{2}{3}\frac{u_0 T}{s^{\frac{4}{3}}}\mathscr{S}[s_1 - s + i(k - k_0)]^{\frac{3}{4}}, \tag{117}$$

from which, using (103), t and $(x - u_0 t)$ may be obtained as functions of s and k. We can then derive the distribution of s and k as a function of $(x - u_0 t)$ for given t. This calculation is quite easy if

$$q = \mathscr{S}\left(\frac{s_1 - s}{s} + i\frac{k - k_0}{s}\right)^{\frac{3}{4}} \tag{118}$$

is used as a parameter, and the results are shown in Fig. 8 for $t = 0·7 t_{max}$ and for $t = t_{max}$, the original form of the wave group (for $t = 0$) being also shown for comparison.

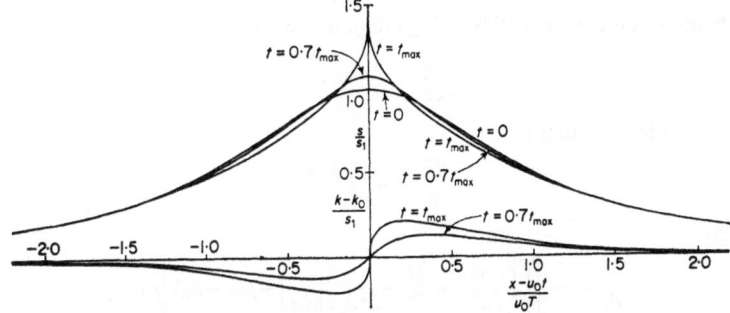

FIG. 8. Variation in the measures of amplitude and wavenumber, s/s_1 and $(k - k_0)/s_1$ respectively, with distance from the centre of the group $(x - u_0 t)$, at times $t = 0$, $0·7 t_{max}$ and t_{max}.

Figure 8 shows that when $t = t_{max}$ the peak of the wave group has become so enhanced that it is actually a cusp; at the same time the gradient of wavenumber k has become infinite. Obviously, the assumptions of the theory cease to be valid before these things happen; but not long before, as we see from the regularity of the curves for $t = 0·7 t_{max}$.

Indeed, the amplitude distribution has not yet changed by much when $t = 0·7 t_{max}$, although it subsequently changes very rapidly. This impression of an "incubation period" before peak amplitude starts to increase significantly is given also by Fig. 7.

The lower curves on Fig. 8 indicate, perhaps, why this is so. The wavenumber k has accomplished a substantially greater proportion of its total change by the time $t = 0·7 t_{max}$ than has the amplitude. The phenomenon can be envisaged, then, as beginning with an increase of wavenumber ahead of, and a decrease of wavenumber behind, the peak of the wave group, and continuing with a process of energy concentration towards the centre of the group, that can be interpreted physically in terms of the tendency for energy associated with greater wavenumbers to be propagated more slowly than that associated with smaller wavenumbers.

The simplest explanation why, as the group propagates, the wavenumber in front of the position of maximum amplitude must increase, is given by equation (45), in the form

$$\frac{\partial k}{\partial t} + f'(k)\frac{\partial k}{\partial x} = -\frac{\partial}{\partial x}[\omega - f(k)].$$

Here, the right-hand side must be positive in front of the position of maximum amplitude for any system where $\omega - f(k)$ increases with amplitude.

We may conclude this section with a direct proof that during the changes of form of the wave group shown in Fig. 8, the energy remains constant. To the approximation considered, the energy at time t is proportional to

$$E(t) = \int_{-\infty}^{\infty} \tfrac{1}{2}s^2 \, dx = -\int (x - u_0 t)s \, ds$$

$$= -\iint \left[\frac{\partial \phi}{\partial k} - \mu(k - k_0)t\right] ds. \tag{119}$$

We obtain the derivative $E'(t)$ by using (103), in the form

$$\frac{\partial \phi}{\partial s} = -\tfrac{1}{2}\mu s t, \tag{120}$$

to show that for constant s

$$\frac{dk}{dt} = -\frac{\tfrac{1}{2}\mu s}{\partial^2 \phi / \partial k \partial s}. \tag{121}$$

This gives

$$\frac{dE}{dt} = -\iint \left[\left(\frac{\partial^2 \phi}{\partial k^2} - \mu t\right)\left(-\frac{\tfrac{1}{2}\mu s}{\partial^2 \phi / \partial k \partial s}\right) - \mu(k - k_0)\right] s \, ds$$

$$= \int d[\tfrac{1}{2}\mu(k - k_0)s^2] = 0, \tag{122}$$

because, for constant t, by (32) and (120),

$$\left(\frac{\partial^2 \phi}{\partial k^2} - \mu t\right) ds = \left(-\frac{\partial^2 \phi}{\partial s^2} - \frac{1}{s}\frac{\partial \phi}{\partial s} - \mu t\right) ds$$

$$= -\left(\frac{\partial^2 \phi}{\partial s^2} + \tfrac{1}{2}\mu t\right) ds = \frac{\partial^2 \phi}{\partial k \partial s} \, dk. \tag{123}$$

This demonstration that the energy of the wave group remains constant is merely a check on the analysis, since the theory was founded on Hamilton's principle for a conservative system; it provides also a check on the calculations leading to Fig. 8, since $E(t)$ has the same value to within graphical accuracy for the three curves in this figure.

16. Discussion of Possible Later Development of the Wave Group

It is clearly interesting to speculate on what may happen to the wave group discussed in the last three sections after $t = t_{max}$, but a definite answer must await experiments, probably using gravity waves on deep water. In the meantime a very tentative discussion of the possibilities is included in this Section.

This discussion is based on the proposition that, since the failure of the theory's assumptions exhibited in Fig. 8 when $t = t_{max}$ is only local, serious errors in what the theory predicts may merely be local, that is, confined to a region (around the centre of the wave group) whose extent may increase with time, but not so as to include most of the group until a rather substantial further time interval has elapsed. In this case, for a range of values of $t > t_{max}$, the solution (110) would continue to have value when $| x - u_0 t |$ is not too small.

For this reason, it is interesting to consider the form of the solution (110) for $t > t_{max}$, even though one must have grave reservations about its value, at least when $| x - u_0 t |$ is small. Now, when the solution (110) appropriate to $k > k_0$ (that is, with the minus sign in front) is extended beyond the curve in the (k,s)-plane corresponding to $t = t_{max}$, we find that points (k,s) corresponding to combinations of all positive values of t with all positive values of $(x - u_0 t)$ lie in the right-hand half of a pear-shaped region drawn approximately in Fig. 9. On its curvilinear boundary BCA, $(x - u_0 t)$ is zero, while $t > t_{max}$; by contrast, we found that the points corresponding to $x - u_0 t = 0$ and $t < t_{max}$ were on the line EB.

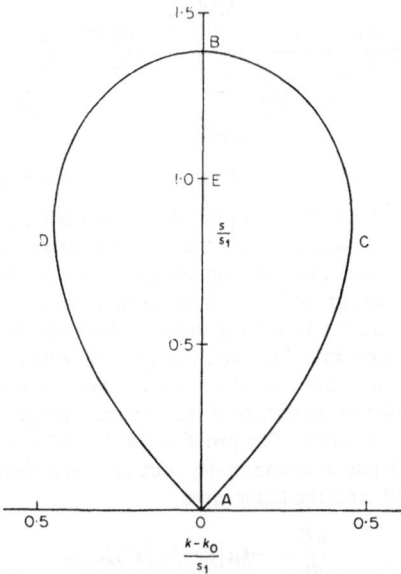

FIG. 9. Pear-shaped region in the (k, s)-plane, into which the whole (x, t)-plane with $t > 0$ is mapped.

Similarly, when the solution (110) appropriate to $k < k_0$ (that is, with the plus sign in front) is extended beyond $t = t_{max}$, we find that points (k,s) corresponding to combinations of all positive values of t with all negative values of $(x - u_0 t)$ lie in the left-hand half of the pear-shaped region. On its curvilinear boundary BDA, we again have $x - u_0 t = 0$ and $t > t_{max}$. Thus, the line $x - u_0 t = 0$, which for $t < t_{max}$ is EB, branches at B, where $t = t_{max}$, and thereafter is both BDA and BCA.

It follows that the solution for $t > t_{max}$ exhibits a discontinuity in k as a function of x at $x = u_0 t$, although s is continuous there. In noting this, however, we must bear in mind the fact that the solution is not expected to be valid for small $| x - u_0 t |$.

The solution, with the limitations that have been made clear, can be calculated approximately from (117), on the assumption that the argument of the hypergeometric function in (110) remains small. When we check *a posteriori* that the argument of F, so calculated, does remain small, the result is less conclusive than in Section 15 since the maximum value of its modulus, which there was 0·07, is in this region 0·25. However, the approximation should still be reasonably accurate, and greater complication is unjustified in the doubtful region $t > t_{max}$.

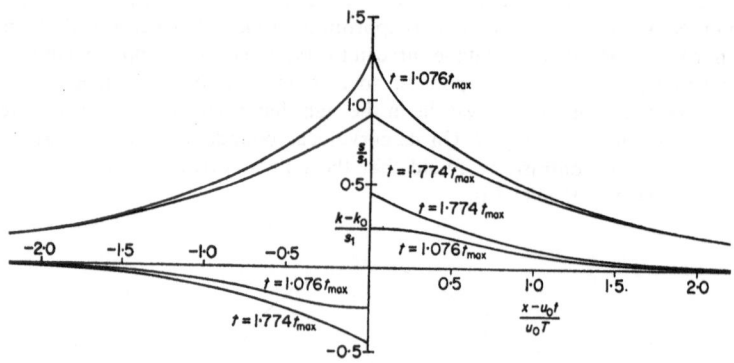

FIG. 10. Continuation of the solution beyond the formation of a discontinuity at $t = t_{max}$.

Solutions so calculated are shown in Fig. 10 for $t = 1·076t_{max}$ and for $t = 1·774t_{max}$. With increasing time the amplitude at the centre is reduced from its peak value $1·390s_1$ attained for $t = t_{max}$, and the cusp becomes a simple discontinuity in slope of the amplitude curve, accompanying the discontinuity in the wavenumber k. These are not features, however, to which too much attention should be paid, since it is the parts of the curves remoter from the centre that should be more reliable.

Some indication of the utility or otherwise of these solutions may be given by calculating the proportion of the energy of the original wave group which they account for. This proportion is $E(t)/E(0)$. The proof in Section 15 that $E(t)$ is constant does not hold good when k has a discontinuity, but all steps therein except the last in equation (122) are valid, and the latter gives

$$\frac{dE}{dt} = -(\mu | k - k_0 | s^2)_{x=u_0 t}. \tag{124}$$

When the values of $| k - k_0 |$ and s at $x = u_0 t$ have been calculated for all t, equation (124) enables $E(t)/E(0)$ to be computed.

The values of $| k - k_0 |/s_1$ and s/s_1 at $x = u_0 t$ are plotted in Fig. 11 as a function of $\mu s_1 t / u_0 T$, from the formulas

$$\frac{\mu s_1 t}{u_0 T} = \frac{1}{8} q^{-3} - \frac{1}{3} q, \quad \frac{| k - k_0 |}{s_1} = \frac{(\frac{1}{4} q^2 - 1 + \frac{1}{4} q^{-2})^{\frac{1}{2}}}{\frac{1}{8} q^{-3} - \frac{1}{3} q}, \quad \frac{s}{s_1} = \frac{\frac{1}{4} q^{-1}}{\frac{1}{8} q^{-3} - \frac{1}{3} q}, \tag{125}$$

using the parameter q defined by equation (118). The quantity $E(t)/E(0)$, thence inferred by (124), is plotted in the same figure. The steady decrease in amplitude at the centre is seen to be accompanied by a steady decrease in total energy, but the discontinuity in wavenumber (which is $2 \mid k - k_0 \mid$) increases to a maximum of $0.90 s_1$ at about $t = 1.92 t_{max}$, and thereafter decreases. Note that the formulas (125) were used also to obtain the shape of the pear-shaped region in Fig. 9, that is, of the curve $x = u_0 t$ in the (k,s)-plane.

Figure 11 shows that the solution plotted in Fig. 10 for $t = 1.076 t_{max}$ accounts for 98.0% of the energy of the original wave group. What has happened to the other 2%? As one extreme possibility one may suppose a special dissipation mechanism that would operate to get rid of energy in a region a few wavelengths long around the discontinuity at $x = u_0 t$, in such a way that the solution remains valid outside that region.

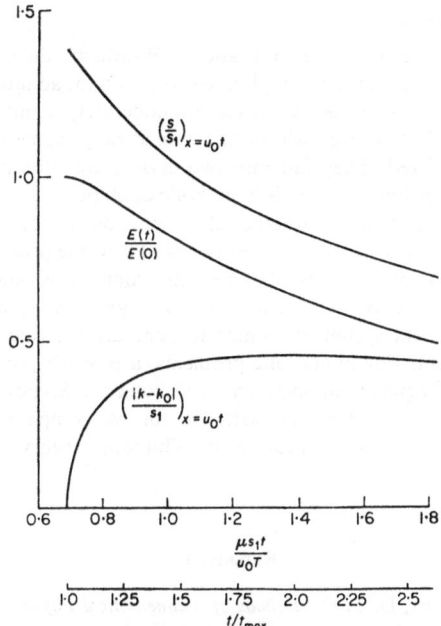

FIG. 11. Variation in central amplitude, wavenumber jump, and group energy, according to the solution for $t > t_{max}$.

Whitham (1965a) pointed out that discontinuous solutions were conceivable in his theory in which either the number of wave crests or the total mechanical energy was conserved, but not both. The former condition requires that the velocity of the discontinuity, here u_0, equals the change in frequency divided by the change in wavenumber, which here can also be shown to be u_0 because s is continuous and $(k - k_0)$ simply changes sign. The discontinuous solution just postulated is therefore one in which precisely enough energy gets removed to maintain conservation of numbers of crests.

An alternative possibility is that a larger region around $x = u_0 t$ differs in character from what is predicted in the theoretical solution. If the solution's deficiency in total energy, namely 2%, is a measure of the discrepancy, it is not great. But clearly, as another extreme possibility, the difference may be a large one over a substantial region and yet not much affect the total energy.

By contrast, the solution plotted in Fig. 10 for $t = 1 \cdot 774 t_{max}$ accounts for only 70·5% of the energy of the original wave group. Departures of the real wave motion from the solution must therefore be substantial, at any rate unless some special mechanism operates at the wavenumber discontinuity to remove energy at just about the required rate.

Only experiment can determine how the wave group really develops. For this purpose, experiments using gravity waves on deep water would appear very suitable.

17. Concluding Remarks

It appears from this survey of the application of Whitham's theory to the competition between frequency dispersion and amplitude dispersion in, admittedly, only a limited range of systems—one-dimensional waves of moderately small amplitude without "pseudo-frequencies"—that even within this limited range some strikingly interesting phenomena are predicted. They fall into two classes, quantitatively quite different, according as the dispersion equation is hyperbolic or elliptic.

The time has come to check whether these phenomena can be realized experimentally; both the more firmly based predictions of how the propagation will develop while the distribution of wave number remains continuous, and the various conjectures regarding its later development. For elliptic problems, gravity waves on deep water form a convenient system on which to compare the calculations of Sections 14–16 with experiment. For hyperbolic problems, a possible procedure would be to use waves of higher frequency on mercury, as discussed in Section 11, to compare the calculations of Section 9 with experiment. Favourable comparisons would act as a major incentive to analysis of implications of Whitham's theory in more complicated systems.

REFERENCES

COURANT, R. & HILBERT, D. 1962 *Methods of Mathematical Physics*, Vol. 2. Interscience, New York.
CRAPPER, G. D. 1957 *J. Fluid Mech.* **2**, 532.
GARABEDIAN, P. R. 1958 *J. Math. Phys.* **36**, 192.
HOWARTH, L. 1953 *Modern Developments in Fluid Dynamics: High Speed Flow*, Vol. 1. Oxford University Press, London.
LAMB, H. 1932 *Hydrodynamics*, 6th ed. Cambridge University Press.
LIGHTHILL, M. J. 1955 In *Naval Hydrodynamics*, Chapter 2. Publication 515 of National Academy of Sciences, National Research Council (Washington).
LIGHTHILL, M. J. 1965 *J. Inst. Maths Applics* **1**, 1.
WARD, G. N. 1955 *Linearised Theory of Steady High-Speed Flow*. Cambridge University Press.
WHITHAM, G. B. 1965a *Proc. R. Soc. A*, **283**, 238.
WHITHAM, G. B. 1965b *J. Fluid Mech.* **22**, 273.
WILTON, J. R. 1915 *Phil. Mag.* (6) **29**, 688.

Proceedings of the Royal Society of London, **A 302**, 529 (1969).

Wavetrains in Inhomogeneous Moving Media

F. P. BRETHERTON and C. I. R. GARRET

When a slowly varying wavetrain of small amplitude propagates in a general medium, changes of frequency and wavenumber are determined along definite paths known as rays. It is shown that, for a wide class of conservative systems in fluid dynamics changes in amplitude along the rays may be computed from conservation of wave action, which is defined as the wave energy divided by the intrinsic frequency. The intrinsic frequency is the frequency which would be measured by an observer moving with the local mean velocity of the medium. This result is the analogue for continuous systems of the adiabatic invariant for a classical simple harmonic oscillator.

If the medium is time dependent or moving with a nonuniform mean velocity the intrinsic frequency is not normally constant, and wave energy is not conserved. Special cases include surface waves on a vertically uniform current in water of finite depth, internal gravity waves in a shear flow at large Richardson number, Alfvèn waves, sound waves, and inertial waves in a homogeneous rotating liquid in geostrophic mean motion.

1. INTRODUCTION

A *wavetrain* is a system of almost sinusoidal propagating waves with a recognizable dominant local frequency ω, vector wavenumber $\mathbf{k}$ and amplitude a. These may vary with position $\mathbf{x}$ and time t, but only slowly, in the sense that appreciable changes are apparent only over many periods and wavelengths. They are defined with a precision which increases the more slowly they vary. The dominant frequency and wavenumber may be derived from a *phase function* $\theta(\mathbf{x}, t)$ by

$$\omega = -\theta_t, \quad k_j = \theta_{x_j} \quad (j = 1, ..., m) \tag{1.1}$$

and the wave crests are surfaces of constant θ. At each point ω, $\mathbf{k}$ are connected by a *dispersion relation*
$$\omega = \Omega(\mathbf{k}, \lambda), \tag{1.2}$$

where the local properties of the medium are for convenience summarized in the parameter $\lambda(\mathbf{x}, t)$ (which may have several components), and are also assumed to be slowly varying. We will confine our attention to linearized waves of small amplitude for which ω, $\mathbf{k}$ are real (i.e. the medium is stable and non-dissipative).

The *group velocity* $\mathbf{c}$ is defined by

$$c_j = \frac{\partial \Omega}{\partial k_j} \quad (j = 1, ..., m) \tag{1.3}$$

and differentiation moving with the group velocity by

$$\frac{\mathrm{d}}{\mathrm{d}t} = \frac{\partial}{\partial t} + c_j \frac{\partial}{\partial x_j}. \tag{1.4}$$

An observer always moving with the local value of $\mathbf{c}$ describes a path in space-time known as a *ray*.

In a uniform time-independent medium ($\lambda = $ constant)

$$\frac{d\omega}{dt} = 0, \quad \frac{dk_j}{dt} = 0, \tag{1·5}$$

i.e. the frequency and wavenumber are constant along a ray. Then from equations (1·2), (1·3) c is also constant, and the ray is straight.

In a nonuniform medium

$$\frac{d\omega}{dt} = \frac{\partial\Omega}{\partial\lambda}\frac{\partial\lambda}{\partial t}, \quad \frac{dk_j}{dt} = -\frac{\partial\Omega}{\partial\lambda}\frac{\partial\lambda}{\partial x_j}, \tag{1·6}$$

so that, if $\lambda(\mathbf{x}, t)$ is known, the frequency, wavenumber and group velocity vary in a predictable manner along a ray. The path of a ray through any given point is determined by integration of equations (1·6), together with

$$\frac{dx_j}{dt} = c_j = \frac{\partial\Omega}{\partial k_j} \tag{1·7}$$

and is in general curved. Equations (1·6), (1·7) are kinematic results, depending only on the existence of a phase function $\theta(\mathbf{x}, t)$ and a local dispersion relation (1·2). They are related to the theory of the Hamilton–Jacobi equation. Their derivation in a fluid dynamical context has been ably explained by Whitham (1960). A general survey of the concept of group velocity has been given by Lighthill (1965).

This paper is concerned with a general procedure for finding changes in the amplitude $a(\mathbf{x}, t)$ in a slowly varying moving or time dependent medium. Unlike the situation in a uniform medium, the maximum displacement of a material particle (for example) will in general vary in a different manner from the maximum velocity of the same particle, because the frequency will change. Either of these would be a suitable measure of wave amplitude, so general formulae directly in terms of a do not exist. However a convenient concept is that of *wave energy density* E. This is discussed in detail in § 3, but in any given problem it is normally straightforward to express it in terms of the local wavenumber, frequency and any convenient measure a of amplitude,

$$E = a^2 F(\omega, \mathbf{k}, \lambda). \tag{1·8}$$

The central result of this paper is that for a wide class of physical systems

$$\frac{d}{dt}\left(\frac{E}{\omega'}\right) + (\nabla \cdot \mathbf{c})\left(\frac{E}{\omega'}\right) = 0, \tag{1·9}$$

where ω' is the frequency relative to a frame of reference in which the mean state of the medium is locally in equilibrium at rest. This result was suggested but not proved by Garrett (1967). The frequency ω, which enters equations (1·2), (1·6) to determine the ray paths, is relative to a fixed observer. It is equal to the *intrinsic frequency* ω', which enters equation (1·9), plus an allowance for the Doppler shift. If the medium at the point under consideration is moving with velocity $\mathbf{U}$ relative to the observer,

$$\omega' = \omega - \mathbf{U} \cdot \mathbf{k}. \tag{1·10}$$

Equation (1·9) describes changes in wave energy density along a ray, in terms of ω' and the spatial divergence $\nabla \cdot \mathbf{c}$ of the rays. Changes in amplitude a then follow from equation (1·8). Ray divergence occurs even in a uniform time-independent medium, for which $\omega' = $ constant, because the group velocities at neighbouring points in a wavetrain are not in general exactly the same, owing to slight differences in $\mathbf{k}$. Because E is an energy *density*, it is not constant down a ray, even if wave energy is conserved. However, in a time dependent and/or nonuniformly moving medium, ω' varies along a ray. If E/ω' is the *wave action density*, total wave action is conserved, whereas total wave energy is not.

The meaning of this is perhaps made clearer by reference to the concept of a *wave packet*. This is a wavetrain of which the amplitude is negligibly small, except within a certain moving region of space V. The dimensions of V are small compared to the scale of variation of the properties of the medium, but large compared to a wavelength. The dominant wavenumber is effectively uniform over the packet, so that it moves as a whole with a well-defined group velocity. Owing to the divergence of the rays, the volume occupied by V may change. However, viewed from the scale of the medium, a wave packet appears as a point associated with a definite position $\mathbf{x}(t)$ and a definite wavenumber $\mathbf{k}(t)$ which moves along a path predetermined by the variation of λ with $\mathbf{x}$ and t, and by the values of $\mathbf{x}$, $\mathbf{k}$ at one time t_0. In a uniform medium the path of every packet is a straight line, and the total wave energy associated with it is constant. In a slowly varying medium the path is in general curved, and the total wave energy is proportional to the instantaneous intrinsic frequency. This result is the analogue for a continuous system of the classical adiabatic invariant for a single discrete oscillator which is subject to a slow change in its defining characteristics (e.g. a pendulum consisting of a bob on a string of varying length). In the latter case the total energy divided by the frequency is constant (Einstein 1911).

The theory here is an approximate one analogous to the W.K.B. approximation and the relation of geometrical optics to electrodynamics. Such theories never describe partial reflexions at abrupt changes in the medium. Their mathematical basis is an asymptotic expansion in powers of a small parameter ϵ, which is the ratio of a typical wavelength to the length scale of the changes. Formally, the fractional changes in ω, $\mathbf{k}$, a over a period and over a wavelength are all proportional to ϵ. The above equations are for the lowest order terms. Although in practice some partial reflexion always occurs, it is exponentially small (e.g. $\exp\{-\epsilon^{-1}\}$) and does not emerge even from higher order corrections.

To establish our result we draw freely on ideas due to Whitham (1965). In a fundamental paper he suggested that, if the equations governing a dynamical system can be derived from a variational principle of a certain type, changes of amplitude in a slowly varying wavetrain (linear or nonlinear) are governed by a conservation equation. The conserved quantity, called by Whitham the adiabatic invariant, is obtained from a local average over a period of the integrand of the governing variational principal. We are concerned here mainly with its physical interpretation for waves which are small perturbations about a mean state, particularly when the latter is in motion. In a section on linear systems, Whitham came close to a statement of equation (1·9) with ω' replaced by ω. However, for linearized, as opposed to linear, systems great care is necessary over the definition of E, otherwise fallacious results may be obtained. This is discussed in §3.

Two further difficulties must also be overcome before equation (1·9) may be established for general systems. Whitham gave no general method whereby a governing variational principle of the required type may be written down. In the example he discussed, long waves on water of finite depth, the principle was in terms of an Eulerian description of the instantaneous motion, and although it is correct, its genesis is not obvious. For our present purposes, Hamilton's principle may often be used, to obtain the required starting point. How this may be done for a wide class of fluid dynamical problems is discussed in §4.

A second difficulty is that Whitham (1965) gave no justification for his averaged variational principle, other than by showing that it gave the correct averaged equations in one special case. This gap has been partly filled by Luke (1966). A complete treatment for linear systems will be given in the following paper (Bretherton 1968). In many problems the fields describing the wave motion are approximately sinusoidal only in some (longitudinal) directions. In others (the lateral directions) they may have a complex normal mode structure. It is necessary that the averaged quantities in the variational principle be integrated over the lateral coordinates, including possibly contributions from the lateral boundary conditions. For example, in waves on water of a finite depth, height is a lateral coordinate, the wave energy is the total average energy per unit horizontal area, including the integrated kinetic energy (which is distributed over depth) and the net potential energy (which is intimately connected with the free surface).

Before discussing these general arguments we shall first illustrate the concepts and methods involved by a simple, rather trivial, example, and then in the remainder of the paper show how these may be applied to a wider class of problems.

2. The stretched string

2·1. Hamilton's principle

We consider an infinitely long string of mass $\rho(x)$ per unit length, under tension $T(t)$. If $\eta(x,t)$ is the transverse displacement of each point of the string from its equilibrium position, the linearized equation of motion is

$$\rho\eta_{tt} - T\eta_{xx} = 0. \tag{2·1}$$

It is important that the mean state $\eta = 0$ is a possible solution of the dynamical equations of motion, even when $\rho(x)$, $T(t)$ are not constant. This is so if the modulus of elasticity of the string (which controls the speed of propagation of longitudinal vibrations) is taken to be infinite.

Equation (2·1) may be derived from the variational principle

$$\delta \int \{\tfrac{1}{2}\rho\eta_t^2 - \tfrac{1}{2}T\eta_x^2\}\,dx\,dt = 0 \tag{2·2}$$

subject to all suitably differentiable infinitesimal variations $\delta\eta$ which vanish for sufficiently large $|x|$, $|t|$. Equation (2·2) is clearly Hamilton's principle for a system defined by a continuum $\eta(x)$ of generalized coordinates. $\int \tfrac{1}{2}\rho\eta_t^2\,dx$ and $\int \tfrac{1}{2}T\eta_x^2\,dx$ are respectively the kinetic and potential energy for small perturbations.

2·2. *Local solutions*

If T, ρ are constant, equation (2·1) has solutions periodic in a phase function $\theta(x,t)$

$$\eta = a\sin\theta \tag{2·3}$$

if

$$\omega = \pm\sqrt{(T/\rho)}\,k, \tag{2·4}$$

where

$$\omega = -\theta_t, \quad k = \theta_x. \tag{2·5}$$

If $T(t)$, $\rho(x)$ are slowly varying equations (2·3) to (2·5) are still valid locally, the problem is to connect values of a, ω, k at widely different points (x,t).

2·3. *The averaged Lagrangian*

We define an averaged Lagrangian density $\mathscr{L}$ by substituting the elementary solution (2·3) into the Lagrangian density in equation (2·2), remembering that

$$\frac{\partial}{\partial t} = -\omega\frac{d}{d\theta} \quad \text{and} \quad \frac{\partial}{\partial x} = k\frac{d}{d\theta}$$

and integrating with respect to θ over a period

$$\mathscr{L} = \frac{1}{2\pi}\int_0^{2\pi}(\tfrac{1}{2}\rho\omega^2\eta_\theta^2 - \tfrac{1}{2}Tk^2\eta_\theta^2)\,d\theta$$

$$= \tfrac{1}{4}(\rho\omega^2 - Tk^2)a^2. \tag{2·6}$$

This is a function of parameters a, ω, k and also (via ρ, T) of x, t. Whitham (1965) suggested that for a slowly varying wavetrain, the dominant local amplitude, frequency, and wavenumber are governed by the variational principle

$$\delta\int\mathscr{L}(a, -\theta_t, \theta_x)\,dx\,dt = 0 \tag{2·7}$$

subject to infinitesimal variations $\delta a(x,t)$, $\delta\theta(x,t)$ which vanish at infinity. Variation with respect to a yields

$$\partial\mathscr{L}/\partial a = \tfrac{1}{2}a(\rho\omega^2 - Tk^2) = 0. \tag{2·8}$$

This is equivalent to the dispersion relation (2·4). Variation with respect to θ yields

$$\frac{\partial}{\partial t}\left(\frac{\partial\mathscr{L}}{\partial\omega}\right) - \frac{\partial}{\partial x}\left(\frac{\partial\mathscr{L}}{\partial k}\right) = 0. \tag{2·9}$$

This is a conservation equation for the quantity $\partial\mathscr{L}/\partial\omega$, subject to the flux $-\partial\mathscr{L}/\partial k$.

For linear waves $\mathscr{L}$ is proportional to a^2, so that the dispersion relation (2·8) is equivalent to

$$\mathscr{L} = 0. \tag{2·10}$$

The group velocity is then

$$c = -\mathscr{L}_k/\mathscr{L}_\omega, \tag{2·11}$$

so that equation (2·9) becomes

$$\frac{d}{dt}(\mathscr{L}_\omega) + \frac{\partial c}{\partial x}\mathscr{L}_\omega = 0. \tag{2·12}$$

Whitham justified his simplified variational principle only by showing for one particular case that it was equivalent to simplified differential equations which could be obtained by averaging in a certain way (also heuristic) the exact differential equations obtained from the exact variational principle. His main interest was in nonlinear waves of finite amplitude. However, Luke (1966) has shown how wave-train solutions of the second order nonlinear Klein–Gordon equation

$$\eta_{tt} - \eta_{xx} + F(\eta) = 0$$

may be obtained as an asymptotic expansion in powers of a small parameter ϵ, the lowest order term being governed by the appropriate forms of equations (2·8), (2·9). He also partially justified this procedure for a general nonlinear second order partial differential equation derived from a variational principle, including equation (2·1) as a special case. It will be shown in the following paper (Bretherton 1968) how Luke's results may be generalized for a much wider class of systems of equations, and for linear systems a complete asymptotic expansion may be constructed.

2·4. Wave energy

We arrive at the concept of wave energy by considering the response of a string, which is initially in equilibrium, to arbitrary small external forces $f(x, t)$ per unit length applied transversely to it. The governing equation is

$$\rho \eta_{tt} - T \eta_{xx} = f. \tag{2·13}$$

To obtain the rate at which work is done by these forces, we multiply by the velocity η_t of the particles to which they are applied, and integrate. After a little manipulation

$$W = \int_{x_1}^{x_2} f \eta_t \, dx = \frac{\partial}{\partial t} \int_{x_1}^{x_2} \{ \tfrac{1}{2}\rho \eta_t^2 + \tfrac{1}{2}T \eta_x^2 \} \, dx + [-T\eta_t \eta_x]_{x_1}^{x_2} - \int_{x_1}^{x_2} \frac{1}{2} \frac{\partial T}{\partial t} \eta_x^2 \, dx. \tag{2·14}$$

If T is constant, this is a conservation equation, changes in

$$\int_{x_1}^{x_2} \{ \tfrac{1}{2}\rho \eta_t^2 + \tfrac{1}{2}T \eta_x^2 \} \, dx$$

being associated only with work done by external forces or with a flux $-T\eta_t \eta_x$ across the points $x = x_1, x_2$. Thus we unhesitatingly identify $\tfrac{1}{2}\rho \eta_t^2 + \tfrac{1}{2}T \eta_x^2$ with a perturbation energy per unit length, for any perturbation $\eta(x, t)$ however produced. If η is sinusoidal, according to equation (2·3), the average of the perturbation energy over a period is the wave energy density

$$E = \tfrac{1}{4}a^2(\rho\omega^2 + Tk^2) \tag{2·15}$$

and the wave energy flux is $\quad\quad \tfrac{1}{2}Ta^2\omega k. \tag{2·16}$

It is easy to see that the latter is equal to the group velocity $c(=\pm \sqrt{(T/\rho)})$ times the wave energy density.

If $T(t)$ is not constant, however, the third term on the right-hand side of equation (2·14) is nonzero, and no amount of manipulation will turn the equation into a conservative form. $\tfrac{1}{2}(\partial T/\partial t) \eta_x^2$ thus describes the interaction between the waves and the mean state. If $T'(t)$ is slowly varying, it is still meaningful to use the local perturbation energy given by equation (2·15). It is still approximately conserved for times of a few wave periods, but over longer times the interaction terms cannot be ignored.

2·5. *The interpretation of* $\mathscr{L}_\omega$

Equation (2·12) does describe conservation of something, but it is not wave energy. To see this, we set up equation (2·14) again, but in terms of the Lagrangian density L.

The string may be regarded as a classical dynamical system; the position of every material particle in it being specified by the instantaneous values of a set of generalized coordinates $\{q_i\}$. In this case the value of η at each point is a coordinate q_i, and the set $\{q_i\} = \eta(x)$ forms a continuum. The set of possible configurations is limited by the requirement that $\eta(x)$ be differentiable. The present problem is typical of many in continuum mechanics in that the kinetic and potential energies of the complete string may be written as integrals of explicit expressions in η and its first derivatives only. Their difference is the Lagrangian

$$L_s(q_i, \dot{q}_i, t) = \int L(\eta, \eta_t, \eta_x; \lambda(x, t))\, \mathrm{d}x \qquad (2\cdot17)$$

$$= \int \{\tfrac{1}{2}\rho(x)\,\eta_t^2 - \tfrac{1}{2}T(t)\,\eta_x^2\}\, \mathrm{d}x. \qquad (2\cdot18)$$

If external forces $\{Q_i\} = f(x)$ (not included in the potential energy above) are applied to the system, their effect may be computed in the manner customary in the derivation of Lagrange's equations by considering an arbitrary infinitesimal virtual displacement $\{\delta q_i\} = \delta\eta(x)$ in which $\eta_t(x)$ is not varied and $\delta\eta$ vanishes for $|x|$ sufficiently large. The virtual work δW done by the external forces is then

$$\delta W = \sum_i Q_i\,\delta q_i$$

$$= \sum_i \left\{\frac{\mathrm{d}}{\mathrm{d}t}\left(\frac{\partial L_s}{\partial \dot{q}_i}\right) - \frac{\partial L_s}{\partial q_i}\right\}\delta q_i$$

$$= \int \left\{\frac{\partial}{\partial t}\left(\frac{\partial L}{\partial \eta_t}\right) + \frac{\partial}{\partial x}\left(\frac{\partial L}{\partial \eta_x}\right) - \frac{\partial L}{\partial \eta}\right\}\delta\eta\, \mathrm{d}x. \qquad (2\cdot19)$$

In this identification $\partial L_s/\partial \dot{q}_i$ is, of course, the partial derivative in which all the remaining $\dot{q}$'s and all the q's and t are held constant. For each i, $(\partial L_s/\partial \dot{q}_i)\,\delta \dot{q}_i$ is equivalent to the change in L_s due to a change $\delta\eta_t$ which is confined to a small unit neighbourhood of x; i.e. to $(\partial L/\partial \eta_t)\,\delta\eta_t$ $(\eta, \eta_x, \lambda$ being held constant). Summation over i is replaced by integration with respect to x. The partial derivative $\partial/\partial t$ in $(\partial/\partial t)\,(\partial L/\partial \eta_t)$ implies that all the arguments η, η_t, η_x, λ of $\partial L/\partial \eta_t$ are regarded as functions of x and t, only x is held constant. $\sum_i \dfrac{\partial L_s}{\partial q_i}\delta q_i$, on the other hand, is the change in L_s due to $\delta\eta(x)$ holding η_t and λ constant. This is

$$\delta L_s = \int \left\{ \frac{\partial L}{\partial \eta} \delta\eta + \frac{\partial L}{\partial \eta_x} \delta\eta_x \right\} dx$$

$$= \int \left\{ \frac{\partial L}{\partial \eta} - \frac{\partial}{\partial x} \left(\frac{\partial L}{\partial \eta_x} \right) \right\} \delta\eta \, dx,$$

where we have used the vanishing of $\delta\eta(x)$ at infinity.

It follows from equation (2·19) that the rate of working by any external forces on the string when it is moving transversely in any manner is

$$W = \int \left\{ \frac{\partial}{\partial t} \left(\frac{\partial L}{\partial \eta_t} \right) + \frac{\partial}{\partial x} \left(\frac{\partial L}{\partial \eta_x} \right) - \frac{\partial L}{\partial \eta} \right\} \eta_t \, dx. \tag{2·20}$$

By simple manipulations, this becomes

$$W = \int \left\{ \frac{\partial}{\partial t} \left(\eta_t \frac{\partial L}{\partial \eta_t} - L \right) + \frac{\partial}{\partial x} \left(\eta_t \frac{\partial L}{\partial \eta_x} \right) + \frac{\partial L}{\partial \lambda} \lambda_t \right\} dx. \tag{2·21}$$

This corresponds term by term to equation (2·14). In particular, in the absence of external forces and when $(\partial L/\partial \lambda)\lambda_t = 0$ it reduces to a conservation equation. Hence $\eta_t(\partial L/\partial \eta_t) - L$ is the perturbation energy per unit length of the string.

The purpose of this discussion is to emphasize the stages in the line of argument leading to the identification of $\eta_t(\partial L/\partial \eta_t) - L$. The basic postulate is equation (2·19), i.e. that, if

$$\frac{\delta L}{\delta \eta} = \frac{\partial L}{\partial \eta} - \frac{\partial}{\partial t} \left(\frac{\partial L}{\partial \eta_t} \right) - \frac{\partial}{\partial x} \left(\frac{\partial L}{\partial \eta_x} \right)$$

is the functional derivative of L with respect to η, the virtual work done by the external forces on the system under any admissible variation $\delta\eta$ is given by

$$\delta W = - \int \frac{\delta L}{\delta \eta} \delta\eta \, dx. \tag{2·22}$$

This is the property of Hamilton's principle which distinguishes it from other variational principles governing other nonphysical systems. It enables a specific interpretation to be given to functions derived from L. This interpretation does not depend on L being obtained as the difference between the kinetic energy density and the potential energy density. In systems including, for example, electromagnetic fields, the distinction between kinetic and potential energy is obscure. Equation (2·22) is basic to the derivation of an expression for the total energy density of the system. This cannot be defined (as in Whitham (1965)) simply as the quantity which is conserved when no external forces are acting and the Lagrangian does not depend explicitly on time, although such conservation is fundamental to the significance of the energy. With such a definition the Lagrangian could be multiplied everywhere by an arbitrary constant scalar without affecting the analysis at all, but the numerical values of the derived quantities would be altered. Furthermore, even when the right-hand side of equation (2·22) cannot be equated to the virtual work (as in systems linearized about a moving mean state, see §3) a 'pseudo energy' may still formally be defined by $\eta_t L\eta_t - L$ and under certain circumstances it is conserved, but its interpretation must be examined very carefully.

When the perturbation has the sinusoidal form of equation (3·3), the Lagrangian density may be regarded as a function of the variable θ, with parameters ω, k, a, λ;

$$L = L(\eta, -\omega\eta_\theta, k\eta_\theta; \lambda), \quad \text{where} \quad \eta = a\sin\theta. \tag{2·23}$$

The frequency ω enters only via $-\omega\eta_\theta$ in lieu of the variable η_t so the average perturbation energy over a period is

$$E = \frac{1}{2\pi}\int_0^{2\pi}\left\{(-\omega\eta_\theta)\frac{\partial L}{\partial(-\omega\eta_\theta)} - L\right\}d\theta$$

$$= \frac{1}{2\pi}\int_0^{2\pi}\left(\omega\frac{\partial L}{\partial\omega} - L\right)d\theta \tag{2·24}$$

$$= \omega\mathscr{L}_\omega - \mathscr{L}. \tag{2·25}$$

But since for waves of small amplitude $\mathscr{L} = 0$, (equation (2·10)), we have finally

$$\mathscr{L}_\omega = E/\omega. \tag{2·26}$$

Because the string is basically at rest, the observed frequency ω and the intrinsic frequency ω' are in this case equal.

3. Perturbation energy and wave energy

3·1. General dynamical systems

The dynamical system of §2 was described by a single field $\eta(x,t)$, the transverse displacement of a material particle in the string. For more general systems more fields are required, and they will be functions of three space variables $\mathbf{x} = (x,y,z)$ and time t. Let them be the set

$$\{\Phi_\alpha + \phi_\alpha\} \quad (\alpha = 1, ..., N),$$

where $\Phi_\alpha(\mathbf{x},t)$ describes the basic state, and $\phi_\alpha(\mathbf{x},t)$ is a perturbation. The following discussion was conceived in the context of a class of problems in classical inviscid fluid dynamics, illustrated by the examples of §6, for which the required fields are the components of the displacement $\Xi + \boldsymbol{\xi}$ of fluid particles, together with the pressure $p + \pi$ which enters as a Lagrange multiplier associated with incompressibility. The treatment envisages a Lagrangian rather than an Eulerian specification of the system, i.e. the position of every material particle is known when the values of the fields are given. It may be extended to include the effects of a 'frozen in' magnetic field as in nondissipative magnetohydrodynamics. It will be assumed that Hamilton's principle can be written down exactly, in terms of the total fields and their first derivatives, and that from this an approximate version may be obtained. In the latter the basic state $\Phi_\alpha(x,t)$ is regarded as known and for convenience is formally subsumed into the composite parameter λ, but by considering variations $\delta\phi_\alpha(\mathbf{x},t)$ linearized equations for $\phi_\alpha(\mathbf{x},t)$ may be derived. Before variation the approximate Lagrangian density $L(\phi_\alpha, \phi_{\alpha x}, \phi_{\alpha t}; \lambda)$ is homogeneous and quadratic in the ϕ_α's. Techniques for obtaining such an approximation to Hamilton's principle are described in §4. The ideas of §3 may be applicable in a wider context (e.g. relativistic situations) but some statements will require modification.

The objective of the remainder of §3 is to define closely the concept of wave energy. For such a definition to be satisfactory, it must be widely applicable, and yield demonstrably unique results when applied to different formulations of the same problem. For example, an arbitrary divergence can be added to the Lagrangian density L in §2. This must not affect the numerical value of E. Or in general fluid dynamical systems the linearized equations may be derived from an Eulerian, rather than a Lagrangian specification of the motion. The wave energy must be calculable and the same in either case. Finally, it must not involve consideration of second order corrections to linearized theory. These requirements are stringent, but in certain circumstances the authors believe they can be met, in a way which in fact corresponds to normal practice.

3·2. *In a medium in equilibrium at rest*

If the state of a system can meaningfully be described as a perturbation about some basic state, it is possible to define the perturbation energy as the difference between the total energy of the system and that of the basic state. This definition is only useful, however, if the basic state is prescribed with adequate precision. As the perturbation energy is a quadratic function of the amplitude a, an adequate description of the relation between the perturbed state and the basic state normally requires consideration of equations correct to second order in a.

There is one set of circumstances in which this is not necessary, and an adequate computation may be made from first order (linearized) theory alone. This is when the basic state is in approximate equilibrium at rest; i.e. the rates of change $\Phi_{\alpha t}$ (including particle velocities), and also the external generalized forces necessary to maintain them, are at most of order a^2. If now external forces f of order a are applied to particles in the system, the particles will move with velocity $\dot{\xi}$ also of order a. The work done in setting up a given perturbation is correctly given (to order a^2) as the integral of $f.\dot{\xi}$ over space and time.

We shall restrict our attention to perturbations which can be excited from equilibrium by mechanical body or surface forces only (i.e. by generalized forces corresponding to displacements $\delta\xi$). The generalized forces associated with other fields are assumed to vanish. For fluid dynamical problems this is not an unreasonable limitation (for an incompressible liquid the generalized force corresponding to the pressure perturbation $\delta\pi$ is the dilatation) and the set of permissible perturbations is very wide, if not completely general. For a frozen-in magnetic field, however, the flux linked with any material circuit is restricted to be equal to the equilibrium value.

Then, just as in §2·5, the rate of working is

$$W = \int f . \dot{\xi} dx$$

$$= -\int \frac{\delta L}{\delta \phi_\alpha} \phi_{\alpha t} dx, \tag{3·1}$$

then

$$W = \iint \left\{ \frac{\partial}{\partial t} \left(\phi_{\alpha t} \frac{\partial L}{\partial \phi_{\alpha t}} - L \right) + \frac{\partial}{\partial x_j} \left(\phi_{\alpha t} \frac{\partial L}{\partial \phi_{\alpha x_j}} \right) - L_\lambda \lambda_t \right\} dx \, dt. \tag{3·2}$$

In a uniform medium in equilibrium at rest

$$L_\lambda \lambda_t = 0. \tag{3.3}$$

The quantity

$$\phi_{\alpha t}(\partial L / \partial \phi_{\alpha t}) - L, \tag{3.4}$$

where summation over α is understood, may be interpreted as a perturbation energy per unit volume. For a strictly sinusoidal wavetrain in a normal mode

$$\phi_\alpha = \mathcal{R}\{a\hat{\phi}_\alpha \, e^{i\theta}\}, \tag{3.5}$$

where $\hat{\phi}_\alpha(\mathbf{k}; \lambda)$ are constants which depend on the mode under consideration. Then the Lagrangian density L may be expressed as a function of the single variable θ and of the parameters ω, $\mathbf{k}$, λ. Derivatives $\partial/\partial t$ are replaced by $-\omega \, \partial/\partial\theta$ and $\partial/\partial x_j$ by $k_j \, \partial/\partial\theta$. We *define* the wave energy density as the average of (3.4) over a period,

$$E = \frac{1}{2\pi} \int_0^{2\pi} \left\{ -\omega\phi_{\alpha\theta} \frac{\partial L}{\partial(-\omega\phi_{\alpha\theta})} - L \right\} d\theta \tag{3.6}$$

$$= \omega \mathcal{L}_\omega - \mathcal{L}. \tag{3.7}$$

This clearly has the form

$$E = a^2 F(\omega, \mathbf{k}; \lambda). \tag{3.8}$$

For a slowly varying wavetrain, the parameters a, $\mathbf{k}$, ω all vary slightly (by order ϵ) over a period, but if ϵ is sufficiently small the averaging procedure (3.6) may be applied with arbitrary accuracy (Bretherton 1968), and equation (3.8) applied locally at each point in the wavetrain.

Several points about this argument should be noted. The perturbation energy density (3.4) is not a well defined quantity, because the distribution between terms in equation (3.2) is not unique. If $\partial(\frac{1}{2}\eta^2)/\partial x$ is added to the Lagrangian density L the equations derived from Hamilton's principle are unaffected, but $\eta\eta_x$ must be subtracted from the perturbation energy density, and $\eta\eta_t$ added to the energy flux.

However, the wave energy obtained by averaging over a period is unique, provided that $L_\lambda \lambda_t = 0$. For then perturbation energy is conserved, and if

$$E = a^2 F(\omega, \mathbf{k}; \lambda); \quad E' = a^2 F'(\omega, \mathbf{k}; \lambda) \tag{3.9}$$

are two different expressions for the perturbation energy density of the same system, but obtained by different methods (e.g. with a different Lagrangian L or with an Eulerian representation), we may consider the work W which must be done by external forces to set up from rest a wave packet of slowly varying amplitude $a(\mathbf{x}, t)$, but with effectively constant ω, $\mathbf{k}$, λ over the volume for which $a \neq 0$. This work is the same for either method of computation, for it depends only on products of first order quantities, and cannot involve second order corrections. Then

$$W = F(\omega, \mathbf{k}, \lambda) \int a^2 \, d\mathbf{x}$$

$$= F'(\omega, \mathbf{k}, \lambda) \int a^2 \, d\mathbf{x},$$

where the integration is over the complete volume V occupied by the packet. Hence

$$F = F'. \tag{3.10}$$

Conservation of perturbation energy is essential to this argument. If interaction terms are admitted in equation (3·2), then, for example, $\partial(\frac{1}{2}\eta^2 t)/\partial t$ may be added to L, implying subtraction of $\frac{1}{2}\eta^2$ from the perturbation energy, and the addition of $\eta\eta_t$ to the terms $L_\lambda \lambda_t$. The average value of $\frac{1}{2}\eta^2$ does not vanish, so the value of E would be affected. However, if the basic state is in equilibrium at rest, it is possible to group terms in equation (3·2) in such a way that the interaction terms are identically zero. If the computation for W is reworked in terms of different fields, presumably a similar conservation equation may also be found, and an expression given for the perturbation energy. Having found it, the wave energy for an approximately sinusoidal wavetrain may be computed in the form (3·9) by averaging, and equation (3·10) shows that the result will be independent of the actual linearized fields used. This remark is very important in applications of equation (1·9), because the wave energy is frequently most easily computed from an Eulerian specification of the motion (cf. §4·2), whereas equation (3·7), which is vital for the proof of equation (1·9), can only be derived by the methods used here in terms of a Lagrangian specification.

The passage from equation (3·6) to (3·7), which is a generalization of equation (2·24), appears to rely on the structure constants $\hat{\phi}_\alpha$ in equation (3·5) being independent of ω, so that the only dependence of L on ω in equation (3·6) is multiplied by $\phi_{\alpha\theta}$ in precisely those places where η_t occurs in (3·4). In each normal mode, the dispersion relation $\omega = \Omega(\mathbf{k}; \lambda)$ may always be used formally to eliminate any such dependence $\hat{\phi}_\alpha(\omega)$ and, for given $\mathbf{k}$, equation (3·5) can be treated as a *definition* of $\phi_\alpha(\theta; \mathbf{k}, \omega, a; \lambda)$ for values of $\omega \neq \Omega$. This device enables $\mathscr{L}$ to be defined even in the physically uninteresting situations when the wavenumber and frequencies are assigned values incompatible with the existence of a solution to the governing equations. Such definition is necessary before meaning can be attached to the partial derivatives $\mathscr{L}_\omega$, $\mathscr{L}_{k_i}$ although of course these are only used when $\omega = \Omega$. This device is simple but arbitrary. Whitham (1965) allowed the wave period to be flexible in order to achieve the same object. The difference in approach is immaterial, because in fact the values of $\mathscr{L}_\omega$ and $\mathscr{L}_{k_i}$ when $\omega = \Omega(\mathbf{k})$ are quite insensitive to the definition of $\phi_\alpha(\theta, \mathbf{k}, \omega)$ off the surface $\omega = \Omega$, provided only that it is differentiable and correct on the surface (Bretherton 1968; Lighthill 1965). The averaged variational principle may be justified in either case.

If, as in §§4·1, 4·2, the fields $\{\phi_\alpha(x, y, z, t)\}$ are sinusoidal only in (x, y, t) but depend also on a lateral coordinate z, the averaged Lagrangian $\mathscr{L}$ is obtained by integrating over z as well as over a wavelength, using the appropriate lateral structure for a normal mode, and including contributions from the lateral boundary conditions. The wave energy density E is then the average perturbation energy per unit horizontal area, integrated with respect to depth, but the formula (3·7) still applies.

The wave energy flux may similarly be defined as the average of $\phi_{\alpha t}\,\partial L/\partial \phi_{\alpha x_j}$ and is given by the expressions

$$G_j = a^2 H_j(\omega, \mathbf{k}, \lambda)$$

$$= -\omega\mathscr{L}_{k_j}. \tag{3·11}$$

The uniqueness of $G_j(\omega, \mathbf{k}; \lambda)$ follows from consideration of a wavetrain which occupies a volume V which is small compared to the scale of variation of ω, $\mathbf{k}$, λ but large compared to a wavelength. For this wavetrain, unlike a wave packet, the amplitude a does not vanish on the boundary S of V. The rate of working by external forces, and the rate of change of wave energy within V are uniquely defined, so from equations (3·2), (3·3),

$$\int_S a^2(\mathbf{x}, t)\,(H_j - H_j')\,\mathrm{d}S_j = 0 \tag{3·12}$$

$H_j - H_j'$ may be taken outside the integral sign, and $\partial a^2/\partial x_j$ is arbitrary, so

$$H_j(\omega, \mathbf{k}; \lambda) - H_j'(\omega, \mathbf{k}; \lambda) = 0. \tag{3·13}$$

For this uniqueness it is essential that the energy flux be in the form (3·11). For Rossby waves, Longuet-Higgins (1964) found that the average $\overline{\pi \dot{\boldsymbol{\xi}}}$ of the perturbation energy flux for a strictly sinusoidal wavetrain is not equal to the group velocity times the wave energy density. Indeed it is in a different direction. However, for a slowly varying wavetrain an additional term of comparable magnitude appears in the expression for $\overline{\pi \dot{\boldsymbol{\xi}}}$ proportional to the gradient of a^2. The difference between this revised expression and $\mathbf{c}E$ is a non-divergent vector, so there are two genuinely alternative forms for the energy flux. However, only the latter does not involve derivatives of a.

Finally, it should be remarked that computations of 'wave energy' as the average value of the term proportional to a^2 in an expansion of the total energy of the system in powers of a, naïvely ignoring second order corrections to linearized theory, can yield different answers if applied in slightly different ways. For example, given a certain field with basic value Φ,

$$\tfrac{1}{2}\overline{(\Phi + \phi)^2} = \tfrac{1}{2}\Phi^2 + \Phi\bar{\phi} + \tfrac{1}{2}\overline{\phi^2}.$$

If the perturbation ϕ is taken to be sinusoidal, the contribution to the 'wave energy' would be

$$E = \tfrac{1}{2}\overline{\phi^2}. \tag{3·14}$$

But suppose
$$\Phi + \phi = (\Psi + \psi)^2$$
so that
$$\phi = 2\Psi\psi$$
to first order in a, then

$$\tfrac{1}{2}\overline{(\Psi + \psi)^4} = \tfrac{1}{2}\Psi^2 + 2\Psi^3\bar{\psi} + 3\Psi^2\overline{\psi^2} + O(a^3).$$

If now ψ is assumed to be sinusoidal, the 'wave energy' is

$$E' = 3\Psi^2\overline{\psi^2} = \tfrac{3}{2}E. \tag{3·15}$$

In practice, there would often be no reason other than prejudice for preferring to develop the linearized theory in terms of ϕ rather than ψ. The quantities ϕ^2 and $(2\Psi\psi)^2$ which are obtained from products of first order quantities are, of course, strictly comparable, at least to $O(a^2)$. The paradox depends on the $O(a^2)$ difference between the assumed mean states in the two cases.

The definition (3·6) of wave energy is restricted to a basic state in approximate equilibrium at rest. We are interested in basic states which are moving.

3·3. *Difficulties in a general moving basic state*

In §4, we will show how to obtain an approximate form of Hamilton's principle which describes linearized perturbations about any changing basic state (uniform or nonuniform) which is an exact solution of the equations of motion (i.e. $\phi_\alpha = 0$ is a dynamically consistent state). However it is not necessary that this standard solution be exact; an error of order a^2 in it would not affect the total Lagrangian in equation (4·16) to order a^2, because such an error may be regarded as a variation about an exact solution. The mean state of a system with waves on it is not normally known more accurately than this. Thus correctly linearized equations of motion may be derived from equation (4·16), even though the basic mean state is not precisely specified. Difficulties arise, however, with the energy.

These difficulties may be illustrated by the example of §2. If the mean state of the stretched string could only be maintained by application of an external transverse force F per unit length, or if in the mean state the particles have a transverse velocity V, the total rate of working by the external forces is not given by equation (2·14) but is

$$\int_{x_1}^{x_2} (F+f)(V+\eta_t)\,\mathrm{d}x,$$

compared with

$$\int_{x_1}^{x_2} F V \,\mathrm{d}x$$

for the basic state. The additional energy supplied can be evaluated correctly to order a^2 from a theory which is linear in f and η_t only if F and V are themselves of order a^2 or smaller. Otherwise a second order theory has to be considered.

Thus although the term of order a^2 in an expansion of Hamilton's principle for the total fields yields a variational principle which gives the correct linearized equations for the perturbations, if the basic state is one of motion the right-hand side of equation (3·1) can no longer be equated to the increment of the rate of working by the external forces. Although the external force **F** necessary to maintain the basic state $\Phi_\alpha(\mathbf{x}, t)$ is negligible (because to order a^2 the basic state is a solution of the equations of motion), the particle velocities $\mathbf{U}(\mathbf{x}, t)$ are not.

A further complication arises because the perturbation velocity $\dot{\boldsymbol{\xi}}$ is not the partial derivative $\partial\boldsymbol{\xi}(\mathbf{x}, t)/\partial t$, but is rather the derivative $([\partial/\partial t] + \mathbf{U}\,.\,[\partial/\partial\mathbf{x}])\,\boldsymbol{\xi}$ following a fluid particle with the velocity of the basic state. It is still formally possible to define quantities like expression (3·4) and the right-hand side of equation (3·7), but the ideal generalization of equation (2·20) has been lost, and their significance is obscure.

3·4. *Definitions in a slowly varying mean state*

However, if the mean state is to a sufficient approximation in uniform constant motion **U** relative to a fixed observer and otherwise independent of time, it is still possible to make meaningful calculations of perturbation energy and wave energy, but only relative to a frame of reference which is itself moving with velocity **U**. In this frame, the rate of working by any external forces is $\mathbf{f}\,.\,\dot{\boldsymbol{\xi}}$ per unit volume, and we may proceed exactly as in §3·2. However, any attempt to relate the wave energy so computed to a difference between the perturbed and mean states in a stationary frame founders because the momentum of the mean state is not known from first order theory with adequate precision.

If the mean velocity is not uniform, no frame of reference can be found in which the mean state of all parts of the system is one of rest. However, if the mean velocity $U(x, t)$ varies only slightly over a large number of wavelengths and periods of a wave packet centred at the point x_0, t_0, this may be achieved *locally* in space and time over a region sufficiently large to envisage the wave packet being set up from relative rest within the region by suitable external forces f of magnitude of order a. The relevant properties of the basic state must also be approximately constant over the time it takes to do this, which will be a large number of wave periods. Over the region $L_\lambda \lambda$ is small, and perturbation energy in a frame moving with velocity $U(x_0, t_0)$ is approximately conserved. Thus we are led to *the wave energy E being the average over a period or wavelength of the perturbation energy in a locally co-moving frame of reference*. Uniqueness follows as in §3·2. No valid interpretation can be made of global integrals of wave energy, but locally it is well defined, asymptotically as the parameter ϵ which measures both the ratio of a wavelength to the scale of variation of the basic state, and of a wave period to the basic time scale tends to zero.

To view this analytically we note that

$$\iint f \cdot \dot{\xi} \, dx \, dt = \iint \frac{\delta L}{\delta \phi_\alpha} \{\phi_{\alpha t} + U_j \phi_{\alpha x_j}\} \, dx \, dt$$

$$= \iint \left[\left\{ \frac{\partial}{\partial t} \left(\phi_{\alpha t} \frac{\partial L}{\partial \phi_{\alpha t}} - L \right) + \frac{\partial}{\partial x_j} \left(\phi_{\alpha t} \frac{\partial L}{\partial \phi_{\alpha x_j}} \right) + \frac{\partial L}{\partial \lambda} \lambda_t \right\} \right.$$

$$\left. + U_j \left\{ \frac{\partial}{\partial t} \left(\phi_{\alpha x_j} \frac{\partial L}{\partial \phi_{\alpha t}} \right) + \frac{\partial}{\partial x_i} \left(\phi_{\alpha x_j} \frac{\partial L}{\partial \phi_{\alpha x_i}} \right) + \frac{\partial L}{\partial \lambda} \lambda_{x_j} \right\} \right] dx \, dt. \quad (3 \cdot 16)$$

In a medium of which the properties are independent of space and time, i.e. $\lambda_t = \lambda_{x_j} = 0$, and in which there are no lateral coordinates, the perturbation equations have sinusoidal solutions like equation (3·5). The averages over a period of

$$\phi_{\alpha t} \frac{\partial L}{\partial \phi_{\alpha t}} - L, \quad \phi_{\alpha t} \frac{\partial L}{\partial \phi_{\alpha x_j}}, \quad \phi_{\alpha x_j} \frac{\partial L}{\partial \phi_{\alpha t}}, \quad \phi_{\alpha x_j} \frac{\partial L}{\partial \phi_{\alpha x_i}},$$

are respectively $\quad \omega \mathscr{L}_\omega - \mathscr{L}, \quad -\omega \mathscr{L}_{k_j}, \quad -k_j \mathscr{L}_\omega, \quad k_j \mathscr{L}_{k_i}.$

Integrating equation (3·16) by parts to include time and space derivatives of U_j and replacing the integrand everywhere by its averaged value, we have for a slowly varying wavetrain

$$W = \iint f \cdot \dot{\xi} \, dx \, dt$$

$$= \iint \left[\frac{\partial}{\partial t} \{(\omega - k_j U_j) \mathscr{L}_\omega - \mathscr{L}\} - \frac{\partial}{\partial x_i} \{(\omega - k_j U_j) \mathscr{L}_{k_i}\} \right.$$

$$\left. + \left\{ \mathscr{L}_\lambda (\lambda_t + U_j \lambda_{x_j}) + \frac{\partial U_j}{\partial t} k_j \mathscr{L}_\omega - \frac{\partial U_j}{\partial x_i} k_i \mathscr{L}_{k_j} \right\} \right] dx \, dt \quad (3 \cdot 17)$$

to an approximation which is arbitrarily good as $\epsilon \to 0$. The approximation involved in replacing the integral with respect to space and time of an approximately sinusoidal function by the integral of its average over a period is discussed in Bretherton (1968).

As $\epsilon \to 0$ the last group of three terms in equation (3·17) becomes small compared to the first two. Thus, as in § 3·2, we identify the wave energy density

$$E = \omega' \mathscr{L}_\omega - \mathscr{L}, \tag{3·18}$$

and the wave energy flux in direction x_j as

$$G_j = -\omega' \mathscr{L}_k, \tag{3·19}$$

where ω' is the intrinsic frequency

$$\omega' = \omega - U_j k_j. \tag{3·20}$$

Alternatively, these results can be obtained by considering a local application of the definition of § 3·2 and equations (3·7), (3·11) in a frame of reference moving with velocity $\mathbf{U}(x_0, t_0)$ and showing that $\mathscr{L}$, $\mathscr{L}_\omega$, $\mathscr{L}_{k_i}$ are unaltered under Galilean transformations.

According to the averaged variational principle, for linear waves the dispersion relation is

$$\mathscr{L} = 0$$

so that we have finally that the adiabatic invariant may be equated with the wave action density

$$\mathscr{L}_\omega = E/\omega'. \tag{3·21}$$

Using the dispersion relation, $c_j = -\mathscr{L}_{k_j}/\mathscr{L}_\omega$

so that the wave energy flux $-\omega' \mathscr{L}_{k_j}$ is indeed equal to the group velocity times the wave energy density. From Whitham's conservation equation

$$\frac{\partial}{\partial t} \mathscr{L}_\omega - \frac{\partial}{\partial x_j} \mathscr{L}_{k_j} = 0$$

we derive equation (1·9).

3·5. *Summary of § 3*

This definition of wave energy by reference to the apparent work done by virtual localized external forces moving with the perturbation velocity of fluid particles is a natural one in many fluid dynamical problems, and the authors believe that it corresponds to normal practice. It is important, however, to recognize the conditions which are necessary for it to be unambiguous. The first is that the external forces be mechanical ones associated with specific fluid particles of which the perturbation velocity is well defined, and no constraints on the system are violated.

Secondly, there must exist an equation expressing conservation of perturbation energy, at least over a local region (cf. discussion in § 3·2). However, equation (3·17) shows that a conservation equation without interaction terms does exist in a suitably moving frame to an arbitrarily good approximation over a local region, provided that the governing equations for the perturbation may be derived from an approximate form of Hamilton's principle, and provided that the variations in the mean state are sufficiently small over the region. Granted its existence, the wave energy may then be calculated from any convenient form of the linearized equations, in particular from an Eulerian formulation (cf. §4·2). The Lagrangian specification assumed here is, however, necessary to establish equation (3·18), because the perturbation velocity must be identified with $\phi_{\alpha l} + U_j \phi_{\alpha x_j}$ for the relevant values of α.

Two more conditions concern the assumption that the basic state is slowly varying. If the wavetrain is slowly varying only in horizontal directions (x, y) and in time but depends on a lateral coordinate z, it is essential that $U(x, y, t)$ be uniform in z, with error no larger than $O(\epsilon)$. Otherwise, the passage from equation (3·16) to (3·17) fails, because U cannot be taken outside the integral sign when $\phi_{\alpha l}(\partial L/\partial \phi_{\alpha l}) - L$ etc. are integrated with respect to z as part of the averaging procedure. Thus we cannot discuss surface waves on a current which varies with depth within the range where there are significant velocity perturbations. In such a case there would be no unambiguous value of the intrinsic frequency ω'.

Finally, the statement that the basic state is slowly varying does not mean that variations of every Φ_α are small; only those fields and those derivatives are restricted which enter into the specification of the parameter λ which sums up the basic state from the point of view of the perturbation. In particular, normally only the particle velocities U are relevant, not their absolute displacements. However, when ϵ is small, these velocities must become the same to within an error which tends to zero with ϵ for all particles in a region which contains a number which tends to infinity of wavelengths and wave periods. This condition may be very restrictive, as for example for Rossby waves, of which the period is a decreasing function of wavelength. The only type of basic flow on a β-plane for which the length scale is very much larger than a wavelength and the time scale very much longer than a period appears to be one which is everywhere round circles of latitude. For internal gravity waves in a uniformly stratified medium the period is independent of wavelength, but depends only on the direction of the wavenumber. The condition then appears to imply that the ratio of the horizontal to the vertical scale of variation of the basic flow is much larger than the ratio of a horizontal to a vertical wavelength. Otherwise the basic flow itself must be changing over a time comparable with a wave period. For sound waves, on the other hand, the relevant condition seems to be simply that the wavelength is small compared to the spatial scale of the medium, the temporal condition is then implied. These restrictions are necessary if $\dot{\lambda}$, $\partial U_j/\partial t$ and $\partial U_j/\partial x_i$ in equation (3·17) are to be negligibly small (say $O(\epsilon)$). It is possible to envisage a wave packet containing $\epsilon^{\frac{1}{2}}$ wavelengths being set up by nearly sinusoidal external forces in a time of $\epsilon^{\frac{1}{2}}$ periods. Over the volume occupied by the packet the error involved in neglecting the interaction terms in equation (3·17) is then $O(\epsilon^{\frac{1}{2}})$ smaller than the rate of change of wave energy and the energy flux separately.

4. HAMILTON'S PRINCIPLE

4·1. *Surface waves on water*

Before the arguments of the previous section can be seen to apply to any particular problem, an appropriate approximation to Hamilton's principle must be found, from which the complete linearized equations for the system may be derived without any approximation about the basic state being slowly varying. As in equation (2·2) the integrand must be a quadratic function of the fields $\{\phi_\alpha\}$ and their first derivatives, and must be integrated with respect to space coordinates and time. Some of the fields $\{\phi_\alpha\}$ must be the displacements $\boldsymbol{\xi}$ of material particles. For the stretched string, such a principle was easily written down. A similar version is also known for small displacements in an elastic solid (Morse & Feshbach 1953, p. 322). For electromagnetic radiation in free space (Morse & Feshbach 1953, p. 327) the corresponding principle does not involve particle displacements. However, when the equations for the waves involve linearization about a non-uniform state of mean motion more sophisticated treatment is required. In this subsection we will illustrate with the problem of surface waves of small amplitude on water of depth $h(x, y, t)$ which is moving with mean velocity $\mathbf{U}(x, y, z) = (U, V, W)$. Here (x, y, z) are rectangular coordinates with z vertically upwards. This is a good prototype for problems in classical fluid dynamics, because it involves lateral boundary conditions, at the free surface and at the bottom, and also the condition that the water be incompressible. This last is a non-holonomic constraint on possible displacements.

The basic idea of this section is due to Eckart (1963). He showed how the equations of motion for small oscillations about a moving mean state for an unbounded, inviscid, adiabatic, compressible fluid in a gravitational field may be obtained from Hamilton's principle. However, here we have an incompressible liquid with a rigid boundary below and a free surface above, and the limit of incompressible motion is not trivial. Also Eckart's treatment is for our purpose unnecessarily complicated by his simultaneously transforming into curvilinear coordinates. We start by writing down Hamilton's principle for fully nonlinear incompressible motions in a Lagrangian specification, the boundary conditions being included. To see how this may be done for a compressible fluid see Herivel (1955). The pressure then no longer enters as an independent field. For the inclusion of electromagnetic fields see Lundgren (1963).

In the usual Lagrangian specification of the motion of a fluid, the instantaneous position vector of each material particle $\mathbf{X} = (X, Y, Z)$ is expressed as a function of its initial position $\mathbf{X}_0$ and time t

$$\mathbf{X} = \mathbf{X}(\mathbf{X}_0, t). \tag{4·1}$$

The kinetic energy is $\frac{1}{2}|\dot{\mathbf{X}}|^2$ per unit mass, the gravitational potential energy is gZ. If the fluid is incompressible, there is an additional constraint that

$$\frac{\partial(X, Y, Z)}{\partial(X_0, Y_0, Z_0)} = 1. \tag{4·2}$$

If it is bounded below by a rigid boundary $b(x, y)$, then

$$Z(\mathbf{X}_0, t) = b(X, Y) \quad \text{if} \quad Z_0 = b(X_0, Y_0). \tag{4·3}$$

Otherwise there are no constraints on the motion, in particular the upper surface

$$Z_0 = s(X_0, Y_0) \tag{4·4}$$

is free.

The exact form of Hamilton's principle is then

$$\delta \int dX_0\, dY_0\, dt \left[\int_b^s \left\{ \tfrac{1}{2}\rho\,|\dot{\mathbf{X}}|^2 - \rho g Z + \lambda_1 \left(\frac{\partial(X,Y,Z)}{\partial(X_0,Y_0,Z_0)} - 1 \right) \right\} dZ_0 \right.$$
$$\left. + \lambda_2 \{ Z - b(X,Y) \}_{Z_0 = b(X_0,\,Y_0)} \right] = 0 \quad (4\cdot5)$$

for all variations $\delta\mathbf{X}(X_0,t)$, $\delta\lambda_1$, $\delta\lambda_2$ which vanish for sufficiently large $|X_0|$, $|Y_0|$, $|t|$. In this equation the integration over the initial coordinates X_0, Y_0, Z_0 corresponds to a summation over material particles to form the Lagrangian. The further integration over t forms the action. λ_1, λ_2 are Lagrange multipliers, their variation implying the incompressibility condition (4·2) and the bottom condition (4·3) respectively. However, as usual with constraints in Lagrange's equations, λ_1 and λ_2 may be interpreted as the associated generalized forces. Thus λ_1 is the generalized force corresponding to a change in volume of a fluid element, i.e. the pressure P. λ_2 is associated with the vertical component of the stress on the bottom; in fact

$$\lambda_2 = \left\{ P\,\frac{\partial(X,Y)}{\partial(X_0,Y_0)} \right\}_{Z_0 = b(X_0,\,Y_0)}, \quad (4\cdot6)$$

a condition which also follows from the variational principle by varying Z. The Jacobian in equation (4·6) is evaluated on $Z_0 = b(X_0,Y_0)$ and is thus given by

$$\left. \frac{\partial(X,Y)}{\partial(X_0,Y_0)} \right|_{Z_0 = b} = \begin{vmatrix} \dfrac{\partial X}{\partial X_0} & \dfrac{\partial X}{\partial Y_0} \\[2mm] \dfrac{\partial Y}{\partial X_0} & \dfrac{\partial Y}{\partial Y_0} \end{vmatrix} + \frac{\partial b}{\partial X_0} \begin{vmatrix} \dfrac{\partial X}{\partial Z_0} & \dfrac{\partial X}{\partial Y_0} \\[2mm] \dfrac{\partial Y}{\partial Z_0} & \dfrac{\partial Y}{\partial Y_0} \end{vmatrix} + \frac{\partial b}{\partial Y_0} \begin{vmatrix} \dfrac{\partial X}{\partial X_0} & \dfrac{\partial X}{\partial Z_0} \\[2mm] \dfrac{\partial Y}{\partial X_0} & \dfrac{\partial Y}{\partial Z_0} \end{vmatrix}. \quad (4\cdot7)$$

Besides equation (4·6), variation of Z gives the Z equation of motion and the free surface condition

$$P(X_0, Y_0, s(X_0, Y_0)) = 0. \quad (4\cdot8)$$

The other equations of motion follow simply from variation of X, Y. However, since each particle has to be identified by its initial position, the equations derived from the variational principle are not in a very convenient form.

We now envisage one standard solution of these equations

$$\mathbf{X} = \mathbf{x}(\mathbf{X}_0, t), \quad P = p(\mathbf{X}_0, t) \quad (4\cdot9)$$

and express the *independent* variables in terms of it

$$\mathbf{X}_0 = \mathbf{X}_0(\mathbf{x}, t). \quad (4\cdot10)$$

This is merely a convenient way of relabelling each material particle in terms, not of its original position, but of the position $\mathbf{x}$ it would have had, had it moved for time $t - t_0$ according to the particle paths of the standard solution. We denote the standard particle velocity

$$(\partial\mathbf{x}/\partial t)_{\mathbf{X}_0} = \mathbf{U} \quad (4\cdot11)$$

and regard all fields as functions of $\mathbf{x}, t$.

From equation (4·2)

$$\frac{\partial(x,y,z)}{\partial(X_0,Y_0,Z_0)} = 1, \quad (4\cdot12)$$

and equation (4·5) becomes

$$\delta \int dx\, dy\, dt \left[\int_b^s \left\{ \tfrac{1}{2}\rho\, |\dot{\mathbf{X}}|^2 - \rho g Z + P\left(\frac{\partial(X,Y,Z)}{\partial(x,y,z)} - 1\right)\right\} dz \right.$$

$$\left. + \left\{ P\frac{\partial(X,Y)}{\partial(x,y)}(Z - b(X,Y))\right\}_{z=b(x,y)}\right] = 0, \quad (4\cdot13)$$

where
$$\dot{\mathbf{X}} = \left(\frac{\partial}{\partial t} + U\frac{\partial}{\partial x} + V\frac{\partial}{\partial y} + W\frac{\partial}{\partial z}\right)\mathbf{X}. \quad (4\cdot14)$$

Equation (4·13) holds for any solution $\mathbf{X}(\mathbf{x},t)$, $P(\mathbf{x},t)$ of the equations of motion. In particular we may put
$$\mathbf{X} = \mathbf{x} + \boldsymbol{\xi}, \quad P = p + \pi, \quad (4\cdot15)$$

where $\boldsymbol{\xi}$, π are small, and consider infinitesimal variations $\delta\boldsymbol{\xi}$, $\delta\pi$. If equation (4·13) is expanded in powers of $\boldsymbol{\xi}$, π and then varied, the lowest order term in the expansion does not involve $\boldsymbol{\xi}$, π at all and hence has no variation. The linear term involves $\mathbf{x}$, p and $\delta\boldsymbol{\xi}$, $\delta\pi$ only, and it vanishes precisely because $\mathbf{x}$, p is a solution of the equations of motion. The quadratic term yields a variational principle governing $\boldsymbol{\xi}$, π. This gives homogeneous Euler equations for perturbations of small magnitude about the standard flow. Picking out these quadratic terms, (4·9) becomes

$$\delta \int dx\, dy\, dt \left[\int_b^s \{\tfrac{1}{2}\rho\,|\dot{\boldsymbol{\xi}}|^2 + \pi(\xi_x + \eta_y + \zeta_z) + pq(\boldsymbol{\xi})\}\, dz \right.$$

$$\left. + \{\pi + p(\xi_x + \eta_y + b_x\,\xi_z + b_y\eta_z)\}\{(\zeta - b_x\xi - b_y\eta) - pr(\boldsymbol{\xi})\}_{z=b}\right] = 0, \quad (4\cdot16)$$

where
$$q(\boldsymbol{\xi}) = \eta_y\zeta_z - \zeta_y\eta_z + \zeta_z\xi_x - \xi_z\zeta_x + \xi_x\eta_y - \eta_x\xi_y, \quad (4\cdot17)$$

and
$$r(\boldsymbol{\xi}) = \tfrac{1}{2}(b_{xx}\xi^2 + 2b_{xy}\xi\eta + b_{yy}\eta^2). \quad (4\cdot18)$$

The integration is strictly over the region occupied by the standard flow, and the representation is quasi-Lagrangian in that the displacements $\boldsymbol{\xi}$ are of fluid particles from the instantaneous value of a moving reference position. $\mathbf{U}$, p, s, b are regarded as known.

Judicious integration by parts and careful consideration of the boundary terms enables one to cast equation (4·16) into the form

$$\delta \int dx\, dy\, dt \left[\int_b^s \{\tfrac{1}{2}\rho\,|\dot{\boldsymbol{\xi}}|^2 + \pi'(\xi_x + \eta_y + \zeta_z) - \tfrac{1}{2}(p_{xx}\xi^2 + p_{yy}\eta^2 + p_{zz}\zeta^2 \right.$$

$$+ 2p_{yz}\eta\zeta + 2p_{zx}\zeta\xi + 2p_{xy}\xi\eta)\}\, dz + \{\tfrac{1}{2}p_z(\zeta - \xi s_x - \eta s_y)^2\}_{z=s}$$

$$\left. + \{\tfrac{1}{2}p_z(\zeta - \xi b_z - \eta b_y)^2 + \pi'(\zeta - \xi b_x - \eta b_y)\}_{z=b}\right] = 0, \quad (4\cdot19)$$

where
$$\pi' = \pi - \xi p_x - \eta p_y - \zeta p_z. \quad (4\cdot20)$$

The basic flow is thus relevant to the perturbation only through the convection terms $\mathbf{U}.\partial/\partial\mathbf{x}$ in the time derivative, and through the pressure gradients. The position $z = s(x,y,t)$ of the free surface is also determined by the fact that the basic flow is a solution of the equations of motion. The volume integral in equation (4·19) is made up of the perturbation kinetic energy $\tfrac{1}{2}\rho\,|\dot{\boldsymbol{\xi}}|^2$ minus two forms of potential energy. The first, $\pi'(\xi_x + \eta_y + \zeta_z)$ is purely virtual; variation of the perturbation pressure π' at any point implies incompressibility, so for real perturbations there is no contribution from this term. The second form is the potential energy of a displaced particle in the basic pressure field, gradients of pressure at the equilibrium position being balanced against basic mass accelerations. For the present problem this term is negligibly small if the basic flow is slowly varying. The surface integral over the free surface describes the only non-zero perturbation potential energy in a slowly varying flow. Then, the basic vertical accelerations are negligible and p_z may be replaced by $-\rho g$. Variation of π' also implies the bottom boundary condition, and contributions to the energy from the bottom integral are purely virtual.

Equation (4·16), subject to infinitely differentiable variations $\delta\xi$, $\delta\pi$ which vanish as $|x|, |y|, |t| \to \infty$ but which are otherwise unrestricted, is the required approximate form of Hamilton's principle, to which the averaging procedure and the arguments of §3 may be applied. Alternatively equation (4·19) could be used; the two are entirely equivalent. The explicit appearance in equation (4·19) of a perturbation potential energy is not an essential part of the argument, merely a comforting check on the calculations.

4·2. Wave energy for surface waves

To use equation (1·9), rather than to prove it, this procedure is unnecessary.

If $\mathbf{U}$, h are uniform ($W = 0$), the governing equations are separable and we may find sinusoidal solutions in which the vertical displacement of a fluid particle is given by

$$\zeta = a \frac{\sinh |\mathbf{k}| (z-b)}{\sinh |\mathbf{k}| h} \cos (kx + ly - \omega t + \beta) \qquad (4·21)$$

provided the dispersion relation

$$\omega = \mathbf{U}.\mathbf{k} \pm \{g |\mathbf{k}| \tanh |\mathbf{k}| h\}^{\frac{1}{2}} \qquad (4·22)$$

is satisfied. Here the bottom is at $z = b$, so the mean position of the free upper surface is
$$s = b + h.$$

The wavenumber $\mathbf{k}$ has two components (k, l) and there are for each $\mathbf{k}$ two normal modes corresponding to the choice of sign in equation (4·22). The state of motion of the system is completely determined when we specify the mode, the wavenumber $\mathbf{k}$, the amplitude a, and the arbitrary constant β in the phase. If the total velocity at the point (x, y, z, t) is $(U + u, V + v, w)$ and ρ is the density of the water (assumed constant), the wave energy density is the average over a wavelength of

$$\int_b^s \tfrac{1}{2}\rho(u^2 + v^2 + w^2) \, dz + \tfrac{1}{2}\rho g \zeta^2 |_{z=s}, \qquad (4·23)$$

which is easily shown to be $E = \tfrac{1}{2}\rho g a^2$.   (4·24)

The intrinsic frequency is $\omega' = \omega - \mathbf{U}\mathbf{k}$
$$= \pm \{g |\mathbf{k}| \tanh |\mathbf{k}| h\}^{\frac{1}{2}}, \qquad (4·25)$$

and the wave action density is $\tfrac{1}{2}\rho g a^2 / \omega'$.

If $\mathbf{U}$, h are not precisely uniform, but vary substantially in a known manner in a horizontal direction over a scale of many wavelengths (but with $W \ll U, V$, and with U, V varying only slightly with z), then we may still determine the motion locally in the same way. However, the values of $\mathbf{k}$, a, β at different places will be different but interconnected. The frequency and wavenumber are connected along rays according to equations (1·6) and (1·7) and the dispersion relation (4·22), and the amplitude is determined by conservation of wave action, according to equations (1·9), (4·24) and (4·25). The phase β requires analysis to a higher order of approximation than considered in this paper.

4·3. *Hamilton's principle for a frozen-in magnetic field*

A statement of Hamilton's principle for general magnetohydrodynamic flows has been given by Lundgren (1963). He extends the classical non-magnetic treatment due to Herivel (1955) to include Galilean invariant electric and magnetic fields in a system consisting of perfectly conducting fluid, vacuum and perfectly conducting solid parts. This is accomplished by subtracting the magnetic energy from the integrand in the statement of Hamilton's principle for non-magnetic systems, and Lundgren shows that this leads to the correct equations of motion for the fluid, and boundary conditions at the fluid-vacuum interface.

The approach of § 4·1 may then be used to derive a variational principle describing small perturbations about any state which is an exact solution of the governing equations. It should be noted that perturbations in the electromagnetic field in any nonconducting regions can only be induced by particle displacements in conductors. The methods of § 3 can only be used to define the wave energy density by reference to the velocities of these conductors. This situation will only arise when adjacent conducting and nonconducting regions are acting as a waveguide implying integration over lateral coordinates.

5. FURTHER REMARKS

The class of problems for which equation (1·9) may be considered established is circumscribed in three ways. First it is necessary that the appropriate approximate form of Hamilton's principle may be obtained. The methods of § 4 appear to be generally applicable to nondissipative problems in nonrelativistic fluid dynamics and magnetohydrodynamics with conservative boundary conditions, and yield a formulation explicitly in terms of particle displacements together with other fields such as the pressure and the magnetic field as appropriate. For compressible fluids the internal energy per unit mass should be expressed as an explicit function of the dilatation, and the pressure no longer appears as an independent field. External gravity fields and nonuniformities in initial density are all easily included.

Secondly, it is necessary that the averaged variational principle may be applied to the approximate form of Hamilton's principle to describe slowly varying wavetrains. The mathematical requirements for this are discussed fully in Bretherton (1968). Here it suffices to remark that the locally valid normal modes, in which an almost sinusoidal oscillation must be found, are required to be well-behaved functions of position and wavenumber, discrete and nondegenerate. This ensures that a wavetrain can indeed propagate between two widely separated points with a continuously varying but well defined internal structure imposed by the appropriate mode at each point along its path. It excludes the possibility of energy being transferred en route to another mode. The normal modes at each point are not required to form a complete set, although a weaker condition of this general type is needed. With a few minor qualifications, the averaging procedure can then be made precise as the zero order approximation in an asymptotic sense as $\epsilon \to 0$, and for linearized equations an infinite sequence of higher order approximations in powers of ϵ can be constructed.

Thirdly, even if the adiabatic invariant $\mathscr{L}_\omega$ can be obtained, it cannot always be identified with the wave action density E/ω'. The discussion of §3 indicates the conditions under which this appears to be possible. Noteworthy are the restrictions implied by the basic state being slowly varying. The train of argument is full of pitfalls for the unwary. The present authors fell at some stage into all of them, some of them several times.

6. APPLICATION TO PARTICULAR PROBLEMS

6·1. *Surface gravity waves*

An important application of the result of this paper is to the propagation of surface gravity waves on a slowly varying nonuniform current, which is approximately independent of the vertical coordinate. This is the prototype problem of §4.

Longuet-Higgins & Stewart (1961, 1964) have investigated special cases of this problem, and show that the results of their detailed perturbation analyses in these cases may be interpreted in terms of the interaction between the rate of strain of the basic current and the radiation stress of the waves. They thus infer from these special cases a general equation governing the energy propagation of surface waves on a nonuniform current

$$\frac{\mathrm{d}E}{\mathrm{d}t} + \nabla \cdot \mathbf{c}\, E + \tfrac{1}{2}S_{ij}\left(\frac{\partial U_i}{\partial x_j} + \frac{\partial U_j}{\partial x_i}\right) = 0, \tag{6·1}$$

where $E = \tfrac{1}{2}\rho g a^2$, a is the amplitude of the surface elevation, and S_{ij} is the radiation stress tensor. Whitham (1962) derived equation (6·1) for special cases by a different method.

It is shown in the appendix that (6·1) may be cast into the form of equation (1·9) using equations (1·6) and the continuity equation of the basic flow. The same equivalence may be demonstrated if the effects of capillarity are included, though the wave energy density is then given by $E = \tfrac{1}{2}\rho g a^2(1 + T\mathbf{k}^2/\rho g)$, where T is the surface tension.

We have also shown the equivalence of equation (6·1) and (1·9) for many other types of wave motion in fluid dynamics; some of these are discussed further in the ensuing paragraphs. This equivalence suggests certain general properties of radiation stress which are still under consideration.

6·2. *Sound waves*

Blokhintsev (1946) considers the problem of sound propagation in a slowly varying steady nonuniform moving medium. Using a W.K.B. approximation he derives an energy equation (his (2·30)) which may be written with our notation

$$\frac{\mathrm{d}}{\mathrm{d}t}\left(\frac{\omega E}{\omega'}\right) + (\nabla \cdot \mathbf{c})\left(\frac{\omega E}{\omega'}\right) = 0. \tag{6·2}$$

This agrees with our result (1·9) when the medium is not changing with time, as we then have $\mathrm{d}\omega/\mathrm{d}t = 0$. Extension of Blokhintsev's analysis gives (1·9) as the correct energy equation even when the basic state is not steady. The same result may also be reached from consideration of the radiation stress for sound waves.

6·3. *Internal gravity waves*

Both types of wave motion discussed so far in this section have their group velocity and phase velocity parallel. Internal gravity waves provide an example for which this is not so. Bretherton (1966) has considered the problem of internal wave propagation in a shear flow $(U(z), V(z), 0)$ with Brunt–Väisälä frequency $N(z)$ and large Richardson number $N^2/(U_z^2 + V_z^2)$. He derives equation (1·9) both by using a W.K.B. approximation, and also by considering the work done by the relevant Reynolds stresses. Hines & Reddy (1966) also arrive at (1·9) for this problem, though by a different method.

6·4. *Alfvèn waves*

From the extension of Hamilton's principle to include magnetic fields, mentioned in §4·3, we see that equation (1·9) applies also to the propagation of Alfvèn and magneto-acoustic waves in a compressible or incompressible fluid with a frozen-in magnetic field, provided that the slowly varying approximation is satisfied. For Alfvèn waves this result has also been verified by consideration of the effect of radiation stress.

6·5. *Inertial waves*

We have also considered the action of radiation stress for wave motions which do not have equipartition between kinetic and potential energy. Inertial waves afford the prime example of this, the wave energy in this case being entirely kinetic. For inertial waves in an incompressible homogeneous liquid rotating about Oz with a constant angular velocity, and with a superimposed geostrophic shear flow $(U(x,y), V(x,y), 0)$, equation (6·1) may again be shown to be equivalent to equation (1·9). For waves of this type, the frequency is independent of the magnitude of the wavenumber (though not its direction). The conditions for slow variation, both spatial and temporal must be satisfied separately. It is not sufficient merely to take the limit of small wavelength.

6·6. *Standing surface waves*

Taylor (1962) has investigated the effect of slowly changing currents of uniform divergence on standing surface waves. In each of the three cases he considers, his results are consistent with the statement that the energy of an area of standing waves expanding or contracting with the current varies in proportion to the frequency of the waves. If we consider a standing wave train to be made up of two progressive wave trains travelling in opposite directions with the same amplitude and frequency, then this result is only to be expected in view of the reversibility of equations (1·6) and (1·9).

6·7. *Wave action density*

It is noteworthy that the quantity E/ω', designated here the 'wave action density', is also of fundamental importance in nonlinear wave interactions. If the resonance conditions are satisfied, a unit of wave action density from one wave train may react with a unit of wave action density from another wave train to produce one unit of wave action density in each of one or more further wave trains, very much in the manner of chemical reaction between different molecules. This is discussed further by Hasselmann (1963) who calls E/ω' the 'wave number density', and by Bretherton (1964).

6·8. *Further generalizations*

The present analysis is essentially confined to nonrelativistic classical dynamical systems. However, a result similar to equation (1·9) is known for gravitational waves in general relativity (Isaacson 1967) and electromagnetic waves in a relativistic gravitational field (Kristian & Sachs 1966).

One of the authors (C. J. R. Garrett) is indebted to the Science Research Council for a maintenance grant. We are also grateful to Professor M. J. Lighthill, Sec. R.S., for constructive criticisms of this manuscript.

APPENDIX

We wish to show that the two energy propagation equations (1·9) and (6·1) are equivalent for surface gravity waves on water of depth h. We thus require

$$\tfrac{1}{2}S_{ij}\left(\frac{\partial U_i}{\partial x_j} + \frac{\partial U_j}{\partial x_i}\right) = -\frac{E}{\omega'}\frac{d\omega'}{dt}.$$

Now
$$\frac{d\omega'}{dt} = \frac{d\omega}{dt} - \frac{d}{dt}(\mathbf{U}.\mathbf{k}),$$

and equations (1·6) hold equally well for a moving medium as for a stationary medium, as they are merely a consequence of the definitions $\omega = -\theta_t$, $k_i = \theta_{x_i}$ in (1·1). Here we have $\omega = \mathbf{U}.\mathbf{k} + \Omega'(\mathbf{k}, h)$.

Thus
$$\frac{d\omega}{dt} = \mathbf{k}.\frac{\partial \mathbf{U}}{\partial t} + \frac{\partial \Omega'}{\partial h}\frac{\partial h}{\partial t},$$

$$\frac{dk_j}{dt} = -k_i\frac{\partial U_i}{\partial x_j} - \frac{\partial \Omega'}{\partial h}\frac{\partial h}{\partial x_j},$$

$$\frac{dU_j}{dt} = \frac{\partial U_j}{\partial t} + U_i\frac{\partial U_j}{\partial x_i} + c_i'\frac{\partial U_j}{\partial x_i},$$

where
$$c_i' = \frac{\partial \Omega'}{\partial k_i}.$$

Hence
$$\frac{d\omega'}{dt} = \frac{\partial \Omega'}{\partial h}\left(\frac{\partial h}{\partial t} + \mathbf{U}.\nabla h\right) - k_j c_i'\frac{\partial U_j}{\partial x_i}$$

$$= -h\frac{\partial \Omega'}{\partial h}\nabla.\mathbf{U} - k_j c_i'\frac{\partial U_j}{\partial x_i},$$

16*

using the continuity equation for the basic flow

$$\frac{\partial h}{\partial t} + \mathbf{U} . \nabla h + h\nabla . \mathbf{U} = 0.$$

If $\mathbf{U}$ is arbitrary, we thus require, using the dispersion relation (4·25),

$$S_{ij} = E\left(\frac{c_i' k_l}{\omega'} - \frac{1}{2}\right)\delta_{ij} + E\frac{c_i' k_j}{\omega'},$$

which is the general form of the radiation stress tensor derived by Longuet-Higgins & Stewart.

References

Blokhintsev, D. I. 1946 Acoustics of a nonhomogeneous moving medium. *N.A.C.A. T.M.* 1399 (1956).

Bretherton, F. P. 1964 Resonant interactions between waves. The case of discrete oscillations. *J. Fluid Mech.* **20**, 457–479.

Bretherton, F. P. 1966 The propagation of groups of internal waves in a shear flow. *Quart J. R. Met. Soc.* **92**, 466–480.

Bretherton, F. P. 1968 Propagation in slowly varying waveguides. *Proc. Roy. Soc.* A **302**, 555.

Eckart, C. 1963 Some transformations of the hydrodynamic equations. *Phys. Fluids* **6**, 1037–1041.

Einstein, A. 1911 *La theorie du rayonnement et les quanta* (ed. P. Langevin and M. de Broglie). Paris: Gauthier-Villars. (1912) (Report of First Solvay Conference) p. 450.

Garrett, C. J. R. 1967 Discussion: the adiabatic invariant for wave propagation in a non-uniform moving medium. *Proc. Roy. Soc.* A **299**, 26–27.

Hasselmann, K. 1963 On the non-linear energy transfer in a gravity wave spectrum. Part 2. Conservation theorems; wave-particle analogy; irreversibility. *J. Fluid Mech.* **15**, 273–281.

Herivel, J. W. 1955 Equations of motion of an ideal fluid. *Proc. Camb. Phil. Soc.* **51**, 344–349.

Hines, C. O. & Reddy, C. A. 1967 On the propagation of atmospheric gravity waves through regions of wind shear. *J. Geophys. Res.* **72**, 1015–1034.

Isaacson, R. A. 1967 Ph.D. thesis. Department of Physics, University of Maryland.

Kristian, J. & Sachs, R. K. 1966 Observations of cosmology. *Astrophys. J.* **143**, 379–399.

Lighthill, M. J. 1965 Group velocity. *J. Inst. Appl. Math. Applic.* **1**, 1–28.

Longuet-Higgins, M. S. 1964 On group velocity and energy flux in planetary wave motions. *Deep Sea Res.* **11**, 35–42.

Longuet-Higgins, M. S. & Stewart, R. W. 1961 The changes in amplitude of short gravity waves on steady non-uniform currents. *J. Fluid Mech.* **10**, 529–549.

Longuet-Higgins, M. S. & Stewart, R. W. 1964 Radiation stress in water waves; a physical discussion, with applications. *Deep Sea Res.* **11**, 529–562.

Luke, J. C. 1966 A perturbation method for nonlinear dispersive wave problems. *Proc. Roy. Soc.* A **292**, 403–412.

Lundgren, T. S. 1963 Hamilton's variational principle for a perfectly conducting plasma continuum. *Phys. Fluids.* **6**, 898–904.

Morse, P. M. & Feshbach, H. 1953 *Methods of theoretical physics*, Part I. New York: McGraw-Hill.

Taylor, Sir Geoffrey 1962 Standing waves on a contracting or expanding current. *J. Fluid Mech.* **13**, 182–192.

Whitham, G. B. 1960 A note on group velocity. *J. Fluid Mech.* **9**, 347–352.

Whitham, G. B. 1962 Mass, momentum and energy flux in water waves. *J. Fluid Mech.* **12**, 135–147.

Whitham, G. B. 1965 A general approach to linear and non-linear dispersive waves using a Lagrangian. *J. Fluid Mech.* **22**, 273–283.

Proceedings of the Royal Society of London, **A 299**, 28 (1967).

Some Special Cases Treated by the Whitham Theory

M. J. Lighthill

The theory of the preceding paper (by G. B. Whitham) may be valuable in problems of non-linear dispersive wave systems governed by equations too complicated for analytical or numerical treatment. It suggests, in fact, that the development of wave groups, whose parameters vary gradually enough, can be represented by much less complicated approximate equations, derived from the assumption that in each separate small region the waves are closely like plane periodic ones. Nevertheless, the effort demanded to solve even those approximate equations would normally be very substantial, and therefore it is important (see §1) to get a preliminary idea of whether the solutions would correspond well with reality or not, by calculating the implications of the theory in detail for some rather simple system or systems and comparing them with experiment.

It is argued (§2) that unidirectional waves on deep water are a suitable system for this purpose. Their propagation, as an analysed by the Whitham theory, is governed by how the Lagrangian density $\mathscr{L}$ depends on the frequency ω and the wavenumber k. In fact, $\mathscr{L}k^2/\rho g$ (where ρ is the density of water) is a function of $z = \omega^2/gk$ alone; and, by interpolating between known values for low waves ($z \to 1$) and for the wave of greatest height ($z = 1\cdot20$), a polynomial approximation

$$\mathscr{L}k^2/\rho g = \tfrac{1}{4}[(z-1)^2 - (z-1)^3 - (z-1)^4],$$

expected to be good throughout the interval $1 \leqslant z \leqslant 1\cdot20$, is found (§4).

It is deduced from this (§5) that, when a wavemaker produces a slightly modulated train of waves of constant frequency on deep water, the exponential rate of increase of modulation with distance from the wavemaker, shown by Brooke Benjamin later in this Discussion to be proportional to amplitude for waves of moderate amplitude, attains a maximum for a wave of about half the greatest height, and then falls to zero, no increase of modulation being predicted in the case of waves of more than three-quarters of the greatest height, which should exhibit instead Whitham's 'splitting of the group velocity' (figure 2). Brooke Benjamin & Feir's agreement with experiment is improved (figure 3) by this modification, the remaining discrepancy being readily attributable to dissipation. However, additional experiments are needed for large values of the amplitude.

The theory is worked out and compared with experiment also for groups within which the wave amplitude varies by a large fraction of itself, although only gradually on a scale of wavelengths and subject to an upper limit on magnitude (§7). The predicted change with time, previously worked out (Lighthill 1965 b) for one special form of group (reciprocal-quadratic; see §8), is calculated in §9 for a general group of uniform wavenumber whose amplitude varies smoothly and symmetrically about its maximum; and detailed results are given for a sinusoidal modulation with arbitrary depth of modulation.

The theory predicts that the group remains symmetrical, and that the peak amplitude increases, first according to a hyperbolic-cosine law but then still faster, until a critical condition is reached when the distribution of wave amplitude has a cusp at the centre of the group, where also the wavenumber exhibits a discontinuity. Beyond this critical time the assumptions of the theory break down and a changed type of propagation is expected.

Experiments by Feir (described below, p. 54), in which the spatial period of modulation was about 12 wavelengths, showed just such a catastrophic alteration at a time close to that predicted (see §10), and wavenumber changes close to those given by the theory. Dissipation of wave energy (e.g. by friction at the side walls) was again too great to permit an accurate check on the theory's amplitude predictions, but the value of the critical time (after which the observed wave group becomes markedly asymmetric) is encouraging evidence that the theory is reliable so long as the variations it predicts remain smooth and slow enough to satisfy its own basic assumptions.

Earlier, in §§3 and 6, the way (not essentially more complicated) in which the theory would be used to study a two dimensional pattern of stationary waves on a uniform stream of deep water is outlined.

1. Preliminary critique of the Whitham theory

I said in my remarks introducing this meeting that the basic principle underlying the theory which Professor Whitham has just expounded, and the determination of its range of applicability, are key matters for investigation. I believe that this investigation can be conducted best by working out the detailed implications of the theory in particular cases where comparison with experiment is possible.

The principle whose validity needs to be investigated is Whitham's fundamental assumption that, if parameters characterizing the waves vary sufficiently gradually, on a scale of wavelengths, then locally the waves must closely approximate to plane periodic ones. If this is correct, then relations between the parameters characterizing the waves must to close approximation take the same form as for plane periodic waves, so that their determination involves the solution only of ordinary differential equations.

The relations thus obtained between the parameters can then be used, with other equations, to determine how the waves vary over large distances. Different methods of doing that analysis can be employed; either starting, in the elegant way Professor Whitham has just expounded, from a variational principle of Hamilton's type; or by the more *ad hoc* approach given in his first paper (Whitham 1965 a) or, for example, by an approximate method, mentioned in Whitham's paper (§4·3, first six equations, culminating in equation (49)) and discussed further below (§6). But whatever mathematical method is used, the results can at most be as accurate as is the principle on which all the work is based.

In favour of this principle it must be noted that it is identical to the principle used when the waves satisfy linear equations, namely, that if the parameters characterizing them vary slowly enough on a scale of wavelengths, then we can use the geometrical optics approximation. In this case the local form of the waves must closely approximate to plane periodic waves; in other words, since the equations are linear, to sinusoidal waves. This implies immediately that the relations between parameters (particularly, frequency and wavenumber) can be taken the same as for sinusoidal waves, which leads at once (Lighthill 1965 a, §5) to the equations governing the motion of individual wave packets.

It seems natural to assume, as Whitham does, that the same method will work when the equations governing the waves are nonlinear. Plane periodic solutions still exist, even though they are no longer sinusoidal in form, and any wave pattern exhibiting gradual variation over a scale of wavelengths might be supposed, locally, to approximate to some plane periodic wave.

However, any calculations based on that assumption are subject to one inherent limitation, not present in the linear geometrical optics case. There, it is well known that solutions one-valued in certain regions of space-time may become many-valued in others. In fact, wherever a caustic appears, the solution on one side of the caustic is at least two-valued, with two separate values of the wavenumber vector, for example, at each point. But although the geometrical optics approximation is not accurate in the immediate neighbourhood of the caustic, the many-valuedness itself, in the region away from the caustic, does not prevent the solution from giving good results. This is because in linear problems solutions can be linearly superposed, and hence two separate waves with different wavenumbers and amplitudes can locally coexist in linear combination.

Unfortunately, there is no such easy way out in the nonlinear case. Whitham's equations, as we shall see, show just as much tendency, if not more, for solutions that are one-valued in certain regions of space-time to become many-valued in others. But, with linear superposition impossible, there is no easy interpretation of a solution with more than one value of the wavenumber vector at each point. Whitham (1965a) suggested that, as in gas dynamics, only one of the solutions would apply at any one point, while a discontinuous transition from one to the other would occur at a surface to be determined—although the rules for its determination are by no means clear. It is possible that in some cases the pattern of waves approximates to this sort of arrangement and that in others, nearer a linear situation, it approximates to a superposition of different wavetrains; but the determination of regions of applicability of the two approaches and the nature of any transition has not yet been attempted.

Two separate problems of evaluation of the Whitham theory exist, then. First, when the solutions it gives remain one-valued, we must ask how accurate they are, or more precisely how slowly the parameters must vary on a scale of wavelengths for the solutions to have tolerable accuracy. Secondly, we must ask what happens physically when the predicted solutions become many-valued. What does it signify, and can any modification of the theory be found that will correspond roughly to what happens?

I should emphasize that the Whitham theory would still be well worth having even if it could be used only in regions of space-time where the predicted solutions remain one-valued, since these are quite extensive. Furthermore, if there is some recognizable physical phenomenon which corresponds to the prediction of many-valuedness, then the fact that the Whitham theory could be used to determine its onset would be valuable in itself.

I shall, in fact, conclude that this is the present position. In the examples I discuss, no modification of the theory which corresponds well with experiment in the many-valued region has yet been found, but a particular characteristic behaviour seems to be associated with the many-valued solution, and its onset is predicted reasonably well.

The whole motivation of the work I shall describe, then, is to obtain detailed results from the Whitham theory in particular examples, and to compare them with the results of experiment. Until this has been done in both cases (one-valued and many-valued solutions) the theory can be of no value, in the sense that formulas are useless if no one knows when they can be applied.

2. Introductory remarks on problems in two independent variables without pseudo-frequencies, including waves on deep water

With the motives described in §1, it seemed essential to study rather simple situations, where various solutions of the Whitham equations could be obtained in a relatively flexible way. Accordingly, I make no apology for having confined my attention to problems with two independent variables, and also to problems where Whitham's pseudo-frequencies (such as the γ in his equation (29), p. 12 above) are absent, essentially because second order partial differential equations in two independent variables are so much more tractable than equations of higher order or equations with more independent variables.

A good example of such systems without pseudo-frequencies is provided by waves on deep water. It is worth noting, incidentally, that if surface waves are considered only in the deep-water case, the difficulties mentioned by Whitham regarding the formulation of an appropriate variational principle disappear.

As he said, the classical Lagrangian (kinetic energy T minus potential energy V) is suitable in any problem where 'Lagrangian' (as opposed to Eulerian) dependent variables are used to specify the motion. The deep-water wave problem is precisely of this kind, because a completely suitable dependent variable, of a Lagrangian character, for specifying the motion exists, namely, the height of the free surface. In the case of unidirectional waves, with motions in the (x, y) plane only (where the y axis is vertical), this height is (say) $\eta(x, t)$, and knowledge of the function $\eta(x, t)$ completely determines the whole motion in the infinite region $y \leqslant \eta(x, t)$, subject to the boundary condition that the velocity tends to zero as $y \to -\infty$, because, at each time t, a velocity potential ϕ, which is unique except for an arbitrary constant, satisfies $\nabla^2 \phi = 0$ in the domain, $|\nabla \phi| \to 0$ as $y \to -\infty$, and the boundary condition on its normal derivative at $y = \eta(x, t)$, namely

$$\frac{\partial \phi}{\partial y} - \frac{\partial \eta}{\partial x} \frac{\partial \phi}{\partial x} = \frac{\partial \eta}{\partial t}.$$

Accordingly, the classical Lagrangian $T - V$ can be used if both T and V are expressed in terms of the function $\eta(x, t)$ which describes the deformation of the surface. (This is not true for waves in shallow water, since uniqueness disappears when a bottom boundary condition replaces $|\nabla \phi| \to 0$; indeed, any multiple of the solution representing steady flow through the channel bounded by the bottom and the instantaneous free-surface shape can then be added on to ϕ.)

Besides possessing this simplifying feature, waves on deep water are especially suitable for comparison of the theory with experiment, since the whole theory neglects viscosity, and motions near a bottom are certain to involve very vigorous viscous action, whose influence can spread and involve the whole motion (Longuet-Higgins 1953). Comparison with experiment is more likely to be favourable, therefore, if motions near solid surfaces are avoided as far as possible, as in waves on water deep enough for motions near the bottom to be negligible.

Because experiments on deep water are convenient and suitable, then, much of what follows will be concerned with them, although the bulk of the theory will be formulated so that it can be applied very widely—in fact, to any problem in two independent variables without pseudo-frequencies. Lighthill (1965b) gave a convenient mathematical method for treating such problems if the independent variables are x and t (as above).

For periodic waves the Lagrangian density $\mathscr{L}$ (e.g. the Lagrangian per unit horizontal area in the case of deep water waves) is a function of the wavenumber k and of a measure of amplitude a. However, the frequency ω is also determinate if we know the wavenumber k and the amplitude a. It is possible, therefore, to eliminate a from these two equations and regard $\mathscr{L}$ as a function

$$\mathscr{L} = \mathscr{L}(\omega, k). \tag{1}$$

For a more general wavetrain, a phase function $\theta(x, t)$ is used, following Whitham (1965 b and preceding paper), to specify its development, where θ varies smoothly and takes successive integer values on successive wave crests. Then the local frequency and wavenumber are defined as

$$\omega = -\frac{\partial\theta}{\partial t}, \quad k = \frac{\partial\theta}{\partial x}, \tag{2}$$

and the Euler equation for the variational principle

$$\delta \iint \mathscr{L}\left(-\frac{\partial\theta}{\partial t}, \frac{\partial\theta}{\partial x}\right) dt\, dx = 0 \tag{3}$$

specifying the function $\theta(x, t)$ can be written

$$\frac{\partial}{\partial t}\left(\frac{\partial\mathscr{L}}{\partial\omega}\right) = \frac{\partial}{\partial x}\left(\frac{\partial\mathscr{L}}{\partial k}\right). \tag{4}$$

Equations (1) and (2) show that this can be written out as a second-order quasi-linear partial differential equation for θ, namely,

$$\frac{\partial^2\mathscr{L}}{\partial\omega^2}\frac{\partial^2\theta}{\partial t^2} - 2\frac{\partial^2\mathscr{L}}{\partial\omega\,\partial k}\frac{\partial^2\theta}{\partial t\,\partial x} + \frac{\partial^2\mathscr{L}}{\partial k^2}\frac{\partial^2\theta}{\partial x^2} = 0. \tag{5}$$

Such an equation can be transformed into a linear equation by a Legendre transformation, which uses the first derivatives (2) as new independent variables. The new dependent variable and its first derivatives are

$$\phi(\omega, k) = kx - \omega t - \theta, \quad \frac{\partial\phi}{\partial\omega} = -t, \quad \frac{\partial\phi}{\partial k} = x, \tag{6}$$

and ϕ satisfies the linear equation

$$\frac{\partial^2\mathscr{L}}{\partial\omega^2}\frac{\partial^2\phi}{\partial k^2} - 2\frac{\partial^2\mathscr{L}}{\partial\omega\,\partial k}\frac{\partial^2\phi}{\partial\omega\,\partial k} + \frac{\partial^2\mathscr{L}}{\partial k^2}\frac{\partial^2\phi}{\partial\omega^2} = 0. \tag{7}$$

All solutions of (5) are solutions of (7) (Courant & Hilbert 1962) except those very simple solutions, easily obtained in other ways, for which ω is a fixed function of k.

The characteristics of a linear equation are fixed curves. Those of equation (7) are, in fact, the 'asymptotic lines' covering the surface $\mathscr{L} = \mathscr{L}(\omega, k)$. These are given by

$$\frac{\partial^2\mathscr{L}}{\partial\omega^2} d\omega^2 + 2\frac{\partial^2\mathscr{L}}{\partial\omega\,\partial k} d\omega\, dk + \frac{\partial^2\mathscr{L}}{\partial k^2} dk^2 = 0, \tag{8}$$

and their directions at any point are those in which the plane tangent to the surface at that point cuts the surface. Thus, they are real, so that the equation is hyperbolic, only where the surface has principal curvatures of opposite sign.

Lighthill (1965 b) showed that, if the dispersion relation for infinitesimal amplitude is $\omega = f(k)$, then for moderate amplitudes this condition for the equation to be hyperbolic is satisfied provided that

$$[\omega - f(k)]f''(k) \tag{9}$$

is positive, that is, provided that the point (ω, k) for moderate amplitude lies on the side of the curve $\omega = f(k)$ towards which that curve is concave. Conversely, if (9) is negative, so that (ω, k) for moderate amplitude is in the region towards which the infinitesimal-amplitude curve is convex, then (7) is elliptic. This is the appropriate condition because the mean Lagrangian vanishes (to order amplitude squared) for infinitesimal amplitudes, and therefore the surface $\mathscr{L} = \mathscr{L}(\omega, k)$ touches the plane $\mathscr{L} = 0$ all along the curve $\omega = f(k)$. Evidently, then, it can have double curvature only on the side towards which that curve is concave.

For waves on deep water, the infinitesimal-amplitude curve is convex upwards ($f(k)$ is $\sqrt{(gk)}$, in fact,) while for finite amplitudes ω increases *above* $\sqrt{(gk)}$. Accordingly, the curve $\omega = f(k)$ is convex to the point (ω, k) (and (9) is negative), in agreement with Whitham's conclusion in the preceding paper that the equations are elliptic in this case.

Two kinds of experimental set-up seem especially suitable for comparison with theory for waves on deep water. Lighthill (1965 b) concentrated on the case where a wavemaker at one end of a long tank is used to create a group of long-crested waves, whose progress down the tank is then observed in sufficient detail to show up any nonlinear interactions. It seems that, independently and simultaneously, Dr Brooke Benjamin and I came to the conclusion that this was a key problem for investigation. It was not the first time that we had embarked on similar work simultaneously, and I believe that as on previous occasions the combination of both investigations uncovered more than either separately would have done!

I shall go into some detail on these wavemaker cases, but I want first to describe another kind of experimental set-up that has not previously been noticed as a suitable field of application for the Whitham theory, but which may prove to be one of the most important. This is the kind of stationary wave pattern that can be generated on a steady uniform stream.

3. APPLICATION OF THE WHITHAM THEORY TO STATIONARY WAVES ON A STEADY FLOW

I shall discuss this mainly in relation to the sort of wave pattern that can be set up on the surface of a deep-water stream by an obstacle in the flow, or, what is the same thing, set up on deep water at rest by a uniformly moving obstacle on or near the surface. However, what I shall say is applicable in any two-dimensional homogeneous fluid system in which plane periodic waves with arbitrary wavenumber (k_1, k_2) are possible for a range of values of amplitude, and in which there is relative motion between an obstacle (acting as a steady source of disturbances) and the fluid. The deep-water case is the obvious one for experimental comparisons, but it seems likely that other applications, such as some kind of plasma waves, may be found.

The study of stationary waves on a steady flow may, in fact, turn out to be more useful than the study of wavemaker problems. In the familiar, non-dispersive, acoustic case, the nonlinear propagation theory turned out to be rather less widely useful in one-dimensional unsteady-flow cases than in two-dimensional steady-flow cases, which include the supersonic aerofoil theory. In general, moreover, in this or any other kind of fluid flow, the study of a steady motion can be experimentally more convenient. With water waves it is arguable, furthermore, that an experimental set-up in which wave crests lie in a range of different directions can be investigated more easily than one in which waves of accurately unidirectional character must be produced and maintained. There is, lastly, the possibility, further discussed below, of practical application to problems of the wave pattern generated by uniform motion of a ship through deep water.

The mathematical approach to the study whose usefulness is suggested by these arguments is very simple indeed, if Whitham's basic variational idea is applied directly. To do this, suppose first that

$$\mathscr{L}_s(k_1, k_2) \tag{10}$$

is the Lagrangian density (per unit horizontal area) for a periodic stationary wave with wavenumber $\mathbf{k} = (k_1, k_2)$ on a stream flowing with uniform velocity $(U, 0)$. In this Lagrangian the kinetic energy which appears is that of the motions relative to the uniform stream.

Consider now a general stationary wave pattern, in which the phase function θ, being independent of time, satisfies

$$\theta = \theta(x_1, x_2), \quad \partial\theta/\partial x_i = k_i. \tag{11}$$

Then Hamilton's principle gives

$$\delta \iint \mathscr{L}_s \left(\frac{\partial\theta}{\partial x_1}, \frac{\partial\theta}{\partial x_2} \right) dx_1 dx_2 = 0, \tag{12}$$

whose Euler equation is simply

$$\frac{\partial}{\partial x_1} \left(\frac{\partial\mathscr{L}_s}{\partial k_1} \right) + \frac{\partial}{\partial x_2} \left(\frac{\partial\mathscr{L}_s}{\partial k_2} \right) = 0. \tag{13}$$

This deduction of the basic equation, for a wave pattern with wave crests in all directions, is so surprisingly simple that it may be thought desirable to check it by an alternative method. To do this, suppose that, in the fluid at rest, a wave of frequency ω and wavenumber $\mathbf{k}$ has Lagrangian density $\mathscr{L}(\omega, k_1, k_2)$ and consider a wave pattern set up by a steady forcing effect travelling through the fluid at a velocity $(-U, 0)$. Such a pattern must have a phase function of the form

$$\theta = \theta(x_1 + Ut, x_2), \tag{14}$$

and evidently the function on the right must be the same as the phase function for the problem, just considered, with stationary waves on the flowing stream.

But on Whitham's theory, any system of waves in the fluid satisfies

$$\frac{\partial}{\partial t} \left(\frac{\partial\mathscr{L}}{\partial \omega} \right) = \frac{\partial}{\partial x_1} \left(\frac{\partial\mathscr{L}}{\partial k_1} \right) + \frac{\partial}{\partial x_2} \left(\frac{\partial\mathscr{L}}{\partial k_2} \right). \tag{15}$$

Furthermore, for functions of the form (14), $\partial/\partial t = U\partial/\partial x_1$. Two special cases of this are the facts that

$$\omega = -\frac{\partial\theta}{\partial t} = -U\frac{\partial\theta}{\partial x_1} = -Uk_1, \tag{16}$$

and that (15) becomes

$$\frac{\partial}{\partial x_1}\left(\frac{\partial\mathscr{L}}{\partial k_1} - U\frac{\partial\mathscr{L}}{\partial\omega}\right) + \frac{\partial}{\partial x_2}\left(\frac{\partial\mathscr{L}}{\partial k_2}\right) = 0. \tag{17}$$

But by (16) the Lagrangian density $\mathscr{L}_s$ for waves satisfying (14) is

$$\mathscr{L}_s(k_1, k_2) = \mathscr{L}(-Uk_1, k_1, k_2), \tag{18}$$

and (17) states that this is subject to the simple equation (13), previously derived directly.

In practical use of (13) it is convenient to use (x, y) as coordinates in place of (x_1, x_2) and to use $(l, -m)$ as wavenumber components in place of (k_1, k_2), so that periodic waves specified by l and m depend on $lx - my$, and their slope in the (x, y) plane is l/m. The equation (13) then becomes

$$\frac{\partial}{\partial y}\left(\frac{\partial\mathscr{L}_s}{\partial m}\right) = \frac{\partial}{\partial x}\left(\frac{\partial\mathscr{L}_s}{\partial l}\right), \tag{19}$$

where

$$m = -\frac{\partial\theta}{\partial y}, \quad l = \frac{\partial\theta}{\partial x}. \tag{20}$$

A useful feature of equations (19) and (20) is that they are absolutely identical with (4) and (2) after the substitutions

for

$$\left.\begin{array}{ccccc} y & x & m & l & \mathscr{L}_s \\ t & x & \omega & k & \mathscr{L} \end{array}\right\} \tag{21}$$

have been made. Accordingly, any piece of general theory done on the 'wavemaker' problem in the (x, t) plane can be immediately taken over and applied to the 'stationary waves on a stream' problem in the (x, y) plane, by just applying the substitutions (21). This fact adds substantially to the utility of any work done on the wavemaker problem.

To obtain the function $\mathscr{L}_s(l, m)$, equation (18) shows that we merely have to take the Lagrangian density $\mathscr{L}$ for plane waves of frequency ω and wavenumber $(l, -m)$, and put $\omega = -Ul$. For waves on deep water Lighthill (1965b) obtained $\mathscr{L}$ in the form (see also §4 below)

$$\mathscr{L} = \frac{g}{8k^2}\left[\left(\frac{\omega^2}{gk} - 1\right)^2 - \left(\frac{\omega^2}{gk} - 1\right)^3 + \ldots\right], \tag{22}$$

where k, the magnitude of the wavenumber, must here be replaced by $\sqrt{(l^2 + m^2)}$. Replacing also ω by $-Ul$, we obtain

$$\mathscr{L}_s(l, m) = \frac{g}{8(l^2 + m^2)}\left[\left(\frac{U^2l^2}{g\sqrt{(l^2 + m^2)}} - 1\right)^2 - \left(\frac{U^2l^2}{g\sqrt{(l^2 + m^2)}} - 1\right)^3 + \ldots\right] \tag{23}$$

for this problem.

Similarly the curve in the (l, m) plane corresponding to waves of infinitesimal amplitude is obtained by putting $\omega = -Ul$ in the relation between ω, l and m for infinitesimal amplitude. The curve, naturally, is that on which (23) vanishes, namely

$$U^2l^2 = g\sqrt{(l^2 + m^2)}, \qquad (24)$$

and it is shown as a plain line in figure 1. Furthermore, for finite amplitude ω^2 increases $above \; gk = g\sqrt{(l^2 + m^2)}$, so given any value of $\sqrt{(l^2 + m^2)}$ in figure 1 the value of U^2l^2 for finite amplitude is $greater$ than the value on the curve. It is, therefore, the

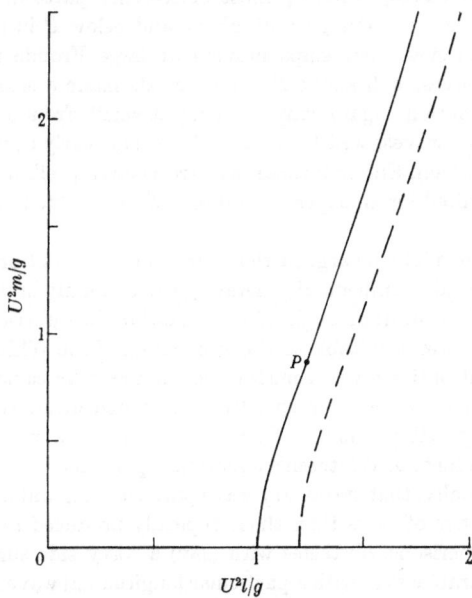

FIGURE 1. The wavenumber (l, m) of waves that are stationary on deep water moving with velocity $(U, 0)$ is represented by a point lying between the plain line (representing waves of infinitesimal amplitude) and the broken line (representing waves of maximum amplitude).

region to the right of the curve that corresponds to real waves; more precisely, the region between it and the broken line (corresponding to waves of maximum height, with $\omega^2 = 1\cdot20 \, gk$ as in (25) below).

On the infinitesimal-amplitude curve in figure 1, the points below P correspond to the long 'transverse' waves downstream of the obstacle. In this region, finite-amplitude waves lie on the side of the curve towards which it is concave. Hence these long transverse waves, generated preferentially when the obstacle is of large length L (more precisely, when the Froude number U^2/gL is small), are governed by a hyperbolic equation.

By contrast, the points above P correspond to the short 'lateral' waves trailing behind the obstacle. In this region, finite-amplitude waves lie on the side of the curve towards which it is convex. Hence these short lateral waves, generated preferentially when the obstacle is small in relation to its speed (so that U^2/gL is large), are governed by an elliptic equation.

It may be urged that the 'ship wave' problem is inherently unsuitable to the application of the Whitham theory, owing to the 'many-valuedness' difficulties mentioned in §1. Indeed, the whole of Kelvin's well known ship-wave pattern is doubly covered with waves, of the 'transverse' and 'lateral' systems, which meet in cusps on the lines of 'caustic' character that trail at an angle $\sin^{-1}\frac{1}{3} = 19\frac{1}{2}°$ behind the obstacle.

This would be a fatal objection for a ship or obstacle which generated a really broad spectrum of waves, filling up most of Kelvin's pattern, and including in particular significant wave energy both above and below P in figure 1. We have noted already, however, that ships moving at large Froude number generate mainly 'lateral' waves, well above P, whose crests make a small angle with the direction of ship motion. Again, ships moving at small Froude number generate mainly 'transverse' waves, well below P, with crests nearly at right angles to the direction of ship motion. Either of these cases (respectively, elliptic and hyperbolic) would seem well suited for an experimental test of 'one-valued' solutions of Whitham's equations.

Less plausibly, it might be argued that particular forms of obstacle, travelling at intermediate Froude number and generating waves mainly in the neighbourhood of the caustic, with crests at an angle of around 55° to the direction of ship motion, might give wave patterns suitable to the application of the Whitham theory. Certainly, geometrical optics gives a misleading picture near caustics, where exact linear theory predicts rather pronouncedly long-crested waves (Ursell 1960); non-linear effects may possibly enhance this tendency, and it cannot be ruled out that this would make a form of Whitham's equations applicable.

We may note, finally, that stationary wave patterns on a uniform stream include a much greater range of cases than those typically produced by ship forms. For experimental comparisons, obstacles with (say) a wavy side surface, that would preferentially generate waves with a particular longitudinal wavenumber, might be convenient, in addition to shapes generating a broader spectrum.

4. Form of Lagrangian for gravity waves on deep water

Gravity waves on deep water may, as we have argued, be particularly useful for comparison of the Whitham theory with experiment. It is desirable, therefore, to calculate the form of the Lagrangian density $\mathscr{L}(\omega, k)$ as accurately as possible for them throughout the range of amplitudes for which periodic waves exist. A beginning on this was made by Lighthill (1965 b), leading to the incipient series (22) quoted above.

It seems most unlikely that the way to satisfactory accuracy throughout the range lies in calculating higher terms in the Taylor expansion in powers of $(\omega^2/gk) - 1$. The algebra involved in obtaining Taylor expansions of high order for water waves is well known to be extremely complex, and the difficulty and likelihood of error increases enormously with each new term, while the interval in which the approximation is good increases only very slowly.

But, if a function is to be approximated in an interval, a Taylor series centred on one end of the interval is inherently inefficient. Instead of extending the length of the series, one may obtain much greater improvements by using knowledge of the function at the other end of the interval. In this section, the problem of approximating the Lagrangian is thus approached from both ends.

The maximum amplitude for a period wave is that for which the surface makes Stokes's 120° angle at the crest, where the fluid velocity is actually equal to the phase velocity of the wave. The remarkable calculation of the shape of this wave by Michell (1893) has stood up excellently to critical study over the intervening years, and can be confidently used. So can his result on the relation between frequency and wavenumber:

$$\omega^2 = 1 \cdot 20 gk. \tag{25}$$

Actually, Michell obtains this as a relation between phase velocity c and wavelength L, namely $c^2 = 0 \cdot 191 gL$. Mistakenly, he describes the result as showing that the velocity is $1 \cdot 2$ times its value for the infinitesimal wave, and Lamb (1932) quotes this statement, but it is really the square of c that is $1 \cdot 2$ times its value (namely $gL/2\pi$) for the infinitesimal wave, and this gives (25).

The Lagrangian density $\mathscr{L}$ as a function of ω and k will be derived from the expression for the potential energy V per unit horizontal area, using the result (equation (85) of Lighthill 1965 b, with a misprint ω^2 for ω^3 corrected) that

$$\frac{\partial(\mathscr{L}/\omega^2)}{\partial \omega} = \frac{2V}{\omega^3}. \tag{26}$$

This follows from Whitham's expression $\omega \partial \mathscr{L}/\partial \omega - \mathscr{L}$ for the total energy density $\mathscr{L} + 2V$ in problems without pseudo-frequencies. When $Vk^2/\rho g$ is known as a function of

$$z = \omega^2/gk \tag{27}$$

from $z = 1$ to $1 \cdot 20$ (see (25)), then $\mathscr{L} k^2/\rho g$ can be deduced, with the aid of (26) in the integrated form

$$\mathscr{L} k^2/\rho g = z \int_1^z (Vk^2/\rho g) z^{-2} dz. \tag{28}$$

The first two terms,

$$Vk^2/\rho g = \tfrac{1}{4}(z-1) - \tfrac{1}{4}(z-1)^2 + O(z-1)^3, \tag{29}$$

in the expansion of the potential energy about $z = 1$ can be obtained by several different methods, and are implicit in equations (90) and (91) of Lighthill (1965 b). Also, the value of the potential energy for $z = 1 \cdot 2$ is easily obtained from Michell's coordinates of the surface, by plotting it as a curve $y = y(x)$ and using

$$V = \frac{1}{\lambda} \int_0^\lambda \tfrac{1}{2}\rho g(y - \bar{y})^2 dx, \quad \text{where} \quad \bar{y} = \frac{1}{\lambda} \int_0^\lambda y \, dx. \tag{30}$$

This gives

$$(Vk^2/\rho g)_{z=1 \cdot 20} = 0 \cdot 0334, \tag{31}$$

where at most the last figure is uncertain.

The first term in (29) would give $0 \cdot 05$ at $z = 1 \cdot 2$, and the parabolic approximation (first two terms) $0 \cdot 04$. Evidently, the deviation from a parabola is not so very great, and the possible forms of curve (which must osculate the parabola at $z = 1$)

are limited. Now, the value of $\mathscr{L}k^2/\rho g$ at $z = 1\cdot2$ calculated from (28) is $0\cdot0041$ if the parabolic approximation to V is used, and is reduced to $0\cdot0038$ if any rather smooth curve, such as a cubic, osculating the parabola at $z = 1$ and satisfying (31) is substituted for V. The correction is small, and almost certainly correct to two figures. Taking $\mathscr{L}k^2/\rho g = 0\cdot0038$ at $z = 1\cdot2$ we obtain from (26) that $\mathrm{d}(\mathscr{L}k^2/\rho g)/\mathrm{d}z = 0\cdot031$.

A simple polynomial approximation to $\mathscr{L}$ satisfying these two conditions at the end $z = 1\cdot2$ of the interval exactly, as well as two conditions at $z = 1$ (namely, possessing the known values of its second and third derivatives there) is

$$\mathscr{L}k^2/\rho g = \tfrac{1}{8}[(z-1)^2 - (z-1)^3 - (z-1)^4]. \tag{32}$$

Such a polynomial approximation satisfying equal numbers of conditions at both ends of an interval is usually particularly efficient, and it is put forward as suitable for use in all applications of the Whitham theory to waves on deep water.

When it is necessary to relate z to the height H between crest and trough it is suggested that the relation

$$z = 1 + (\pi H/\lambda)^2 \tag{33}$$

be regarded as an adequate approximation, since it incorporates the first two terms of the expansion in powers of H/λ correctly and is practically exact also in Michell's case $(z = 1\cdot2, H = 0\cdot142\lambda)$.

5. Weak slow modulations to a train of deep-water waves

Benjamin describes in his paper (p. 59 below) the experiments by Feir and himself in which a train of water waves of fixed frequency is created, with a constant average amplitude on which a weak modulation of amplitude is superimposed. He also describes the development of the modulation, by means of a theory that can be expected to be valid (i) *while the modulation remains small*, provided (ii) *that the initial amplitude is moderate*. Finally, he shows that, when these conditions are satisfied, the experiments agree rather well with the theory.

In this paper, the Whitham theory will be used to calculate the development in two different cases, namely when either (i) or (ii) above is relaxed. However, the basic assumption of the Whitham theory demands (iii) *that the modulation is slow*.

Thus, Benjamin's theory requires (i) and (ii) but not (iii). A theory requiring (ii) and (iii) but not (i) will be given in §9; while this section gives a theory requiring (i) and (iii) but not (ii), and compares it with Benjamin & Feir's experiments, including those for larger initial amplitudes. Specifically, using the Lagrangian for arbitrary amplitude calculated in §4, we calculate the development of weak slow modulations to a train of deep-water gravity waves during the period while they remain weak.

If the undisturbed values of ω and k are ω_0 and k_0 then small departures of the phase θ from the undisturbed value $k_0 x - \omega_0 t$ satisfy equation (5) with the coefficients replaced by constants, namely their undisturbed values $\mathscr{L}_{\omega\omega0}$, $\mathscr{L}_{\omega k0}$ and $\mathscr{L}_{kk0}$. At the wavemaker $x = 0$, they satisfy

$$\theta = -\omega_0 t, \quad \partial\theta/\partial x = k_0 + \epsilon\,\mathrm{e}^{\mathrm{i}xt}, \tag{34}$$

so that the frequency $\omega = -\partial\theta/\partial t$ is fixed at ω_0 but the wavenumber (and therefore also the amplitude) is modulated. Solution of equation (5) subject to these conditions gives

$$\theta = k_0 x - \omega_0 t + \epsilon\, e^{i\alpha(t - x/c_0)} \frac{\sinh \beta x}{\beta}, \tag{35}$$

where c_0 is an 'effective group velocity' (velocity of propagation of the modulation along the tank), given by

$$c_0 = -\mathscr{L}_{kk0}/\mathscr{L}_{\omega k0}, \tag{36}$$

and β is a rate of exponential increase of modulation with distance given by

$$\beta = \alpha(\mathscr{L}_{\omega\omega0}\mathscr{L}_{kk0} - \mathscr{L}_{\omega k0}^2)^{\frac{1}{2}}/\mathscr{L}_{kk0}, \tag{37}$$

being real if the equation is elliptic (as we know that it is, from §2, at any rate for moderate amplitude).

If β were imaginary, the solution (35) would still be valid but the interpretation would be different. No exponential increase would occur but the modulation would propagate at two speeds c_1 and c_2 given by

$$\frac{\alpha}{c_1} = \frac{\alpha}{c_0} - |\beta|, \quad \frac{\alpha}{c_2} = \frac{\alpha}{c_0} + |\beta|. \tag{38}$$

This is the 'splitting of the group velocity' in the hyperbolic case which Whitham (1965 a) first noted.

It is easy to compute c_0 and β from the value of $\mathscr{L}$ given by (32). Such a computation shows that although β is real for moderate amplitude it becomes imaginary for $z_0 > 1\cdot115$, where $z_0^2 = \omega_0^2/gk_0$. This means that for amplitudes exceeding about three-quarters of the maximum amplitude the equation is hyperbolic.

The results of the computations are given in figure 2, all speeds being related to the undisturbed phase velocity ω_0/k_0. Results of calculating β are expressed by means of the fraction $2\pi c_0\beta/\alpha$, which is the logarithmic increase in depth of modulation in the distance $2\pi c_0/\alpha$ between successive maxima of wave amplitude. For height/length ratios between $0\cdot06$ and $0\cdot09$, this fraction is close to $0\cdot7$, so that the depth of modulation associated with a particular (travelling) amplitude peak will have doubled approximately by the time the next amplitude peak is generated at the wavemaker. For small height/length ratios it is

$$(\pi^2\sqrt{2})(H/\lambda) = 14H/\lambda;$$

all theories, including Brooke Benjamin's, give this result when conditions (i), (ii) and (iii) all hold.

For height/length ratios exceeding $0\cdot108$, however, there is no increase in depth of modulation. By contrast, the splitting of the group velocities becomes very pronounced, and one of them increases to values considerably in excess of the phase velocity.

Other papers in this Discussion demonstrate that the basic instability of long-crested gravity waves on deep water under slow modulation is subject to a cut-off when the depth is reduced below a certain value (Whitham) or when the frequency of modulation is increased above a certain value (Benjamin). The present section has added to these results the information that a third cutoff appears when the amplitude exceeds a certain value.†

Experimental results on rate of logarithmic increase of depth of modulation of a periodic wave are summarized in figure 4 of Benjamin's paper (p. 73 below), for the case $\alpha = 0.1\omega_0$. This means that the time period of amplitude modulation is ten cycles, while its *spatial* period is about five wavelengths (rather low in relation to Whitham's main assumption (iii)).

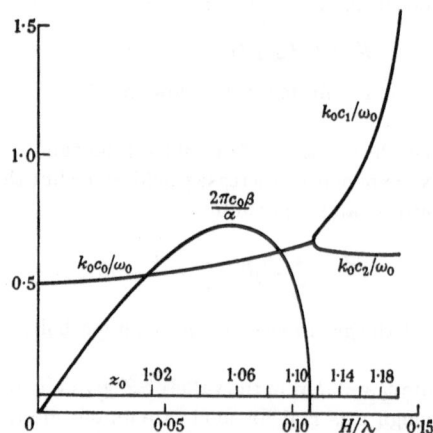

FIGURE 2. For unidirectional periodic waves on deep water, with frequency ω_0 and wavenumber k_0, slightly modulated at a frequency $\alpha \ll \omega_0$, the rate β of exponential increase of depth of modulation with distance, and the velocity c_0 of propagation of amplitude peaks, are plotted as a function of H/λ (vertical distance between crest and trough divided by wavelength). When $H/\lambda > 0.108$, β is zero but the propagation velocity splits into two (c_1 and c_2).

The present figure 3 compares those experimental points with three theoretical curves: the dash-dotted line, based on all the assumptions (i) (ii) and (iii) above; the broken line, from Benjamin & Feir's theory, assuming only (i) and (ii) and expected to be good except at the higher values of H/λ; and the full line, based on the theory of this section, assuming only (i) and (iii).

The error in using Whitham's assumption (iii) when $\alpha = 0.1\omega_0$ may be estimated, by comparing the broken and dash-dotted lines, as being substantial only for the smaller values of H/λ (say, less than 0.05).‡ The error in using the moderate-amplitude assumption (ii) may be estimated, by comparing the plain and dash-dotted lines, as being substantial only for the larger values of H/λ (say, greater than 0.04).

† A fourth extension of the basic theory is contributed by Phillips's paper, with its consideration of three-dimensional modes of instability.

‡ This is because Whitham's assumption requires the relaxation time, during which energy exchange between the fundamental and its harmonics brings them into the proportions characteristic of periodic waves of a given amplitude, to be small compared with the period. For given period this requires (since the relaxation time increases with decreasing amplitude) that the amplitude be not too small.

These estimates suggest that the development of a modulation, while it remains a small one (assumption (i)), should be given to reasonable accuracy, independently of assumptions (ii) and (iii), by the dotted curve which makes a transition between the broken and plain lines at about $H/\lambda = 0.045$.

Evidently, the neglect of viscous dissipation in all the theories would lead one to expect the experimental points to lie below even this curve, but it is encouraging that they lie considerably closer to it than to any of those based on assumptions (ii) or (iii). More results, however, are needed for higher values of H/λ.

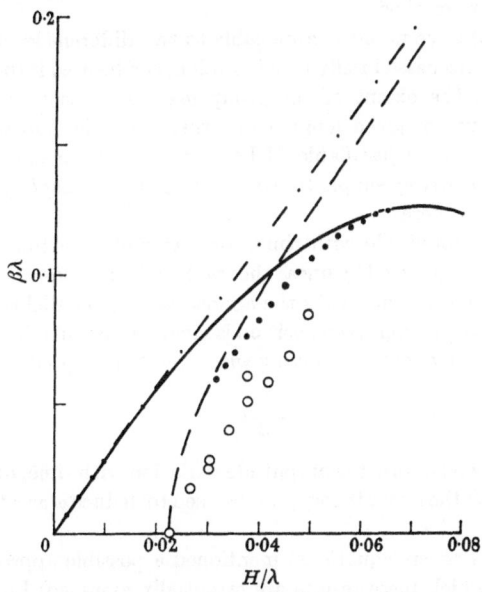

FIGURE 3. Rate of exponential increase in depth of modulation per unit wavelength, $\beta\lambda$, plotted against H/λ. ———, Theory of §5 (and figure 2); ————, Theory of Benjamin & Feir; —·—·—, Limit of plain line for small H/λ or of broken line for large H/λ; ...., tentative interpolation between plain and broken line; O, experimental results of Benjamin & Feir.

6. DEVELOPMENT OF WAVE GROUP OF MODERATE AMPLITUDE OVER AN EXTENDED PERIOD

We turn now to problems when conditions (ii) and (iii) of §5, but not (i), are satisfied. The object is to study modulations after they have grown in depth and become comparable with the initial amplitude; or, more generally, to study any group of waves (not necessarily a modulated periodic wave) if the parameters vary slowly through the group and the amplitude everywhere remains moderate.

It might be thought that the restriction to moderate amplitude would prevent the theory from obtaining any results that could not just as easily be obtained by

17*

mode-interaction methods. This supposition would be incorrect, because the theory gives the development of a group of waves over a time comparable with the duration of the group divided by a nondimensional measure of amplitude. This is a time long enough for each piece of wave energy to have been subjected to a very large number of mode interactions. Their individual calculation would be very difficult, but the smoothed-out result of them should be given correctly by the present theory.

Similarly, in the nondispersive case, as I recalled in my introductory remarks, small nonlinear terms produce over long distances of propagation a cumulative influence whose effects are rather large. This influence, also, cannot be calculated by mode-interaction methods.

Lighthill (1965 b) gave methods applicable to two different kinds of group. The first method, to be discussed briefly in this section, can be used if the spread of wave frequency through the extent of the group greatly exceeds any differences in frequency (for waves of given length) resulting from the spread in amplitude. Approximations are then justifiable, if both spreads are smooth enough, which makes the theory relatively simple to work out (Lighthill 1965 b, §5), and only the results will be quoted here.

The variation of amplitude with time, for waves of a particular length, turns out to be the same as predicted by linear theory; but their *position* is rather different; and this combined statement recalls the almost exactly parallel result in the non-dispersive case. More precisely, the position is given by the rule that waves of wave-number k are found in a location moving with time at the speed

$$\left(\frac{\partial \omega}{\partial k}\right)_l, \tag{39}$$

whose value is known because the amplitude variation with time, obtained for each k in accordance with the linear theory, can be used to deduce ω as a function of t and k.

Whitham (§4·3, first six equations) mentioned a possible approximate method of calculation, in which these results are essentially assumed; his third equation says that energy propagates as in the linear theory while his fifth, which follows from (2) above, can be written

$$0 = \frac{\partial k}{\partial t} + \frac{\partial \omega}{\partial x} = \frac{\partial k}{\partial t} + \left(\frac{\partial \omega}{\partial k}\right) \frac{\partial k}{\partial x}, \tag{40}$$

and states therefore that k is constant for changes dx and dt which are in the ratio (39). This method was mentioned by him as an approximate alternative to the variational method, but Lighthill's inference of it from the variational equations, when amplitudes are moderate and frequency spreads are large in comparison, indicates that those are the conditions for the approximation to be good.

The corresponding result in the ship-wave problem (§3) is that waves of given longitudinal wavenumber l have their amplitude varying with distance y from the ship's track as in linear theory, but their position satisfies

$$\left(\frac{\partial x}{\partial y}\right)_l = \left(\frac{\partial m}{\partial l}\right)_y. \tag{41}$$

This problem only will be further treated here, as problems in the (x, t) plane were fully discussed by Lighthill (1965 b).

On linear theory, the point (l, m) lies on the plain-line curve shown in figure 1, whose equation (24) can be written

$$m = \left(\frac{U^4 l^4}{g^2} - l^2\right)^{\frac{1}{2}} = f(l). \tag{42}$$

For finite amplitudes the dependence of m on l is different, and this alters the slope (41). In particular, for moderate amplitudes, the difference

$$w = m - f(l) \tag{43}$$

varies as the square of the amplitude, in a manner obtained by putting $\omega = -Ul$ in the relation between ω, l, m and the amplitude.

On linear theory, if waves of longitudinal wavenumber l are at $x = x_0(l)$ where $y = 0$, then for greater values of y they are at

$$x = x_0(l) + yf'(l), \tag{44}$$

whereas waves of longitudinal wavenumber $l + \delta l$ are at

$$x = x_0(l) + x_0'(l)\,\delta l + y[f'(l) + f''(l)\,\delta l]. \tag{45}$$

Hence, the longitudinal separation between the waves of wavenumbers l and $l + \delta l$ increases proportionally to $1 + \alpha(l)y$, where

$$\alpha(l) = f''(l)/x_0'(l). \tag{46}$$

Furthermore, the value of w, proportional to the square of the amplitude and hence to the y component of energy flux, must vary inversely as the space available so these waves, namely as

$$w = \frac{w_0(l)}{1 + \alpha(l)y}, \tag{47}$$

where $w_0(l)$ is the value for $y = 0$.

For moderate amplitudes, the variation of w with y for given l is approximately the same as in the linear case (47), but the position of the waves is governed by (41), giving

$$\left(\frac{\partial x}{\partial y}\right)_l = f'(l) + \left(\frac{\partial w}{\partial l}\right)_y$$

$$= f'(l) + \frac{w_0'(l)}{1 + \alpha(l)y} - \frac{w_0(l)\,\alpha'(l)}{[1 + \alpha(l)y]^2}. \tag{48}$$

Accordingly, equation (44) is replaced by

$$x = x_0(l) + yf'(l) + \frac{w_0'(l)}{\alpha(l)} \ln\left[1 + \alpha(l)y\right] - \frac{w_0(l)\,\alpha'(l)\,y}{1 + \alpha(l)y}. \tag{49}$$

Evidently, fairly substantial departures from the linear-theory positions can build up as y increases.

For small values of y, the initial change in slope (41) due to finite amplitude can be found most conveniently by plotting the l, m relation for $y = 0$ by use of figure 1. Evidently changes of several degrees are possible if the amplitude varies substantially with wavelength. The wedge of waves of given l is widened (because $\partial x/\partial y$ is reduced) if the amplitude is an increasing function of l.

7. CASE OF SMALL FREQUENCY SPREAD

I shall return now to the wavemaker problem, and consider it in the second case foreshadowed in §6, namely the case of small frequency spread. This complicates rather than simplifying the moderate-amplitude problem, because it means that changes in position of waves due to amplitude variations can completely upset the changes that would have occurred due to frequency variations alone. The two influences are in evenly matched competition and the former cannot, as in §6, be treated as a perturbation on the latter.

In elliptic cases, which alone will be treated here, and which include as we have seen the moderate-amplitude deep-water case, the analysis of this competition is greatly assisted by the introduction of a new variable, defined in general as

$$s = 2\left[-\frac{\omega - f(k)}{f''(k)}\right]^{\frac{1}{2}}, \tag{50}$$

so that it varies directly as the amplitude for moderate amplitude. For waves of amplitude a on deep water, for example,

$$s \doteq 2k^2a \sqrt{2} \tag{51}$$

for moderate values of the maximum wave slope ka.

Lighthill (1965b) showed that, if amplitudes remain only very moderate, the leading terms in equation (7), as transformed by taking s and k as independent variables, are

$$\frac{\partial^2\phi}{\partial k^2} + \frac{\partial^2\phi}{\partial s^2} + \frac{1}{s}\frac{\partial\phi}{\partial s} = 0. \tag{52}$$

This is a valuable transformation because the resulting approximate equation is a rather simple one (the axisymmetric Laplace equation) and because it is the same for all one-dimensional wave systems of moderate amplitude without pseudo-frequencies. Any work that is done in solving boundary-value problems for equation (52) is applicable equally to all such systems, and also (after the substitutions (21)) to the analogous systems of stationary waves on a stream.

Boundary conditions appropriate to (52) are easily obtained from the relations

$$t = -\frac{2}{\mu s}\frac{1}{\partial s}\frac{\partial\phi}{\partial s}, \quad x - u_0 t = \frac{\partial\phi}{\partial k} + 2\frac{k - k_0}{s}\frac{\partial\phi}{\partial s} \tag{53}$$

between the original independent variables (x, t) and those obtained after the Legendre transformation (50). In (53), k_0 is a reference wavenumber about which variations of k are only small, while

$$f'(k_0) = u_0, \quad f''(k_0) = -\mu. \tag{54}$$

Evidently, the boundary conditions appropriate to any problem of wave group development are of initial-value type; two conditions are given on some initial curve in the (s, k) plane. They may actually be conditions describing the group at $t = 0$; they state then that along a certain curve in the (s, k) plane t vanishes and x is a known function. Alternatively, for waves created at a fixed wavemaker, x vanishes and t is a known function on the curve.

The combination of an elliptic differential equation (52) with boundary conditions of this initial-value or hyperbolic type means that the problem is what used to be called an 'incorrectly set' one. Although such combinations have been increasingly used in recent years, and it has become recognized that their derivation and solution can play a role of value in the discussion of physically important questions, there are still perhaps some philosophical questions to answer when such a problem arises. Especially there are the difficulties that, unless the initial values satisfy 'smoothness' conditions, no solution may exist at all, and that, even if they do, a solution is unlikely to be free of singularities throughout the plane.

The philosophical difficulties are not, in reality, substantial in the case of Whitham's theory. The theory states only what the equations are *if variations are smooth enough*. Solution of the equations determines subsequent development *so long as* variations remain smooth. If singularities appear, this simply means that after a finite time, which can be calculated, the parameters governing the waves (wavelength and amplitude) cease to vary smoothly. There is no physical reason why this should not happen, and a theory which predicts when it will happen has evident physical interest.

An excellent method for solving elliptic equations subject to hyperbolic boundary conditions is the method of characteristics. Generally speaking, the 'type' of the boundary condition, more than that of the equation, determines the best method of solution. Riemann's methods of characteristics is excellently suited to initial-value problems. Several authors (see, for example, Garabedian 1958) have shown that the imaginary nature of the characteristics for an elliptic equation does not create insuperable difficulties in the application of the method. The value of ϕ at any point can be expressed as an integral along the initial curve, between the points where the two characteristics through (k, s) intersect it. These points are imaginary; but the integral can be evaluated provided that the initial values can be continued analytically into the complex plane as far as those points. The form of the integral for equation (52) is particularly simple because that equation has a particularly simple Riemann function.

8. Case of Initially Uniform Wavelength

A particular problem that illustrates well the general situation described in §7 is that of calculating the development in time of a group of waves of uniform length but variable amplitude, so that

$$k = k_0, \quad s = s_0(x) \quad \text{at} \quad t = 0. \tag{55}$$

Equations (53) then give

$$\phi = 0, \quad \partial\phi/\partial k = x_0(s) \quad \text{on} \quad k = k_0, \tag{56}$$

where the arbitrary constant from integrating $\partial\phi/\partial s = 0$ has been taken as zero and where

$$x = x_0(s) \quad \text{is the inverse of} \quad s = s_0(x). \tag{57}$$

Lighthill (1965 b) shows that two other problems of interest lead to the same boundary conditions (56) within the same order of approximation as gave equation (52). First, if the initial conditions are given not at $t = 0$ but at a fixed wavemaker $x = 0$, $s = s_1(t)$, with inverse $t = t_1(s)$, then to that approximation we still obtain (56), with $x_0(s) = -u_0 t_1(s)$. Secondly, if the wavemaker operates so as to keep the frequency fixed and equal to $f(k_0)$ (instead of k fixed and equal to k_0), the difference in the boundary condition (56) produces small changes in the solution, which to the same order of approximation can be neglected.

The solution of (52) by means of its known Riemann function, subject to the boundary conditions (56), is

$$\phi = -\tfrac{1}{2}i \int_{s-i(k-k_0)}^{s+i(k-k_0)} x_0(S) \left(\frac{S}{s}\right)^{\frac{1}{2}} F\left(\tfrac{1}{2}, \tfrac{1}{2}; 1; \frac{(k-k_0)^2 + (s-S)^2}{-4sS}\right) dS, \tag{58}$$

where F is a hypergeometric function and the limits of integration are the values of s where the characteristics through (k, s) intersect the initial curve $k = k_0$. The integral (58) can be calculated if only $x_0(s)$ can be analytically continued into the complex plane.

Now, wherever the initial amplitude variation $s = s_0(x)$ is analytic in a domain including the real axis, the inverse function $x = x_0(s)$ has locally just the same property, except at stationary values of $s_0(x)$. We are interested, for example, in groups whose amplitude variation $s_0(x)$ possesses a maximum, say $s_0(x_m) = s_m$. The inverse function $x_0(s)$ in such a case is analytic in the complex s plane only if this is cut from s_m to ∞. The integral (58) must then be taken along a path that does not cross the cut. Lighthill (1965 b) shows that, provided this is done, no singularities in the solution (58) are introduced by the presence of the cut, even for values of $s > s_m$.

He considers a particular form of initial variation of amplitude in the group,

$$s = s_0(x) = \frac{s_m}{1 + (x/X)^2} \tag{59}$$

(here x_m, the value where $s_0(x)$ is a maximum, is zero, and X, the distance in which the amplitude falls to $\tfrac{1}{2}s_m$, must be large on a scale of wavelengths). In this case he shows that the solution does not involve any singularity in the (s, k) plane; nevertheless, after a certain critical time

$$t_{\text{crit.}} = 0.69 X/\mu s_m \tag{60}$$

inversely proportional to the maximum amplitude s_m, a singularity in the transformation between the (x, t) and (k, s) planes appears, and the condition of smoothness breaks down in the physical plane. The distance $u_0 t_{\text{crit.}}$ travelled by the group during the critical time takes, for gravity waves, the form $0.49X/ka$.

Up to $t = t_{\text{crit.}}$ the variation of amplitude in the group remains smooth, and symmetrical about a point moving with velocity u_0. This fact, that s is an even

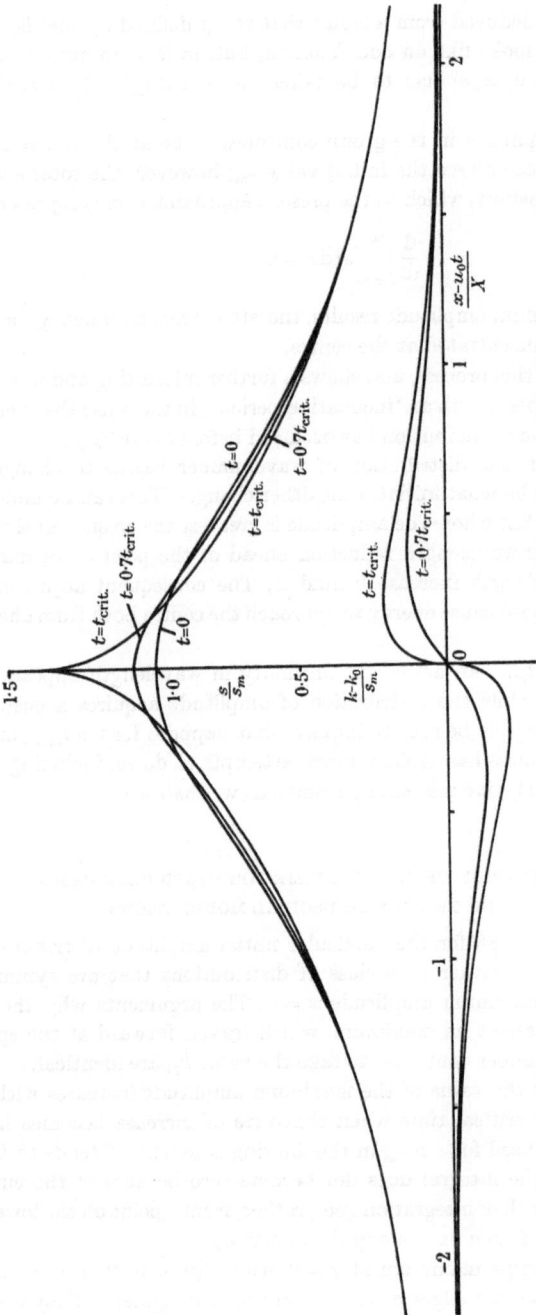

FIGURE 4. Change with time in the distribution of the wavenumber k and the amplitude variable s (see equation (51)) in a wave group initially of uniform wavenumber k_0 and with s satisfying (59). Here $u_0 = f'(k_0)$ and the critical time $t_{\text{crit.}}$ is defined by (60), where $u = -f''(k_0)$, and for waves on deep water $f(k) = \sqrt{(gk)}$.

function of $x - u_0 t$ is deduced from a result that the ϕ defined by (58) is an even function of $k - k_0$ (it looks like an odd function, but, in fact, to obtain solutions corresponding to $t > 0$, $x_0(s)$ has to be taken as $\pm X \sqrt{(s_m/s - 1)}$ according as $k \gtrless k_0$).

The maximum amplitude in the group continues to be at the centre $x = u_0 t$, and is found to increase above the initial value s_m; however, the total energy in the group remains constant, which to the present approximation is represented by the fact that

$$\frac{d}{dt} \int_{-\infty}^{\infty} s^2 \, dx = 0. \tag{61}$$

The increase in maximum amplitude results, therefore, from a tendency for energy to become actually concentrated at the centre.

Figure 4 illustrates this process, and shows a further interesting and unexpected feature of it. It is a process with an 'incubation period', in the sense that very little change in the amplitude distribution has occurred before $t = 0.7 t_{\text{crit.}}$.

On the other hand, the distribution of wavenumber begins to change more quickly, and seems to be what initiates the other changes. This can be understood physically by saying that where the amplitude is greatest the crests travel forward fastest. This produces wavelength reduction ahead of the position of maximum amplitude, and wavelength increase behind it. The consequent adjustments in energy propagation speed cause energy to approach the centre both from ahead and from behind.

Ultimately, at $t = t_{\text{crit.}}$, an actual discontinuity in wavelength appears at the centre of the group, while the distribution of amplitude acquires a cusp there. The theory cannot properly be used to inquire what happens for $t > t_{\text{crit.}}$, after the assumption of smoothness has broken down; attempts to do so, including that in § 16 of Lighthill (1965 b), give misleading results, as we shall see.

9. DETERMINATION OF CRITICAL TIME FOR PARTICULAR CASES, INCLUDING THAT OF SINUSOIDAL MODULATIONS

The results quoted in § 8 for the particular initial amplitude distribution (59) are obtained also for a rather wise class of distributions that are symmetrical about a position of maximum amplitude $x = 0$. The arguments why the group remains symmetrical about its maximum, which travels forward at the speed u_0 and where the wavenumber continues to take the value k_0, are identical.

In order to see how the value of the maximum amplitude increases with time, and to determine the critical time when that rate of increase becomes infinite, equation (58) may be used for $s > s_m$ in the limiting case when k tends to k_0 from above. In this limit the integral does not become zero because of the cut from $s = s_m$ to $s = \infty$. The path of integration goes, rather, from a point on the lower edge of the cut to $s = s_m$ and then back along the upper edge.

If $s = s_0(x)$ has a simple maximum at $x = 0$, with $s''(0) < 0$, then the values of its inverse $x_0(S)$ on the two edges of the cut are pure imaginary. They are most easily expressed in terms of the inverse $\xi_0(s)$ of the function $s = s_0(i\xi)$. In terms of this,

$\alpha_u(S) = i\xi_0(S)$ on the lower edge of the cut and $-i\xi_0(S)$ on the upper, so that (58) becomes

$$\phi = -\int_{s_m}^{s} \xi_0(S) \left(\frac{S}{s}\right)^{\frac{1}{4}} F\left(\tfrac{1}{2}, \tfrac{1}{2}; 1; \frac{(s-S)^2}{-4sS}\right) dS, \tag{62}$$

which together with expression (53) for the time, namely

$$t = -\frac{2}{\mu s}\frac{1}{\partial s}\frac{\partial \phi}{\partial s}, \tag{63}$$

shows how the amplitude s at the centre varies with t.

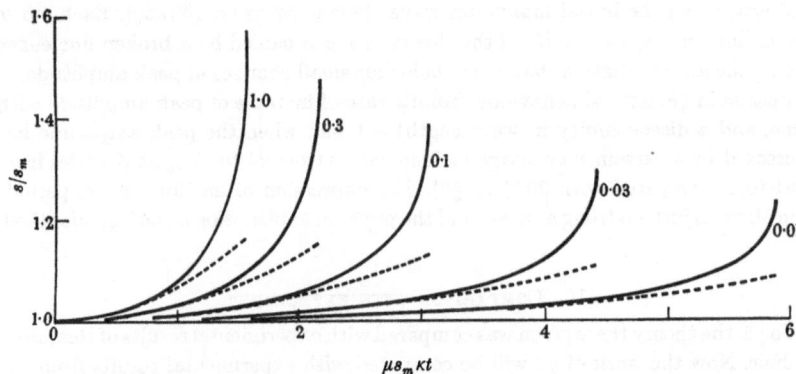

FIGURE 5. Variation in maximum amplitude with time for a wave group which initially possesses uniform wavelength and the slow sinusoidal modulation (64), where A takes the values 1·0, 0·3, 0·1, 0·03 and 0·01 marked on the curves. ———, theory of §9; ----, theory for small $s - s_m$ (equation (68)).

Some interesting cases are those for which

$$s_0(x) = s_a(1 + A \cos \kappa x), \tag{64}$$

so that the average amplitude s_a is modulated between a maximum

$$s_m = s_a(1 + A) \tag{65}$$

and a minimum $s_a(1 - A)$, where $0 \leqslant A \leqslant 1$. The assumption (i) of §5 would require that A is small, but the present method can be used for any A and applies provided only assumptions (ii) and (iii) hold; that is, if s_m is moderately small and $\kappa \ll k_0$.

Evidently, $\xi_0(s)$, the inverse of $s_0(i\xi)$, is

$$\xi_0(s) = \frac{1}{\kappa}\cosh^{-1}\left(\frac{s - s_a}{As_a}\right), \tag{66}$$

which with (62) and (63) determines the build-up of amplitude at the centre in these cases. This is given in figure 5 for $A = 1, 0·3, 0·1, 0·03$, and 0·01.

In each case the initial build-up of amplitude is exponential, as expected from §5. For when $s - s_m$ is small, (62) and (63) give

$$\tfrac{1}{2}\mu s_m t \doteqdot -\frac{\partial \phi}{\partial s} \doteqdot \xi_0(s), \tag{67}$$

which when (66) holds gives

$$s = s_a[1 + A \cosh (\tfrac{1}{2}\mu s_m t)]. \tag{68}$$

This agrees with results obtained by Benjamin's and other methods when conditions (i), (ii) and (iii) of §5 all hold; the argument $\tfrac{1}{2}\mu s_m t$ of the hyperbolic cosine increases by $(\pi \sqrt{2}) ka$ (which, in terms of height H between crest and trough, is $(\pi^2 \sqrt{2}) (H/\lambda)$ as in §5) when the distance $u_0 t$ travelled by the amplitude peak becomes equal to the distance $2\pi/\kappa$ between successive amplitude peaks.

In figure 5 the ordinate s/s_m describes the increase in peak amplitude relative to its initial value, while the abscissa $\mu s_m \kappa t$ measures time; another form of this abscissa, when the initial maximum wave slope is ka, is $(ka \sqrt{2}) \kappa u_0 t$. Each curve (plain line) giving the results of the theory is accompanied by a broken-line curve giving the approximation (68) which holds for small changes in peak amplitude.

Just as in §8, critical behaviour (infinite rate of increase of peak amplitude with time, and a discontinuity in wavelength) is found when the peak amplitude has increased by a certain percentage (taking values from 24 to 57 % as A varies from 0·01 to 1, compared with 39 % in §8). The impression of an 'incubation period' is in all cases just as strong as in §8, and the physical explanation is, no less, identical.

10. COMPARISON WITH EXPERIMENT

In §5, the theory there given was compared with experimental results of Benjamin & Feir. Now the work of §9 will be compared with experimental results from the same laboratory.

The first public description of part of the theory in this paper was made in April 1965, and consisted mainly of the calculations of §8, relating to a wave group of the particular form (59), with a single amplitude maximum at the centre of the group. Soon afterwards Mr Feir decided to do an experiment that would test the predictions of the theory in a case as similar to that one as could be realized in an apparatus allowing only sinusoidal modulation of amplitude. To this end, he caused his wavemaker to create a group closely satisfying (64) with $A = 1$ for $|x| < \pi/\kappa$, but satisfying $s_0(x) = 0$ for $|x| > \pi/\kappa$. Thus, the amplitude variation consisted of a single sinusoidal lobe only; its second but not its first derivative was discontinuous at $x = \pm \pi/\kappa$.

Comparison of his observed value for the critical time gave reasonable agreement with the theory of §8 if the value of κ was chosen so that the wave groups matched each other as closely as possible (say, in a least-squares sense). But it is obviously more satisfactory to compare the experimental findings with theoretical results for actual sinusoidal amplitude variation, as became possible when the theory of §9 had been worked out.

Evidently, the detailed comparison of the theoretical predictions when $A = 1$ in (64) could best be made with experiments (not so far done) in which (64) represented the amplitude variation not just for $|x| < \pi/\kappa$ but for a much wider range of values of x. The lack of smoothness represented by the discontinuous second derivative of amplitude variation at $x = \pm \pi/\kappa$ causes a deviation from theory, which would be present even for infinitesimal amplitude, when according to linear analysis it results in the local frequency exhibiting a narrow peak whose height increases with distance from the wavemaker.

In fact, Feir's results (see p. 54 below) exhibit just such a narrow frequency peak at the two ends of the group. However, this result of the initial lack of smoothness there seems to remain rather localized (as linear analysis, probably quite accurate at such a location where the amplitude is small, would predict) and the changes in the centre of the group up to the critical time seem to proceed independently of it.

These were measured at a distance of 8·5 m from the wavemaker for different values of the amplitude and of the modulation period. The results when the time period of amplitude modulation is 25 times the wave period may be expected, and are found, to be rather closely comparable with the Whitham theory. For some (low) amplitudes the distance of 8·5 m is less than the predicted critical distance and the observed group is then still accurately symmetrical, with its centre proceeding at the group velocity $u_0 = 31$ cm/s associated with the wave frequency 2·5 c/s. For other (greater) amplitudes, the distance of 8·5 m exceeds the critical distance and the observed group has become markedly asymmetrical.

Exact comparison of the variation of maximum amplitude with theory is impossible because the different sources of dissipation that are present (including losses at the side walls of the tank) reduce the initial energy of the group to about half in the course of 8·5 m. However, the critical distance can be compared. Theoretically, in terms of the initial maximum wave slope ka, it is

$$\frac{0\cdot54}{ka}\text{ metres.} \tag{69}$$

Experimentally, it appears to be equal to 8·5 m when $ka = 0\cdot07$, because then the wave group at 8·5 m is still symmetrical but has acquired a sharp-pointed appearance strongly reminiscent of the theoretical prediction for $t = t_{\text{crit.}}$. (The amplitude variation in Feir's figure 2 is for $ka = 0\cdot078$ and shows asymmetry just beginning to appear.)

Actually, the theoretical expression (69) equals 8·5 m for $ka = 0\cdot063$. The reason why this is less than the experimentally observed value is probably that the energy dissipation already mentioned causes the maximum wave slope to be reduced from about 0·07 to about 0·06 in the distance in question.

At the same time, the calculations described in §8 suggest (see figure 4) that wavenumber increases of about

$$0\cdot2s_m \doteqdot (0\cdot56ka)\,k \tag{70}$$

will occur in the front of the group, with corresponding decreases behind. Feir's figure 2 shows that changes of about $0\cdot033k$ in fact occur, which would agree with (70) if $ka = 0\cdot06$. In this respect, as before, the experimental evidence can be regarded as provisionally confirming the Whitham theory up to the critical time.

The post-critical state of the group is shown to be markedly asymmetric (as in Feir's figure 3). Although the assumptions underlying the Whitham theory break down in this region, Lighthill (1965 b) made a tentative adaptation of the theory for use in post-critical conditions. This adaptation must be rejected, as it predicts that the group remains symmetrical. At present no theoretical explanation of the onset of asymmetry has been given.†

A certain analogy is, nevertheless, suggestive. A symmetrical two-dimensional aerofoil, possessing also fore-and-aft symmetry, gives rise, in a steady uniform inviscid subsonic stream, to a flow with complete fore-and-aft symmetry at all Mach numbers up to a certain critical value, when a discontinuity (that is, a shock wave) first appears. Nevertheless, above that Mach number, the flow is highly asymmetrical. In the present problem the appearance of a discontinuity (in wavelength) at the centre of the group similarly triggers off a high degree of asymmetry in the amplitude distribution.

The provisional conclusion of this paper from limited comparisons with experiment for simple systems is that the Whitham theory gives reliable results in regions of space time where its assumptions are satisfied, and (in particular) where the solutions remain one-valued. Work on the analytical problems posed by its application to more complicated systems, although difficult, seems now potentially useful enough to justify fully the effort involved.

REFERENCES (Lighthill)

Courant, R. & Hilbert, D. 1962 *Methods of mathematical physics*, vol. 2, p. 35. New York: Interscience.
Garabedian, P. R. 1958 *J. Math. Phys.* **36**, 192.
Lamb, H. 1932 *Hydrodynamics*, 6th ed. Cambridge University Press.
Lighthill, M. J. 1965*a* *J. Inst. Math. Applic.* **1**, 1.
Lighthill, M. J. 1965*b* *J. Inst. Math. Applic.* **1**, 269.
Longuet-Higgins, M. S. 1953 *Phil. Trans.* A **245**, 535.
Michell, J. H. 1893 *Phil. Mag.* (5), **36**, 430.
Ursell F. 1960 *J. Fluid Mech.* **8**, 418.
Whitham, G. B. 1965*a* *Proc. Roy. Soc.* A **283**, 238.
Whitham, G. B. 1965*b* *J. Fluid Mech.* **22**, 273.

† In the preceding paper, §4·5 (added after the meeting), Whitham indicates how, perhaps, such an explanation might be initiated.

J. Fluid Mech. (1968), *vol.* 32, *part* 4, *pp.* 779–789

Phase Jumps

M. S. Howe

An earlier paper (Howe 1967) considered a non-linear theory of open-channel steady flow of deep water past a slowly modulated wavy wall. The wave pattern on the free surface of the water was obtained as the solution of a stably posed elliptic Cauchy problem, the main feature of the solution being the appearance of a 'shock' across which there is an abrupt change of phase. Such phase jumps can occur in a wide range of similar problems, but the advantage of the present case is that it is rather well suited to experimental investigation. This paper is therefore a lead-in to the more general problem of phase jumps, and uses the principle of conservation of energy in conjunction with the earlier solution to predict the possible position of the discontinuity on the free surface of the water. The possible nature of the free surface in the vicinity of the phase jump is also discussed (figure 4). This is a region where the width of the wave troughs becomes dramatically shorter than that of the neighbouring troughs. An approximate method of determining the line along which the phase jump occurs, not depending on a knowledge of the solution of the Cauchy problem, is also presented.

1. Introduction

In an earlier paper (Howe 1967) a non-linear theory of steady flow of deep water past a slowly modulated wavy wall was considered. The discussion was based on a theory proposed by Whitham (1965a, b) which describes the dispersion of slowly varying wave trains of large amplitude. Briefly, the method consists in first supposing the wave train to be locally a close approximation to an exactly periodic solution of the full non-linear equations of motion, from which an average Lagrangian is calculated in terms of the wave parameters. The dispersion equations governing the slow variation of these parameters are then obtained by an application of Hamilton's Principle.

Previous calculations of the dispersion of large amplitude wave-groups (Lighthill 1965, 1967; Whitham 1967a) had shown that in the particular case in which the dispersion equations formed an elliptic system a certain instability in the solutions of the initial value problems arose. Experimental confirmation of such instability, in the case of the development in time and space of a one-dimensional finite amplitude wave-group on deep water, has been reported by Benjamin & Feir (1967). However, because of the experimental difficulties involved in making accurate observations of moving wave trains, it appeared desirable to do a non-linear calculation for a case in which the wave pattern is steady. Such steady wave patterns could in principle be observed from a ship moving at constant velocity relative to still water but, because of difficulties mentioned by Howe (1967), it is not immediately clear how to apply the Whitham theory to the general ship-wave problem. The simpler problem of steady flow of deep water past a wavy wall of locally sinusoidal shape, whose amplitude decays slowly from a central maximum, was therefore considered, as a possible introduction to the ship-wave case. As explained in some detail in the earlier paper, the dispersion equation describing the wave-field on the free surface of the water is of elliptic type for moderate wall amplitudes, provided the wavelength of the wall is not too large.

The result of the earlier calculation is shown in figure 1, which is a map of the curves of constant phase on the free surface of the water (corresponding to the wave-crests and troughs). The x-axis is taken from left to right along the mean surface of the wall, the origin being at the point of maximum wall amplitude, and the positive direction of the y-axis is outwards from the wall. All distances and wavelengths are quoted with respect to U^2/g as the unit of length, where U is the undisturbed free stream speed which is in the positive x-direction. The amplitude of the wall is exaggerated by a factor of 8 in the figure. The wave pattern was obtained as the solution of an initial value problem, the wave-number κ and the phase θ being prescribed along the line $y = 1$.

It can be seen from the figure that there is a region of the (x, y)-plane within which the solution is indeterminate. This will be referred to as the 'gap region'. It is argued in the earlier paper that this is a manifestation of a shock-like phenomenon, across which there are discontinuous jumps in the wave-number and phase, the direction and spacing of the wave-crests changing abruptly through the shock. The shock, or 'phase jump', forms largely as a result of amplitude dispersion, whereby the larger amplitude waves have the larger phase velocity. The present paper aims to complete the solution of the problem by predicting the position of the discontinuity using the solution obtained and shown in figure 1 on either side of the shock. This is described in §3, and involves an application of the principle of conservation of energy. Out of all possible positions of the shock only one ensures that this is not violated.

Comparison with the analogous case in gas dynamics suggests the possibility of error in the solution which is being used behind (i.e. to the right of) the gap region. Such errors are third order effects in gas dynamics, and can often be neglected in the case of weak shocks; however, misleading results may be obtained in the strong shock case. Hence, the possible need for the solution behind the gap region to be modified, with possible consequential changes in shock position, must not be overlooked.

The present calculation assumes therefore that the shock is sufficiently weak for the method to be valid. Support for this view comes from the observations of Benjamin & Feir (1967), which indicate that the instability of finite amplitude waves does not result in turbulent dissipation of energy, but rather a redistribution of the energy in the spectrum of the wave-group.

The relevant conservation equations are derived in the next section. In §3 the predicted position of the shock is found, and the predicted amount of phase jump across the shock is used to estimate the difference in the number of wavecrests entering and leaving the shock. The possible form of the free surface in the neighbourhood of the phase jump is obtained by taking cross-sections of the free surface at right-angles to the shock. Observations on this would form a rather precise test of the theory. Finally, in §4 a crude physical argument is given which interestingly enough enables one to predict rather well the line along which the phase jump lies, although the method does not give its location on the line correctly.

2. Conservation equations and the dispersion equation

Let $\mathscr{L}(\kappa, \omega)$ denote the average Lagrangian per unit horizontal area for finite amplitude waves on deep water, where $\kappa = (l, m)$ is the wave-number and ω the time frequency. The system is assumed to be non-dissipative, since in an almost uniform wave train velocity gradients are smoothly varying functions of position and so the effect of viscosity may be ignored. In this case Lighthill (1967) has given an explicit expression for $\mathscr{L}$ approximately valid for all possible amplitudes, and this has been used in the computations of Howe (1967).

The wave-field is described in terms of a phase function $\theta(x_i, t)$ (where t is the time), which is smoothly varying and takes successive integral multiple values of 2π on successive wave-crests. In terms of θ the wave-number and frequency are given by

$$\kappa_i = \partial\theta/\partial x_i, \quad \omega = -\partial\theta/\partial t. \tag{2.1}$$

Hamilton's Principle,
$$\delta \iint \mathscr{L}(\kappa_i, \omega)\, d\mathbf{x}\, dt = 0, \tag{2.2}$$

then leads to the Euler equation

$$\frac{\partial}{\partial t}(\mathscr{L}_\omega) = \frac{\partial}{\partial x_i}(\mathscr{L}_{\kappa i}) \tag{2.3}$$

which, as pointed out by Whitham (1967b), is the conservation equation representing the balance between changes in the space-like adiabatic invariant $\mathscr{L}_{\kappa i}$ and the time-like adiabatic invariant $\mathscr{L}_\omega$. Other conservation equations may be derived from (2.2) by applying Noether's Theorem (Bogoliubov & Shirkov 1957, p. 20). In particular the invariance of $\mathscr{L}$ with respect to arbitrary time and space translations lead respectively to the energy equation

$$\frac{\partial}{\partial t}(\omega\mathscr{L}_\omega - \mathscr{L}) - \frac{\partial}{\partial x_i}(\omega\mathscr{L}_{\kappa i}) = 0 \tag{2.4}$$

and the momentum equation,

$$\frac{\partial}{\partial t}(\kappa_j \mathscr{L}_\omega) - \frac{\partial}{\partial x_i}(\kappa_j \mathscr{L}_{\kappa i} - \mathscr{L}\delta_{ij}) = 0. \tag{2.5}$$

The definitions (2.1) give the two further conservation equations

$$\frac{\partial\omega}{\partial x_i} + \frac{\partial\kappa_i}{\partial t} = 0, \quad \text{curl } \kappa = 0. \tag{2.6}$$

The first of (2.6) represents the conservation of wave-crests, and the second says that wave-crests originate and terminate only on the boundaries of the wave-

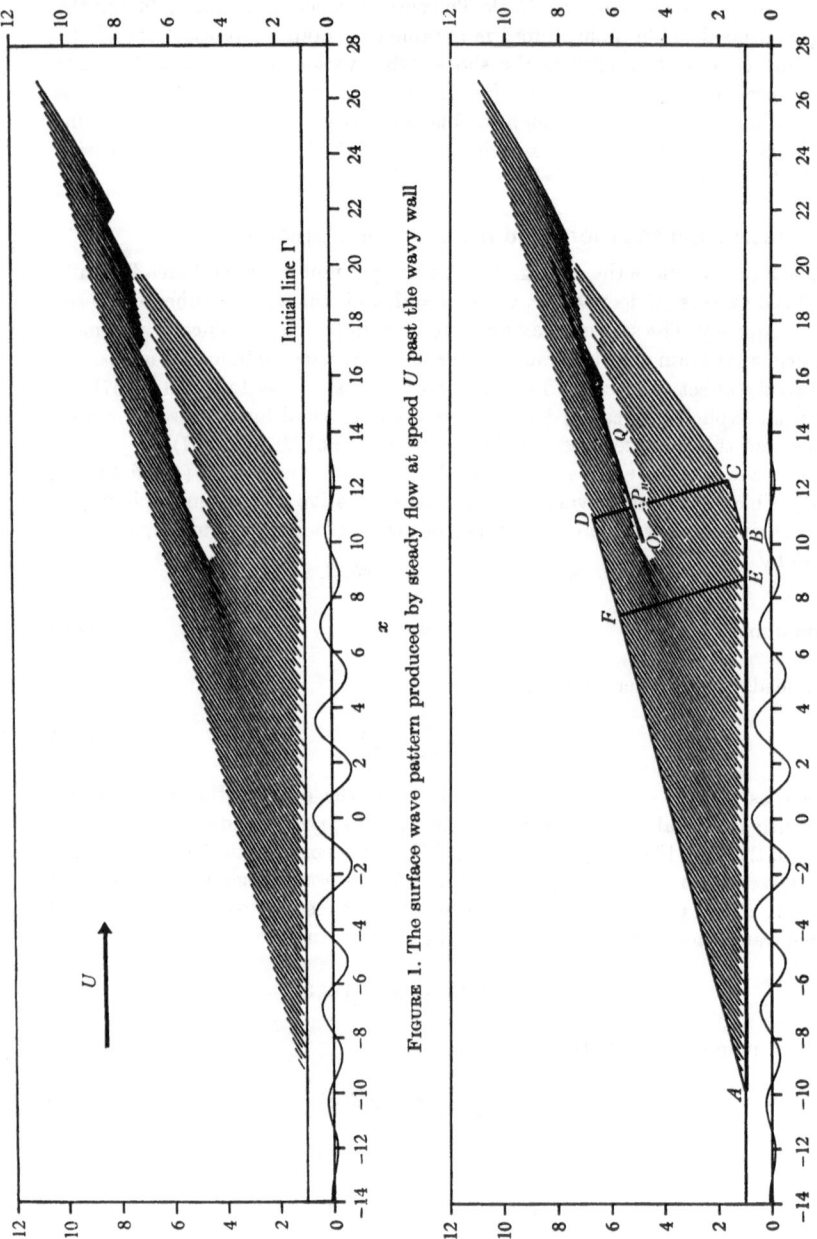

FIGURE 1. The surface wave pattern produced by steady flow at speed U past the wavy wall

FIGURE 2. The position of the phase jump OQ as determined by the principle of conservation of energy. $ABCDA$ is a typical contour used to predict the P_m, and $AEFA$ is a typical test-integral contour.

field. However, (2.3) and (2.6) are valid only in those regions where the wave-field varies slowly on a scale of wavelength, whereas the energy and momentum equations (2.4), (2.5) are valid under all circumstances when the average Lagrangian $\mathscr{L}$ is replaced by the exact Lagrangian density.

The wavy wall problem concerns the steady-state distribution of waves on the free surface, and it is shown in the earlier paper how Lighthill's Lagrangian is expressed in terms of axes fixed relative to the wall, with respect to which the wave pattern is steady, so that in (2.1) and (2.6) $\omega = 0$ and $\partial/\partial t = 0$. The dispersion equation obtained from (2.1) and (2.3) becomes in this case

$$a(l,m)\theta_{xx} + 2b(l,m)\theta_{xy} + c(l,m)\theta_{yy} = 0, \tag{2.7}$$

where $\qquad a(l,m) = \mathscr{L}_{ll}, \quad b(l,m) = \mathscr{L}_{lm}, \quad c(l,m) = \mathscr{L}_{mm}.$

For the particular wall considered (2.7) is a quasi-linear elliptic equation, and the associated initial value problem was solved by the method of imaginary characteristics, the data l, m, θ being specified along the line $y = 1$.

3. The phase jump in the wavy wall problem

In this section the position of the phase jump in the wavy wall problem is predicted using the data obtained in the earlier paper on either side of the gap region. Discontinuous jumps across a shock are usually treated in terms of the physical conservation equations of the problem. Whitham (1967 b) has pointed out that in non-linear problems of the present type there are always more conservation equations than the number of required shock conditions. Essentially one must distinguish between those conservation equations which remain valid in an un-averaged form, i.e. through the shock, and those which are true only for slowly varying wave trains. Thus, for example, it would be wrong to choose the wave conservation equations (2.6). However, the energy and momentum equations (2.4) and (2.5) may be used.

Consider in particular the energy conservation equation

$$\frac{\partial}{\partial t}(\omega\mathscr{L}_\omega - \mathscr{L}) - \frac{\partial}{\partial x_i}(\omega\mathscr{L}_{\kappa i}) = 0. \tag{3.1}$$

Here $\mathscr{L} = \mathscr{L}(\kappa_i, \omega)$ is the Lighthill average Lagrangian, and is in terms of axes fixed relative to the mean motion of the water. In order that (3.1) may be written in terms of axes fixed relative to the wall, for which the wave pattern is steady, one first computes the derivatives $\mathscr{L}_\omega$ and $\mathscr{L}_{\kappa i}$ in (3.1) and then replaces ω by $-Ul$ and $\partial/\partial t$ by $U_i(\partial/\partial x_i)$ where $U_i = (U, 0)$ is the undisturbed stream velocity, to obtain,

$$U_i\frac{\partial}{\partial x_i}[Ul\mathscr{L}_\omega + \mathscr{L}] - \frac{\partial}{\partial x_i}[Ul\mathscr{L}_{\kappa i}] = 0. \tag{3.2}$$

Thus about any contour S in the (x, y)-plane

$$\oint_S [(Ul\mathscr{L}_\omega + \mathscr{L})U_i - Ul\mathscr{L}_{\kappa i}]n_i ds = 0, \tag{3.3}$$

where n_i is the outward normal of S and ds the curvilinear length element. Equation (3.3) represents the balance of the energy fluxes into and out of S.

18*

The position of the phase jump is determined by choosing a suitable sequence of contours S_n which pass through the gap region of the wavy wall solution. Since the solution is known on either side of this region one then extrapolates the integrand along the contour from either side of the gap region to a point P_n, say. At P_n there will be a discontinuous jump in the value of the integrand, and the precise location of P_n on the contour is adjusted to make the integral about the whole contour vanish. One may note that an integral relation of the form (3.3) can also be derived from the momentum equation (2.4). The same procedure applied in this case would lead to the same prediction of the position of P_n.

Figure 2 reproduces the solution of the wavy wall problem and shows also the position of the phase jump as determined by the above method. A typical integration contour is also shown, consisting of a segment AB from $x = -10$ to $+10$ of the initial line $y = 1$, a portion BC of the 'output curve' from the point $(10, 1)$ (see §4), a transversal CD cutting the shock approximately at right-angles, and the segment DA of the output curve from $(-10, 1)$. Before attempting to find the position of the jump, the reliability of the numerical solution obtained by Howe (1967) in satisfying (3.3) was examined by integrating around several closed contours not passing through the gap region, e.g., $AEFA$. The data available from the solution gave the wave-number κ to four decimal places, and when used in the test integrals the latter were found to differ from zero only in the fifth decimal place (the contributions along AE and EF separately were of order 1); in no case was the error in excess of ± 0.00005.

The sequence of contours S_n was generated by moving the component CD of the contour shown in figure 2 along AD and BC. The points P_n determined in this way were found to lie closely along a straight line, and the location of the phase jump in figure 2 is on a segment of the line

$$y = 0.3372x + 1.4875, \tag{3.4}$$

which is the least squares fit to the points P_n. The standard deviation of the fit is 0.012. The tip of the shock is rather ill-defined, but is approximately at the point O where $x = 9.17$, and data for the line (3.4) extends as far as $x = 18.61$.

The difference in the number of wave-crests which enter the segment OQ of the shock from the left (1) and which leave it on the right (2) is given by

$$\delta = \frac{1}{2\pi} \int_0^Q [\kappa] \, .d\mathbf{s} . \tag{3.5}$$

The integration is taken along the discontinuity, and the square bracket denotes the jump in wave-number. Since $\kappa = \operatorname{grad} \theta$, and $\theta_1 = \theta_2$ at O,

$$\delta = \frac{1}{2\pi} \int_0^Q [d\theta] = \frac{1}{2\pi} (\theta_2 - \theta_1)_Q . \tag{3.6}$$

The variation of $2\pi\delta$ with distance s of Q along the jump from O is shown in figure 3, for possible future comparison with experiment.

One may illustrate the possible nature of the free surface of the water in the neighbourhood of the phase jump by plotting the surface elevation along a line such as CD in figure 2, cutting the jump at right angles. The free surface elevation is obtained from the solution of the earlier paper by substituting the amplitude and phase into the Stokes expansion for finite amplitude surface waves. Possible forms for the surface in the gap region may then be inferred by graphical interpolation between the curves of the known elevation on either side. By taking a large number of such cross-sections and plotting them on the same diagram a picture of the surface which seems almost three-dimensional is obtained.

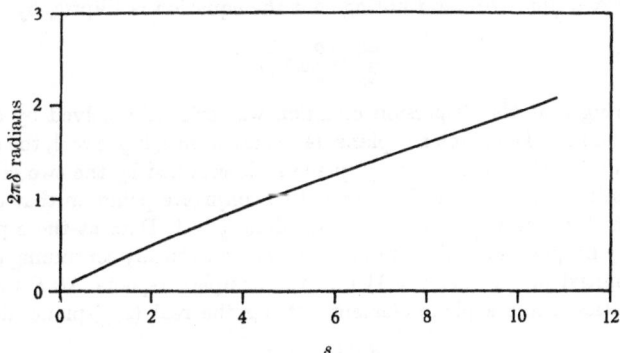

FIGURE 3. $2\pi\delta$ radians is the increase in phase in passing across the discontinuity in the positive x-direction at a distance s from the tip O of the phase jump.

This has been done in figure 4. Each wavy curve represents a cross-section of the free surface (magnified by a factor of 5) along a line such as CD in figure 2. Such a line is taken as the abscissa and the positive direction of the ordinate is in the sense OQ of figure 2. Thus figure 4 should be viewed by looking from the left along the direction of the phase jump line, which is also shown. With respect to this orientation the left-hand tip of each curve occurs where the wave energy is very small and so lies approximately at the undisturbed level of the free surface. The broken-line segments indicate a possible form for the free surface in the gap region.

The effect of the phase jump is rather dramatically emphasized by the shortening of the widths of those troughs shown entering the jump from below. Comment may also be made on the change in form of the crest which passes just to the left of the jump origin. The wave is of characteristic trochoidal section which becomes more peaked as it nears the tip of the jump. However, this tendency towards breaking is apparently arrested by the appearance of the jump which effectively acts as a means of transfer of wave energy from that crest, which moves on above the jump, to the waves in the region beneath the jump, so that the wave crest emerges on the other side much reduced in amplitude.

4. An approximate method for locating the line along which the jump lies

This discussion of the wavy wall problem is conveniently concluded by a consideration of an approximate method of predicting the line along which the phase jump is formed, which gives rather more physical insight than the purely numerical approach.

Consider the dispersion equation (2.7)

$$a\theta_{xx} + 2b\theta_{xy} + c\theta_{yy} = 0.$$

The characteristics of this equation are given by

$$ady^2 - 2bdxdy + cdx^2 = 0,$$

or, setting $\Delta = \sqrt{(ac - b^2)}$, and noting that the equation is elliptic, by

$$\frac{dx}{dy} = \frac{b}{c} \pm i\frac{\Delta}{c}. \tag{4.1}$$

Remembering that the dispersion equation was originally solved by analytic continuation into the complex x-plane, (4.1) states that, if y is real, the solution at any point P, say, in the real (x, y)-plane is determined by the two conjugate characteristics through P originating at two conjugate points in that complex x-plane which meets the real (x, y)-plane along $y = 1$. Data at these points is obtained by keeping the real part of x fixed and analytically continuing the data from the initial line Γ (i.e. $y = 1$) into the complex x-plane. At P these two characteristics define a plane element cutting the real (x, y)-plane along the curve

$$dx/dy = b/c. \tag{4.2}$$

As a first approximation (4.2) may be regarded as defining the rays along which energy is propagated. Indeed in the case of infinitesimal amplitude they *do* degenerate into the steady-state analogue of the group velocity lines, since

$$\frac{dx}{dy} = \frac{b}{c} = \frac{\mathscr{L}_{lm}}{\mathscr{L}_{mm}} = \frac{\partial(\mathscr{L}_m, m) . \partial(m, l)}{\partial(l, m) . \partial(\mathscr{L}_m, l)} = -\left(\frac{\partial m}{\partial l}\right)_{\mathscr{L}_m}.$$

For finite amplitudes the neglect of the imaginary part of (4.1) implies that the propagation of changes along the approximate 'rays' (4.2) would be to some extent blurred.

Suppose now that the wall has constant amplitude. The solution of the dispersion equation is then trivially a plane wave, and the energy rays (4.2) become the family of parallel straight lines

$$x = \frac{b}{c}y + \text{const.} \tag{4.3}$$

Here b/c is a function of κ which is constant. But when the amplitude of the wall varies, κ will vary along Γ. Provided that this variation is slow enough, in the sense of Whitham's approximation, the energy rays would be expected to remain straight for some distance from Γ, but the slope b/c would vary in accordance with the variation of κ along the initial line. It is conceivable that under these circumstances the family (4.3) would envelope a caustic along which the solution of the dispersion equation would be indeterminate. Physically one then expects to find a shock in the solution.

These ideas may be illustrated by referring to the results of the wavy wall calculation. The dispersion equation was solved by reducing it to a set of five

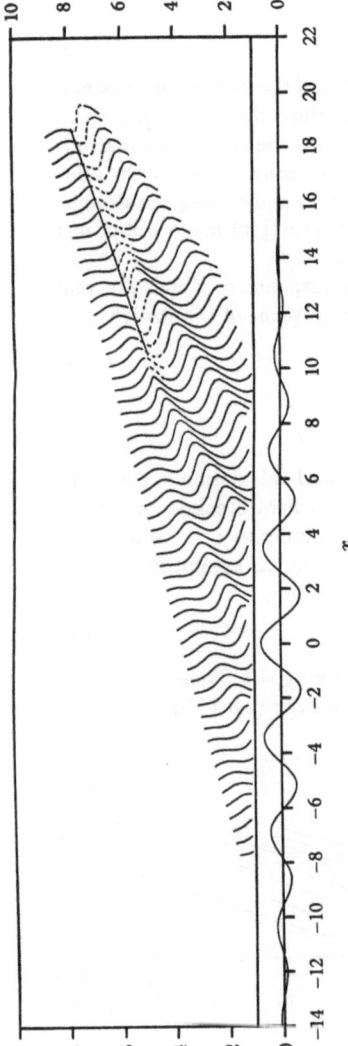

FIGURE 4. Each wavy curve is a cross-section of the free surface along a direction perpendicular to the phase jump. The figure should be viewed by looking from the left along the direction of the discontinuity. The broken-line segments indicate a possible form for the free surface in the neighbourhood of the phase jump.

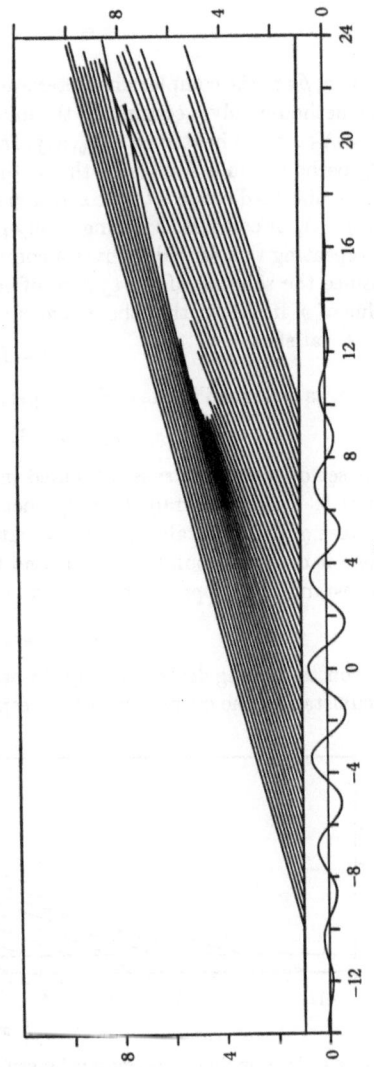

FIGURE 5. The 'output' curves obtained by Howe (1967). The method of imaginary characteristics used in solving the dispersion equation gives the values of l, m and θ along these rays.

first order partial differential equations in which the unknowns (x, y, l, m, θ) were to be determined in terms of co-ordinates (ξ, η) defined by

$$\xi = \frac{\alpha + \beta}{2}, \quad \eta = \frac{\alpha - \beta}{2i},$$

where α, β are the complex characteristic co-ordinates of the dispersion equation. The method involves continuing the initial data into the complex η-plane, where if $\eta = \lambda + i\sigma$, λ is held fixed at λ_0, say. In the (ξ, σ)-plane the system of equations is hyperbolic and is soluble by the method of characteristics. At $\sigma = 0$ the solution so obtained reduces to the solution in the real (ξ, η)-plane along the straight line $\eta = \lambda_0$. A complete covering of any portion of the real (ξ, η)-plane is obtained by repeating this procedure over a complete interval of λ_0.

Since the solution of the system of equations corresponds to a constant real value of η, it follows that the image curve in the (x, y)-plane of the straight line $\eta = \lambda_0$ satisfies

$$d\alpha - d\beta = 0, \tag{4.4}$$

where, as described in the earlier paper,

$$x = \lambda_0, \quad y = 1, \quad \text{on} \quad \alpha + \beta = 0. \tag{4.5}$$

This set of 'output curves' obtained in the wavy wall calculation is shown in figure 5. They are straight except near the region where breakdown occurs. Because of this it is valid to approximate to each by giving to κ the value it has where it cuts the boundary $y = 1$, and to solve (4.4) on the assumption that κ is constant. This approximation readily yields the solution

$$x = (b/c)y + \text{const.}, \tag{4.6}$$

the constant being determined by the starting-point on $y = 1$. Thus under these circumstances the output curves correspond to the energy rays (4.3).

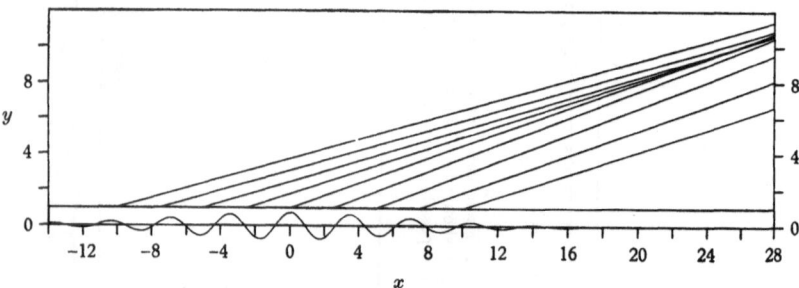

FIGURE 6. The family of 'energy rays' given by equation (4.3). The effect of finite amplitude is to blur the propagation of changes along these rays.

Figure 5 reveals that the output curves of the wavy wall calculation *do* in fact tend to form an envelope along the front of the gap region. This may be compared with figure 6 in which are drawn the energy rays (4.3) originating from the same points on $y = 1$ as the output curves. The energy rays form a cusp at (23·37, 9·21). This is quite a lot farther from the initial line than the start of the envelope in figure 5.

However, the line (3.4) along which the phase jump occurs is rather well determined by that along which the cusp forms in the case of the energy rays. Taking our definition of the latter line as that ray which touches the cusp at its tip, one obtains for its equation,

$$y = 0·3235x + 1·6609. \tag{4.7}$$

This is negligibly different from (3.4) within the range $x = 10, 20$ of validity of the phase-jump calculation. Thus the blurring of the energy rays may alter the distance to formation of the discontinuity but apparently does not alter the direction in which the phase-jump is formed.

It is again a great pleasure to acknowledge the generous help and guidance of Professor M. J. Lighthill, Sec. R.S., throughout the preparation of this work. Support for this research was provided by a grant from the Central Electricity Generating Board, which is gratefully acknowledged.

REFERENCES

BENJAMIN, T. B. & FEIR, J. E. 1967 *J. Fluid Mech.* **27**, 417.
BOGOLIUBOV, N. N. & SHIRKOV, D. V. 1957 *Introduction to the Theory of Quantum Fields.* Moscow.
HOWE, M. S. 1967 *J. Fluid Mech.* **30**, 497.
LIGHTHILL, M. J. 1965 *J.I.M.A.* **1**, 269.
LIGHTHILL, M. J. 1967 *Proc. Roy. Soc.* A, **299**, 28.
WHITHAM, G. B. 1965a *Proc. Roy. Soc.* A, **283**, 238.
WHITHAM, G. B. 1965b *J. Fluid Mech.* **22**, 273.
WHITHAM, G. B. 1967a *J. Fluid Mech.* **27**, 399.
WHITHAM, G. B. 1967b *Proc. Roy. Soc.* A, **299**, 6.

Math. Annalen **162**, 228–236 (1966).

Équations Hyperboliques Non-Stricts: Contre-Exemples, du Type De Giorgi, aux Théorèmes d'Existence et d'Unicité

J. Leray

Introduction

1. Considérons dans $\mathbf{R}^l$ un problème de Cauchy, hyperbolique non strict, d'inconnue $u(x)$:

$$(1.1) \qquad \begin{cases} a_1(x, D) \ldots a_p(x, D)\, u(x) = b(x, D)\, u(x) + v(x) \\ D^{m-1} u \,|\, S_0 \ \text{donné}; \end{cases}$$

$D = \dfrac{\partial}{\partial x}$; $a_1, \ldots, a_p$ sont p opérateurs strictement hyperboliques relativement à S_0. Notons

$$\text{ordre } (a_1, \ldots, a_p) = m; \ \text{ordre } (b) \leq m - p + q, \quad \text{où} \quad 0 \leq q.$$

Supposons que S_0 est un hyperplan; notons $\gamma^{n,\,(\alpha)}$ la classe des fonctions $\mathbf{R}^l \to \mathbf{C}$ dont les dérivées $f^{(n)}$ d'ordres $\leq n$ ont des restrictions aux hyperplans S_t parallèles à S_0 qui vérifient uniformément par rapport à t la condition d'appartenir à la classe α de Gevrey:

$$\sup_{\substack{x \in S_t \\ |\beta| \leq s}} |D^\beta f^{(n)}| \leq (\text{const.})^s \, (s!)^\alpha$$

où D^β est une dérivée, d'ordre $|\beta|$, sur S_t.

On sait ceci (pour l'énoncé précis voir [2], nº 23,24): si les données du problème de Cauchy (1.1) appartiennent à la classe $\gamma^{n,\,(\alpha)}$, alors *ce problème possède une solution unique u et $u \in \gamma^{n,\,(\alpha)}$*, quand on a:

$$n \geq m + p, \quad 1 \leq \alpha < \frac{p}{q}.$$

Si $1 \leq \alpha = \dfrac{p}{q}$, ces théorèmes d'existence et d'unicité valent sous certaines restrictions (existence locale, c'est-à-dire au voisinage de S_0; unicité sous l'hypothèse $u \in \gamma_2^{m+p,\,(\alpha)}$).

Un exemple DE GIORGI montre que ces théorèmes deviennent faux quand on supprime l'hypothèse $\alpha \leq \dfrac{p}{q}$; plus précisément, DE GIORGI montre que cette hypothèse est nécessaire dans le cas $m = p = 8$, $q = 4$.

Nous allons construire, par un procédé simplifiant[1]), celui qu'emploie DE GIORGI, des contre-exemples prouvant que, quels que soient[2]) $m \geqq p \geqq 1$ et $q \geqq 1$, *l'hypothèse* $\alpha \leqq \dfrac{p}{q}$ *est nécessaire à la validité des théorèmes d'existence et d'unicité*[3]) qu'énonce [2] (n⁰ 23, 24, 25 et 26).

Cependant, si l'on impose à $a_1, \ldots, a_p, b$ d'être *réels*, nous ne prouvons la nécessité de cette hypothèse $\alpha \leqq \dfrac{p}{q}$ que dans le cas où q est *pair*.

§ 1. Préliminaires

2. RÉDUCTION AU CAS : $l = 2, m = p$. — Le théorème d'existence implique le théorème d'unicité, d'après HOLMGREN : voir [2], n⁰ 24. Il suffit donc de construire un contre-exemple au théorème d'unicité. Nous choisissons ce contre-exemple fonction de deux des variables indépendantes, ce qui nous ramène au cas où $\mathbf{R}^l = \mathbf{R}^2$.

Supposons que l'équation, à coefficients indéfiniment différentiables,

$$(2.1) \qquad \frac{\partial^p u}{\partial t^p} = b\left(t, x, \frac{\partial}{\partial x}\right) u \quad \text{(ordre } (b) \leqq q)$$

possède une solution, indéfiniment différentiable, contredisant le théorème d'unicité, c'est-à-dire s'annulant p fois avec t; on voit que toutes ses dérivées s'annulent avec t. Par suite u est un contre-exemple au théorème d'unicité pour l'équation

$$\prod_{k=1}^{m-p} \left(\frac{\partial}{\partial t} - k\frac{\partial}{\partial x}\right) \frac{\partial^p u}{\partial t^p} = \prod_{k=1}^{m-p} \left(\frac{\partial}{\partial t} - k\frac{\partial}{\partial x}\right) b u$$

qui est du type (1.1), avec

$$a_1 = \prod_{k=0}^{m-p} \left(\frac{\partial}{\partial t} - k\frac{\partial}{\partial x}\right), \ a_j = \frac{\partial}{\partial t} \ (1 < j \leqq p).$$

Pour traiter le cas (m, p, q) quelconque, il nous suffit donc de construire, pour tout (p, q, α) tel que $\dfrac{p}{q} < \alpha$, un contre-exemple au théorème d'unicité concernant une équation du type (2.1); pour ce type d'équation, $m = p$.

3. QUASI-NORMES FORMELLES. — Nous notons (t, x) les coordonnées de $\mathbf{R}^2$ et S_t la droite d'abscisse t. Etant donnée une fonction $u(t, x)$, définie sur une bande $T_0 \leqq t \leqq T_1$, nous définissons sa quasi-norme

$$|u, S_t| = \sup_x |u(t, x)|$$

et sa quasi-norme formelle

$$(3.1) \qquad |D^{h, \infty} u, S_t, \varrho| = \sum_{s=0}^{\infty} \frac{\varrho^s}{s!} \sup_j \left|\frac{\partial^{j+s} u}{\partial t^j \partial x^s}, S_t\right|, \text{ où } \quad 0 \leqq j \leqq h \, ;$$

c'est une série formelle en ϱ, qui peut être une fonction de ϱ holomorphe à l'origine.

---

[1]) Là où nos § 2 et § 3 emploient 5 bandes, DE GIORGI en emploie 7.

[2]) Aucune hypothèse n'est faite sur p/q.

[3]) Et aussi à la validité de théorèmes de G. TALENTI [3] apparentés à ceux-ci.

Soit une série formelle

$$\Phi(\varrho) = \sum_{s=0}^{\infty} \frac{\varrho^s}{s!}\ \Phi_s;$$

$$\Phi(\varrho) \gg 0 \quad \text{signifie} \quad \Phi_s \geqq 0\ ,\quad \forall s\ ;$$

on dit que $\Phi \in \Gamma^{(\alpha)}$ *(classe de Gevrey formelle)* quand il existe une constante c, dépendant de Φ, telle que

$$(3.2) \qquad\qquad\qquad \Phi_s \leqq c^s (s!)^\alpha\ ;$$

on dit que $u \in \gamma^{h,\,(\alpha)}$ *(classe de Gevrey)* quand il existe une série formelle $\Phi(\varrho)$, indépendante de t, telle que

$$(3.3) \qquad\qquad |D^{h,\,\infty} u, S_t, \varrho| \leqq \Phi(\varrho) \in \Gamma^{(\alpha)}\ .$$

Etant donné un opérateur différentiel

$$b\left(t, x, \frac{\partial}{\partial x}\right) = \sum_{j=0}^{q} b_j(t, x) \left(\frac{\partial}{\partial x}\right)^q,$$

nous notons

$$|D^{h,\,\infty} b, S_t, \varrho| = \sum_{j=0}^{q} |D^{h,\,\infty} b_j, S_t, \varrho|;$$

nous disons que $b \in \gamma^{h,\,(\alpha)}$ quand $b_j \in \gamma^{h,\,(\alpha)}$, $\forall j$.

4. Le contre-exemple à construire est, d'après le n° 2, le suivant: Etant donnés (p, q, α) tels que

$$p \geqq 1,\quad q \geqq 1,\quad \frac{p}{q} < \alpha\ ,$$

construire, sur une bande $0 \leqq t \leqq T$ de $\mathbf{R}^2$, une équation linéaire homogène

$$(4.1) \qquad\qquad \frac{\partial^p u}{\partial t^p} = b\left(t, x, \frac{\partial}{\partial x}\right) u \quad \text{(ordre } b \leqq q)$$

possèdant une solution $u(t, x) \neq 0$, telle que

$$(4.2) \qquad\qquad \frac{\partial^h u}{\partial t^h}(0, x) = 0,\quad \forall h;$$

$$(4.3) \qquad\qquad u \in \gamma^{h,\,(\alpha)},\quad b \in \gamma^{h,\,(\alpha)},\quad \forall h\ .$$

Note. — u et b sont indépendants de h.

De Giorgi construit un tel contre-exemple en résolvant d'abord le problème non homogène que voici.

5. Enoncé d'un problème non homogène. — Nous nous donnons (p, q, α), tels que

$$p \geqq 1,\quad q \geqq 1,\quad \frac{p}{q} < \alpha\ ,$$

un nombre l_1 et un paramètre $l \leqq l_1$; nous cherchons sur la bande

$$0 \leqq t \leqq 1$$

de $\mathbf{R}^2$ une équation linéaire homogène

$$(5.1) \qquad \frac{\partial^p u}{\partial t^p} = b\left(t, x, \frac{\partial}{\partial x}\right) u \quad \text{(ordre } (b) \leqq q; \ b \text{ dépend de } l)$$

et une solution u de cette équation telles que:

(5.2)
$$\begin{cases} u(t, x) = e^l, & b = 0 \quad \text{pour } t \text{ voisin de } 0 , \\ u(t, x) = e^{l'(l)}, & b = 0 \quad \text{pour } t \text{ voisin de } 1 . \end{cases}$$

(5.3)
$$\begin{cases} |D^{h,\infty} u, S_t, \varrho| \ll \theta(l) \, \Phi(\varrho), & \forall h , \\ |D^{h,\infty} b, S_t, \varrho| \ll \theta(l) \, \Phi(\varrho), & \forall h , \end{cases}$$

où: l', θ, Φ dépendent de h; Φ ne dépend pas de l; $\Phi \in \Gamma^{(\alpha)}$; l' et θ sont des fonctions de l, ayant les propriétés suivantes:

$$l'(l) < l ;$$

si nous définissons les suites $l_1, l_2, \ldots, \theta_1, \theta_2, \ldots$ par la loi de récurrence:

$$l_{k+1} = l'(l_k) , \quad \theta_k = \theta(l_k)$$

alors

(5.4)
$$\lim_{k \to \infty} k^c \, \theta_k = 0 \quad \text{pour toute constante } c .$$

6. Construction[4]) du contre-exemple $u(t, x)$ ayant les propriétés qu'exige le n° 4. — Supposons résolu le problème non homogène qu'énonce le n° 5; sa solution, pour $l = l_k$, sera notée $b_k\left(t, x, \dfrac{\partial}{\partial x}\right)$, $u_k(t, x)$.

Définissons $T_1, T_2, \ldots$ par la loi de récurrence:

$$T_1 = 0, \ T_{k+1} - T_k = \frac{1}{k^2} ;$$

soit

$$T = \lim_{k \to \infty} T_k = \sum_{k=1}^{\infty} \frac{1}{k^2} < \infty .$$

Définissons

$$b\left(x, t, \frac{\partial}{\partial x}\right) = k^{2p} b_k\left(\frac{t - T_k}{T_{k+1} - T_k}, x, \frac{\partial}{\partial x}\right)$$

$$u(x, t) = u_k\left(\frac{t - T_k}{T_{k+1} - T_k}, x\right) \quad \text{pour} \quad T_k \leq t \leq T_{k+1} .$$

Vu (5.2), b et u sont indéfiniment dérivables sur la bande $0 \leq t < T$; vu (5.1), sur cette bande, (4.1) est vérifiée.

Vu (5.3):
$$|D^{h,\infty} u, S_t, \varrho| \ll k^{2h} \theta_k \Phi(\varrho)$$
$$|D^{h,\infty} b, S_t, \varrho| \ll k^{2(h+p)} \theta_k \Phi(\varrho) , \quad \text{où} \quad \Phi \in \Gamma^{(\alpha)}$$

D'où, vu (5.4):
$$|D^{h,\infty} u, S_t, \varrho| \ll \varepsilon(t) \, \Phi(\varrho) ,$$
$$|D^{h,\infty} b, S_t, \varrho| \ll \varepsilon(t) \, \Phi(\varrho) ,$$

où $\lim\limits_{t \to T} \varepsilon(t) = 0$; bien entendu, $\varepsilon(t)$ dépend de h.

Donc $u \in \gamma^{h, (\alpha)}$, $b \in \gamma^{h, (\alpha)}$, $\forall h$; toutes les dérivées de u et des coefficients de b s'annulent pour $t = T$.

---

[4]) Je remercie K. Jörgens d'avoir rectifié cette partie de mon exposé.

Nous avons construit le contre-exemple qu'exige le n⁰ 4, à la permutation près de 0 et T.

7. CONCLUSION DU § 1. — Ce qu'affirme l'introduction, à savoir *la nécessité de l'hypothèse $\alpha \leq p/q$ dans les théorèmes d'existence et d'unicité concernant l'équation hyperbolique non stricte, sera donc prouvé quand nous aurons résolu le problème non homogène*, qu'énonce le n⁰ 5.

§ 2. Résolution du problème non homogène (n⁰ 5)

Il faut évidemment supposer u et b fonctions de x; il suffira de prendre u linéaire en $e^{i\omega x}$, où $\omega = \omega(l)$. Le terme de u indépendant de x est une fonction de t qui sera constante près des bords de la bande; le coefficient de $e^{i\omega x}$ aura pour coefficient, dans u, une fonction de t qui sera constante au centre de la bande. Cette bande ne sera pas la bande $0 \leq t \leq 1$, comme l'annonce le n⁰ 5, mais la bande

$$0 \leq t \leq 5 .$$

Notation. c désignera divers nombres, fonctions de (h, p, q), mais indépendants de l.

8. INTRODUCTION DU TERME EN $e^{i\omega x}$ DANS u. —

Lemme 1. *Donnons-nous des nombres*

$$m < l, \quad \omega > 1 .$$

On peut construire sur la bande

$$0 \leq t \leq 1$$

une équation du type (5.1) admettant une solution u, telle que

$$u(t, x) = e^l, b = 0 \quad \text{pour } t \text{ voisin de } 0 ;$$

$$u(t, x) = e^l + e^{m + i\omega x}, b = 0 \quad \text{pour } t \text{ voisin de } 1 ;$$

$$|D^{h, \infty} u, S_t, \varrho| \ll c e^{l + \omega \varrho} ;$$

$$|D^{h, \infty} b, S_t, \varrho| \ll c e^{m - l + \omega \varrho} .$$

Notation. — $f(t)$ désignera une fonction fixe, indéfiniment dérivable, telle que

$$f(t) = 0 \text{ pour } t \text{ voisin de } 0, \quad f(t) = 1 \text{ pour } t \text{ voisin de } 1 .$$

Preuve. — La fonction u et l'opérateur b que voici vérifient (5.1):

$$u = e^l + e^{m + i\omega x} f(t)$$

$$b = e^{m - l + i\omega x} \frac{d^p f(t)}{dt^p} \left(\frac{i}{\omega} \frac{\partial}{\partial x} + 1 \right) .$$

9. AUGMENTATION DU COEFFICIENT DE $e^{i\omega x}$ DANS u. —

Lemme 2. *Donnons-nous des nombres*

$$m < l < n, \omega \quad \text{tels que} \quad n - m > 1, \omega > 1 .$$

On peut construire sur la bande

$$0 \leq t \leq 2$$

une équation du typo (5.1) *admettant une solution u, telle que*

$$u(t, x) = e^l, \ b = 0 \quad pour \ t \ voisin \ de \ 0;$$

$$u(t, x) = e^l + e^{n+i\omega x}, \ b = 0 \quad pour \ t \ voisin \ de \ 2;$$

$$|D^{h, \infty} u, S_t, \varrho| \ll c(n-m)^h e^{n+\omega\varrho}$$

$$|D^{h, \infty} b, S_t, \varrho| \ll c \, e^{m-l+\omega\varrho} + c \, \frac{(n-m)^p}{\omega^q}.$$

Preuve. — Définissons b et u par le lemme 1 pour $0 \leqq t \leqq 1$. Pour $1 \leqq t \leqq 2$, la fonction u et l'opérateur b que voici vérifient (5.1):

$$u = e^l + e^{nf+m(1-f)+i\omega x} \quad \text{où} \quad f = f(t-1);$$

$$b = e^{-nf-m(1-f)} \frac{d^p e^{nf+m(1-f)}}{dt^p} \frac{1}{(i\omega)^q} \frac{\partial^q}{\partial x^q}.$$

Or

$$e^{-nf-m(1-f)} \frac{d^p e^{nf+m(1-f)}}{dt^p}$$

est un polynome en $n-m$ de degré p, dont les coefficients sont des fonctions fixes de t.

10. MODIFICATION DU TERME DE u INDÉPENDANT DE x. —

Lemme 3. *Donnons-nous des nombres*

$$m < l' < l < n, \quad \omega \ tels \ que \ n - m > 1, \ \omega > 1.$$

On peut construire sur la bande

$$0 \leqq t \leqq 3$$

une équation du type (5.1) *admettant une solution u, telle que*

$$u(t, x) = e^l, \quad b = 0 \ pour \ t \ voisin \ de \ 0;$$

$$u(t, x) = e^{l'} + e^{n+i\omega x}, \quad b = 0 \ pour \ t \ voisin \ de \ 3;$$

$$|D^{h, \infty} u, S_t, \varrho| \ll c(n-m)^h e^{n+\omega\varrho}$$

$$|D^{h, \infty} b, S_t, \varrho| \ll c(n-m)^{h+p} e^{l-n+\omega\varrho} + c \, e^{m-l+\omega\varrho} + c \, \frac{(n-m)^p}{\omega^q}.$$

Preuve. — Définissons b et u par le lemme 2 pour $0 \leqq t \leqq 2$. Pour $2 \leqq t \leqq 3$, la fonction u et l'opérateur b que voici vérifient (5.1):

$$u = e^{l(1-f)+l'f} + e^{n+i\omega x}, \quad \text{où} \quad f = f(t-2);$$

$$b = e^{-n-i\omega x} \frac{d^p e^{l(1-f)+l'f}}{dt^p} \frac{1}{i\omega} \frac{\partial}{\partial x}.$$

11. FIN DE LA CONSTRUCTION DE b ET u. — Pour $0 \leqq t \leqq 3$, définissons b et u par le lemme 3; pour $3 \leqq t \leqq 5$, définissons b et u par le lemme 2, où l'on remplace

$$0 \leqq t \leqq 2 \quad par \quad 5 \geqq t \geqq 3$$

$$m < l < n \quad par \quad m < l' < n.$$

Il vient:

Lemme 4. *Donnons-nous des nombres*

(11.1) $$m < l' < l < n, \omega \quad tels \ que \quad n - m > 1 \quad et \quad \omega > 1.$$

On peut construire sur la bande

$$0 \leq t \leq 5$$

une équation du type (5.1) *admettant une solution* u, *telle que*

(11.2) $\begin{cases} u(t, x) = e^l, b = 0 & \text{pour } t \text{ voisin de } 0; \\ u(t, x) = e^{l'}, b = 0 & \text{pour } t \text{ voisin de } 5; \end{cases}$

(11.3) $\begin{cases} |D^{h, \infty} u, S_t, \varrho| \ll c(n-m)^h e^{n+\omega\varrho}; \\ |D^{h, \infty} b, S_t, \varrho| \ll c(n-m)^{h+p} e^{l-n+\omega\varrho} + c e^{m-l'+\omega\varrho} + c \dfrac{(n-m)^p}{\omega^q}. \end{cases}$

12. Choix de l', m, n, ω en fonction de l. — Soient un paramètre $L > 1/4$ et un nombre fixe $\alpha \geq 1$.

Choisissons, en accord avec (11.1):

$$m = -8L, \quad l' = -6L, \quad l = -4L, \quad n = -2L, \quad \omega = l^\alpha;$$

définissons

(12.1) $$\theta = |l|^{p - \alpha q}$$

Puisque $\sup_L L^e e^{-L} < \infty$, (11.3) donne

(12.2) $\begin{cases} |D^{h, \infty} u, S_t, \varrho| \ll c\theta e^{-L + L^\alpha \varrho} \\ |D^{h, \infty} b, S_t, \varrho| \ll c\theta [e^{-L + L^\alpha \varrho} + 1]. \end{cases}$

Le n⁰ 13 va prouver le lemme suivant:

Lemme 5. — *Il existe une série formelle* $\Phi(\varrho) \in \Gamma^{(\alpha)}$, *indépendante de* L, *telle que*

$$e^{-L + L^\alpha \varrho} \ll \Phi(\varrho), \quad \forall L \geq 0.$$

Donc (12.2) implique (5.3): le problème non homogène qu'énonce le n⁰ 5 est résolu, quand (5.4) a lieu. Or:

$$l'(l) = \frac{3}{2} l;$$

d'où, en choisissant $l_1 = -\dfrac{3}{2}$, vu (12.1):

$$l_k = -\left(\frac{3}{2}\right)^k; \quad \theta_k = \left(\frac{2}{3}\right)^{(\alpha q - p)k};$$

d'où (5.4), si, comme le suppose le n⁰ 5:

$$\alpha > \frac{p}{q}.$$

Le problème non homogène (n⁰ 5) a donc une solution; vu le n⁰ 7, ce qu'affirme l'introduction est prouvé; mais b a été choisi *non réel*.

13. Preuve du lemme 5. — On a

(13.1) $$e^{-L + L^\alpha \varrho} = \sum_{s=0}^{\infty} \frac{\varrho^s}{s!} L^{\alpha s} e^{-L}.$$

Or

(13.2) $$\sup_{L>0} (L^\beta e^{-L}) = \left(\frac{\beta}{e}\right)^\beta, \quad \text{si} \quad \beta \geqq 0,$$

car ce sup est atteint pour $L = \beta$.

Rappelons[5]) que

$$\left(\frac{s}{e}\right)^s < s!;$$

de (13.2) résulte donc

$$\sup_{L>0} (L^{\alpha s} e^{-L}) = \left(\frac{\alpha s}{e}\right)^{\alpha s} \leqq \alpha^{\alpha s} (s!)^\alpha.$$

En portant cette inégalité dans (13.1), nous obtenons

$$e^{-L + L^\alpha \varrho} \ll \sum_{s=0}^\infty (\alpha^\alpha \varrho)^s (s!)^{\alpha-1} \in \Gamma^{(\alpha)}.$$

Voici prouvé le lemme 5.

14. Conclusion du § 2. — Ce qu'affirme l'introduction, à savoir *la nécessité de l'hypothèse $\alpha \leq p/q$ dans les théorèmes d'existence et d'unicité concernant l'équation hyperbolique non-stricte, est donc prouvé.*

Mais b a été choisi non réel.

§ 3. Choix d'un b réel

Si q est *pair*, on peut faire pour u et b un autre choix, *réel*, pour lequel subsistent les majorations des quasi-normes formelles employées ci-dessus et par suite les conclusions prouvées.

Indiquons rapidement ce choix.

15. Modifications a apporter au lemme 1. — *Modification à son énoncé.*

$$u(t, x) = e^l + e^m \sin(\omega x), \, b = 0 \text{ pour } t \text{ voisin de } 1.$$

Modification à sa preuve. —

$$u = e^l + e^m f(t) \sin(\omega x)$$

$$b = e^{m-l} \frac{d^p f}{dt^p} \sin(\omega x) \left[\frac{1}{\omega^2} \frac{\partial^2}{\partial x^2} + 1\right].$$

16. Modification au lemme 2. —

$$u(t, x) = e^l + e^n \sin(\omega x), \quad b = 0 \text{ pour } t \text{ voisin de } 2.$$

Modification à sa preuve. —

$$u = e^l + e^{nf + m(1-f)} \sin(\omega x)$$

$$b = e^{-nf - m(1-f)} \frac{d^p e^{nf + m(1-f)}}{dt^p} \frac{1}{(i\omega)^q} \frac{\partial^q}{\partial x^q},$$

en supposant q *pair*.

17. Modification au lemme 3.

$$u(t, x) = e^{l'} + e^n \sin(\omega x), \quad b = 0 \text{ pour } t \text{ voisin de } 3.$$

---

[5]) car $\dfrac{x^s}{s!} < e^x$.

Modification à sa preuve. —

$$u = e^{l(1-f)+l'f} + e^n \sin(\omega x)$$

$$b = e^{-n} \frac{d^p \, e^{l(1-f)+l'f}}{dt^p} \left[\frac{1}{\omega} \cos(\omega x) \frac{\partial}{\partial x} - \frac{1}{\omega^2} \sin(\omega x) \frac{\partial^2}{\partial x^2} \right].$$

Bibliographie

[1] DE GIORGI: Un esempio di non-unicità della soluzione del problema di Cauchy. Università di Roma. Rend. Mat. 14, 382—387 (1955).

[2] LERAY, J., et Y. OHYA: Systèmes linéaires, hyperboliques non-stricts. Colloque C.B.M., Louvain (1964).

[3] TALENTI, G.: Sur le problème de Cauchy pour les équations aux dérivées partielles. C. R. Acad. Sci. 259, 1932—1933 (1964).

(Reçu le 2 juillet 1965)

Centre Belge de Reehuchs Mathematiques. Deuxième Colloque sur l'Analyse Fonctionnelle, p. 105.

Systèmes Linéaires, Hyperboliques Non-Stricts

J. LERAY et Y. OHYA

Introduction

1. HISTORIQUE

Les systèmes strictement hyperboliques se résolvent dans des espaces de fonctions ayant un nombre donné, fini, de dérivées (Petrowsky [10], Leray [5], [6], Gårding [1]). Une équation à caractéristiques multiples ne peut plus être résolue dans de tels espaces (Yamaguti [12]; Mme Lax [4]; Hörmander [3], ch. V, qui réserve le terme « hyperbolique » au strictement hyperbolique). Mais elle peut l'être dans des espaces de fonctions indéfiniment différentiables : les classes de Gevrey $\gamma^{(\alpha)}$ (Hörmander [3], théorème 5.7.3, traite l'équation linéaire à coefficients constants; Ohya [9] traite l'équation linéaire à coefficients variables, dont le *polynôme caractéristique est un produit de polynômes strictement hyperboliques*; *le domaine d'influence existe*. Ce domaine peut s'étudier comme dans le cas strictement hyperbolique, [5], ch VI).

Nous allons étendre ce résultat d'Ohya au système linéaire; nous compléterons ses conclusions en employant une famille plus large de classes de Gevrey : elle s'étend de la classe des fonctions analytiques à celle des fonctions ayant un nombre fini de dérivées bornées ou de carrés sommables.

Notre méthode peut s'appliquer au système non linéaire, grâce à un théorème de L. Waelbroeck [7] sur la composition des fonctions.

2. SOMMAIRE

Nous résolvons le problème de Cauchy, hyperbolique non strict, par approximations successives; ces approximations s'obtiennent en résolvant des problèmes de Cauchy strictement hyperboliques; pour prouver leur convergence il faut employer les normes de toutes leurs dérivées.

A cet effet, suivant une suggestion de L. Waelbroeck, nous employons des séries formelles; la preuve de la convergence se réduit à la résolution d'un problème de Cauchy pour des séries formelles; ce problème se résout en s'aidant du théorème de Cauchy-Kowaleski (problème de Cauchy analytique).

Le problème posé se trouve néanmoins résolu dans des classes de Gevrey non quasi-analytiques : elles contiennent des fonctions à supports compacts; *le domaine d'influence existe*, ce qui, pour nous, caractérise l'hyperbolique.

Notre raisonnement diffère beaucoup de celui d'Ohya; cependant, il est né de son étude.

On trouvera l'énoncé de nos conclusions aux n° 23, 24, 25 et 26, à la fin de l'article.

§ 1. Norme formelle

3. NOTATIONS

$X = R^l$. Étant donnée une fonction $f: X \to C$, nous notons :

$$|f, X|_p = \left[\int_X |f(x)|^p \, dx \right]^{1/p}; \quad |f, X|_\infty = \sup_{x \in X} |f(x)|;$$

$L_p(X)$ l'espace de Banach des f tels que $|f, X|_p < \infty (1 \leqslant p \leqslant \infty)$.
Choisissons des coordonnées $(x_1, \ldots, x_l)$ sur X; notons :

$$D_i f = \frac{\partial f}{\partial x_i},$$

$D^\sigma f = D_1^{\sigma_1} \ldots D_l^{\sigma_l} f$ où $\sigma = (\sigma_1, \ldots, \sigma_l)$, $|\sigma| = \sigma_1 + \ldots + \sigma_l$;

$$|D^m f, X|_p = \sum_{|\sigma| \leqslant m} \frac{1}{\sigma!} |D^\sigma f, X|_p, \quad \text{où} \quad \sigma! = \sigma_1! \ldots \sigma_l!;$$

$L_p^m(X)$ l'espace de Banach des f telles que $|D^m f, X|_p < \infty$; c'est l'espace de Sobolev.

Bien entendu, $|D^\sigma f, X|_p = \infty$ si $D^\sigma f \notin L_p(X)$.

4. Définition et propriétés

Nous nommons *norme formelle* de la fonction $f : X \to C$ la série formelle de la variable ϱ, à coefficients $\geqslant 0$ et $\leqslant +\infty$:

$$|D^\infty f, X, \varrho|_p = \sum_{s=0}^{\infty} \frac{\varrho^s}{s!} \sup_\sigma |D^\sigma f, X|_p, \quad \text{où} \quad |\sigma| = s. \quad (4.1)$$

Notons $[D^\sigma, f]$ le commutateur de D^σ et f, c'est-à-dire l'opérateur tel que

$$[D^\sigma, f]g = D^\sigma(fg) - fD^\sigma g, \quad \text{où} \quad g : X \to C;$$

définissons la série formelle

$$|[D^\infty, f]g, X, \varrho|_p = \sum_{s=0}^{\infty} \frac{\varrho^s}{s!} \sup_\sigma |[D^\sigma, f]g, X|_p. \quad (4.2)$$

Nous allons établir les propriétés suivantes, où

$$\sum_s \frac{\varrho^s}{s!} F_s \leqslant \sum_s \frac{\varrho^s}{s!} G_s$$

signifie $F_s \leqslant G_s \; \forall s$.

Formule du produit. — Si $1/p = 1/q + 1/r$, alors

$$|D^\infty(fg), X, \varrho|_p \leqslant |D^\infty f, X, \varrho|_q \, |D^\infty g, X, \varrho|_r. \quad (4.3)$$

Formule du commutateur. — Si $1/p = 1/q + 1/r$, alors

$$|[D^\infty, f]g, X, \varrho|_p \leqslant [|D^\infty f, X, \varrho|_q - |f, X|_q] \cdot |D^\infty g, X, \varrho|_r. \quad (4.4)$$

Signalons qu'il est aisé d'appliquer (4.6) à (4.4), puisque toute série formelle $F(\varrho) \geqslant 0$ vérifie évidemment

$$F(\varrho) - F(0) \leqslant \varrho \frac{\partial F(\varrho)}{\partial \varrho}. \quad (4.5)$$

Formules de la dérivée

$$\frac{\partial}{\partial \varrho} |D^\infty f, X, \varrho|_p \leqslant \sum_{j=1}^{l} |D^\infty D_j f, X, \varrho|_p; \quad (4.6)$$

$$|D^\infty D_j f, X, \varrho|_p \leqslant \frac{\partial}{\partial \varrho} |D^\infty f, X, \varrho|_p. \quad (4.7)$$

5. Preuve de la formule du produit

Définissons la série formelle en $\xi = (\xi_1, ..., \xi_l)$:

$$|D^\infty f, X; \xi|_p = \sum_\sigma \frac{\xi^\sigma}{\sigma!} |D^\sigma f, X|_p , \qquad (5.1)$$

où

$$\xi^\sigma = \xi_1^{\sigma_1} ... \xi_l^{\sigma_l} .$$

Nous avons

$$|D^\infty (fg), X; \xi|_p \ll |D^\infty f, X; \xi|_q . |D^\infty g, X; \xi|_r \qquad (5.2)$$

($\ll$ signifie $\leqslant$ pour les coefficients homologues).

Preuve de (5.2). — D'après Leibniz

$$\frac{1}{\sigma!} D^\sigma (fg) = \sum_{\lambda, \mu} \frac{1}{\lambda!} (D^\lambda f) \frac{1}{\mu!} (D^\mu g), \quad \text{où} \quad \lambda + \mu = \sigma;$$

d'après Hölder

$$|(D^\lambda f) . (D^\mu g), X|_p \leqslant |D^\lambda f, X|_q |D^\mu g, X|_r;$$

donc

$$|D^\infty (fg), X; \xi|_p \ll \sum_{\lambda, \mu} \frac{\xi^\lambda}{\lambda!} |D^\lambda f, X|_q \frac{\xi^\mu}{\mu!} |D^\mu g, X|_r,$$

ce qui prouve (5.2).

Lemme 5. — Soient

$$\varphi(\xi) = \sum_\sigma \frac{\xi^\sigma}{\sigma!} \varphi_\sigma, \quad \psi(\xi), \quad \theta(\xi)$$

des séries formelles vérifiant

$$\theta(\xi) \ll \varphi(\xi) \psi(\xi) \quad (0 \leqslant \varphi_\sigma \leqslant +\infty, \text{ etc.}); \qquad (5.3)$$

définissons comme suit des séries formelles $\Phi(\varrho), \Psi(\varrho), \Theta(\varrho)$:

$$\Phi(\varrho) = \sum_{s=0}^\infty \frac{\varrho^s}{s!} \sup_\sigma \varphi_\sigma, \quad \text{où} \quad |\sigma| = s. \qquad (5.4)$$

On a

$$\Theta(\varrho) \ll \Phi(\varrho) \Psi(\varrho). \qquad (5.5)$$

Preuve. — Notons $\varrho = \xi_1 + \ldots + \xi_l$; la formule du binôme donne, pour $|\sigma| = s$:

$$\frac{\varrho^s}{s!} = \sum_\sigma \frac{\xi^\sigma}{\sigma!}.$$

La définition (5.4) de $\Phi(\varrho)$ peut donc s'énoncer (*) comme suit : $\Phi(\varrho)$ est la plus petite série formelle en $\varrho = \xi_1 + \ldots + \xi_l$ qui majore $\varphi(\xi)$, c'est-à-dire qui vérifie :

$$\varphi(\xi) \ll \Phi(\varrho).$$

Donc

$$\theta(\xi) \ll \varphi(\xi) \, . \, \psi(\xi) \ll \Phi(\varrho) \, . \, \Psi(\varrho)$$

et, par suite

$$\Theta(\varrho) \ll \Phi(\varrho) \, . \, \Psi(\varrho).$$

Preuve de la formule du produit (4.3). On applique le lemme à (5.2).

6. Preuve de la formule du commutateur

Le calcul établissant (5.2) donne

$$|[D^\infty, f]g, X; \xi|_p \ll [|D^\infty f, X; \xi|_q - |f, X|_q] \, . \, |D^\infty g, X; \xi|_r.$$

On applique le lemme 5.

7. Preuve des formules de la dérivée

Preuve de (4.6). — Aux deux membres, les coefficients respectifs de $\varrho^s/s!$ sont

$$\sup_{\sigma'} |D^{\sigma'}f, X|_p \leqslant \sum_l \sup_\sigma |D^\sigma D_l f, X|_p, \quad \text{où} \quad |\sigma'| - 1 = |\sigma| = s.$$

Preuve de (4.7). — Aux deux membres, les coefficients respectifs de $\varrho^s/s!$ sont

$$\sup_\sigma |D^\sigma D_l f, X|_p \leqslant \sup_{\sigma'} |D^{\sigma'}f, X|_p, \quad \text{où} \quad |\sigma'| - 1 = |\sigma| = s.$$

(*) Cet énoncé est emprunté à L. Gårding [14].

§ 2. **Quasi-norme formelle**

8. Notations

Soient $(x_0, x_1, ..., x_l)$ des coordonnées de R^{l+1}; soit Y la bande
de R^{l+1} d'équation

$$Y : 0 \leqslant x_0 \leqslant |Y|;$$

soit S_t l'hyperplan de cette bande d'équation

$$S_t : x_0 = t, \qquad \text{où} \quad 0 \leqslant t \leqslant |Y|.$$

Étant donnée une fonction

$$f : Y \rightarrow C,$$

nous notons $|f, S_t|_p$ la norme de sa restriction à S_t et $D^\sigma f$ ses
dérivées, où $\sigma = (\sigma_0, \sigma_1, ..., \sigma_l)$; $|\sigma|$ désigne $\sigma_0 + ... + \sigma_l$; $(\sigma) \leqslant (r, s)$
signifie : $\sigma_0 \leqslant r$, $|\sigma| \leqslant r + s$.

Pour $\sigma = (j, 0, ..., 0)$, D^σ est noté D_0^j.

9. Définition

Étant donnée une fonction

$$f : Y \rightarrow C,$$

un nombre $p(1 \leqslant p \leqslant \infty)$ et un entier $n \geqslant 0$, nous nommons
quasi-norme formelle de f la série formelle, dont les coefficients
sont des fonctions de t à valeurs $\geqslant 0$ et $\leqslant \infty$:

$$|D^{n,\infty}f, S_t, \varrho|_p = \sum_\gamma \frac{1}{\gamma!} |D^\infty D^\gamma f, S_t, \varrho|_p, \quad \text{où} \quad |\gamma| \leqslant n,$$

$$\tag{9.1}$$

$$= \sum_{\gamma, s} \frac{1}{\gamma!} \frac{\varrho^s}{s!} \sup_\sigma |D^{\gamma+\sigma}f, S_t|_p \quad \text{où} \quad |\gamma| \leqslant n, 0 \leqslant s, (\sigma) = (0, s).$$

Nous emploierons la *norme formelle*

$$|D^{n,\infty}f, Y, \varrho|_\infty = \sum_{\gamma, s} \frac{1}{\gamma!} \frac{\varrho^s}{s!} \sup_{\sigma, x} |D^\gamma D^\sigma f| \tag{9.2}$$

où $|\gamma| \leqslant n$, $0 \leqslant s$, $(\sigma) = (0, s)$, $x \in Y$;
autrement dit

$$|D^{n,\infty}f, Y, \varrho|_\infty = \sup |D^{n,\infty}f, S_t, \varrho|_\infty \quad \text{où} \quad 0 \leqslant t \leqslant |Y|, \tag{9.3}$$

en convenant que

$$\sup_t \sum_s \varrho^s \Phi_s(t) = \sum_s \varrho^s \sup_t \Phi_s(t).$$

Soit $a(x, D)$ un opérateur différentiel; ses quasi-norme et norme formelles

$$|D^{n,\infty} a, S_t, \varrho|_p \qquad \text{et} \qquad |D^{n,\infty} a, Y, \varrho|_\infty$$

sont les sommes de celles de ses coefficients; nous notons

$$|[D^{n,\infty}, a]f, S_t, \varrho|_p = \sum_{\gamma, s} \frac{1}{\gamma!} \frac{\varrho^s}{s!} \sup_\sigma |[D^\gamma D^\sigma, a]f, S_t|_p \qquad (9.4)$$

où $0 \leqslant |\gamma| \leqslant n, 0 \leqslant s, (\sigma) = (0, s), [D^\gamma D^\sigma, a]f = D^\gamma D^\sigma af - a D^\gamma D^\sigma f$.

10. Propriétés

On a les formules suivantes :

Formule du produit. — Si $1/p = 1/q + 1/r$, alors

$$|D^{n,\infty}(fg), S_t, \varrho|_p \ll |D^{n,\infty}f, S_t, \varrho|_q |D^{n,\infty}g, S_t, \varrho|_r. \qquad (10.1)$$

Preuve. — D'après la formule de Leibniz

$$|D^{n,\infty}(fg), S_t, \varrho|_p \ll \sum_{\alpha, \beta} \frac{1}{\alpha! \beta!} |D^\infty(D^\alpha f)(D^\beta g), S_t, \varrho|_p$$

pour $|\alpha + \beta| \leqslant n$; donc, *a fortiori*, pour $|\alpha| \leqslant n, |\beta| \leqslant n$; il suffit alors d'appliquer au second membre la formule du produit (4.3).

Formule de la dérivée. — De (4.6) et (4.7) résultent les formules, où $j \neq 0$:

$$\frac{\partial}{\partial \varrho} |D^{n,\infty}f, S_t, \varrho|_p \ll \sum_{j=1}^l |D^{n,\infty}D_j f, S_t, \varrho|_p; \qquad (10.2)$$

$$|D^{n,\infty}D_j f, S_t, \varrho|_p \ll \frac{\partial}{\partial \varrho} |D^{n,\infty}f, S_t, \varrho|_p \quad (j > 0). \qquad (10.3)$$

Le n° 11 va prouver la formule suivante, que les § 3 et 4 emploieront :

Formule du commutateur. — Soit $1/p = 1/q + 1/r$; soit $a(x, D)$ un opérateur *normal* ([1]) d'ordre m; on a, si $n \geqslant 1$:
$$|[D^{n,\infty}, a]f, S_t, \varrho|_p \ll \qquad\qquad\qquad\qquad\qquad\qquad (10.4)$$

$$\ll c(m, n) |D^{n,\infty} a, S_t, \varrho|_q \left(1 + \varrho \frac{\partial}{\partial \varrho}\right) |D^{m+n-1,\infty}f, S_t, \varrho|_r,$$

où $c(m, n)$ ne dépend que de l, m et n.

[1] Le coefficient de D_0^m (qui est nommé *premier coefficient*) vaut 1.

11. Preuve de la formule du commutateur

On a

$$[D^\gamma D^\sigma, a]f = D^\sigma[D^\gamma, a]f + [D^\sigma, a]D^\gamma f;$$

d'où, vu les définitions (9.1) et (9.4) :

$$|[D^{n,\infty}, a]f, S_t, \varrho|_p \ll$$

$$\ll \sum_\gamma |D^{0,\infty} b_\gamma f, S_t, \varrho|_p + \sum_\gamma \frac{1}{\gamma!} |[D^{0,\infty}, a]D^\gamma f, S_t, \varrho|_p,$$

où

$$|\gamma| \leqslant n; \quad b_\gamma = \frac{1}{\gamma!}[D^\gamma, a]; \quad \text{ordre } (b_\gamma) \leqslant m + n - 1;$$

les coefficients de b_γ sont des dérivées de ceux de a d'ordres $\leqslant n$. Donc, vu la formule du produit (4.3) :

$$|D^{0,\infty} b_\gamma f, S_t, \varrho|_p \ll c(m, n) \, |D^{n,\infty} a, S_t, \varrho|_q \, |D^{m+n-1} f, S_t, \varrho|_r.$$

Pour établir la formule du commutateur (10.4), il suffit donc de prouver que

$$|[D^{0,\infty}, a]D^\gamma f, S_t, \varrho|_p \ll \tag{11.1}$$

$$\ll c(m, n) \, |D^{n,\infty} a, S_t, \varrho|_q \left(1 + \varrho \frac{\partial}{\partial \varrho}\right) |D^{m+n-1,\infty} f, S_t, \varrho|_r$$

si $|\gamma| \leqslant n$. Il suffit de le prouver quand a est monome :

$$a(x, D) = a_\alpha(x)D^\alpha, \qquad \text{où} \quad |\alpha| \leqslant m.$$

Quand $|\alpha| \leqslant m - 1$, la formule (4.4), où l'on remplace X par S_t et où l'on majore $-|f, X|_q$ par 0, donne

$$|[D^{0,\infty}, a]D^\gamma f, S_t, \varrho|_p \ll c |D^{n,\infty} a, S_t, \varrho|_q \, |D^{m+n-1,\infty} f, S_t, \varrho|_r$$

et prouve donc (11.1).

Quand $(\alpha) = (m, 0)$, alors $a_\alpha = 1$, puisque a est normal; donc, vu la définition (9.4),

$$|[D^{0,\infty}, a]D^\gamma f, S_t, \varrho|_p = 0.$$

Supposons enfin $(\alpha) \leqslant (m - 1, 1)$, c'est-à-dire $a(x, D) = a_\alpha(x)D_i D^\beta$ où $1 \leqslant i \leqslant l$, $|\beta| \leqslant m - 1$; remarquons que les formules (4.4), (4.5), (4.6), (4.7) donnent

$$|[D^\infty, f]D_i g, X, \varrho|_p \ll \sum_{j=1}^{l} |D^\infty D_j f, X, \varrho|_q \, \varrho \frac{\partial}{\partial \varrho} |D^\infty g, X, \varrho|_r;$$

d'où en remplaçant f, g, X par a_α, $D^{\beta+\gamma} f$, S_t :

$$|[D^{0,\infty}, a] D^\gamma f, S_t, \varrho|_p \ll c |D^{1,\infty} a, S_t, \varrho|_q \varrho \frac{\partial}{\partial \varrho} |D^{m+n-1} f, S_t, \varrho|_r ,$$

ce qui achève la preuve de (11.1), donc de la formule du commutateur.

§ 3. Approximations successives

12. L'ÉQUATION STRICTEMENT HYPERBOLIQUE a les propriétés que voici (voir [1], [5], [6]) :

Sur la bande Y, soit un opérateur $a(x, D)$ d'ordre m, normal et régulièrement hyperbolique ([2]) pour les hyperplans S_t;

on suppose donné un entier $n \geqslant 1$ tel que le premier terme $|D^n a, Y|_\infty$ de la norme formelle $|D^{n,\infty} a, Y, \varrho|_\infty$ soit fini;

alors l'opérateur a se *minore* comme suit : si $|D^{m+n-1} f, S_t|_2$ est une fonction sommable de t, alors

$$|D^{m+n-2} f, S_t|_2 \leqslant c \int_0^t |D^{n-1} af, S_{t'}|_2 \, dt' + c |D^{m+n-2} f, S_0|_2 ;$$
$$(12.1)$$

les nombres c ne dépendent que de $|D^{n-1} a, Y|_\infty$ et, si $n = 1$, de $|D^1 a, Y|_\infty$.

Cette minoration implique évidemment le

Théorème d'unicité. — Le problème de Cauchy d'inconnue u

$$au = v, \quad D^{m-1} u | S_0 \text{ donné } ([3]) \qquad (12.2)$$

a au plus une solution telle que $|D^m u, S_t|_2$ soit fonction sommable de t.

La fin de ce n° 12 va montrer que la minoration (12.1) de a permet de majorer comme suit la solution (12.2) du problème de Cauchy (12.2) :

([2]) Soit $g(x, p)$ le polynome caractéristique de a, c'est-à-dire la partie principale du polynome $a(x, p)$ de p. On suppose qu'il a les propriétés suivantes pour $p_1, \ldots, p_l$ réels :

$g(x, p)$ a toutes ses racines p_0 finies, réelles (hyperbolicité) et distinctes (stricte hyperbolicité);

tout polynome $g(\infty, p)$, qui est une limite pour $|x| \to \infty$ de $g(x, p)$, a lui aussi ses racines p_0 finies et distinctes (hyperbolicité régulière).

([3]) C'est-à-dire : $u | S_0, \ldots, D_0^{m-1} u | S_0$ donnés.

Lemme 12. — On a

$$|D^{m+n-1,\infty}u, S_t, \varrho|_2 \ll \varphi(t, \varrho), \qquad (12.3)$$

en notant $\varphi(t, \varrho)$ la série formelle que définit le problème de Cauchy

$$\begin{cases} \left[\dfrac{\partial}{\partial t} - C_1(\varrho) \left(1 + \varrho \, \dfrac{\partial}{\partial \varrho} \right) \right] \varphi(t, \varrho) = \psi(t, \varrho) \\ \varphi(0, \varrho) = \theta(\varrho), \end{cases} \qquad (12.4)$$

où C_1, ψ et θ sont des séries formelles en ϱ telles que

$$\begin{cases} c_0 |D^{n,\infty}a, Y, \varrho|_\infty \ll C_1(\varrho) \\ c_0 |D^{n,\infty}v, S_t, \varrho|_2 \ll \psi(t, \varrho), \quad c_0 |D^{m+n-1,\infty}u, S_0, \varrho|_2 \ll \theta(\varrho) \end{cases} \qquad (12.5)$$

les c_0 étant des nombres dépendant de $|D^n a, Y|_\infty$.

Note 12.1. — Il est immédiat de calculer $D^{m+n-1}u\,|\,S_0$, donc de majorer $|D^{m+n-1,\infty}u, S_0, \varrho|_2$ en fonction des données. Par exemple, si

$$D^{m-1}u\,|\,S_0 = 0 \quad \text{et} \quad D^{n-1}v\,|\,S_0 = 0, \quad \text{alors} \quad D^{m+n-1}u\,|\,S_0 = 0$$

et l'on peut prendre $\theta(\varrho) = 0$.

Note 12.2. — La résolution de (12.4) est élémentaire : les coefficients $\varphi_s(t)$ de

$$\varphi(t, \varrho) = \sum_s \frac{\varrho^s}{s!} \, \varphi_s(t)$$

s'obtiennent par des quadratures portant sur des fonctions $\geqslant 0$; ces formules gardent un sens quand les coefficients des données C_1, ψ, θ ne sont pas sommables [4].

Théorème d'existence. — Le problème de Cauchy (12.2) possède une solution u, telle que $|D^{m+n-1}u, S_t|_2$ est une fonction bornée de t, quand les premiers termes

$$|D^n v, S_t|_2, \ |D^{m+n-1}u, S_0|_2$$

des séries formelles figurant dans (12.5) sont respectivement sommables et finis; c'est-à-dire, *quand $\varphi(t, 0)$ est une fonction bornée de t.* Cette solution vérifie (12.1); c'est-à-dire :

$$|D^{m+n-1}u, S_t|_2 \leqslant c \int_0^t |D^n v, S_{t'}|_2 \, dt' + c|D^{m+n-1}u, S_0|_2. \quad (12.6)$$

---

[4] Nous pourrions aisément nous limiter au cas où elles le sont.

Preuve du lemme 12. — Supposons les r premiers termes de la série formelle $\varphi(t, \varrho)$ bornés; c'est-à-dire les r premiers termes des séries formelles

$$|D^{n,\infty}v, S_t, \varrho|_2, \quad |D^{m+n-1,\infty}u, S_0, \varrho|_2, \quad |D^{n,\infty}a, Y, \varrho|_\infty$$

respectivement sommables et finis; on sait que les r premiers termes de $|D^{m+n-1,\infty}u, S_t, \varrho|_2$ sont alors bornés. Soient γ et σ tels que $|\gamma| \leqslant n, \ (\sigma) \leqslant (0, r)$.

Puisque

$$aD^\gamma D^\sigma u = -[D^\gamma D^\sigma, a]u + D^\gamma D^\sigma v,$$

on a d'après (12.1) :

$$|D^{m-1}D^\gamma D^\sigma u, S_t|_2 \ll c \int_0^t |[D^\gamma D^\sigma, a]u, S_{t'}|_2 \, dt' +$$

$$+ \ c \int_0^t |D^\gamma D^\sigma v, S_{t'}|_2 \, dt' + |D^{m-1}D^\gamma D^\sigma u, S_0|_2,$$

les c étant indépendants de r; chacun des termes de cette relation est une fonction bornée de t. Appliquons

$$\sum_{\gamma, s} \frac{1}{\gamma!} \frac{\varrho^s}{s!} \sup_\sigma,$$

où $|\gamma| \leqslant n, |\sigma| = s \leqslant r$; il vient, en modifiant les c (qui restent indépendants des r) :

$$|D^{m+n-1,\infty}u, S_t, \varrho|_2 \ll c \int_0^t |[D^{n,\infty}, a]u, S_{t'}, \varrho|_2 \, dt' +$$

$$+ \ c \int_0^t |D^{n,\infty}v, S_{t'}, \varrho|_2 \, dt' + |D^{m+n-1,\infty}u, S_0, \varrho|_2 \quad \text{mod. } \varrho^r.$$

Majorons le second membre par la formule du commutateur (10.4); il vient :

$$|D^{m+n-1,\infty}u, S_t, \varrho|_2 \ll$$

$$\ll c|D^{n,\infty}a, Y, \varrho|_\infty \left(1 + \varrho \frac{\partial}{\partial \varrho}\right) \int_0^t |D^{m+n-1,\infty}u, S_{t'}, \varrho|_2 \, dt' +$$

$$+ \ c \int_0^t |D^{n,\infty}v, S_{t'}, \varrho|_2 \, dt' + |D^{m+n-1,\infty}u, S_0, \varrho|_2 \text{ mod. } \varrho^r.$$

L'intégration de cette inégalité est classique; elle donne ceci, en prenant dans (12.4) et (12.5) des c_0 assez grands, mais indépendants de r : on a

$$|D^{m+n-1}u, S_t, \varrho|_2 \ll \varphi(t, \varrho) \mod \varrho^r \qquad (12.7)$$

pour $0 \leqslant t \leqslant T_r(T_r \leqslant |Y|)$, si pour ces valeurs de t les coefficients de $\varrho^j(j = 0, ..., r-1)$ dans φ sont des fonctions bornées de t.

Or la définition de $\varphi(t, \varrho)$, par (12.4) et la Note 12.2, montre que si le coefficient de ϱ^{r-1} dans φ vaut ∞ pour $t = T$, alors tous les coefficients de $\varrho^s(s \geqslant r-1)$ valent ∞ pour $T \leqslant t \leqslant |Y|$. Donc (12.7) implique (12.3), c'est-à-dire le lemme 12.

13. Équation décomposable en équations strictement hyper-boliques

Sur la bande Y, soit

$$a = a_1 \ldots a_p$$

un opérateur d'ordre m, produit de p opérateurs normaux, régu-lièrement hyperboliques pour les hyperplans S_t; soient $m_1, ..., m_p$ leurs ordres :

$$m = m_1 + \ldots + m_p.$$

Posons-nous sur Y le problème de Cauchy d'inconnue u :

$$au = v, \quad D^{m-1}u|S_0 = 0 \qquad (13.1)$$

quand on se donne un entier $n \geqslant p$ tel que :

$$D^{n-1}v|S_0 = 0, \qquad (13.2)$$

ce qui impliquera $D^{m+n-p+q-1}u|S_0 = 0$ dans le lemme ci-dessous;

$$|D^{m_1+...+m_j-j+n}a_{j+1}, Y|_\infty < \infty \qquad (0 \leqslant j < p). \qquad (13.3)$$

Soient C_1, C_2 et ψ des séries formelles en ϱ telles que

$$\begin{cases} c_0|D^{m_1+...+m_j-j+n,\infty}a_{j+1}, Y, \varrho|_\infty \ll C_1(\varrho) \quad \text{pour} \quad j = 0, ..., p-1, \\ c_0[1 + |D^{n-p+q,\infty}a, Y, \varrho|_\infty]^q \ll C_2(\varrho) \\ c_0|D^{n,\infty}v, S_t, \varrho|_2 \ll \psi(t, \varrho), \end{cases} \qquad (13.4)$$

les c_0 étant des nombres dépendant de $|D^{m_1+...+m_j-j+n}a_{j+1}, Y|_\infty$.

Soit $\varphi(t, \varrho)$ la série formelle en ϱ que définit le problème de Cauchy (à données de Cauchy évidemment nulles) :

$$\begin{cases} \left[\dfrac{\partial}{\partial t} - C_1(\varrho)\left(1 + \varrho\dfrac{\partial}{\partial \varrho}\right)\right]^p \varphi(t, \varrho) = \psi(t, \varrho), \\ \left[\dfrac{\partial}{\partial t} - C_1(\varrho)\left(1 + \varrho\dfrac{\partial}{\partial \varrho}\right)\right]^j \varphi(t, \varrho) = 0 \quad \text{pour} \quad t = 0, j = 0, ..., p-1. \end{cases} \qquad (13.5)$$

Lemme 13.1. — Si $|D^n v, S_t|_2$ est une fonction sommable de t, en particulier si $\varphi(t, 0)$ *est une fonction bornée de t* pour $0 \leqslant t \leqslant |Y|$, alors le problème de Cauchy (13.1) possède une et une seule solution telle que $|D^{m+n-p}u, S_t|_2$ soit une fonction bornée de t. Cette solution vérifie :

$$|D^{m+n-p,\infty}u, S_t, \varrho|_2 \ll \varphi(t, \varrho); \qquad (13.6)$$

et aussi

$$|D^{m+n-p+q,\infty}u, S_t, \varrho|_2 \ll C_2(\varrho) \left(1 + \frac{\partial}{\partial t} + \frac{\partial}{\partial \varrho} \right)^q \varphi(t, \varrho), \quad (13.7)$$

si q satisfait la condition $0 < q < p$,

La note 12.2 s'applique au problème (13.5).

Preuve. — Notons $v_0 = v$ et envisageons les problèmes de Cauchy d'inconnues $v_j (j = 1, ..., p)$:

$$a_j v_j = v_{j-1}, \quad D^{m_j-1} v_j | S_0 = 0.$$

Le n° 12 et une récurrence sur j montrent qu'ils définissent sans ambiguïté des v_j tels que $|D^{m_1+...+m_j+n-j} v_j, S_t|_2$ est une fonction bornée de t. On a :

$$D^{m_1+...+m_j+n-j-1} v_j | S_0 = 0$$

et, vu le lemme 12 :

$$|D^{m_1+...+m_j+n-j,\infty} v_j, S_t, \varrho|_2 \ll \psi_j(t, \varrho) \qquad (13.8)$$

les ψ_j étant les séries formelles en ϱ que définissent les problèmes de Cauchy :

$$\left[\frac{\partial}{\partial t} - C_1(\varrho) \left(1 + \varrho \frac{\partial}{\partial \varrho} \right) \right] \psi_j(t, \varrho) = \psi_{j-1}(t, \varrho), \quad \psi_j(0, \varrho) = 0 \quad (13.9)_j$$

où $\psi_0 = \psi$, $j = 1, ..., p$. Vu le n° 12, $u = v_p$ est la solution unique du problème (13.1); (13.8) donne (13.6), car ψ_p est la solution φ du problème (13.5).

La preuve de (13.7) est la suivante : il s'agit de prouver que

$$|D^{\infty} D^{\gamma}u, S_t, \varrho|_2 \ll C_2(\varrho) \left(1 + \frac{\partial}{\partial t} + \frac{\partial}{\partial \varrho} \right)^q \varphi(t, \varrho), \quad (13.10)$$

pour tout γ tel que

$$|\gamma| \leqslant m + n - p + q, \qquad \text{où} \quad q < p.$$

Nous le prouverons en montrant ceci :

$$D^\gamma u = e_\gamma(x, D)u + f_\gamma(x, D)D_0^{n-p}v, \qquad (13.11)$$

où e_γ et f_γ sont des opérateurs différentiels ayant les propriétés que voici :

$$\text{ordre } (e_\gamma) \leqslant (m+n-p, q), \text{ ordre } (f_\gamma) \leqslant q$$

les coefficients de e_γ et f_γ sont des polynomes en les dérivées d'ordres $\leqslant n - p + q$ des coefficients de a; ces polynomes sont de degré q.

2°) On a

$$|D^\infty e_\gamma(x, D)u, S_t, \varrho|_2 \ll C_2(\varrho) \left(1 + \frac{\partial}{\partial\varrho}\right)^q \varphi(t, \varrho); \quad (13.12)$$

3°) On a de même :

$$|D^\infty f_\gamma(x, D)D_0^{n-p}v, S_t, \varrho|_2 \ll C_2(\varrho) \left(1 + \frac{\partial}{\partial t} + \frac{\partial}{\partial\varrho}\right)^q \varphi(t, \varrho). \quad (13.13)$$

La preuve de (13.7) sera alors achevée.

Preuve de (13.11) — Cette relation est évidente si $(\gamma) \leqslant \leqslant (m+n-p, q)$: on prend $e_\gamma u = D^\gamma u$; $f_\gamma = 0$. Il suffit donc de le prouver quand

$$D^\gamma = D_0^{m+n-p} D^\beta, \quad |\beta| = q > 1,$$

en la supposant vraie pour $|\beta| < q$.

Appliquons $D_0^{n-p} D^\beta$ à l'équation $au = v$; nous obtenons

$$D^\gamma u = g(x, D)u + \sum_{j=1}^{q-1} h_j(x, D) D_0^{m+n-p+j} u + D^\beta D_0^{n-p}v,$$

où

$$\text{ordre } (g) \leqslant (m+n-p, q), \text{ ordre } (h_j) \leqslant (0, q-j).$$

Remplaçons dans la relation précédente les $D_0^{m+n-p+j} u$ par leurs expressions (13.11) : nous obtenons (13.11) pour la valeur donnée de γ.

Preuve de (13.12). — La formule du produit (4.3), de la dérivée (4.7) et la définition (9.1) de la quasi-norme formelle donnent

$$|D^\infty e_\gamma u, S_t, \varrho|_2 \ll C_2(\varrho) \left(1 + \frac{\partial}{\partial\varrho}\right)^q |D^{m+n-p,\infty} u, S_t, \varrho|_2;$$

on majore le second membre par (13.6).

Preuve de (13.13).   On prouve de même (13.13), en employant la formule

$$c_0 |D^\infty D_0^{n-p+j} v, S_t, \varrho|_2 \ll \left(\frac{\partial}{\partial t}\right)^j \varphi(t, \varrho), \text{ pour } j \leqslant q, \quad (13.14)$$

dont voici la preuve. Puisque $D_0^{n-p+q} v | S_0 = 0$, on a

$$D_0^{n-p+j} v(x, t) = \int_0^t \frac{(t-t')^{p-j-1}}{(p-j-1)!} D_0^n v(x, t') \, dt' ;$$

donc

$$|D^\infty D_0^{n-p+j} v, S_t, \varrho|_2 \ll \int_0^t \frac{(t-t')^{p-j-1}}{(p-j-1)!} |D^\infty D_0^n v, S_{t'}, \varrho|_2 \, dt' ;$$

c'est-à-dire, vu la définition (13.4) de $\psi = \psi_0$:

$$c_0 |D^\infty D_0^{n-p+j} v, S_t, \varrho|_2 \ll \int_0^t \frac{(t-t')^{p-j-1}}{(p-j-1)!} \psi_0(t', \varrho) \, dt' . \quad (13.15)$$

Or, puisque $\psi_0(t, \varrho) \geqslant 0$, l'équation $(13.9)_1$ donne

$$\psi_1(t, \varrho) \geqslant 0, \frac{\partial \psi_1}{\partial t} \geqslant \psi_0 ;$$

l'équation $(13.9)_2$ donne alors

$$\psi_2(t, \varrho) \geqslant 0, \frac{\partial \psi_2}{\partial t} \geqslant \psi_1 ;$$

d'où, en appliquant $\partial/\partial t$ à $(13.9)_2$:

$$\frac{\partial^2 \psi_2}{\partial t^2} \geqslant \frac{\partial \psi_1}{\partial t} ;$$

finalement :

$$\frac{\partial^j \psi_p}{\partial t^j} \geqslant \frac{\partial^{j-1} \psi_{p-1}}{\partial t^{j-1}} \geqslant \ldots \geqslant \psi_{p-j} \geqslant 0 \quad \text{pour } j \leqslant p. \quad (13.16)$$

D'où, puisque d'après (13.9) $\psi_j(0, \varrho) = 0$ si $j > 0$:

$$\int_0^t \frac{(t-t')^{p-j-1}}{(p-j-1)!} \psi_0(t', \varrho) \, dt' \ll \int_0^t \frac{(t-t')^{p-j-1}}{(p-j-1)!} \frac{\partial \psi_1(t', \varrho)}{\partial t'} \, dt' -$$

$$\int_0^t \frac{(t-t')^{p-j-2}}{(p-j-2)!} \psi_1(t', \varrho) \, dt' \ll \ldots \ll \psi_{p-j}(t, \varrho) \ll \frac{\partial^j \psi_p}{\partial t^j} = \frac{\partial^j \varphi(t, \varrho)}{\partial t^j}.$$

En portant cette inégalité dans (13.15), on obtient (13.14), ce qui achève la preuve de (13.7) et celle du lemme 13.1.

Notons que (13.16) résulte de l'inégalité $\psi_0(t, \varrho) \geqslant 0$ et des relations $(13.9)_j$, où l'on peut remplacer la condition $\psi_j(0, \varrho) = 0$

par $\psi_j(0, \varrho) \gg 0$; on peut donc énoncer le lemme suivant, où

$$L_1 = C_1 \left(1 + \varrho \, \frac{\partial}{\partial \varrho}\right):$$

Lemme 13.2. — Supposons $C_1 \gg 0$,

$$\left[\frac{\partial}{\partial t} - L_1\right]^p \varphi(t, \varrho) \gg 0,$$

$$\left[\frac{\partial}{\partial t} - L_1\right]^j \varphi(t, \varrho) \gg 0 \quad \text{pour} \quad t = 0, \, j = 0, 1, \dots, p-1.$$

Alors

$$0 \ll \varphi, \dots, 0 \ll \frac{\partial^p \varphi(t, \varrho)}{\partial t^p}.$$

14. Le système dont la partie principale est diagonale

Sur la bande Y, soient des opérateurs différentiels $a(x, D)$ et $b_\mu^\nu(x, D)$ $(\mu, \nu = 1, 2, \dots, N)$ du type suivant :

$a(x, D) = a_1(x, D) \dots a_p(x, D)$ est le produit de p opérateurs normaux, régulièrement hyperboliques pour les hyperplans S_t;

$$\text{ordre } (a_i) = m_i; \quad \text{ordre } (a) = m = m_1 + \dots + m_p;$$
$$\text{ordre } (b_\mu^\nu) = m + n^\mu - n^\nu - p + q,$$

où n^ν et q sont des entiers tels que ([5])

$$0 \leqslant q < p \leqslant n^\nu.$$

Posons-nous, sur Y, *le problème de Cauchy* d'inconnue $u^\nu(x)$. :

$$\begin{cases} a(x, D)u^\nu(x) = \displaystyle\sum_\mu b_\mu^\nu(x, D)u^\mu(x) + v^\nu(x) \\ D^{m-1} u^\nu \mid S_0 = 0 \end{cases} \tag{14.1}$$

([5]) Et même tels que

$$0 \leqslant q \leqslant p \leqslant n^\nu,$$

si b_μ^ν ne contient pas

$$\left(\frac{\partial}{\partial t}\right)^{m+n^\mu-n^\nu}$$

On retrouve ainsi un théorème de L. A. Lednev [13] [14], par un procédé dû à G. Talenti.

Signalons que la Note [11] de G. Talenti complète certains de nos résultats; elle emploie des méthodes analogues aux nôtres, puisque son opérateur

$$\left(\frac{\partial}{\partial t}\right)^m$$

est une puissance de l'opérateur strictement hyperbolique $\partial/\partial t$.

en supposant

$$D^{n^{\nu}-1} v^{\nu} | S_0 = 0 \qquad (14.2)$$

Notons $C_1(\varrho)$, $C_2(\varrho)$, $C_3(\varrho)$, $\psi(t, \varrho)$ des séries formelles en ϱ telles qu'on ait :

$$\begin{cases} c_0 |D^{m_1+\ldots+m_j-j+n^{\nu},\infty} a_{j+1}, Y, \varrho|_{\infty} \ll C_1(\varrho) \\ c_0 [1 + |D^{n^{\nu}-p+q,\infty} a, Y, \varrho|_{\infty}]^q \ll C_2(\varrho) \\ c_0 |D^{n^{\nu},\infty} b^{\nu}_{\mu}, Y, \varrho|_{\infty} \ll C_3(\varrho) \\ c_0 |D^{n^{\nu},\infty} v^{\nu}, S_t, \varrho|_2 \ll \psi(t, \varrho), \end{cases} \qquad (14.3)$$

les c_0 étant des nombres dépendant de $|D^{m_1+\ldots+m_j-j+n^{\nu}} a_{j+1}, Y|_{\infty}$; nous poserons [6] :

$$\begin{cases} C_1(\varrho) \left(1 + \varrho \dfrac{\partial}{\partial \varrho}\right) = L_1\left(\varrho, \varrho \dfrac{\partial}{\partial \varrho}\right) \\ C_2 C_3 \left(1 + \dfrac{\partial}{\partial t} + \dfrac{\partial}{\partial \varrho}\right)^q = L_q\left(\varrho, \dfrac{\partial}{\partial t}, \dfrac{\partial}{\partial \varrho}\right); \end{cases} \qquad (14.4)$$

L_1 et L_q sont donc des opérateurs différentiels d'ordres 1 et q.

15. Résolution de (14.1) par approximations successives

Définition de ces approximations

   Cette définition supposera seulement ceci : le premier terme de chacune des séries formelles (14.3) est fini ou sommable. Rappelons que t varie de 0 à $|Y|$.

   Vu le lemme 13.1, le problème de Cauchy

$$a(x, D) u^{\nu}_0(x) = v^{\nu}(x), \qquad D^{m-1} u^{\nu}_0 | S_0 = 0 \qquad (15.1)_0$$

possède une solution unique u^{ν}_0 telle que $|D^{m+n^{\nu}-p+q} u^{\nu}_0, S_t|_2$ est une fonction bornée de t; (13.6) et (13.7), où l'on fait $n = n^{\nu}$, la majorent.

   Soit un entier $k > 0$; supposons u^{ν}_{k-1} défini pour tout ν et $|D^{m+n^{\nu}-p+q} u^{\nu}_{k-1}, S_t|_2$ fonction sommable de t; vu le lemme 13.1, le problème de Cauchy

$$a(x, D) u^{\nu}_k(x) = \sum_{\mu} b^{\nu}_{\mu}(x, D) u^{\mu}_{k-1}(x), \qquad D^{m-1} u^{\nu}_k | S_0 = 0 \quad (15.1)_k$$

[6] Si $p = q$, on pose :

$$C_2 C_3 \left(1 + \frac{\partial}{\partial t} + \frac{\partial}{\partial \varrho}\right)^{p-1} \left(1 + \frac{\partial}{\partial \varrho}\right) = L_p\left(\varrho, \frac{\partial}{\partial t}, \frac{\partial}{\partial \varrho}\right).$$

possède une solution unique u_k^ν telle que $|D^{m+n^\nu-p+q} u_k^\nu, S_t|_2$ est une fonction bornée de t; (13.6) et (13.7) la majorent; en effet :

$$|D^{n^\nu} b_\mu^\nu u_{k-1}^\mu, S_t|_2 \text{ est sommable et } D^{n^\nu-1} b_\mu^\nu u_{k-1}^\mu |S_0 = 0,$$

vu l'hypothèse précédente, complétée par le théorème du produit, et le lemme que voici :

Lemme 15.1. — On a $D^{m+n^\nu-p+q-1} u_{k-1}^\nu |S_0 = 0$, si u_k^ν est défini.

Preuve. — Si $k = 1$, cela résulte de (14.2) et (15.1)$_0$. Soit $k \geqslant 2$; supposons prouvé que

$$D^{m+n^\mu-p+q-1} u_{k-2}^\mu |S_0 = 0;$$

puisque ordre $(b_\mu^\nu) \leqslant m + n^\mu - n^\nu - p + q$, on a

$$D^{n^\nu-1}(b_\mu^\nu u_{k-2}^\mu)|S_0 = 0,$$

donc, vu (15.1)$_k$ et puisque $q < p$:

$$D^{m+n^\nu-p+q-1} u_{k-1}^\nu |S_0 = 0.$$

Définitions de séries formelles

Soient $\varphi_k(t, \varrho)$ des séries formelles vérifiant les inégalités

$$\begin{cases} \left[\dfrac{\partial}{\partial t} - L_1\right]^p \varphi_0(t, \varrho) \gg \psi(t, \varrho) \\ \left[\dfrac{\partial}{\partial t} - L_1\right]^j \varphi_0(t, \varrho) = 0 \text{ pour } t = 0, j = 0, ..., p-1; \end{cases} \quad (15.2)_0$$

$$\begin{cases} \left[\dfrac{\partial}{\partial t} - L_1\right]^p \varphi_k(t, \varrho) \gg L_q \varphi_{k-1}(t, \varrho) \\ \left[\dfrac{\partial}{\partial t} - L_1\right]^j \varphi_k(t, \varrho) = 0 \text{ pour } t = 0, j = 0, ..., p-1 \end{cases} \quad (15.2)_k$$

où $k = 1, 2, ...$. Soit

$$\varphi(t, \varrho) = \sum_{k=0}^{\infty} \varphi_k(t, \varrho).$$

La note 12.2 s'applique à (15.2); les coefficients des séries formelles φ, φ_k peuvent prendre la valeur $+\infty$; ils sont $\geqslant 0$, car, vu le lemme 13.2 :

$$\frac{\partial^j \varphi_k}{\partial t^j} \gg 0, \quad \frac{\partial^j \varphi}{\partial t^j} \gg 0 \quad \text{pour} \quad j = 0, ..., p. \quad (15.3)$$

Lemme 15.2. — Pour tous les k tels que $\varphi_k(t, 0)$ est une fonction bornée de t, $u_k^\nu(x)$ existe et vérifie ([7])

$$|D^{m+n^\nu-p,\infty} u_k^\nu, S_t, \varrho|_2 \ll \varphi_k(t, \varrho) \qquad (15.4)_k$$

$$|D^{m+n^\nu-p+q,\infty} u_k^\nu, S_t, \varrho|_2 \ll C_2(\varrho) \left(1 + \frac{\partial}{\partial t} + \frac{\partial}{\partial \varrho}\right)^q \varphi_k(t, \varrho). \qquad (15.5)_k$$

Preuve. — Pour $k = 0$, ce lemme est le lemme 13.1. Supposons-le établi pour $0, 1, ..., k-1$: u_{k-1}^ν existe et vérifie $(15.5)_{k-1}$; or nous supposons $\varphi_k(t, 0)$ bornée; donc le second membre de $(15.2)_k$ est sommable en t pour $\varrho = 0$; donc celui de $(15.5)_{k-1}$; donc

$$|D^{m+n^\nu-p+q} u_{k-1}^\nu, S_t|_2 \ ;$$

donc u_k^ν existe; le lemme 13.1 donne $(15.4)_k$ et $(15.5)_k$ puisque, vu $(15.5)_{k-1}$ ([8]) et la formule du produit (10.1), on a :

$$c_0 |D^{n^\nu} b_\mu^\nu u_{k-1}^\mu, S_t, \varrho|_2 \ll L_q \varphi_{k-1}(t, \varrho).$$

Complétons le lemme précédent par trois remarques évidentes : $\sum_k \varphi_k(t, 0)$, qui est d'après (15.2) une série de fonctions croissantes

$\geqslant 0$, ne peut converger qu'uniformément (à l'intérieur de son intervalle de convergence);

Si u_k^ν existe quel que soit k et si la série

$$\sum_{k=0}^\infty |D^{m+n^\nu-p} u_k^\nu, S_t|_2$$

converge uniformément, alors le problème de Cauchy (14.1) possède la solution

$$u^\nu(x) = \sum_{k=0}^\infty u_k^\nu(x) ;$$

l'émission ([9]) du support de v contient les supports de $u_0^\nu, ..., u_k^\nu, ...$ ..., u^ν.

([7]) Si $p = q$, alors $(15.5)_k$ s'écrit :

$$|D^{m+n^\nu-1,\infty} u_k^\nu, S_t, \varrho|_2 \ll C_2 \left(1 + \frac{\partial}{\partial t} + \frac{\partial}{\partial \varrho}\right)^{p-1} \varphi_k(t, \varrho).$$

([8]) ... la formule de la dérivée (10.3), si $p = q$, ...
([9]) ... ou domaine d'influence.

Nous obtenons la conclusion suivante :

Proposition 15. — Si $\varphi(t, 0)$ est une fonction bornée de t pour $0 \leqslant t \leqslant |Y|$, alors le problème de Cauchy (14.1) possède une solution $u^{\nu}(x)$ telle que :

$$|D^{m+n^{\nu}-p,\infty} u^{\nu}, S_t, \varrho|_2 \ll \varphi(t, \varrho), \tag{15.6}$$

$$|D^{m+n^{\nu}-p+q,\infty} u^{\nu}, S_t, \varrho|_2 \ll C_2(\varrho) \left(1 + \frac{\partial}{\partial t} + \frac{\partial}{\partial \varrho}\right)^q \varphi(t, \varrho) ; \tag{15.7}$$

le support de u^{ν} appartient à l'émission de celui de v.

16. Remarque : Système strictement hyperbolique

Si $q = 0$ la conclusion précédente se précise comme suit : (15.2) permet de calculer $\varphi_k(t, 0)$ $(k = 0, 1, ...)$ à partir de $\psi(t, 0)$, qui est sommable par hypothèse ;

$$\sum_k \varphi_k(t, 0) = \varphi(t, 0)$$

est la solution du problème de Cauchy :

$$\begin{cases} \left[\dfrac{\partial}{\partial t} - C_1(t, 0)\right]^p \varphi(t, 0) = c_0 \varphi(t, 0) + \psi(t, 0) \\ \left[\dfrac{\partial}{\partial t} - C_1(t, 0)\right]^j \varphi(t, 0) = 0 \quad \text{pour} \quad j = 0, ..., p-1 . \end{cases}$$

On complète la définition de $\varphi(t, \varrho)$ arbitrairement; par exemple en prenant tous les coefficients des séries formelles $\varphi_k(t, \varrho)$, égaux à $+ \infty$, à l'exception du premier, qui est $\varphi_k(t, 0)$.

La conclusion du n° 15 est alors celle-ci :

Proposition 16. — Supposons $q = 0$ et le premier terme des séries formelles (14.3) fini ou sommable. Alors le problème de Cauchy (14.1) possède une solution $u^{\nu}(x)$ vérifiant l'inégalité

$$|D^{m+n^{\nu}-p} u^{\nu}, S_t|_2 \leqslant \varphi(t, 0),$$

dont le second membre est borné. Le support de u^{ν} appartient à l'émission de celui de v.

Quand $q = 0$, le système (14.1) est *strictement hyperbolique*; la proposition 18 est un théorème d'existence classique.

17. Des problèmes de Cauchy formels vont nous permettre de choisir commodément des φ_k vérifiant (15.2).

Définissons deux séries formelles en ϱ, $\Theta(t, \varrho)$ et $\Omega(t, \varrho)$ par les deux problèmes de Cauchy formels :

$$\frac{\partial}{\partial t} \Theta(t, \varrho) = \varrho\, C_1(\varrho) \frac{\partial}{\partial \varrho} \Theta(t, \varrho), \quad \Theta(0, \varrho) = \varrho\,; \qquad (17.1)$$

$$\frac{\partial \Omega}{\partial t} = \varrho\, C_1 \frac{\partial \Omega}{\partial \varrho} + C_1, \quad \Omega(0, \varrho) = 0. \qquad (17.2)$$

La résolution de ces problèmes est élémentaire : les coefficients successifs de ϱ se calculent par quadratures; ces coefficients sont des exponentielles-polynômes; on a :

$$\Theta(t, 0) = 0\,;\ \varrho \ll \Theta(t, \varrho),\ 0 \ll \frac{\partial^j}{\partial t^j} \Theta(t, \varrho)\ \text{pour tout } j\,; \qquad (17.3)$$

$$\Omega(t, 0) = t\, C_1(0)\,;\ 0 \ll \frac{\partial^j}{\partial t^j} \Omega(t, \varrho)\ \text{pour tout } j\,. \qquad (17.4)$$

La propriété de Θ et Ω que nous emploierons est la suivante : soit $\Phi(t, \theta)$ une série formelle en θ, à coefficients fonctions de t; définissons :

$$\varphi(t, \varrho) = e^{\Omega(t,\varrho)}\, \Phi(t, \Theta(t, \varrho)), \qquad (17.5)$$

ce qui a un sens car $\Theta(t, 0) = 0$; on a évidemment, vu la définition (14.4) de L_1 :

$$\left[\frac{\partial}{\partial t} - L_1\right]^j \varphi(t, \varrho) = e^{\Omega(t,\varrho)} \left[\frac{\partial^j}{\partial t^j} \Phi(t, \theta)\right]_{\theta=\Theta(t,\varrho)} \qquad (17.6)$$

Notons $M_q(t, \varrho, \partial/\partial t, \partial/\partial \theta)$ l'opérateur différentiel [10] d'ordre q, à coefficients séries formelles en ϱ, tel que tout Φ vérifie :

$$L_q[e^{\Omega(t,\varrho)}\, \Phi(t, \Theta(t, \varrho))] =$$

$$= e^{\Omega(t,\varrho)} \left[M_q\left(t, \varrho, \frac{\partial}{\partial t}, \frac{\partial}{\partial \theta}\right) \Phi(t, \theta)\right]_{\theta=\Theta(t,\varrho)}\,; \qquad (17.7)$$

les coefficients de M_q sont linéaires en ceux de L_q, polynomiaux en les dérivées de Θ et Ω; ils sont $\gg 0$.

Choisissons les φ_k comme suit :

$$\varphi_k(t, \varrho) = e^{\Omega(t,\varrho)}\, \Phi_k(t, \Theta(t, \varrho)),$$

[10] ..., ne contenant pas $(\partial/\partial t)^p$, si $p = q$, ...

les $\Phi_k(t, \theta)$ étant définis par les problèmes de Cauchy formels, dont la résolution est élémentaire :

$$\begin{cases} \dfrac{\partial^p}{\partial t^p}\, \Phi_0(t, \theta) \,=\, \psi(t, \theta), \\[2mm] \dfrac{\partial^j}{\partial t^j}\, \Phi_0(t, \theta) \,=\, 0 \ \text{pour}\ t = 0, j = 0, ..., p-1; \end{cases} \quad (17.8)_0$$

$$\begin{cases} \dfrac{\partial^p}{\partial t^p}\, \Phi_k(t, \theta) \,=\, M_q\left(t, \theta, \dfrac{\partial}{\partial t},\ \dfrac{\partial}{\partial \theta}\right) \Phi_{k-1}(t, \theta) \\[2mm] \dfrac{\partial^j}{\partial t^j}\, \Phi_k(t, \theta) \,=\, 0 \ \text{pour}\ t = 0, j = 0, ..., p-1, \end{cases} \quad (17.8)_k$$

où $k = 1, 2, ...$. Soit

$$\Phi(t, \theta) \,=\, \sum_{k=0}^{\infty} \Phi_k(t, \theta). \qquad (17.9)$$

La note 12.2 s'applique à (17.8); les coefficients des séries formelles Φ, Φ_k peuvent prendre la valeur $+\infty$; ils sont évidemment $\geqslant 0$; plus précisément :

$$\frac{\partial^j}{\partial t^j}\, \Phi_k(t, \theta) \geqslant 0, \ \frac{\partial^j}{\partial t^j}\, \Phi(t, \theta) \geqslant 0 \ \text{pour}\ j = 0, ..., p. \ (17.10)$$

De (17.6), où $\Omega \geqslant 0$, et de (17.8)$_0$, où l'on fait $\theta = \Theta \geqslant \varrho$, résulte que φ_0 vérifie (15.2)$_0$. De (17.6), (17.7) et (17.8)$_k$, où l'on fait $\theta = \Theta \geqslant \varrho$, résulte que φ_k vérifie (15.2)$_k$.

On a

$$\varphi(t, \varrho) \,=\, e^{\Omega(t, \varrho)}\, \Phi(t, \Theta(t, \varrho)); \qquad (17.11)$$

en particulier, vu (17.3) et (17.4) :

$$\varphi(t, 0) \,=\, e^{t C_1(0)}\, \Phi(t, 0).$$

La proposition 15 prouve donc ceci :

Proposition 17. — Si $\Phi(t, 0)$ est une fonction bornée de t pour $0 \leqslant t \leqslant |Y|$, alors le problème de Cauchy (14.1) possède une solution $u^v(x)$, qui satisfait (15.6) et (15.7) et dont le support appartient à l'émission de celui de v.

18. $\Phi(t, \theta)$ Peut être caractérisé par un problème de Cauchy formel, dont la solution n'est pas élémentaire, comme l'était celle des problèmes (17.1), (17.2), (17.8) :

Proposition 18.1. — Soit un opérateur $M_q(t, \varrho, \partial/\partial t, \partial/\partial \varrho)$, d'ordre $q \leqslant p$ ne contenant pas $(\partial/\partial t)^p$, dont les coefficients sont des séries formelles $\gg 0$ en ϱ, ayant elles-mêmes pour coefficients des fonctions bornées de t $(0 \leqslant t \leqslant |Y|)$. Soit $\psi(t, \varrho)$ une série formelle $\gg 0$, à coefficients fonctions sommables de t. Définissons Φ_k et Φ par (17.8) et (17.9), où nous remplaçons θ par ϱ.

1) Si les coefficients de $\partial^j/\partial t^j\, \Phi(t, \varrho)$ $(j = 0, ..., q)$ sont des fonctions sommables de t, alors $\Phi(t, \varrho)$ est une solution du problème de Cauchy formel :

$$\begin{cases} \dfrac{\partial^p}{\partial t^p}\, \Phi(t, \varrho) = M_q \Phi(t, \varrho) + \psi(t, \varrho), \\[2mm] \dfrac{\partial^j \Phi}{\partial t^j} = 0 \text{ pour } t = 0,\, j = 0, ..., p-1. \end{cases} \qquad (18.1)$$

2) On a

$$\frac{\partial^j}{\partial t^j}\, \Phi(t, \varrho) \ll \frac{\partial^j}{\partial t^j}\, \Psi(t, \varrho) \text{ pour } j = 0, ..., p, \qquad (18.2)$$

quelle que soit la série formelle $\Psi(t, \varrho)$ vérifiant les inégalités :

$$\begin{cases} \dfrac{\partial^p}{\partial t^p}\, \Psi(t, \varrho) \gg M_q \Psi(t, \varrho) + \psi(t, \varrho), \\[2mm] \dfrac{\partial^j}{\partial t^j}\, \Psi(t, \varrho) \gg 0 \text{ pour } j = 0, ..., p-1. \end{cases} \qquad (18.3)$$

Note : Autrement dit : Φ est la plus petite des solutions de (18.3).

Preuve de 1). — Vu (17.10), les coefficients de la série formelle

$$\sum_k \frac{\partial^j \Phi_k}{\partial t^j}\ (j = 0, ..., p)$$

sont des séries de fonctions croissantes $\geqslant 0$; elles convergent donc uniformément, sauf peut-être au voisinage de $t = |Y|$; (18.1) résulte donc de (17.8).

Preuve de 2). — Notons

$$\Delta_K(t, \varrho) = \Psi(t, \varrho) - \sum_{k=0}^{K-1} \Phi_k(t, \varrho) \text{ si } K > 0$$

$$\Delta_0(t, \varrho) = \Psi(t, \varrho).$$

Les relations (18.1) et (17.8) donnent, si $K > 0$:

$$\begin{cases} \dfrac{\partial^p}{\partial t^p}\, \Delta_K(t, \varrho) \gg M_q\, \Delta_{K-1} \\[2mm] \dfrac{\partial^j \Delta_K}{\partial t^j} \gg 0 \text{ pour } t = 0,\, j = 0, ..., p-1; \end{cases}$$

d'autre part, vu (18.3),

$$\frac{\partial^j \Delta_0}{\partial t^j} \geqslant 0 \quad \text{pour} \quad j = 0, ..., p, \ M_q \Delta_0 \geqslant 0.$$

D'où

$$\frac{\partial^j \Delta_K}{\partial t^j} \geqslant 0 \ (j = 0, ..., p) \,,$$

successivement pour $K = 1, 2, ...$; d'où (18.2).

En résumé, le § 3 a réduit *le problème de Cauchy* (14.1) à la majoration de la solution minimum du *problème de Cauchy formel* (18.1), c'est-à-dire à la recherche *d'une solution des inégalités* (18.3).

§ 4. Résolution du problème de Cauchy formel

Il s'agit de montrer que le problème (18.1) possède une solution à coefficients finis, sous des hypothèses à préciser. Il suffit de construire une solution des inégalités (18.3); nous le ferons à l'aide du théorème de Cauchy-Kowalewski et des opérateurs que voici :

19. OPÉRATEURS SUR LES SÉRIES FORMELLES

Soit une suite de nombres > 0 :

$$\lambda = (\lambda_0 = 1, \ \lambda_1, ..., \lambda_s, ...);$$

nous ferons opérer λ comme suit sur une série formelle $\Phi(\varrho)$:
si

$$\Phi(\varrho) = \sum_s \frac{\varrho^s}{s!} \Phi_s,$$

alors

$$\lambda \Phi(\varrho) = \sum_s \lambda_s \frac{\varrho^s}{s!} \Phi_s. \tag{19.1}$$

Le produit des deux opérateurs $\lambda' = (..., \lambda'_s, ...)$, $\lambda'' = (..., \lambda''_s, ...)$ est évidemment $\lambda' \lambda'' = (..., \lambda'_s \lambda''_s, ...)$; si $\lambda = (1, \lambda_1, ..., \lambda_1^s, ...)$ alors $\lambda \Phi(\varrho) = \Phi(\lambda_1 \varrho)$; il nous suffira donc de nous limiter au cas où $\lambda_1 = 1$.

Évidemment

$$\lambda \left(\varrho \frac{\partial}{\partial \varrho} \Phi \right) = \varrho \frac{\partial}{\partial \varrho} (\lambda \Phi) \tag{19.2}$$

Si Φ et $\Psi \gg 0$, alors la condition nécessaire et suffisante pour que

$$\lambda(\Phi\Psi) \ll (\lambda\Phi)\,(\lambda\Psi) \tag{19.3}$$

est que

$$\lambda_{r+s} \leqslant \lambda_r\,\lambda_s. \tag{19.4}$$

(Il suffit de le prouver quand $\Phi(\varrho) = \varrho^r$, $\Psi(\varrho) = \varrho^s$; c'est alors évident).

Nous supposerons désormais (19.4), qui implique $\lambda_{r+1} \leqslant \lambda_r$, donc :

$$1 = \lambda_0 = \lambda_1 \geqslant \lambda_2 \geqslant \lambda_3 \geqslant \ldots \geqslant \lambda_s \geqslant \ldots . \tag{19.5}$$

D'où, si $\Phi \gg 0$:

$$\frac{\partial}{\partial\varrho}\,(\lambda\Phi) \ll \lambda\left(\frac{\partial\Phi}{\partial\varrho}\right).$$

L'inégalité en sens opposé que voici est celle que nous aurons à employer : soit $\Phi \gg 0$; pour que

$$\lambda\left(\frac{\partial}{\partial\varrho}\right)^j \Phi \ll \left(\frac{\partial}{\partial\varrho}\right)^j \left(\eta + \varepsilon\varrho\,\frac{\partial}{\partial\varrho}\right)^r (\lambda\Phi), \quad \text{si } j \leqslant q, \tag{19.6}$$

(q, ε et η : constantes), il faut et suffit que

$$\lambda_{s-q} \leqslant (\eta + \varepsilon s)^r\,\lambda_s, \quad \text{quel que soit } s. \tag{19.7}$$

(Il suffit de le prouver quand $\Phi(\varrho) = \varrho^s$; c'est alors immédiat, vu (19.2) et (19.5)).

Pour satisfaire (19.7), il suffit de choisir $(s\,!)^\delta\,\lambda_s$ croissant, δ étant un nombre tel que $0 \leqslant q\delta \leqslant r$;

$$\begin{cases} \text{si } q\delta = r, \text{ on prend } \varepsilon = 1, \eta = 0; \\ \text{si } 0 < q\delta < r, \text{ on prend } \varepsilon > 0 \text{ et } \eta = \varepsilon^{p'}, \\ p' \text{ étant le nombre } < 0 \text{ tel que } \dfrac{r}{q\delta} + \dfrac{1}{p'} = 1; \\ \text{si } \delta = 0, \text{ on prend } \eta = 1, \varepsilon = 0. \end{cases} \tag{19.8}$$

Preuve. — Si $(s\,!)^\delta\,\lambda_s$ est croissant, alors (19.7) est vérifié quand

$$\frac{s\,!^\delta}{(s-q)\,!^\delta} \leqslant (\eta + \varepsilon s)^r, \quad \text{quel que soit } s;$$

puisque

$$\frac{s\,!}{(s-q)\,!} \leqslant s^q,$$

il suffit d'avoir

$$s^{q\delta} \leqslant (\eta + \varepsilon s)^r, \text{ quel que soit } s.$$

Il est nécessaire que

$$0 \leqslant q\delta \leqslant r.$$

Notons $p = q\delta/r$. Si $p = 1$, on prend évidemment $\eta = 0$, $\varepsilon = 1$. Si $0 < p < 1$, la concavité de s^p montre qu'on peut prendre

$$\eta + \varepsilon s = s_0^p + (s - s_0) \frac{d(s_0^p)}{ds_0}, \quad \text{quel que soit } s_0 > 0;$$

c'est-à-dire

$$\varepsilon = p s_0^{p-1}, \quad \eta = (1-p)s_0^p;$$

il suffit donc de prendre

$$\varepsilon = s_0^{p-1}, \quad \eta = s_0^p = \varepsilon^{p'}.$$

Nous satisferons (19.4), (19.5) et (19.7) en prenant

$$\lambda_s = (s!)^{-\delta}, \tag{19.9}$$

δ étant tel que $0 \leqslant q\delta \leqslant r$.

Résumons ce qui précède :

Lemme 19. — Étant donnée une série formelle

$$\Phi(\varrho) = \sum_{s=0}^{\infty} \frac{\varrho^s}{s!} \Phi_s$$

et un nombre δ tel que $0 \leqslant q\delta \leqslant r$, nous notons

$$\lambda\Phi(\varrho) = \sum_{s=0}^{\infty} \frac{\varrho^s}{(s!)^{1+\delta}} \Phi_s. \tag{19.10}$$

Nous avons alors (19.2), (19.3) et (19.6), où ε et η sont des constantes définies par (19.8).

20. Classe de gevrey formelle

Définition. — Étant donnés un entier $p \geqslant 0$ et un nombre $\alpha \geqslant 1$, nous nommons classe de Gevrey formelle $\Gamma^{p,(\alpha)}(|Y|)$ l'ensemble des séries formelles en ϱ, à coefficients fonctions de t ($0 \leqslant t \leqslant |Y|$),

$$\Phi(t, \varrho) = \sum_{s=0}^{\infty} \frac{\varrho^s}{s!} \Phi_s(t), \tag{20.1}$$

vérifiant la condition suivante :

$$\lambda \frac{\partial^j}{\partial t^j} \Phi(t, \varrho) = \sum_{s=0}^{\infty} \frac{\varrho^s}{(s\,!)^\alpha} \frac{d^j}{dt^j} \Phi_s(t) \quad (j = 0, \ldots, p) \quad (20.2)$$

sont des fonctions de ϱ holomorphes à l'origine, uniformément par rapport à t : il existe un voisinage de $\varrho = 0$, indépendant de t, où elles ont une borne, indépendante de t.

Cette condition peut donc s'énoncer :

$$\sup_{s,t} \frac{1}{[1+s]^\alpha} \left| \frac{d^j \Phi_s(t)}{dt^j} \right|^{1/s} < \infty. \quad (20.3)$$

Définition. — $\Gamma^{(\alpha)}$ est l'ensemble des $\Phi(\varrho) \in \Gamma^{p,(\alpha)}$.
Décrivons les propriétés de $\Gamma^{p,(\alpha)}$:

Lemme 20.1. — $\Gamma^{p,(\alpha)}$ est une algèbre, stable pour $\partial/\partial\varrho$.

Preuve : (19.3), puis (19.2).

Voici un lemme qui sera appliqué à la série formelle composée (17.5) :

Lemme 20.2. — Les hypothèses

$$0 \ll \Phi(t,\varrho) \in \Gamma^{p,(\alpha)}, \; 0 \ll \Theta(t, \varrho) \in \Gamma^{p,(\alpha)}, \; \Theta(t, 0) = 0$$

impliquent

$$\Phi(t, \Theta(t, \varrho)) \in \Gamma^{p,(\alpha)}.$$

Preuve. — Par hypothèse :

$$\Theta(t, \varrho) = \varrho\Psi(t, \varrho); \; \lambda\Phi(t, \varrho) = \varphi(t, \varrho) \text{ et } \lambda\Psi(t, \varrho) = \psi(t, \varrho)$$

sont des fonctions de ϱ holomorphes à l'origine. Vu (19.3)

$$\lambda[\Theta(t, \varrho)]^s \ll [\psi(t, \varrho)]^s \, \lambda\varrho^s;$$

d'où, vu (19.3)

$$\lambda[\Phi(t, \Theta(t, \varrho))] = \sum_s \frac{\lambda[\Theta(t, \varrho)]^s}{s\,!} \Phi_s(t) \ll$$

$$\sum_s \frac{(\lambda\varrho^s)}{s\,!} [\psi(t, \varrho)]^s \Phi_s(t) = \varphi(t, \varrho\,\psi(t, \varrho));$$

or $\varphi(t, \varrho\psi(t, \varrho))$ est holomorphe en ϱ, à l'origine, uniformément par rapport à t; donc $\lambda[\Phi(t, \Theta(t, \varrho))]$ aussi : le lemme est prouvé.

Voici un lemme que nous appliquerons aux problèmes de Cauchy (17.1) et (17.2) :

Lemme 20.3. — Soit $\Phi(t, \varrho)$ la série formelle solution du problème de Cauchy

$$\begin{cases} \dfrac{\partial \Phi(t, \varrho)}{\partial t} = \left[\varrho\, C_1(\varrho)\, \dfrac{\partial}{\partial \varrho} + C_2(\varrho) \right] \Phi(t, \varrho) + C_3(\varrho) \\ \Phi(0, \varrho) = C_4(\varrho) \end{cases} \qquad (20.4)$$

où

$$0 \ll C_i(\varrho) \in \Gamma^{(\alpha)} \; ;$$

on a

$$0 \ll \Phi(t, \varrho) \in \Gamma^{p,(\alpha)}, \text{ quel que soit } p.$$

Preuve pour $p = 0$. — Les coefficients de $\Phi(t, \varrho)$ se calculent successivement, par quadratures : ce sont des exponentielles-polynomes $\geqslant 0$; donc (20.4) a une solution unique $\Phi(t, \varrho)$; $\Phi(t, \varrho) \gg 0$.

Notons $c_i(\varrho) = \lambda C_i(\varrho)$. Soit $\varphi(t, \varrho)$ la solution du problème de Cauchy-Kowalewski, qui s'intègre par quadratures :

$$\begin{cases} \dfrac{\partial}{\partial t}\, \varphi(t, \varrho) = \left[\varrho c_1(\varrho)\, \dfrac{\partial}{\partial \varrho} + c_2(\varrho) \right] \varphi(t, \varrho) + c_3(\varrho) \\ \varphi(0, \varrho) = c_4(\varrho) \, ; \end{cases} \qquad (20.5)$$

ces quadratures montrent que $\varphi(t, \varrho)$ est holomorphe dans un bicylindre : $|t| \leqslant |Y|$, $|\varrho| < \text{const.}$

Les formules (19.2) et (19.3) donnent

$$\begin{cases} \dfrac{\partial}{\partial t}\, \lambda \Phi(t, \varrho) \ll \left[\varrho c_1(\varrho)\dfrac{\partial}{\partial \varrho} + c_2(\varrho) \right] \lambda \Phi(t, \varrho) + c_3(\varrho) \\ \lambda \Phi(0, \varrho) = c_4(\varrho). \end{cases} \qquad (20.6)$$

La comparaison de (20.5) et (20.6) montre que les coefficients successifs de $\varphi(t, \varrho) - \lambda \Phi(t, \varrho)$ sont $\geqslant 0$; d'où

$$\lambda \Phi(t, \varrho) \ll \varphi(t, \varrho),$$

ce qui prouve que $\lambda \Phi(t, \varrho)$ est une fonction de ϱ holomorphe à l'origine, uniformément par rapport à t : le lemme est prouvé pour $p = 0$.

Preuve pour $p > 0$. — On déduit de (20.4) que $\partial^j \Phi/\partial t^j$ vérifie un problème de Cauchy du même type; le raisonnement précédent prouve donc que $\lambda\, \partial^j \Phi/\partial t^j$ est une fonction de ϱ holomorphe à l'origine.

Lemme 20.4. — Considérons le problème de Cauchy formel

$$\begin{cases} \dfrac{\partial^p}{\partial t^p}\, \Phi(t, \varrho) = M_q\left(t, \varrho, \dfrac{\partial}{\partial t},\ \dfrac{\partial}{\partial \varrho}\right) \Phi(t, \varrho) + \psi(t, \varrho) \\ \dfrac{\partial^j}{\partial t^j}\, \Phi(t, \varrho) = 0 \text{ pour } t = 0,\ j = 0, ..., p-1 \end{cases} \quad (20.7)$$

où M_q est un opérateur différentiel d'ordre $q \leqslant p$; ne contenant pas $(\partial/\partial t)^p$; par hypothèse $\psi(t, \varrho)$ et les coefficients de M_q sont des séries formelles en $\varrho \in \Gamma^{0,(\alpha)}\,(|Y|)$; elles sont $\geqslant 0$. Ce problème possède au moins une solution Φ telle que

$$0 \ll \Phi \in \Gamma^{p,(\alpha)}(|Z|),$$

si $|Z|$ est un nombre > 0 suffisamment petit et si

$$1 \leqslant \alpha \leqslant \frac{p}{q}.$$

$$|Z| = |Y|, \text{ si } 1 \leqslant \alpha\ < \frac{p}{q}.$$

Preuve. — Vu la proposition 18.1, il suffit de trouver

$$\Phi \in \Gamma^{p,(\alpha)}\,(|Z|)$$

tel que

$$\begin{cases} \dfrac{\partial^p \Phi(t, \varrho)}{\partial t^p} \gg M_q\left(t, \varrho, \dfrac{\partial}{\partial t},\ \dfrac{\partial}{\partial \varrho}\right) \Phi(t, \varrho) + \psi(t, \varrho), \\ \dfrac{\partial^j \Phi(t, \varrho)}{\partial t^j} \gg 0 \quad \text{pour}\quad j = 0, ..., p-1. \end{cases} \quad (20.8)$$

Appliquons λ à (20.8), en notant :

$$\delta = \alpha - 1\,;$$
$$\varphi(t, \varrho) = \lambda\Phi(t, \varrho)\,;$$

$c_1(\varrho)$ une série formelle majorant $\lambda\psi(t, \varrho)$ (c'est-à-dire : $\lambda\psi \ll c_1$ pour $0 \leqslant t \leqslant |Y|$); $m_q(\varrho, \partial/\partial t, \partial/\partial \varrho)$ un opérateur différentiel dont les coefficients majorent les transformés par λ de ceux de $M_q(t, \varrho, \partial/\partial t, \partial/\partial \varrho)$. Pour que (20.8) ait lieu, il suffit, vu (19.3) et (19.6), qu'on ait [11]

$$\frac{\partial^p}{\partial t^p}\, \varphi(t, \varrho) \gg m_q\left(\varrho, \frac{\partial}{\partial t},\ \frac{\partial}{\partial \varrho}\right) \left(\eta + \varepsilon\varrho\, \frac{\partial}{\partial \varrho}\right)^{p-q} \varphi(t, \varrho) + c_1(\varrho), \quad (20.9)$$

$$\frac{\partial^j}{\partial t^j}\, \varphi(t, \varrho) \gg 0 \quad \text{pour}\quad j = 0, ..., p. \quad (20.10)$$

Rappelons que (19.6) exige $q\,\delta \leqslant p - q$, c'est-à-dire $q\alpha \leqslant p$.

[11] Rappelons que si $p = q$, alors $\delta = 0, \alpha = 1, \eta = 1, \varepsilon = 0$.

Pour réaliser (20.10) et la condition $\Phi \in \Gamma^{p,(\alpha)}(|Z|)$, il suffit de choisir $\varphi(t, \varrho)$ holomorphe en (t, ϱ) dans le bicylindre

$$|t| \leqslant |Z|, \quad |\varrho| < \text{const.}$$

ses coefficients de Taylor étant $\geqslant 0$. Par hypothèse m_q et c_1 sont holomorphes au voisinage de $\varrho = 0$.

Nous prendrons

$$\varphi(t, \varrho) = \varphi[\tau], \quad \text{où} \quad \tau = t + \varrho/\varepsilon_0, \ \varepsilon_0 = \text{const.} > 0,$$

$\varphi[\tau]$ étant défini par le problème de Cauchy, à données holomorphes $\geqslant 0$:

$$\begin{cases} \dfrac{d^p \varphi[\tau]}{d\tau^p} = m_q\left(\varepsilon_0\tau, \dfrac{d}{d\tau}, \dfrac{1}{\varepsilon_0}\dfrac{d}{d\tau}\right)\left(\eta + \varepsilon\tau\dfrac{d}{d\tau}\right)^{p-q}\varphi[\tau] + c_1(\varepsilon_0\tau) \\[2mm] \dfrac{d^j\varphi}{d\tau^j} = 0 \quad \text{pour} \quad \tau = 0, j = 0, ..., p-1\, ; \end{cases}$$

nous choisissons ε_0 assez petit pour que $m_q(\varepsilon_0\tau, ...)$, $c_1(\varepsilon_0\tau)$ soient holomorphes pour $|\tau| < 2|Y|$. Puisque $q \leqslant p$, le théorème de Cauchy-Kowalewski montre que $\varphi[\tau]$ est holomorphe pour $|\tau| < 2|Z|$, $|Z|$ étant suffisamment petit ; évidemment $\varphi[\tau] \geqslant 0$ (c'est-à-dire a ses coefficients de Taylor $\geqslant 0$) et (20.9) est vérifié pour $|t| \leqslant |Z|$, $\varrho < \varepsilon_0|Z|$.

Supposons $q\alpha < p$, c'est-à-dire $q\delta < p-q$; vu (19.8), nous pouvons prendre ε voisin de 0 ; nous le prenons assez petit pour que $\varphi[\tau]$ soit holomorphe [12] pour $|\tau| < 2|Y|$: on peut donc prendre $|Z| = |Y|$.

21. Application au problème de Cauchy (14.1)

Supposons que les seconds membres $C_i(\varrho)$ et $\psi(t, \varrho)$ de (14.3) vérifient :

$$C_i(\varrho) \in \Gamma^{(\alpha)}, \ \psi(t, \varrho) \in \Gamma^{0,(\alpha)}(|Y|). \tag{21.1}$$

Alors, d'après le lemme 20.3, les fonctions Θ et Ω, que définit le n° 17, vérifient

$$\Theta(t, \varrho), \ \Omega(t, \varrho) \in \Gamma^{p,(\alpha)}(|Y|).$$

Donc, vu les lemmes 20.1 et 20.2 :

---

[12] Car les solutions d'une équation différentielle ordinaire et *normale* sont holomorphes dans le même domaine que ses coefficients.

la fonction φ que définit (17.11), vérifie

$$\varphi(t, \varrho) \in \Gamma^{p,(\alpha)}(|Z|), \quad \text{si} \quad \Phi(t, \varrho) \in \Gamma^{p,(\alpha)}(|Z|);$$

les coefficients de l'opérateur M_q, que définit la proposition 18.1, appartiennent à $\Gamma^{0,(\alpha)}(|Y|)$.

Donc, vu le lemme 20.4, les inégalités (18.3) [et même les égalités en résultant par substitution de $=$ à $\geqslant$] possèdent une solution $\in \Gamma^{p,(\alpha)}(|Z|)$.

Donc, vu la proposition 18.1, la fonction $\Phi(t, \varrho)$ que définit le n° 17 vérifie $\Phi(t, \varrho) \in \Gamma^{p,(\alpha)}(|Z|)$, quand on ne fait varier t que de 0 à $|Z|$.

La proposition 17 prouve donc ceci :

Proposition 21. — Faisons les hypothèses (21.1) et supposons $1 \leqslant \alpha \leqslant p/q$; dans une bande suffisamment étroite

$$Z : 0 \leqslant x_0 \leqslant |Z|,$$

le problème de Cauchy (14.1) possède une solution $u^{\nu}(x)$ telle que

$$|D^{m+n^{\nu}-p+q,\infty} u^{\nu}, S_t, \varrho|_2 \in \Gamma^{0,(\alpha)}(|Z|).$$

$$Z = Y, \text{ si l'on a } 1 \leqslant \alpha < p/q$$

Le support de u^{ν} appartient à l'émission de celui des données.

La conclusion des § 3 et 4 est *le théorème d'existence* que constitue la proposition ci-dessus.

§ 5. Fin de l'étude du système dont la partie principale est diagonale

Explicitons la proposition 21, qui est un théorème d'existence, et le théorème d'unicité qui en résulte.

22. CLASSES DE GEVREY

Définition. — Soit $\alpha \geqslant 1$, $p \geqslant 1$; $\gamma_p^{(\alpha)}(S_0)$ désignera l'ensemble des fonctions

$$f : S_0 \to C$$

telles que

$$\sup_\sigma \frac{1}{[1 + |\sigma|]^\alpha} [|D^\sigma f, S_0|_p]^{1/|\sigma|} < \infty, \quad \text{où} \quad \sigma_0 = 0;$$

c'est la classe de Gevrey classique si $p = \infty$.

Y étant la bande $0 \leqslant x_0 \leqslant |Y|$, $\gamma_p^{n,(\alpha)}(Y)$ désignera l'ensemble des fonctions

$$f : Y \to C$$

telles que

$$\sup_{\beta,\sigma,t} \frac{1}{[1 + |\sigma|]^\alpha} [|D^{\beta+\sigma} f, S_t|_p]^{1/|\sigma|} < \infty,$$

où $|\beta| \leqslant n$, $\sigma_0 = 0$, $0 \leqslant t \leqslant |Y|$.

Evidemment, vu (20.3), on a :

Lemme 22. — $f \in \gamma_p^{n,(\alpha)}(Y)$ équivaut à

$$|D^{n,\infty} f, Y, S_t, \varrho|_p \in \Gamma^{0,(\alpha)}$$

Note. — Nous appliquerons la définition de $\gamma^{(\alpha)}$ avec $\alpha = \infty$, en convenant que dans ce cas

$$\frac{1}{[1 + |\sigma|]^\alpha} = 1 \quad \text{si} \quad |\sigma| = 0, \quad \frac{1}{[1 + |\sigma|]^\alpha} \dots = 0 \quad \text{si} \quad |\sigma| > 0.$$

$\gamma_p^{(\infty)}(S_0)$ est donc l'espace $L_p(S_0)$ des fonctions sur S_0 dont la puissance $p^{\text{ième}}$ est sommable; $\gamma_p^{n,(\infty)}(Y)$ est l'espace $L_{\infty,p}^n(Y)$ des fonctions sur Y dont les quasi-normes [13] $|D^n f, S_t|_p$ sont fonctions bornées de $t(0 \leqslant t \leqslant |Y|)$.

Propriétés des classes de Gevrey γ. — (Voir : Gevrey [2]). Evidemment : ces classes croissent avec α; si $\beta_0 = 0$, D^β les applique en elles-mêmes.

La formule du produit (10.1), les lemmes 20.1 et 22 prouvent ceci :

$\gamma_\infty^{(\alpha)}(S_0)$ est une algèbre et $\gamma_p^{(\alpha)}(S_0)$ un module sur cette algèbre; $\gamma_\infty^{n,(\alpha)}(Y)$ est une algèbre et $\gamma_p^{n,(\alpha)}(Y)$ un module sur cette algèbre.

Pour $\alpha = 1$, ces classes sont des classes de fonctions analytiques en $(x_1, \dots, x_l)$; *pour $\alpha > 1$, ce sont des classes de fonctions non quasi-analytiques* : on peut décomposer l'unité en une somme de fonctions leur appartenant et ayant des supports arbitrairement petits (Voir Mandelbrojt [8]).

En composant deux fonctions de l'algèbre $\gamma_\infty^\alpha(S_0)$ ou $\gamma_\infty^{n,(\alpha)}(Y)$ on obtient une fonction de cette classe (Voir : Gevrey [2]; ou un résultat plus précis de Leray-Waelbroeck [7]).

En particulier : cette algèbre contient l'inverse d'un de ses éléments, quand cet inverse est une fonction bornée.

---

[13] Normes des restrictions à S_t de leurs dérivées d'ordres $\leqslant p$.

Note. — Cette dernière propriété prouve qu'il va être superflu de supposer normaux les opérateurs a et a_j, comme nous l'avions fait jusqu'ici.

23. Existence

Sur la bande Y, soient des opérateurs $a(x, D)$ et $b^\nu_\mu(x, D)$ (μ, $\nu = 1, \dots N$) du type suivant :

$a(x, D) = a_1(x, D) \dots a_p(x, D)$ est le produit de p opérateurs régulièrement hyperboliques pour les hyperplans S_t ;

ordre $(a_j) = m_j$; ordre $(a) = m = m_1 + \dots + m_p$;

ordre $(b^\nu_\mu) = m + n^\mu - n^\nu - p + q$, où $0 \leqslant q \leqslant p \leqslant n^\nu$;

b^ν_μ ne contient pas $(\partial/\partial t)^{m+n^\mu-n^\nu}$; notons $n = \sup_\nu n^\nu$;

a_{j+1} a ses coefficients $\in \gamma^{m_1 + \dots + m_j - j + n, (\alpha)}_\infty (Y)$;

a a ses coefficients $\in \gamma^{n, (\alpha)}_\infty (Y)$;

$$\text{où } 1 \leqslant \alpha \leqslant p/q.$$

b^ν_μ a ses coefficients $\in \gamma^{n^\nu, (\alpha)}_\infty (Y)$, où $1 \leqslant \alpha \leqslant p/q$.

On considère le problème de Cauchy d'inconnue u^ν :

$$\begin{cases} a(x, D) u^\nu(x) = \displaystyle\sum_\mu b^\nu_\mu(x, D) u^\mu(x) + v^\nu(x) \\ D^{m-1} u^\nu | S_0 \text{ donné} \end{cases} \tag{23.1}$$

Théorème d'existence. — *Supposons*

$$v^\nu \in \gamma^{n^\nu, (\alpha)}_2 (Y), \ D^j_0 u^\nu | S_0 \in \gamma^{(\alpha)}_2 (S_0) \ (j = 0, \dots, m-1).$$

Alors le problème de Cauchy (23.1) *possède une solution*

$$u^\nu \in \gamma^{m+n^\nu, (\alpha)}_2 (Z) \tag{23.2}$$

qui est définie dans une bande suffisamment étroite

$$Z : 0 \leqslant x_0 \leqslant |Z|,$$

et dont le support est dans l'émission de celui de v et $D^{m-1} u | S_0$.

$Z = Y$, *si $1 \leqslant \alpha < p/q$.*

Note 23.1. — Supposons $q = 0$: c'est le cas strictement hyperbolique. On peut choisir $\alpha = \infty$; il suffit de supposer $|D^{n^\nu} v^\nu, S_t|_2$ sommable ; $|D^{m+n^\nu-p+q} u^\nu, S_t|_2$ est borné ; $|D^{m+n^\nu} u^\nu, S_t|_2$ aussi : voir la preuve de (23.2).

Preuve du théorème quand

$$D^{m-1} u^\nu | S_0 = 0, \quad D^{n^\nu - 1} v^\nu | S_0 = 0. \tag{23.3}$$

Le problème (23.1) est identique au problème (14.1); le lemme 22 montre que le théorème équivaut à la proposition 21 établit l'existance de u^ν vérifiant

$$u^\nu \in \gamma_2^{m+n^\nu - p + q, (\alpha)}(Z). \tag{23.4}$$

Fin de la preuve du théorème. — En appliquant $D_0^j (j = 0, ...,$ $n^\nu - 1)$ à l'équation $au^\nu = \sum_\mu b_\mu^\nu u^\mu + v^\nu$, on constate que (23.1) permet de calculer $D^{m+n^\nu - 1} u^\nu | S_0$ et que

$$D^{m+n^\nu - 1} u^\nu | S_0 \in \gamma_2^{(\alpha)}(S_0). \tag{23.5}$$

On construit $w^\nu \in \gamma^{m+n^\nu, (\alpha)}(Y)$ tel que $D^{m+n^\nu - 1} w^\nu | S_0$ ait les valeurs (23.5). $u^\nu - w^\nu$ est défini par un problème du type (23.1), vérifiant (23.3) : on est ramené au cas précédent.

Nous n'avons pas prouvé (23.2), mais seulement (23.4).

Preuve de 23.2. — En appliquant $D_0^{n^\nu - p + q + 1}, ..., D_0^{n^\nu}$ à l'équation $au^\nu = \sum_\mu b_\mu^\nu u^\mu + v^\nu$ on constate que le premier terme des relations ainsi obtenues vérifie :

$$D_0^{m+n^\nu - p + q + 1} u^\nu \in \gamma_2^{0, (\alpha)}(Z), ..., D^{m+n^\nu} u^\nu \in \gamma_2^{0, (\alpha)}(Z) ;$$

donc

$$u^\nu \in \gamma_2^{m+n^\nu, (\alpha)}(Z).$$

Preuve de la note 23.1. — $q = 0$; le système est strictement hyperbolique; on applique la proposition 16.

24. Unicité

Le problème adjoint à (23.1) a au plus une solution, sous les hypothèses qu'énonce le n° 23, quand

$$1 \leqslant \alpha < \frac{p}{q}.$$

Plus précisément, notons $\bar{a}(x, D)$ l'adjoint de $a(x, D)$; si

$$a(x, D)f = \sum_\beta a_\beta(x) D^\beta f,$$

alors

$$\bar{a}(x, D)f = \sum_\beta (-1)^\beta D^\beta(a_\beta f).$$

Théorème d'unicité. — *Supposons* $1 \leqslant \alpha < p/q$. *Soient* $w_\mu(x)$ *des distributions, définies au voisinage d'un point de* S_0, *vérifiant sur leur domaine de définition*

$$\bar{a}(x, D)w_\mu(x) = \sum_\nu \bar{b}_\mu^\nu(x, D)w_\nu(x) ,$$

et s'annulant hors de Y. *Alors*

$$w_\mu = 0 \text{ au voisinage de } S_0.$$

Note. Ce voisinage contient tous les points ayant, dans Y, une émission rétrograde intérieure au domaine de définition de w.

Preuve (Holmgren). — Permutons les deux bords de Y : w est définie au voisinage d'un point de $S_{|Y|}$ Prenons $\alpha > 1$ et $v^\nu \in \gamma_{\frac{n}{2}}^{\nu,(\alpha)}(Y)$, v ayant un support dont l'émission est intérieure au domaine de définition de w; soit u^ν la solution du problème de Cauchy (23.1), on choisit

$$D^{m-1}u^\nu | S_0 = 0;$$

on a

$$0 = \int_Y \sum_{\mu,\nu} u^\mu[\bar{a}w_\mu - \bar{b}_\mu^\nu w_\nu]dx =$$

$$= \int_Y \sum_{\mu,\nu} w_\nu[au^\nu - b_\mu^\nu u^\mu]dx = \int_Y \sum_\nu w_\nu v^\nu \, dx.$$

Cela suffit à prouver que $w_\nu = 0$ près de $S_{|Y|}$.

§ 6. Système hyperbolique quelconque

Les théorèmes précédents s'appliquent aisément à un système hyperbolique quelconque : par exemple comme ceci :

25. EXISTENCE

Sur une bande de R^{l+1} :

$$Y : 0 \leqslant x_0 \leqslant |Y|,$$

donnons-nous des opérateurs différentiels $a_\mu^\nu(x, D)$ $(\mu, \nu = 1, ..., N)$ tels que :

$$\text{ordre } (a_\mu^\nu) = m^\mu - n^\nu \quad (a_\mu^\nu = 0 \quad \text{si} \quad m^\mu < n^\nu)$$

et que, modulo les opérateurs d'ordre inférieur à

$$m = \sum_\mu (m^\mu - n^\mu),$$

on ait ([14])

$$\text{dét. } (a_\mu^\nu) \equiv a_1(x, D) \, ... \, a_p(x, D),$$

les opérateurs différentiels a_j étant régulièrement hyperboliques sur Y pour les hyperplans $S_t : x_0 = t$. Leurs ordres m_j vérifient donc

$$m = m_1 + ... + m_p.$$

Notons

$$\bar{n} = \sup_\nu n^\nu, \quad n = \inf_\nu n^\nu.$$

Nous considérons *le problème de Cauchy* d'inconnue $u^\mu(x)$:

$$\begin{cases} \sum_\mu a_\mu^\nu(x, D) u^\mu(x) = v^\nu(x) \\ D^{m^\mu - n} u^\mu \,|\, S_0 \text{ donné} \end{cases} \tag{25.1}$$

en supposant les données de Cauchy $(25.1)_2$ *compatibles* avec le système $(25.1)_1$; autrement dit : les restrictions à S_0 des équations qu'on obtient en appliquant $D_0^j (j = 0, ... n^\nu - n)$ à l'équation $\sum_\mu a_\mu^\nu u^\mu = v^\nu$ doivent être vérifiées par ces données de Cauchy.

Voici les hypothèses que nous faisons sur les a_μ^ν et a_j : nous notons r le plus grand entier $\leqslant p$ tel que tous les commutateurs

$$[a_\mu^\lambda, a_\pi^\nu] = a_\mu^\lambda a_\pi^\nu - a_\pi^\nu a_\mu^\lambda$$

vérifient :

$$\text{ordre } [a_\mu^\lambda, a_\pi^\nu] \leqslant \text{ordre } (a_\mu^\lambda) + \text{ordre } (a_\pi^\nu) - r$$

([14]) Nous convenons que le déterminant de a_μ^ν, non commutatifs, est

$$\text{dét}(a_\mu^\nu) = \sum_\pi \pm \, a_{\pi(1)}^1 \, ... \, a_{\pi(N)}^N,$$

pour toutes les permutations π paires $(+)$ et impaires $(-)$.

et que

$$\text{ordre [dét. } (a_\mu^\nu) - a_1(x, D) \dots a_p(x, D)] \leqslant m - r \text{ ;}$$

si $r = 1$, la première de ces deux conditions est vérifiée :

si $r = 0$, elles le sont toutes les deux et la considération des a_j est superflue, mais nous supposons que $(\partial/\partial t)^m$ a pour coefficient dans dét. (a_μ^ν) une fonction bornée ainsi que son inverse.

Nous ajoutons un même entier à tous les entiers m^μ et n^ν de façon que $p \leqslant n$ et nous avons donc

$$0 \leqslant r \leqslant p \leqslant n \leqslant n^\nu \leqslant \bar{n}, \ n \leqslant m^\mu, \ p \leqslant m.$$

Nous envisageons des classes de Gevrey (n° 22), dont l'indice α est un nombre tel que

$$1 \leqslant \alpha \leqslant \frac{p}{p-r} ; \tag{25.2}$$

nous supposons que

a_μ^ν a des coefficients $\in \gamma_\infty^{m+\bar{n}-m^\mu+n^\nu, (\alpha)}(Y)$

a_{j+1} a des coefficients $\in \gamma_\infty^{m_1+\dots+m_j-j+\bar{n}, (\alpha)}(Y)$

a a des coefficients $\in \gamma_\infty^{\bar{n}, (\alpha)}(Y)$,

où $a(x, D) = a_1(x, D) \dots a_p(x, D)$.

Théorème d'existence. — *Supposons*

$$v^\nu \in \gamma_2^{n^\nu, (\alpha)}(Y), \ D^{m^\mu-n} u^\mu | S_0 \in \gamma_2^{(\alpha)}(S_0). \tag{25.3}$$

Alors, dans une bande suffisamment étroite

$$Z : 0 \leqslant x_0 \leqslant |Z|$$

le problème de Cauchy (25.1) possède une solution

$$u^\mu \in \gamma_2^{m^\mu, (\alpha)}(Z), \tag{25.4}$$

dont le support est dans l'émission de celui des données v^ν, $D^{m^\mu-n} u^\mu | S_0$.

$$Z = Y \quad si \quad 1 \leqslant \alpha < \frac{p}{p-r}.$$

Note. — Si $r = p$, ce problème (25.1) est strictement hyperbolique : $Y = Z$; on peut choisir $\alpha = \infty$ et remplacer l'hypothèse

(25.3)$_1$ par l'hypothèse plus générale :

$|D^{n^{\nu}} v^{\nu}, S_t|_2$ est une fonction sommable de t;

la conclusion (25.4) s'énonce :

$|D^{m^{\mu}} u^{\mu}, S_t|_2$ est une fonction bornée de $t (0 \leqslant t \leqslant |Y|)$.

Note. — On peut compléter comme suit ce théorème : si les coefficients de a_{μ}^{ν} sont dans la classe de Gevrey $\gamma_{\infty}^{(\alpha)}(Y)$ et si $v^{\nu} \in \gamma_2^{(\alpha)}(Y)$, alors $u^{\mu} \in \gamma_2^{(\alpha)}(Y)$; voir Y. Ohya [9], proposition 2.4.

Preuve quand les données de Cauchy sont $D^{m^{\mu}-n} u^{\mu}|S_0 = 0$. — Notons $A_{\nu}^{\mu}(x, D)$ le mineur de $a_{\mu}^{\nu}(x, D)$ dans dét. (a_{μ}^{ν}); évidemment :

$$\text{ordre } (A_{\nu}^{\mu}) \leqslant m - m^{\mu} + n^{\nu};$$

$$\text{ordre (dét. } (a_{\mu}^{\nu}) - \sum_{\lambda} a_{\lambda}^{\nu} A_{\nu}^{\lambda}) \leqslant m - r;$$

les coefficients de $\sum_{\lambda} a_{\lambda}^{\nu} A_{\nu}^{\lambda}$ et de dét. $(a_{\mu}^{\nu}) \in \gamma_{\infty}^{\bar{n},(\alpha)}(Y)$;

si $\mu \neq \nu$, ordre $\left(\sum_{\lambda} a_{\lambda}^{\nu} A_{\mu}^{\lambda} \right) \leqslant m + n^{\mu} - n^{\nu} - r$;

les coefficients de $\sum_{\lambda} a_{\lambda}^{\nu} A_{\mu}^{\lambda} \in \gamma_{\infty}^{\bar{n}+n^{\nu}-n^{\mu},(\alpha)}(Y) \subset \gamma_{\infty}^{n^{\nu},(\alpha)}(Y)$.

Soit U^{ν} la solution du problème, traité au n° 23 et auquel s'applique le théorème d'existence :

$$(25.5) \quad \begin{cases} \sum_{\lambda,\mu} a_{\lambda}^{\nu}(x, D) A_{\mu}^{\lambda}(x, D) U^{\mu}(x) = v^{\nu}(x) \\ D^{m-1} U^{\mu}|S_0 = 0; \end{cases}$$

on a :

$$U^{\nu} \in \gamma_2^{m+n^{\nu},(\alpha)}(Z).$$

La condition de compatibilité de (25.1) s'écrit :

$$D^{n^{\nu}-n} v^{\nu}|S_0 = 0;$$

portée dans (25.5), elle donne

$$D^{m+n^{\nu}-n} U^{\nu}|S_0 = 0.$$

Le problème (25.1) admet donc la solution :

$$u^\mu = \sum_\nu A^\mu_\nu U^\nu.$$

Fin de la preuve du théorème. — Soient w^μ des fonctions $\in \gamma_2^{m^\mu, (\alpha)}(Z)$ vérifiant les données de Cauchy : $u^\mu - w^\mu$ vérifie un problème du type (25.1), à données de Cauchy nulles.

26. UNICITÉ. — Le problème adjoint à (25.1) a au plus une solution, sous les hypothèses qu'énonce le n° 25, quand

$$1 \leqslant \alpha < \frac{p}{p-r}.$$

Plus précisément : notons $\bar{a}^\nu_\mu$ l'adjoint de a^ν_μ; on a :

Théorème d'unicité. — *Supposons*

$$1 \leqslant \alpha < \frac{p}{p-r}.$$

Soient $w_\nu(x)$ des distributions, définies au voisinage d'un point de S_0, vérifiant sur leur domaine de définition

$$\sum_\nu \bar{a}^\nu_\mu(x, D)\, w_\nu(x) = 0$$

et s'annulant hors de Y. Alors

$$w_\nu = 0 \ \text{au voisinage de } S_0.$$

Preuve. — Le raisonnement de Holmgren, que cite le n° 24.

27. NÉCESSITÉ DES HYPOTHÈSES

C. Pucci et G. Talenti nous ont signalé *un cas de non-unicité dû à E. de Giorgi* [15], d'où résulte ceci :

Les théorèmes d'unicité et par suite les *théorèmes d'existence* qui précèdent *deviennent faux* quand on prend $\alpha > p/q$, au lieu de prendre $\alpha \leqslant p/q$.

En effet : cet exemple dépend d'un paramètre $\alpha > 2$;

$$p/q = 2 (m = p = 8, \ q = 4);$$

les coefficients des opérations sont indéfiniment différentiables et appartiennent à $\gamma_\infty^{n, (\alpha)}(Y)$, quel que soit n.

BIBLIOGRAPHIE

[1] L. GÅRDING, *Cauchy's problem for hyperbolic equations*, Lecture Notes, University of Chicago, 1957.
Energy inequalities for hyperbolic systems, Colloque international de Bombay, 1964.

[2] M. GEVREY, Sur la nature analytique des solutions des équations aux dérivées partielles, *Annales École norm. sup.*, t. 35 (1917), pp. 129-189.

[3] L. HÖRMANDER, *Linear partial differential operators*, Springer (1963).

[4] A. LAX, On Cauchy's problem for partial differential equations with multiples caracteristics, *Comm. pure appl. math.*, t. 9 (1956), pp. 135-169.

[5] J. LERAY, *Hyperbolic differential equations*, Institute for adv. study, Princeton, 1953.

[6] J. LERAY, *La théorie de Gårding des équations hyperboliques linéaires*, CIME (Centro internazionale matematico estivo), Varenna, 1956.

[7] J. LERAY et L. WAELBROECK, *Normes des fonctions composées* (préliminaires à l'étude des systèmes non linéaires, hyperboliques non stricts); article ci-dessous, p. 145.

[8] S. MANDELBROJT, *Séries adhérentes, régularisations des suites, applications*, Gauthier-Villars (1952). Leçons professées au Collège de France et au Rice Institute.

[9] Y. OHYA, Le problème de Cauchy pour les équations à caractéristiques multiples, *Journal of the Math. Soc. of Japan*, t. 16, pp. 268-296, 1964.

[10] I. PETROWSKY, Uber das Cauchysche Problem für Systeme von partiellen Differential-gleichungen, *Rev. math.* (*Mat. Sbornik*), *Moscou, N.S.* 2, 1937, pp. 814-868; id. N.S. 39, 1956, pp. 267-272.

[11] G. TALENTI, *C.R. Acad. Sciences*, t. 259 (1964), pp. 1932-33, Sur le problème de Cauchy pour les équations aux dérivées partielles.

[12] M. YAMAGUTI, Le problème de Cauchy et les opérateurs d'intégrale singulière, *Mem. Coll. Sc. Univ. Kyoto*, t. 32 (1959), pp. 121-151.

Pour le théorème de N.A. Lednev, voir :

[13] N. A. LEDNEV, Nouvelles méthodes de résolution des équations aux dérivées partielles, *Mat. Sbornik*, t. 22, 1948, pp. 205-266.

[14] L. GÅRDING, *Une variante de la méthode des séries majorantes* (Congrès scandinave, 1964).

Pour le contre-exemple, voir :

[15] E. DE GIORGI, Un esempio di non-unicità della soluzione del problema di Cauchy; *Università di Roma, Rendiconti di Matematica*, t. 14, 1955, pp. 382-387.

Centre Belge de Rechuchs Muthémutiques. Deuxième Colloque sur l'Analyse Fonctionnelle, p. 145.

Norme Formelle d'une Fonction Composée (Préliminaire a l'Étude des Systèmes Non-Linéaires, Hyperboliques Non-Stricts)

J. Leray et L. Waelbroeck

Introduction

1. Relation avec la théorie des équations aux dérivées partielles

J. Leray et Y. Ohya [4], en employant une suggestion de L. Waelbroeck, ont étudié les systèmes hyperboliques non stricts, dans le cas linéaire. Cette méthode s'adapte au cas non linéaire : on opère [5] par approximations successives, comme le fait P. Dionne [2] dans le cas strictement hyperbolique, mais en remplaçant les espaces de Sobolev par des classes de Gevrey, les normes de Sobolev par des normes formelles; la majoration de ces approximations successives résulte de la résolution d'un problème de Cauchy formel, non linéaire, qu'on ramène au problème de Cauchy-Kowalewski ([1]) par des opérateurs transformant les classes de Gevrey formelles en classes de fonctions holomorphes.

C'est possible, parce que le théorème de Sobolev sur la norme d'une fonction composée s'étend aux normes formelles et parce que ces opérateurs respectent l'inégalité exprimant ce théorème; cet article le prouve; il complète donc les n° 4 5 et 19 de [4].

2. Sommaire

Étant donnée une algèbre normée de fonctions, nous définissons la norme formelle de ces fonctions; *nous majorons la norme formelle d'une fonction composée par la composée des normes for-*

([1]) Problème de Cauchy à données holomorphes.

melles : voir (4.3). Nous définissons sur les séries formelles, *des opérateurs, tels que la transformée d'une série composée soit majorée par la composée des séries transformées* : voir (7.2). Parmi ces opérateurs se trouvent en particulier ceux qu'emploie [4] : *les opérateurs de Gevrey.*

Note. — L'une des conséquences évidentes de ces formules est le théorème classique de Gevrey [3] : une classe de Gevrey est une algèbre, contenant les composés de ses éléments.

§ 1. **Norme formelle**

3. Notations

Nous nous donnons : un domaine $X \subset R^l (l < \infty)$; un espace vectoriel $R^m (m < \infty)$;
une algèbre de Banach $A(X)$ de fonctions $a : X \to R$;
l'espace vectoriel $V(X)$ ayant pour éléments les applications

$$v = (v_1, ..., v_m) : X \to R^m \text{ telles que } v_1, ..., v_m \in A(X).$$

L'algèbre $A(X)$ ne contient pas nécessairement d'élément unité; la norme $|a, X|$ de $a \in A(X)$ est une norme d'algèbre :

$$|a_1 . a_2, X| \leqslant |a_1, X| . |a_2, X|;$$

$|v, X|$ est le vecteur à composantes $\geqslant 0$:

$$|v, X| = (|v_1, X|, ..., |v_m, X|).$$

Nous nous donnons en outre :
un domaine $Y \subset R^m$;
un espace vectoriel $B(X, Y)$ de fonctions $b : X \times Y \to C$;
sur cet espace vectoriel B, une quasi-norme [1] $\|b, X \times Y, n\|$ dépendant d'un paramètre $n = (n_1, ..., n_m)$, où $n_1, ..., n_m \geqslant 0$.

Notons $b \circ v$ la composée de $b \in B(X, Y)$ et $v \in V(X)$, c'est-à-dire la fonction qui est définie quand $x \in X$ et $v(x) \in Y$ et qui vaut alors

$$(b \circ v)(x) = b(x, v(x)).$$

[1] C'est une fonction, définie sur B, à valeurs $\geqslant 0$ et $\leqslant +\infty$, telle que :
$$\|\lambda b, X \times Y, n\| = |\lambda| . \|b, X \times Y, n\|, \quad \forall \lambda \in C;$$
$$\|b_1 + b_2, X \times Y, n\| \leqslant \|b_1, X \times Y, n\| + \|b_2, X \times Y, n\|.$$

Nous supposons que ces données satisfont *la condition sui-vante* (^{2}) :

$$\begin{cases} \text{si} \quad b \in B(X, Y), \; v \in V(X) \quad \text{et} \quad ||b, X \times Y, |v, X| \, || < \infty, \\ \text{alors } b \circ v \in A(X) \quad \text{et} \quad |b \circ v, X| \leqslant ||b, X \times Y, |v, X| \, ||. \end{cases} \quad (3.1)$$

Étant donnés

$$\beta = (\beta_1, ..., \beta_l), \; \gamma = (\gamma_1, ..., \gamma_m) \; (\beta_1, ..., \gamma_m : \text{entiers} \geqslant 0),$$

nous notons

$$D_x^\beta = \frac{\partial^{\beta_1 + ... + \beta_l}}{\partial_{x_1}^{\beta_1} \cdots \partial_{x_l}^{\beta_l}}, \; D_y^\gamma = \frac{\partial^{\gamma_1 + ... + \gamma_m}}{\partial_{y_1}^{\gamma_1} \cdots \partial_{y_m}^{\gamma_m}}.$$

Si $a \in A(X)$ et $D_x^\beta a \notin A(X)$, alors nous convenons que

$$|D_x^\beta a, X| = +\infty;$$

de même, si $b \in B(X, Y)$, $D_x^\beta D_y^\gamma b \notin B(X, Y)$, nous convenons que

$$||D_x^\beta D_y^\gamma b, X \times Y, n|| = +\infty.$$

4. Définitions

Introduisons des variables commutatives :

$$\varrho, \eta_1, ..., \eta_m;$$

notons

$$\eta = (\eta_1, ..., \eta_m), \; \eta^\gamma = \eta_1^{\gamma_1} \cdots \eta_m^{\gamma_m}.$$

Nous nommons *normes formelles* de a et b les séries formelles

$$|D^\infty a, X, \varrho| = \sum_s \frac{\varrho^s}{s!} \sup_\beta |D_x^\beta a, X|, \quad \text{où} \quad |\beta| = s; \quad (4.1)$$

$$||D^\infty b, X \times Y, \varrho, \eta, n|| = \sum_{s, \gamma} \frac{\varrho^s}{s!} \frac{\eta^\gamma}{\gamma!} \sup_\beta ||D_x^\beta D_y^\gamma b, X \times Y, n||,$$

où
$$|\beta| = s. \quad (4.2)$$

Si $v = (v_1, ..., v_m) \in V(X)$, nous notons $|D^\infty v, X, \varrho|$ le vecteur, ayant pour composantes des séries formelles, que voici :

$$|D^\infty v, X, \varrho| = (|D^\infty v_1, X, \varrho|, ..., |D^\infty v_m, X, \varrho|).$$

(^{2}) Le théorème de composition de S. Sobolev et les compléments que P. Dionne [^{2}] lui a apportés permettent de satisfaire cette condition.

Ces séries formelles et toutes celles que nous allons considérer sont $\geqslant 0$, c'est-à-dire à coefficients $\geqslant 0$; ces coefficients peuvent valoir $+\infty$.

Donnons-nous une série formelle $\Psi(\varrho, \eta)$ et un vecteur

$$\Phi(\varrho) = (\Phi_1, ..., \Phi_m),$$

dont les composantes $\Phi_1(\varrho), ..., \Phi_m(\varrho)$ sont des séries formelles; supposons leurs premiers coefficients nuls, c'est-à-dire

$$\Phi(0) = 0.$$

Alors *la série formelle composée* $\Psi \circ \Phi = \Psi(\varrho, \Phi(\varrho))$ a une définition évidente; évidemment : elle est $\geqslant 0$ comme Φ et Ψ; ses coefficients sont $< \infty$, si ceux de Φ et Ψ sont $< \infty$.

Nous allons compléter le n° 4 de [4] par la majoration suivante de la norme formelle d'une fonction composée :

Formule de composition. — Sous l'hypothèse (3.1), on a :

$$|D^\infty(b \circ v), X, \varrho| \ll ||D^\infty b, X \times Y, \varrho, |D^\infty v, X, \varrho| - |v, X|, |v, X| \,||.$$
$$(4.3)$$

si $b \in B(X, Y)$ et $v \in V(X)$.

Notons que $|v, X| = |D^\infty v, X, 0|$.

Par exemple, on a *la formule du produit* (voir [4] (4.3)) :

$$|D^\infty(v_1 v_2), X, \varrho| \ll |D^\infty v_1, X, \varrho| \cdot |D^\infty v_2, X, \varrho|.$$

si v_1 et $v_2 \in A(X)$.

5. Preuve de la formule de composition (4.3)

Introduisons des variables $\xi_1, ..., \xi_l$, commutant avec $\eta_1, ...,\eta_m$. Définissons les séries formelles

$$|D^\infty a, X; \xi| = \sum_\beta \frac{\xi^\beta}{\beta!} |D_x^\beta a, X| \qquad (5.1)$$

$$||D^\infty b, X \times Y; \xi, \eta, n|| = \sum_{\beta, \gamma} \frac{\xi^\beta}{\beta!} \frac{\eta^\gamma}{\gamma!} ||D_x^\beta D_y^\gamma b, X \times Y, n||. \qquad (5.2)$$

Notons

$$b_{\beta\gamma}(x, y) = D_x^\beta D_y^\gamma b(x, y)$$

et notons comme suit la formule de dérivation de la fonction composée $b \circ v$:

$$D_x^\alpha (b \circ v) = \sum_{\beta \gamma} (b_{\beta \gamma} \circ v) \cdot (P_\gamma^{\alpha - \beta} \circ D^{|\alpha - \beta|} v), \ (|\beta + \gamma| \leqslant |\alpha|),$$

où $P_\gamma^{\alpha - \beta}$ est un polynome à coefficients $\geqslant 0$, qui ne dépend que de $\alpha - \beta$ et γ; (il est de degré $|\gamma|$; on le compose avec $D^{|\alpha - \beta|} v$; plus précisément avec les dérivées de v d'ordres $\leqslant |\alpha - \beta|$ et > 0).

D'où, vu que $|v_j, X|$ est une norme d'algèbre :

$$|D_x^\alpha (b \circ v), X| \leqslant \sum_{\beta, \gamma} |b_{\beta \gamma} \circ v, X| \cdot (P_\gamma^{\alpha - \beta} \circ |D^{|\alpha - \beta|} v, X|) =$$

$$[D_\xi^\alpha \, \|D^\infty (b \circ v), X; \, |D^\infty v, X; \xi| - |v, X| \,\|]_{\xi=0} \,;$$

d'où, vu la condition (3.1) :

$$|D_x^\alpha (b \circ v), X| \leqslant [D_\xi^\alpha \, \|D^\infty b, X \times Y; \, \xi, \, |D^\infty v, X; \xi| - |v, X|,$$
$$|v, X| \,\|]_{\xi=0}$$

c'est-à-dire

$$|D^\infty (b \circ v), X; \xi| \ll \|D^\infty b, X \times Y; \, \xi, \, |D^\infty v, X; \xi| - |v, X| \,\|. \quad (5.3)$$

Notons $\varrho = \xi_1 + \dots + \xi_l$; la formule du binome donne, pour $|\sigma| = s$

$$\frac{\varrho^s}{s!} = \sum_\sigma \frac{\xi^\sigma}{\sigma!}.$$

D'où, en comparant les définitions (4.1) et (5.1), (4.2) et (5.2) :

$$\|D^\infty b, X \times Y; \, \xi, \, |D^\infty v, X; \xi| - |v, X|, \, |v, X| \,\|$$
$$\ll \|D^\infty b, X \times Y, \varrho, \, |D^\infty v, X, \varrho| - |v, X|, \, |v, X| \,\| \ .$$

L'inégalité (5.3) donne donc :

$$\|D^\infty (b \circ v), X; \xi\| \ll \|D^\infty b, X \times Y, \varrho, \, |D^\infty v, X, \varrho| - |v, X|, \, |v, X| \,\|. \quad (5.4)$$

Or une inégalité du type

$$\theta(\xi) \ll \Omega(\varrho),$$

où

$$\theta(\xi) = \sum_\sigma \frac{\xi^\sigma}{\sigma!} \theta_\sigma \quad ,$$

signifie

$$\theta(\varrho) \ll \Omega(\varrho), \quad \text{si} \quad \theta(\varrho) = \sum_{s=0}^{\infty} \frac{\varrho^s}{s!} \sup_{\sigma} \theta_\sigma, \quad \text{où} \quad |\sigma| = s;$$

car $\theta(\varrho)$ est la plus petite série en $\varrho = \xi_1 + ... + \xi_l$ majorant $\theta(\xi)$ (voir [4], preuve du lemme 5). Donc (5.4) prouve la formule de composition (4.3).

§ 2. Opérateurs sur les séries formelles

Le § 4 de [4], pour employer les normes formelles, leur applique des opérateurs, opérant sur les séries formelles; *Beurling* [1] les a employés depuis longtemps.

6. Définition d'opérateurs

(Voir : [4], n° 19). Donnons-nous une suite de nombres > 0

$$\lambda = (\lambda_0 = 1, \lambda_1, ..., \lambda_s, ...).$$

Étant données des séries formelles

$$\Phi(\varrho) = \sum_{s \geq 0} \frac{\varrho^s}{s!} f_s, \quad \Psi(\varrho, \eta) = \sum_{s, \gamma} \frac{\varrho^s}{s!} \frac{\eta^\gamma}{\gamma!} \Psi_{s, \gamma},$$

nous définissons comme suit des séries formelles $\lambda\Phi$ et $\lambda\Psi$:

$$\lambda\Phi(\varrho) = \sum_s \lambda_s \frac{\varrho^s}{s!} f_s, \quad \lambda\Psi(\varrho, \eta) = \sum_{s, \gamma} \lambda_{s+|\gamma|} \frac{\varrho^s}{s!} \frac{\eta^\gamma}{\gamma!} \Psi_{s, \gamma}.$$

Si $\Phi(\varrho) = (\Phi_1(\varrho), ..., \Phi_m(\varrho))$ est un vecteur ayant pour composantes les séries formelles $\Phi_1(\varrho), ..., \Phi_m(\varrho)$, alors on définit

$$\lambda\Phi(\varrho) = (\lambda\Phi_1(\varrho), ..., \lambda\Phi_m(\varrho)).$$

Il est évident que le produit des deux opérateurs

$$\lambda' = (..., \lambda'_s ...), \quad \lambda'' = (..., \lambda''_s, ...) \text{ est } \lambda = (..., \lambda'_s \lambda''_s, ...).$$

Si $\lambda_s = \lambda_1^s$ on a $\lambda\Psi(\varrho, \eta) = \Psi(\lambda_1 \varrho, \lambda_1 \eta)$; il nous suffira donc de nous limiter au cas où

$$\lambda_1 = 1.$$

7. Propriétés de ces opérateurs

Nous aurons besoin des propriétés que voici.

1) *Propriété du produit* : Si Ψ_1 et $\Psi_2 \geqslant 0$, alors :

$$\lambda[\Psi_1(\varrho, \eta)\, \Psi_2(\varrho, \eta)] \ll [\lambda\Psi_1(\varrho, \eta)] \cdot [\lambda\Psi_2(\varrho, \eta)]. \qquad (7.1)$$

2) *Propriété de la série composée* : Si

$$\theta(\varrho) = \Psi \circ \Phi,$$

où

$$\Phi = (\Phi_1, ..., \Phi_m), \ \Phi_i \geqslant 0, \ \Phi(0) = 0, \ \Psi \geqslant 0,$$

alors

$$\lambda\theta(\varrho) \ll (\lambda\Psi) \circ (\lambda\Phi). \qquad (7.2)$$

Pour que ces deux propriétés aient lieu, il faut et il suffit que λ vérifie les conditions suivantes :

$$\begin{cases} \lambda_0 = \lambda_1 = 1 \geqslant \lambda_2; \\ \lambda_{r+s-1} \leqslant \lambda_r \lambda_s \quad \text{si} \quad r \text{ et } s \geqslant 1. \end{cases} \qquad (7.3)$$

Exemple. — Les conditions (7.3) sont vérifiées quand

$$\lambda_0 = \lambda_1 = 1, \quad \lambda_{s-1} \cdot \lambda_{s+1} \leqslant \lambda_s^2, \qquad (7.4)$$

c'est-à-dire quand $\lambda_s^{-\frac{1}{s}}$ est une fonction de s *logarithmiquement convexe*.

Preuve de (7.3). — Pour que (7.1) ait lieu, il faut et suffit qu'il ait lieu quand Ψ_1 et Ψ_2 sont des monomes :

$$\Psi_1 = \varrho^s \eta^\gamma, \ \Psi_2 = \varrho^{s'} \eta^{\gamma'};$$

donc que

$$\lambda_{r+s} \leqslant \lambda_r \cdot \lambda_s. \qquad (7.5)$$

En faisant $r = 1$, on voit que cette condition implique

$$\lambda_0 = \lambda_1 = 1 \geqslant \lambda_2 \geqslant ... \geqslant \lambda_s \geqslant ... . \qquad (7.6)$$

Pour que (7.2) ait lieu, il faut et suffit que (7.2) ait lieu quand Φ et Ψ sont des monomes :

$$\Phi(\varrho) = (\varrho^{s_1}, ..., \varrho^{s_m}), \ \Psi(\varrho, \eta) = \varrho^{s_0} \eta^\gamma, \ s_1, ..., s_m \geqslant 1;$$

c'est-à-dire que

$$\lambda_{s_0 + \gamma_1 s_1 + ... + \gamma_m s_m} \leqslant \lambda_{s_0 + |\gamma|} (\lambda_{s_1})^{\gamma_1} ... (\lambda_{s_m})^{\gamma_m},$$

si $s_1 ... s_m \geqslant 1$.

22 Hyperbolic Equations and Waves

Cette condition est vérifiée si elle l'est pour $m = 1$, $\gamma_1 = 1$; elle équivaut donc à la condition :

$$\lambda_{r+s-1} \leqslant \lambda_r \cdot \lambda_s \text{ pour } r \text{ et } s \geqslant 1. \tag{7.7}$$

Or (7.7) implique (7.5), vu (7.6); et (7.6) résulte de (7.7), où l'on prend $r = 2$, si $\lambda_2 \leqslant 1$.

Preuve que l'exemple (7.4) vérifie (7.3). — En faisant $s = 1$ dans (7.4), on obtient $\lambda_2 \leqslant 1$. D'autre part, vu (7.4), $\lambda_s / \lambda_{s+1}$ est une fonction croissante de s; donc $\lambda_s / \lambda_{s+r}$ aussi; donc

$$\frac{1}{\lambda_{r+1}} \leqslant \frac{\lambda_s}{\lambda_{s+r}} \quad \text{si} \quad s \geqslant 1.$$

8. Opérateurs de gevrey $\lambda^{(\alpha)}$. — Ces opérateurs dépendent d'un paramètre numérique $\alpha \geqslant 1$; ils se définissent comme suit :

$$\lambda_s = (s!)^{1-\alpha}.$$

Si $\Phi(\varrho) = \sum_{s=0}^{\infty} \frac{\varrho^s}{s!} \Phi_s$, alors $\lambda^{(\alpha)} \Phi(\varrho) = \sum_s \frac{\varrho^s}{(s!)^\alpha} \Phi_s$.

La propriété (7.1) du produit et la propriété (7.2) de la série composée valent pour ces opérateurs.

Preuve. — La vérification de (7.4) est immédiate : d'où (7.3); d'où (7.1) et (7.2).

9. Note. — Si λ vérifie (7.3), alors il possède aussi la propriété suivante, qu'emploie [5] :

$$\lambda \left[\Psi_1(\varrho) \frac{\partial \Psi_2}{\partial \varrho} \right] \ll \left[\lambda \Psi_1(\varrho) \right] \cdot \frac{\partial}{\partial \varrho} \lambda \Psi_2(\varrho) \text{ , si } \Psi_1(0) = 0.$$

BIBLIOGRAPHIE

[1] Beurling, Congrès scandinave, 1938.

[2] P. Dionne, Sur les problèmes de Cauchy hyperboliques bien posés, *Journal d'Analyse math.*, t. 10 (1962), chap. V et VI, pp. 1-90.

[3] M. Gevrey, Sur la nature analytique des solutions des équations aux dérivées partielles, *Annales École norm. sup.*, t. 35 (1917), pp. 129-189.

[4] L. Leray et Y. Ohya, *Systèmes linéaires, hyperboliques non stricts* (exposé précédent).

[5] J. Leray et Y. Ohya, *Systèmes non linéaires, hyperboliques non stricts*, CIME, Varenna (Italie), 1964.

Math. Ann. **170**, 167–205 (1967).

Équations et Systèmes Non Linéaires,
Hyperboliques Non-Stricts

J. LERAY et Y. OHYA

Introduction

0. Historique

Le problème de Cauchy fut étudié d'abord quand les données et les inconnues sont holomorphes (CAUCHY-KOWALESKI; N. A. LEDNEV [8] supprime l'hypothèse d'holomorphie par rapport au «temps», tout en conservant l'hypothèse d'holomorphie par rapport aux coordonnées «d'espace»). Puis ce problème le fut, sous l'hypothèse d'hyperbolicité stricte, quand les données et les inconnues sont des fonctions dérivables jusqu'à un ordre donné ou même des distributions (HADAMARD, PETROWSKY, J. LERAY [9], L. GÅRDING [4], P. DIONNE [3]); alors la solution ne dépend que localement des données; plus précisément, il existe des «domaines d'influence».

Récemment divers auteurs ont étudié des cas intermédiaires: DE GIORGI [6] discute l'unicité, C. PUCCI [14] et G. TALENTI [15] prouvent l'existence quand le cône caractéristique se réduit à des droites parallèles; L. HÖRMANDER [7] (théorème 5.7.3) traite l'équation linéaire à coefficients constants, hyperbolique non stricte[1]; Y. OHYA [13] étudie, en coefficients variables, l'opérateur de Calderon-Zygmund et, en particulier, l'opérateur linéaire hyperbolique, dont le polynome caractéristique est un produit d'opérateurs strictement hyperboliques; nous avons étendu ses conclusions aux systèmes linéaires [10] en formalisant son procédé et en employant une suggestion de L. WAELBROECK, dont l'article [11] va maintenant nous permettre de traiter le cas non linéaire. Tous ces travaux ont des conclusions du type que le n° 1 va énoncer.

1. Énoncé des résultats

Nous résolvons le problème de Cauchy pour un système non linéaire. Nos hypothèses ont pour cas extrêmes les deux cas suivants:

1°) *hyperbolicité stricte;* données et inconnues *indéfiniment dérivables;* (il y a alors *des domaines d'influence*);

2°) *aucune hypothèse d'hyperbolicité;* données et inconnues *holomorphes* par rapport aux coordonnées d'espaces; (il n'y a pas de domaine d'influence).

Hors de ces cas extrêmes, nos hypothèses sont les suivantes:

---

[1] HÖRMANDER réserve le terme «hyperbolique» au strictement hyperbolique. Pour nous, il y a hyperbolicité quand il y a domaine d'influence.

3°) *le polynôme caractéristique est un produit de polynômes strictement hyperboliques;* les données et les inconnues sont indéfiniment différentiables par rapport aux coordonnées d'espace; plus précisément. elles sont dans une *classe de Gevrey, non quasi-analytique;* il existe *des domaines d'influence.*

On trouvera les énoncés précis aux n° 20, 27 et 29, Mme. CHOQUET-BRUHAT les a complétés [2].

Applications. S. S. CHERN et HANS LEWY [1] ont rencontré en géométrie différentielle le problème non linéaire que nous résolvons.

Mme. Y. CHOQUET-BRUHAT [2] et A. LICHNÉROWICZ [12] ont ramené à ce problème la résolution des équations de la magnéto-hydrodynamique relativiste.

2. Sommaire

Nous adaptons au cas non-linéaire le procédé qu'emploie l'article [10], dont la connaissance n'est pas indispensable; ce procédé se simplifie, car l'étude non linéaire est purement locale; cependant il doit employer pour les coefficients des normes un peu moins simples: les normes de Schauder.

Le problème est résolu par approximations successives, que définissent des problèmes de Cauchy linéaires, strictement hyperboliques. L'étude de ces approximations successives emploie leurs normes formelles, c'est-à-dire des séries formelles ayant pour coefficients les normes de toutes leurs dérivées. La majoration des approximations successives résulte de la résolution d'un problème de Cauchy non linéaire formel, c'est-à-dire ayant pour données et inconnue des séries formelles, appartenant à une classe de Gevrey formelle. La convergence des approximations successives résulte de la résolution d'un second problème de Cauchy formel, qui est linéaire.

L'existence des domaines d'influence résulte du théorème d'unicité que nous avons obtenu dans le cas linéaire [10]; la précision de ce théorème d'unicité provient de ce que, dans ce cas linéaire, un théorème d'existence non local peut être obtenu, par ces raisonnements mêmes dont la suppression allège le présent article.

§ 1. Normes formelles

3. Normes

Notons les coordonnées de $\mathbf{R}^{l+1}$

$$(x_0, x_1, \ldots, x_l)$$

et

$$D_x^\beta = \frac{\partial^{|\beta|}}{\partial x_0^{\beta_0} \ldots \partial x_l^{\beta_l}}.$$

Soit X la bande de $\mathbf{R}^{l+1}$ d'équation

$$X : 0 \leqq x_0 \leqq |X| ;$$

soit S_t l'hyperplan de X d'équation

$$S_t : x_0 = t .$$

Notons : K_t les cubes, de côté 1, appartenant à S_t;

$$|f, S_t|_2 - \left[\int_{S_t} |f|^2 \, dx_1 \dots dx_l \right]^{1/2};$$

$$|f, K_t|_2 = \left[\int_{K_t} |f|^2 \, dx_1 \dots dx_l \right]^{1/2}.$$

Etant donné un entier $n \geq 0$, nous nommons *quasi-normes* d'une fonction

$$f : X \to \mathbf{C}$$

les deux fonctions de t :

$$|D^n f, S_t| = c \sup_\beta |D_x^\beta f, S_t|_2$$

$$\|D^n f, S_t\| = c \sup_{\beta, K_t} |D_x^\beta f, K_t|_2; \qquad (|\beta| \leq n)$$

ce sont des normes de $f \bmod (x_0 - t)^n$; $c = c(l, n)$ est une fonction de (l, n), croissante en n et assez grande pour que la propriété (3.1) et la formule (3.2) soient exactes.

DIONNE [3], ch. 1, (6.3.9), déduit des théorèmes de Sobolev ceci, sous l'hypothèse :

$$n > l/2:$$

(3.1)   ces deux normes sont des *normes d'algèbres*;
        leur finitude entraîne la continuité de f;

on a *la formule du produit*:

(3.2) $$|D^n(f \cdot g), S_t| \leq \|D^n f, S_t\| \cdot |D^n g, S_t|.$$

Soit un domaine $Y \subset \mathbf{C}^m$. Nous nommons *quasi-normes* d'une fonction

$$F : X \times Y \to \mathbf{C}$$

les deux fonctions de t, dépendant d'un vecteur $v = (v_1, \dots, v_m)$, à composantes $v_j \geq 0$:

$$|D^n F, S_t \times Y, v| = c \sup_\beta \left| \sup_{y \in Y} \left| D_{x,y}^\beta F(x, y) \right|, S_t \right|_2 (1 + c'|v|)^n$$

$$\|D^n F, S_t \times Y, v\| = c \sup_{\beta, K_t} \left| \sup_{y \in Y} \left| D_{x,y}^\beta F(x, y) \right|, K_t \right|_2 (1 + c'|v|)^n,$$

où

$$D_{x,y}^\beta = \frac{\partial^{|\beta|}}{\partial x_0^{\beta_0} \dots \partial y_m^{\beta_{m+l}}}, \qquad |\beta| \leq n, |v| = v_1 + \dots + v_m,$$

$c' = c'(m)$ suffisamment grand pour avoir (3.3). Soit une application

$$v = (v_1, \dots, v_m) : X \to Y;$$

notons $F \circ v$ la fonction composée

$$(F \circ v)(x) = F(x, v(x));$$

notons $|D^n v, S_t|$ le vecteur de composantes $|D^n v_j, S_t|$ $(j = 1, \dots, m)$. DIONNE [3], théorème 6.4, explicite comme suit le *théorème de composition* de SOBOLEV:

si on a $n > l/2 + 1$,

(3.3) $$\|D^n(F \circ v), S_t\| \leqq \|D^n F, S_t \times Y, |D^n v, S_t|\|;$$

on peut remplacer $\| \dots \|$ par $| \dots |$.

4. Normes formelles

On nomme *quasi-normes formelles* de $f : X \to \mathbf{C}$ les deux séries formelles de ϱ, à coefficients fonctions de t:

$$|D^{n,\infty} f, S_t, \varrho| = \sum_{s=0}^{\infty} \frac{\varrho^s}{s!} \sup_{\sigma} |D^n D_x^\sigma f, S_t|$$

$$= c \sum_{s=0}^{\infty} \frac{\varrho^s}{s!} \sup_{\beta,\sigma} |D_x^{\beta+\sigma} f, S_t|_2,$$

$$\|D^{n,\infty} f, S_t, \varrho\| = \sum_{s=0}^{\infty} \frac{\varrho^s}{s!} \sup_{\sigma} \|D^n D_x^\sigma f, S_t\|$$

$$= c \sum_{s=0}^{\infty} \frac{\varrho^s}{s!} \sup_{\beta,\sigma,K_t} |D_x^{\beta+\sigma} f, K_t|_2$$

où

$$|\beta| \leqq n, \quad \sigma = (0, \sigma_1, \dots, \sigma_l), \quad |\sigma| = \sigma_1 + \cdots + \sigma_l = s.$$

Introduisons des variables commutatives $(\varrho, \eta_1, \dots, \eta_m, v)$; notons $\eta = (\eta_1, \dots, \eta_m)$, $\eta^\tau = \eta_1^{\tau_1} \dots \eta_m^{\tau_m}$; nous définissons de même les *quasi-normes formelles* de $F : X \times Y \to \mathbf{C}$:

$$\|D^{n,\infty} F, S_t \times Y, \varrho, \eta, v\| = \sum_{s,\tau} \frac{\varrho^s}{s!} \frac{\eta^\tau}{\tau!} \sup_{\sigma} \|D^n D_x^\sigma D_y^\tau F, S_t \times Y, v\|$$

$$|D^{n,\infty} F, S_t \times Y, \varrho, \eta, v| = \sum_{s,\tau} \frac{\varrho^s}{s!} \frac{\eta^\tau}{\tau!} \sup_{\sigma} |D^n D_x^\sigma D_y^\tau F, S_t \times Y, v|,$$

où

$$\sigma = (0, \sigma_1, \dots, \sigma_l), \quad |\sigma| = \sigma_1 + \cdots + \sigma_l = s.$$

Une série formelle $\gg 0$ est une série à coefficients $\geqq 0$.

Enonçons les propriétés des quasi-normes formelles; le $n° 5$ les prouvera.

Formule du produit. Si $n > l/2$, on a:

(4.1) $$|D^{n,\infty}(f g), S_t, \varrho| \ll \|D^{n,\infty} f, S_t, \varrho\| \cdot |D^{n,\infty} g, S_t, \varrho|;$$

on peut remplacer $|\dots|$ par $\| \dots \|$.

Formule de la dérivée. Notons $D_j = \dfrac{\partial}{\partial x_j}$; si $j > 0$, on a

(4.2)
$$|D^{n,\infty} D_j f, S_t, \varrho| \ll \frac{\partial}{\partial \varrho} |D^{n,\infty} f, S_t, \varrho| \ll |D^{n+1,\infty} f, S_t, \varrho| \ll$$

$$\ll c'' |D^{0,\infty} D_0^{n+1} f, S_t, \varrho| + c'' \left(1 + \frac{\partial}{\partial \varrho}\right) |D^{n,\infty} f, S_t, \varrho|,$$

où $c'' = c''(l, n)$; on peut remplacer $|\dots|$ par $\| \dots \|$.

Formule du commutateur. Soit $a(x, D)$ un opérateur différentiel linéaire normal [2] d'ordre $m \geq 1$; sa quasi-norme formelle $\|D^{n, \infty} a, S_t, \varrho\|$ sera la somme de celles de ses coefficients; nous définissons

$$|D^n[D^\infty, a] f, S_t, \varrho| = \sum_{s=0}^\infty \frac{\varrho^s}{s!} \sup_\sigma |D^n[D^\sigma, a] f, S_t|$$

$$= c \sum_{s=0}^\infty \frac{\varrho^s}{s!} \sup_{\beta, \sigma} |D^\beta[D^\sigma, a] f, S_t|_2$$

où

$$[D^\sigma, a] f = D^\sigma(a f) - a(D^\sigma f), \quad |\beta| \leq n, \sigma = (0, \sigma_1, ..., \sigma_l), \quad |\sigma| = s.$$

Nous avons, si $n > l/2$:

(4.3)
$$|D^n[D^\infty, a] f, S_t, \varrho| \ll$$

$$\ll [\|D^{n, \infty} a, S_t, \varrho\| - \|D^n a, S_t\|] \left(1 + \frac{\partial}{\partial \varrho}\right) |D^{m+n-1, \infty} f, S_t, \varrho|.$$

Formule de composition. Si $v : X \to Y$, $(F \circ v)(x) = F(x, v(x))$, et $n > l/2 + 1$, nous avons :

(4.4)
$$\|D^{n, \infty}(F \circ v), S_t, \varrho\| \ll$$

$$\ll \|D^{n, \infty} F, S_t \times Y, \varrho, |D^{n, \infty} v, S_t, \varrho| - |D^n v, S_t|, |D^n v, S_t|\| ;$$

on peut remplacer $\| ... \|$ par $| ... |$.

5. Preuves des formules précédentes

[10] montre comment (3.3) implique *la formule de composition* (4.4); (il faut remplacer dans [6] $| ... |$ par $|D^n ...|$, $\| ... \|$ par $\|D^n ... \|$).

La formule de la dérivée (4.2) est facile à prouver.

Prouvons *celle du commutateur*, en prouvant d'abord la suivante [dont il suffit de modifier légèrement la preuve pour établir celle du produit (4.1)]:

Une formule préliminaire. Définissons la série formelle en $\xi = (\xi_1, ..., \xi_l)$:

$$\|D^{n, \infty} f, S_t; \xi\| = \sum_\sigma \frac{\xi^\sigma}{\sigma!} \|D^n D^\sigma f, S_t\|$$

et de même avec $|...|$ au lieu de $\| ... \|$; rappelons que

$$\sigma! = \sigma_1! ... \sigma_l!, \quad \xi^\sigma = \xi_1^{\sigma_1} ... \xi_l^{\sigma_l}.$$

Notons

$$[D^\sigma, f] g = D^\sigma(f g) - f D^\sigma g,$$

(5.1) $$|D^n[D^\infty, f] g, S_t; \xi| = \sum_\sigma \frac{\xi^\sigma}{\sigma!} |D^n[D^\sigma, f] g, S_t|, \qquad \text{où} \quad \sigma_0 = 0;$$

(5.2) $$|D^n[D^\infty, f] g, S_t, \varrho| = \sum_s \frac{\varrho^s}{s!} \sup_\sigma |D^n[D^\sigma, f] g, S_t| \qquad \text{où} \quad \sigma_0 = 0, |\sigma| = s.$$

---

[2] Son premier coefficient, c'est-à-dire celui de D_0^m, vaut 1 ; il suffit de diviser un opérateur par son premier coefficient pour le rendre normal.

D'après la formule de Leibniz de la dérivée d'un produit:

$$|D^n[D^\infty, f]g, S_t; \xi| = \sum_\sigma \frac{\xi^\sigma}{\sigma!} |D^n(D^\sigma(f \cdot g) - f \cdot D^\sigma g), S_t| \ll$$

$$\ll \sum_{\sigma, \tau} \frac{\xi^\sigma}{\sigma!} \frac{\xi^\tau}{\tau!} |D^n(D^\sigma f) \cdot (D^\tau g), S_t|; \quad \text{où} \quad |\sigma| > 0;$$

$$\sigma_0 = \tau_0 = 0,$$

donc, d'après la formule du produit (3.2):

$$|D^n[D^\infty, f]g, S_t; \xi| \ll [\|D^{n, \infty} f, S_t; \xi\| - \|D^n f, S_t\|] |D^{n, \infty} g, S_t; \xi|;$$

d'où, en posant

$$\varrho = \xi_1 + \cdots + \xi_l,$$

ce qui implique

(5.3) $$\frac{\varrho^s}{s!} = \sum_\sigma \frac{\xi^\sigma}{\sigma!} \qquad (|\sigma| = s),$$

$$|D^n[D^\infty, f]g, S_t; \xi| \ll [\|D^{n, \infty} f, S_t, \varrho\| - \|D^n f, S_t\|] |D^{n, \infty} g, S_t, \varrho|.$$

Or, (L. Gårding), vu (5.3), (5.2) est la plus petite série en ϱ qui majore (5.1); l'inégalité précédente signifie donc que

(5.4) $$|D^n[D^\infty, f]g, S_t, \varrho| \ll [\|D^{n, \infty} f, S_t, \varrho\| - \|D^n f, S_t\|] \cdot |D^{n, \infty} g, S_t, \varrho|.$$

Preuve de la formule du commutateur (4.3). Il suffit de prouver cette formule quand $a(x, D)$ est un monôme:

$$a(x, D) = a_\alpha(x)D^\alpha, \quad \text{où} \quad |\alpha| \le m.$$

Si $|\alpha| \le m - 1$, (5.4) donne

$$|D^n[D^\infty, a]f, S_t, \varrho| \ll [\|D^{n, \infty} a, S_t, \varrho\| - \|D^n a, S_t\|] \cdot |D^{m+n-1, \infty} f, S_t, \varrho|.$$

Si $\alpha = (m, 0, \ldots, 0)$, alors $a_\alpha = 1$, puisque a est normal; donc

$$|D^n[D^\infty, a]f, S_t, \varrho| = 0.$$

Enfin si $|\alpha| = m$ et $\alpha_0 < m$, alors $D^\alpha = D^\beta D_j$ où $|\beta| = m - 1$, $1 \le j$ et (5.4) donne:

$$|D^n[D^\infty, a]f, S_t, \varrho| \ll [\|D^{n, \infty} a, S_t, \varrho\| - \|D^n a, S_t\|] |D^{m+n-1, \infty} D_j f, S_t, \varrho| \ll$$

$$\ll [\|D^{n, \infty} a, S_t, \varrho\| - \|D^n a, S_t\|] \frac{\partial}{\partial \varrho} |D^{m+n-1, \infty} f, S_t, \varrho|,$$

vu la formule de la dérivée (4.2).

§ 2. Opérateurs linéaires hyperboliques non stricts

6. *L'opérateur strictement hyperbolique a les propriétés que voici* (Dionne [3])

Sur la bande X soit un opérateur hyperbolique d'ordre m

$$a(x, D) = \sum_{|\beta| \le m} a_\beta(x)D^\beta$$

et une fonction $b(x)$; posons le problème de Cauchy d'inconnue $u(x)$

(6.1) $$u(x, D) u(x) = b(x), \qquad D^{m-1} u \mid S_0 = 0.$$

Nous supposons $a(x, D)$ normal et régulièrement hyperbolique pour les hyperplans S_t; nous notons $\chi(a)$ son caractère de régularité: rappelons qu'il dépend de l'image de X par l'application $\{a_\beta(x)\}$ ($|\beta| = m$), sans dépendre des valeurs des dérivées des $a_\beta(x)$. Nous supposons

$$\|D^{n,\infty} a, S_t, \varrho\| \ll C(t, \varrho), \quad |D^{n,\infty} b, S_t, \varrho| \ll B(t, \varrho)$$

B [et C] étant une série formelle en ϱ, ayant pour coefficients des fonctions bornées [et *croissantes*] de t. Nous supposons enfin:

(6.2) $$D_0^j b \mid S_0 = 0 \quad \text{pour} \quad j < n$$

ce qui impliquera

$$D_0^j u \mid S_0 = 0 \quad \text{pour} \quad j < m + n;$$

(6.3) $$n > \frac{l}{2} + 1.$$

On sait [4], [9] que le problème de Cauchy (6.1) possède une et une seule solution telle que $|D^m u, S_t|$ soit borné; on sait que cette solution vérifie l'inégalité

(6.4) $$|D^{m+n-1} u, S_t| \leq A_0(t) \int_0^t B(t', 0) \, dt',$$

où

$$A_0(t) = c(l, m, \chi, C(t, 0));$$

$c(l, m, \chi, C)$ est une fonction connue, dont toutes les dérivées en C sont ≥ 0.
Précisons comme suit ces résultats:

Lemme 6.1. *On a*

$$|D^{m+n-1,\infty} u, S_t, \varrho| \ll A_0(t) \Phi(t, \varrho) \quad pour \quad 0 \leq t \leq |X|;$$

$A_0(t)$ *vient d'être défini; $\Phi(t, \varrho)$ est la série formelle que définit le problème de Cauchy formel*

(6.5) $$\begin{cases} \left[\dfrac{\partial}{\partial t} - A(t, \varrho) \left(1 + \dfrac{\partial}{\partial \varrho}\right)\right] \Phi(t, \varrho) = B(t, \varrho) \\ \Phi(0, \varrho) = 0, \end{cases}$$

où $A(t, \varrho)$ est la série formelle $\gg 0$, s'annulant avec ϱ:

$$A(t, \varrho) = A_0(t) [C(t, \varrho) - C(t, 0)].$$

Note 6. La résolution du problème de Cauchy (6.5) est élémentaire: le coefficient $\Phi_s(t)$ de

$$\Phi(t, \varrho) = \sum_s \frac{\varrho^s}{s!} \Phi_s(t)$$

s'obtient successivement pour $s = 0, 1, \ldots$ en résolvant (par quadratures: voir lemme 8) le problème de Cauchy

$$(6.6) \qquad \left[\frac{d}{dt} - s a_0(t)\right] \Phi_s(t) = \Psi_{s-1}(t), \qquad\qquad \Phi_s(0) = 0,$$

où $\Psi_{s-1}(t) - B_s(t)$ est une combinaison linéaire de $\Phi_0, \Phi_{s-1}(t)$; les coefficients sont ceux de A; ils sont $\geqq 0$;

$$a_0(t) = \frac{\partial A}{\partial \varrho}(t, 0).$$

Preuve. On peut prouver l'existence de toutes les dérivées $D^\sigma u$, où $\sigma_0 = 0$, en les construisant successivement pour $|\sigma| = m + n, \ m + n + 1, \ldots$ par les problèmes de Cauchy

$$(6.7) \qquad a D^\sigma u = -[D^\sigma, a]u + D^\sigma b, \quad D^{m-1} D^\sigma u | S_0 = 0;$$

elles sont donc telles que $|D^{m+n-1} D^\sigma u, S_t|$ soit une fonction bornée de t. D'après (6.7) et (6.4), on a pour tout σ tel que $\sigma_0 = 0$:

$$|D^{m+n-1} D^\sigma u, S_t| \leqq A_0(t) \int\limits_0^t |D^n[D^\sigma, a]u, S_{t'}| \, dt' + A_0(t) \int\limits_0^t |D^n D^\sigma b, S_{t'}| \, dt';$$

d'où, en appliquant $\sum\limits_s \dfrac{\varrho^s}{s!} \sup\limits_\sigma$, où $|\sigma| = s$:

$$|D^{m+n-1, \infty} u, S_t, \varrho| \ll$$

$$\ll A_0(t) \int\limits_0^t |D^n[D^\infty, a]u, S_{t'}, \varrho| \, dt' + A_0(t) \int\limits_0^t |D^{n, \infty} b, S_{t'}, \varrho| \, dt';$$

d'où, en appliquant la formule du commutateur (4.3) et en notant $|D^{n, \infty} u, S_t, \varrho| = A_0(t) \varphi(t, \varrho)$:

$$\varphi(t, \varrho) \ll \int\limits_0^t A(t', \varrho) \left(1 + \frac{\partial}{\partial \varrho}\right) \varphi(t', \varrho) \, dt' + \int\limits_0^t B(t', \varrho) \, dt'.$$

Explicitons cette inégalité, en posant

$$\varphi(t, \varrho) = \sum\limits_{s=0}^{\infty} \frac{\varrho^s}{s!} \varphi_s(t);$$

puisque $A(t, 0) = 0$, il vient, en posant $a_0(t) = \dfrac{\partial A}{\partial \varrho}(t, 0)$:

$$\varphi_s(t) \leqq \int\limits_0^t s \, a_0(t') \varphi_s(t') \, dt' + \psi_{s-1}(t),$$

où ψ_{s-1} ne dépend que de $\varphi_0, \ldots, \varphi_{s-1}$ et des données A, B; d'où, par une intégration d'inégalité classique:

$$\varphi(t, \varrho) \ll \Phi(t, \varrho),$$

si Φ est défini par l'équation intégrale

$$\Phi(t, \varrho) = \int\limits_0^t A(t', \varrho) \left(1 + \frac{\partial}{\partial \varrho}\right) \Phi(t', \varrho)\, dt' + \int\limits_0^t B(t', \varrho)\, dt',$$

c'est-à-dire par le problème de Cauchy (6.5).   C.Q.F.D.

L'emploi du lemme 6.1 que nous allons faire sera facilité par le lemme suivant:

Lemme 6.2. *Soit Φ^* la solution du problème* (6.5), *quand on y remplace par B^* la donné B. Supposons*

$$0 \ll B(t, \varrho) \ll A_0(t)\, B^*(t, \varrho), \quad \text{où} \quad A_0(t) \text{ est croissant}.$$

Alors

$$\Phi(t, \varrho) \ll A_0(t)\, \Phi^*(t, \varrho).$$

Preuve. Vu la note 6, il suffit de prouver que les solutions $\Phi(t)$ et $\Phi^*(t)$ des problèmes de Cauchy (6.6)

$$\left[\frac{\partial}{\partial t} - s\, c_0(t)\right] \Phi(t) = B(t), \qquad \Phi(0) = 0$$

$$\left[\frac{\partial}{\partial t} - s\, c_0(t)\right] \Phi^*(t) = B^*(t), \quad \Phi^*(0) = 0$$

vérifient $\Phi(t) \leq A_0(t)\, \Phi^*(t)$, si $0 \leq B \leq A_0 B^*$ (A_0 croissant). Or cela résulte immédiatement des solutions explicites (8.3) de ces problèmes.

7. Produit d'opérateurs strictement hyperboliques

Sur la bande X, nous nous donnons à nouveau un opérateur $a(x, D)$ hyperbolique et une fonction $b(x)$; notons

$$m = \text{ordre}\,(a);$$

nous nous posons le problème de Cauchy

$$(7.1) \qquad a(x, D)\, u(x) = b(x), \quad D_0^j u \,|\, S_0 = 0 \quad \text{pour} \quad j < m.$$

Nous supposons que

$$a(x, D) = a_1(x, D) \dots a_p(x, D)$$

est le produit de p opérateurs $a_j(x, D)$ normaux et régulièrement hyperboliques pour les hyperplans S_t; notons $m_j = \text{ordre}\,(a_1) + \dots + \text{ordre}\,(a_j)$; donc $m_p = m$; notons $\chi(a)$ l'ensemble des caractères de régularité des a_j; nous supposons:

$$(7.2) \quad \begin{cases} \|D^{m_j + n - j, \infty} a_{j+1}, S_t, \varrho\| \ll C(t, \varrho) \quad \forall j; \\ \|D^{n - p + k, \infty} a, S_t, \varrho\| \ll C_k(t, \varphi), \quad (k: \text{entier donné tel que } 0 \leq k \leq p); \\ |D^{n, \infty} b, S_t, \varrho| \ll B(t, \varrho); \end{cases}$$

$C(t, \varrho)$, $C_k(t, \varrho)$ et $B(t, \varrho)$ sont des séries formelles, dont chaque coefficient est une fonction bornée de t; nous supposons

$$\frac{\partial^j C(t, \varrho)}{\partial t^j} \gg 0 \quad \text{pour} \quad j = 0, \dots, p.$$

Nous définissons, comme au n° 6:

(7.3) $$A_0(t) = c(l, m, \chi, C(t, 0)).$$

Nous définissons la série formelle, *s'annulant avec* ϱ:

$$A(t, \varrho) = A_0(t) [C(t, \varrho) - C(t, 0)];$$

puis, c_k'' ne dépendant que de l, m, n, p, k:

$$A_k(t, \varrho) = c_k'' A_0(t) [1 + C_k(t, \varrho)]^k.$$

Bien entendu:

$$A_0(t, \varrho) = A_0(t), c_0'' = 1.$$

Nous supposons

(7.4) $$n > \frac{l}{2} + p, D^j b \,|\, S_0 = 0 \quad \text{pour} \quad j < n.$$

Lemme 7. *Le problème de Cauchy* (7.1) *possède une et une seule solution* $u(x)$ *telle que* $|D^m u, S_t|$ *soit borné; on a pour* $0 \le t \le |X|$, $0 \le k \le p$:

(7.5)$_k$ $$|D^{m+n-p+k, \infty} u, S_t, \varrho| \ll A_k(t, \varrho) \left(1 + \frac{\partial}{\partial t} + \frac{\partial}{\partial \varrho}\right)^k \Phi(t, \varrho);$$

$\Phi(t, \varrho)$ *est la série formelle que définit le problème de Cauchy formel*

(7.6) $$\begin{cases} \left[\dfrac{\partial}{\partial t} - A(t, \varrho) \left(1 + \dfrac{\partial}{\partial \varrho}\right)\right]^p \Phi(t, \varrho) = B(t, \varrho) \\ \dfrac{\partial^j \Phi}{\partial t^j} (0, \varrho) = 0 \quad \text{pour} \quad j = 0, ..., p-1. \end{cases}$$

Note. Ce problème (7.6) se résout en calculant successivement les coefficients $\Phi_s(t)$ ($s = 0, 1, ...$) de $\Phi(t, \varrho)$; ce calcul se fait par quadratures.

Preuve de (7.5)$_0$. Le problème (7.1) équivaut à la suite de problèmes de Cauchy:

$$a_j(x, D) u_j(x) = u_{j-1}(x), \quad D^{m_j - 1} u_j \,|\, S_0 = 0,$$

où $j = 1, ..., p, u_0 = b, u_p = u$.

D'où, par application du n° 6, l'existence de u, son unicité et les majorations:

$$|D^{m_1 + \cdots + m_j + n - j, \infty} u_j, S_t, \varrho| \ll c_1(t) \Phi_j(t, \varrho),$$

les $\Phi_j(t, \varrho)$ étant les séries formelles définies par les problèmes de Cauchy formels:

$$\begin{cases} \left[\dfrac{\partial}{\partial t} - A(t, \varrho) \left(1 + \dfrac{\partial}{\partial \varrho}\right)\right] \Phi_j(t, \varrho) = \Phi_{j-1}(t, \varrho) \\ \Phi_j(0, \varrho) = 0 \end{cases}$$

où $\Phi_0 = B$. D'où (7.4) en prenant $\Phi = \Phi_p$, ce qui revient à définir Φ par (7.6).

Preuve de (7.5)$_k$ *pour* $1 \le k \le p$. La formule de la dérivée (4.2) donne

$$|D^{m+n-p+k, \infty} u, S_t, \varrho| \ll$$

$$\ll c'' |D^{0, \infty} D_0^{m+n-p+k} u, S_t, \varrho| + c'' \left(1 + \frac{\partial}{\partial \varrho}\right) |D^{m+n-p+k-1, \infty} u, S_t, \varrho|;$$

or, puisque $a(x, D)u = b$, on a, vu la formule de la dérivée $|D^{0, \infty} D_0^j \quad | \ll$
$\ll |D^{j, \infty} \ldots |$,

$$|D^{0, \infty} D_0^{m+n-p+k} u, S_t, \varrho| \ll |D^{n-p+k, \infty} [a(x, D) - D_0^m]u, S_t, \varrho| + |D^{n-p+k, \infty} b, S_t, \varrho|$$

où $a(x, D) - D_0^m$ a un premier coefficient nul, car a est normal; donc, vu la formule
du produit (4.1), qui s'applique car $n - p + k > l/2$, et la formule de la dérivée
(4.2):

$$|D^{n-p+k, \infty} [a(x, D) - D_0^m]u, S_t, \varrho| \ll$$

$$\ll \|D^{n-p+k, \infty} a, S_t, \varrho\| \left(1 + \frac{\partial}{\partial \varrho}\right) |D^{m+n-p+k-1, \infty} u, S_t, \varrho|.$$

Les trois inégalités précédentes donnent

$$|D^{m+n-p+k, \infty} u, S_t, \varrho| \ll$$

$$\ll c''[1 + \|D^{n-p+k, \infty} a, S_t, \varrho\|] \left(1 + \frac{\partial}{\partial \varrho}\right) |D^{m+n-p+k-1, \infty} u, S_t, \varrho| +$$

$$+ c'' |D^{n-p+k, \infty} b, S_t, \varrho|.$$

D'où, par récurrence sur $k > 0$, la formule, évidente pour $k = 0$:

$$|D^{m+n-p+k, \infty} u, S_t, \varrho| \ll [1 + \|D^{n-p+k, \infty} a, S_t, \varrho\|]^k \left(1 + \frac{\partial}{\partial \varrho}\right)^k |D^{m+n-p, \infty} u, S_t, \varrho|$$

$$+ c'' \sum_{j=1}^k [1 + \|D^{n-p+k, \infty} a, S_t, \varrho\|]^{k-j} \left(1 + \frac{\partial}{\partial \varrho}\right)^{k-j} |D^{n-p+j, \infty} b, S_t, \varrho|;$$

la valeur de c'' a été modifiée; la formule de la dérivée (4.3) a été appliquée
à $\|\ldots\|$.

Pour tirer $(7.5)_k$ de l'inégalité précédente, il suffit évidemment, vu $(7.5)_0$,
de prouver ceci:

$$(7.7) \qquad |D^{n-p+j, \infty} b, S_t, \varrho| \ll \left(\frac{\partial}{\partial t}\right)^j \Phi(t, \varrho) \quad \text{pour} \quad j = 1, \ldots, p.$$

Preuve de (7.7). Puisque $D^{n-1} b \,|\, S_0 = 0$, nous avons

$$D^{\beta + \sigma} b(x) = \int_0^{x_0} \frac{(x_0 - x_0')^{j-1}}{(j-1)!} D_0^j D^{\beta + \sigma} b(x') \, dx_0'$$

pour $x = (x_0, x_1, \ldots, x_l)$, $x' = (x_0', x_1, \ldots, x_l)$,

$$\sigma_0 = 0, \quad 0 < j, \quad j + \beta_0 \leqq n;$$

d'où

$$|D^{\beta + \sigma} b, S_t|_2 \leqq \int_0^t \frac{(t - t')^{j-1}}{(j-1)!} |D_0^j D^{\beta + \sigma} b, S_{t'}|_2 \, dt'$$

et, en appliquant $\sum_s \frac{\varrho^s}{s!} \sup_{\beta, \sigma} \ldots$, où $|\beta| \leqq n - j$ et $\sigma_0 = 0$:

$$
|D^{n-j, \infty} b, S_t, \varrho| \ll \int_0^t \frac{(t - t')^{j-1}}{(j-1)!} |D^{n, \infty} b, S_{t'}, \varrho| \, dt'
$$

(7.8)

$$
\ll \int_0^t \frac{(t - t')^{j-1}}{(j-1)!} B(t', \varrho) \, dt', \quad \text{pour} \quad 0 < j \leqq n,
$$

car

(7.9) $$\qquad\qquad |D^{n, \infty} b, S_t, \varrho| \ll B(t, \varrho).$$

Or le lemme 9.2 va déduire de l'hypothèse

$$
\frac{\partial^j A(t, \varrho)}{\partial t^j} \gg 0 \quad \text{pour} \quad j = 0, \ldots, p-1
$$

que

$$
B(t, \varrho) \ll \frac{\partial^p \Phi(t, \varrho)}{\partial t^p}, \quad \int_0^t \frac{(t - t')^{j-1}}{(j-1)!} B(t', \varrho) \, dt' \ll \frac{\partial^{p-j} \Phi(t, \varrho)}{\partial t^{p-j}} \quad \text{pour} \quad 0 < j \leqq p.
$$
(7.10)

Les majorations (7.8) et (7.9) de b donnent donc:

$$
|D^{n-j, \infty} b, S_t, \varrho| \ll \frac{\partial^{p-j} \Phi(t, \varrho)}{\partial t^{p-j}} \quad \text{pour} \quad 0 \leqq j \leqq p.
$$

Voici prouvé (7.7), donc le lemme 7.

§ 3. Problèmes de Cauchy formels

L'emploi du lemme 7 va introduire des problèmes de Cauchy formels. Etudions leurs propriétés, dont l'une (7.10) vient d'être appliquée.

8. L'inégalité classique pour l'équation différentielle du premier ordre

Lemme 8. *Soit $\Phi(t)$ la solution du problème de Cauchy*

(8.1) $$\qquad\qquad \left[\frac{d}{dt} - a(t)\right] \Phi(t) = b(t), \quad \Phi(0) = 0,$$

où a et b sont des fonctions sommables

$$
a(t) \geqq 0; \quad t \geqq 0.
$$

Alors l'application

$$
(a, b) \to \Phi
$$

est croissante en b et, si b $\geq$ 0, en a, pour les relations d'ordre suivantes :

$$(8.2) \qquad \begin{cases} a \prec a^* & signifie: \quad a(t) \leq a^*(t); \\ b \prec b^* & signifie: \quad b(t) \leq b^*(t); \\ \Phi \prec \Phi^* & signifie: \quad \Phi(t) \leq \Phi^*(t) \ \text{et} \ \dfrac{d\Phi}{dt} \leq \dfrac{d\Phi^*}{dt}, \ \forall t \geq 0. \end{cases}$$

Preuve. C'est évident, car

$$(8.3) \qquad \Phi(t) = \int\limits_0^t b(t') \exp\left[\int\limits_{t'}^t a(t'') \, dt''\right] dt' \ \text{et} \ \frac{d\Phi}{dt} = a\Phi + b.$$

9. Extension de cette inégalité à un problème de Cauchy formel

Donnons-nous une série formelle en ϱ, fonction de $t \geq 0$, $A(t, \varrho)$ telle que

$$A(t, 0) = 0;$$

notons [3] L l'opérateur

$$L\left(t, \varrho, \frac{\partial}{\partial\varrho}\right) = A(t, \varrho)\left(1 + \frac{\partial}{\partial\varrho}\right);$$

soit un entier $p > 0$. Etant donnée $B(t, \varrho)$, série formelle en ϱ, fonction de $t \geq 0$, nous en cherchons une autre, $\Phi(t, \varrho)$, qui soit solution du problème de Cauchy formel

$$(9.1) \left[\frac{\partial}{\partial t} - L\right]^p \Phi(t, \varrho) = B(t, \varrho), \quad \frac{\partial^j \Phi}{\partial t^j}(0, \varrho) = 0 \quad \text{pour} \quad j = 0, \dots, p - 1.$$

Lemme 9.1. *Ce problème (9.1) possède une solution unique; elle s'obtient par quadratures.*

Lemme 9.2. *Supposons*

$$(9.2) \qquad \frac{\partial^j A(t, \varrho)}{\partial t^j} \gg 0 \quad \text{pour} \quad j = 0, \dots, p - 1.$$

Alors l'application $(A, B) \to \Phi$ est croissante en B et, si $B \gg 0$, en A, pour les relations d'ordre suivantes :

$$A(t, \varrho) \prec A^*(t, \varrho) \quad signifie: \quad \left(\frac{\partial}{\partial t}\right)^j A \ll \left(\frac{\partial}{\partial t}\right)^j A^* \quad \text{pour} \quad j = 0, \dots, p - 1;$$

$$B(t, \varrho) \prec B^*(t, \varrho) \quad signifie: \quad B \ll B^*;$$

$$\Phi(t, \varrho) \prec \Phi^*(t, \varrho) \quad signifie: \quad \left(\frac{\partial}{\partial t}\right)^j \Phi \ll \left(\frac{\partial}{\partial t}\right)^j \Phi^* \quad \text{pour} \quad j = 0, \dots, p.$$

---

[3] Ce qui suit est plus généralement vrai pour

$$L\left(t, \varrho, \frac{\partial}{\partial\varrho}\right) = A'(t, \varrho) + A(t, \varrho)\frac{\partial}{\partial\varrho}$$

$A'(t, \varrho)$, $A(t, \varrho)$ étant des séries formelles en ϱ, fonctions de t, vérifiant: $A(t, 0) = 0$; on complète (9.2) par

$$\frac{\partial^j A'(t, \varrho)}{\partial t^j} \gg 0.$$

D'où, en particulier, puisque $0 \prec A$, les inégalités (7.10): si $B \geqslant 0$, alors

$$(9.3) \quad \begin{cases} 0 \ll B(t, \varrho) \ll \dfrac{\partial^p \Phi}{\partial t^p}(t, \varrho), \\[3mm] 0 \ll \displaystyle\int\limits_0^t \dfrac{(t-t')^{j-1}}{(j-1)!} B(t', \varrho)\, dt' \ll \dfrac{\partial^{p-j}\Phi(t, \varrho)}{\partial t^{p-j}} \end{cases}$$

Preuve du lemme 9.1. Notons

$$\Phi_j = \left[\frac{\partial}{\partial t} - L\right]^{p-j} \Phi;$$

le problème (9.1) se décompose en les p problèmes d'ordre 1:

$$(9.4)_j \qquad \left[\frac{\partial}{\partial t} - L\right] \Phi_j(t, \varrho) = \Phi_{j-1}(t, \varrho), \quad \Phi_j(0, \varrho) = 0$$

où

$$j = 1, \dots, p, \quad \Phi_0 = B \quad \text{et} \quad \Phi_p = \Phi.$$

Supposons $\Phi_{j-1}(t, \varrho)$ calculé; il s'agit de résoudre (9.4); les coefficients $\varphi_s(t)$ de

$$\Phi_j(t, \varrho) = \sum_{s=0}^{\infty} \frac{\varrho^s}{s!} \varphi_s(t)$$

se calculent successivement pour $s = 0, 1, 2, \dots$ en résolvant des problèmes de Cauchy du type (8.1):

$$(9.5) \qquad \left[\frac{d}{dt} - s\, a_0(t)\right] \varphi_s(t) = \text{donnée}, \quad \varphi_s(0) = 0$$

où

$$a_0(t) = \frac{\partial A}{\partial \varrho}(t, 0).$$

Preuve du lemme 9.2 *pour* $p = 1$. Les coefficients $\varphi_s(t)$ de $\Phi(t, \varrho)$ se calculent par (9.5), où le second membre donné est une combinaison linéaire, à coefficients positifs, des coefficients de B et des coefficients $\varphi_0, \dots, \varphi_{s-1}$ de Φ. Il suffit donc d'appliquer le lemme 8.

Preuve du lemme 9.2 *pour* $p > 1$. Puisque le lemme vaut pour $p = 1$, $(9.4)_1$ prouve que Φ_1 et $\dfrac{\partial \Phi_1}{\partial t}$ sont croissants[4] et, si $B \geqslant 0$, qu'ils sont $\geqslant 0$. Puisque

---

[4] La croissance de $\left(\dfrac{\partial}{\partial t}\right)^i \Phi_j$ signifie la croissance de l'application

$$(A, B) \to \left(\frac{\partial}{\partial t}\right)^i \Phi_j$$

pour la relation d'ordre suivante:

$$\left(\frac{\partial}{\partial t}\right)^i \Phi_j \prec \left(\frac{\partial}{\partial t}\right)^i \Psi_j \quad \text{signifie} \quad \left(\frac{\partial}{\partial t}\right)^i \Phi_j(t, \varrho) \ll \left(\frac{\partial}{\partial t}\right)^i \Psi_j(t, \varrho).$$

le lemme vaut pour $p = 1$, $(9.4)_2$ prouve donc que Φ_2 et $\dfrac{\partial \Phi_2}{\partial t}$ sont croissants[4]

et, si $B \gg 0$, qu'ils sont $\gg 0$; d'où, en appliquant $\dfrac{\partial}{\partial t}$ à $(9.4)_2$ et en employant

l'hypothèse $\dfrac{\partial A}{\partial t} \gg 0 : \dfrac{\partial^2 \Phi_2}{\partial t^2}$ est croissant[4] et, si $B \gg 0$, est $\gg 0$.

Le raisonnement se poursuit de façon évidente.

Voici un lemme analogue au précédent :

Lemme 9.3. *Soit $\Phi(t, \varrho)$ la série formelle que définit le problème de Cauchy* (9.1). *Supposons*

$$\frac{\partial^j A}{\partial t^j}(0, \varrho) \gg 0 \quad \text{pour} \quad j = 0, \ldots, p + k - 1$$

$$\frac{\partial^j B}{\partial t^j}(0, \varrho) \gg 0 \quad \text{pour} \quad j = 0, \ldots, k$$

Alors

$$\frac{\partial^j \Phi}{\partial t^j}(0, \varrho) \gg 0 \quad \text{pour} \quad j = 0, \ldots, p + k.$$

Preuve pour $p = 1$. On applique $\left(\dfrac{\partial}{\partial t}\right)^j$ $(j = 0, \ldots, k)$ à l'équation $\dfrac{\partial \Phi}{\partial t} = L\Phi + B$, puis l'on fait $t = 0$.

Preuve pour $p > 1$. Puisque le lemme vaut pour $p = 1$,

$$(9.4)_1 \quad \text{donne} \frac{\partial^j \Phi_1}{\partial t^j}(0, \varrho) \gg 0 \quad \text{pour} \quad j = 0, \ldots, k + 1;$$

$$(9.4)_2 \quad \text{donne} \frac{\partial^j \Phi_2}{\partial t^j}(0, \varrho) \gg 0 \quad \text{pour} \quad j = 0, \ldots, k + 2;$$

le raisonnement se poursuit de façon évidente.

10. Énoncé d'un problème de Cauchy formel non-linéaire

Notations. Etant donnée $\Phi(t, \varrho)$, série formelle en ϱ, fonction de $t \geqq 0$, nous notons $D^q \Phi(t, \varrho)$ l'ensemble de ses dérivées $\dfrac{\partial^{i+j}}{\partial t^i \partial \varrho^j} \Phi(t, \varrho)$ d'ordre $i + j \leqq q$; leur nombre est $\dfrac{(q + 1)(q + 2)}{2}$.

Notons : τ un vecteur variable ayant pour composantes $\dfrac{(q + 1)(q + 2)}{2}$ variables numériques $\geqq 0$; θ un vecteur ayant pour composantes $\dfrac{(q + 1)(q + 2)}{2}$ variables formelles commutant entre elles et avec ϱ; $F_q[\tau, \varrho, \theta]$ une série formelle en (ϱ, θ), à coefficients fonctions de τ; $F_q \gg 0$ signifie que ces coefficients sont $\geqq 0$. Notons

$$F_q(D^q \Phi) = F_q[D^q \Phi(t, 0), \varrho, D^q \Phi(t, \varrho) - D^q \Phi(t, 0)];$$

c'est une série formelle en ϱ, s'annulant avec ϱ si $F_q[\tau, 0, 0] = 0$.

Etant donné deux entiers $p \geqq q$ et deux séries formelles, F_0 et F_q, nous considérons *le problème de Cauchy formel* suivant, (il servira à majorer le problème qu'énonce le n° 1): trouver pour $0 \leqq t \leqq T$ (T petit) une série formelle $\Phi(t, \varrho)$ vérifiant

$$(10.1) \quad \left[\frac{\partial}{\partial t} - F_0(\Phi) \left(1 + \frac{\partial}{\partial \varrho} \right) \right]^p \Phi = F_q(D^q \Phi), \quad \frac{\partial^j \Phi}{\partial t^j}(0, \varrho) = 0$$

$$\text{pour} \quad j = 0, ..., p-1$$

et telle que

$$(10.2) \qquad\qquad \frac{\partial^j \Phi(t, \varrho)}{\partial t^j} \gg 0 \quad \text{pour} \quad j = 0, ..., p;$$

nous supposons ceci:

$$(10.3) \quad \begin{cases} F_q(\tau, \varrho, \theta] \gg 0 \\ F_0[\tau, 0, 0] = 0; \quad \dfrac{\partial^j}{\partial \tau^j} F_0[\tau, \varrho, \theta] \gg 0 \quad \text{pour} \quad |j| = 0, ..., p; \end{cases}$$

si $p = q$, alors $\dfrac{\partial^p \Phi}{\partial t^p}$ ne figure pas dans $F_p(D^p \Phi)$.

11. Le théorème de Cauchy-Kowalewski permet de résoudre le problème (10.1) sous les hypothèses suivantes: $F_0[\tau, \varrho, \theta]$ et $F_p[\tau, \varrho, \theta]$ sont des *fonctions holomorphes au point* $(0, 0, 0)$; $p = q$.

En effet (10.1) est du type Cauchy-Kowalewski à un détail près: dans l'équation figure non seulement

$$\frac{\partial^{i+j} \Phi}{\partial t^i \partial \varrho^j}(t, \varrho),$$

mais aussi

$$\frac{\partial^{i+j} \Phi}{\partial t^i \partial \varrho^j}(t, 0);$$

mais ce détail n'altère ni l'énoncé ni la preuve du théorème de Cauchy-Ko-walewski.

Le problème (10.1) possède donc une solution $\Phi(t, \varrho)$ qui est une série de Taylor en ϱ; ses coefficients sont des fonctions de t holomorphes pour $0 \leqq |t| \leqq T$; T est un nombre > 0, dépendant des données.

Prouvons que Φ vérifie (10.2) si *tous les coefficients de Taylor des fonctions holomorphes* $F_0(\tau, \varrho, \theta)$ *et* $F_p(\tau, \varrho, \theta)$ *sont* $\geqq 0$. Notons

$$A(t, \varrho) = F_0(\Phi), \quad B(t, \varrho) = F_p(D^p \Phi);$$

vu $(10.3)_2$, nous avons:

$$A(t, 0) = 0.$$

Supposons prouvé que:

$$(11.1)_{k-1} \qquad \frac{\partial^j \Phi}{\partial t^j}(0, \varrho) \gg 0 \quad \text{pour} \quad j = 0, ..., p+k-1 \ (k \geqq 0),$$

ce qui a lieu, d'après (10.1), pour $k = 0$. Nous avons alors:

$$\frac{\partial^j A}{\partial t^j}(0, \varrho) \gg 0 \quad \text{pour} \quad j = 0, \ldots, p+k-1,$$

$$\frac{\partial^j B}{\partial t^j}(0, \varrho) \gg 0 \quad \text{pour} \quad j = 0, \ldots, k;$$

d'où $(11.1)_k$, vu le lemme 9.3.

Donc $(11.1)_k$ a lieu pour tout k; les coefficients de $\varPhi(t, \varrho)$, dévelopée en série de puissances de ϱ, sont donc des fonctions de t, holomorphes à l'origine, dont tous les coefficients de Taylor sont ≥ 0; ces fonctions et toutes leurs dérivées sont donc ≥ 0 pour $0 \leq t \leq T$; d'où, en particulier (10.2).

En résumé:

Lemme 11. *Adjoignons aux hypothèses* (10.3) *les suivantes:*

$$p = q;$$

$F_0[\tau, \varrho, \theta]$ *et* $F_p[\tau, \varrho, \theta]$ *sont des fonctions holomorphes au point* $(0, 0, 0)$; *leurs coefficients de Taylor en ce point sont tous* ≥ 0.

Alors le problème de Cauchy formel (10.1) *possède pour*

$$0 \leq t \leq T \quad (T \text{ petit}, \ T > 0)$$

au moins une solution vérifiant (10.2).

12. Opérateurs sur les séries formelles

Etant donné un nombre $\alpha \geq 1$, nommons λ l'opérateur qui transforme comme suit les séries formelles:

si $\quad \varPhi(t, \varrho) = \sum\limits_{s=0}^{\infty} \dfrac{\varrho^s}{s!} \varPhi_s(t)$, alors $\quad \lambda \varPhi(t, \varrho) = \sum\limits_{s} \dfrac{\varrho^s}{(s!)^\alpha} \varPhi_s(t)$;

si $\quad F(\tau, \varrho, \theta) = \sum\limits_{s, \gamma} \dfrac{\varrho^s}{s!} \dfrac{\theta^\gamma}{\gamma!} F_{s\gamma}(\tau)$,

où $\quad \gamma = (\gamma_1, \gamma_2, \ldots), \ \theta = (\theta_1, \theta_2, \ldots), \ \theta^\gamma = \theta_1^{\gamma_1} \theta_2^{\gamma_2} \ldots, \ \gamma! = \gamma_1! \gamma_2! \ldots$

alors

$$\lambda F(\tau, \varrho, \theta) = \sum\limits_{s, \gamma} \frac{1}{[(s + |\gamma|)!]^{\alpha - 1}} \frac{\varrho^s}{s!} \frac{\theta^\gamma}{\gamma!} F_{s\gamma}(\tau), \quad \text{où} \quad |\gamma| = \gamma_1 + \gamma_2 + \cdots.$$

L'opérateur λ a les propriétés suivantes, faciles à vérifier (voir [10], n° 19 et [11], n° 6 et 9):

Formule du produit.

(12.1) $$\lambda(\varPhi \cdot \varPsi) \ll (\lambda \varPhi) \cdot (\lambda \varPsi).$$

Formules de la dérivée.

(12.2) $$\begin{cases} \lambda \left(\varPhi \cdot \dfrac{\partial \varPsi}{\partial \varrho} \right) \ll (\lambda \varPhi) \cdot \dfrac{\partial}{\partial \varrho}(\lambda \varPsi), & \text{si} \quad \varPhi(t, 0) = 0. \\[2ex] \lambda \left(\dfrac{\partial}{\partial \varrho} \right)^j \varPhi \ll \left(\dfrac{\partial}{\partial \varrho} \right)^j \left(1 + \varrho \dfrac{\partial}{\partial \varrho} \right)^r \lambda \varPhi, & \text{si} \quad j \ll q, \alpha \ll \dfrac{q+r}{q}. \end{cases}$$

Formule de composition (que [6] note: $\lambda(F \circ \Phi) \ll (\lambda F) \circ (\lambda \Phi)$):

(12.3) $\lambda F(\Phi) \ll f(\lambda \Phi)$,   si $\lambda F = f$.

Appliquons ces formules au problème de Cauchy linéaire, formel (9.1).

Lemme 12. Considérons le problème (9.1) et le problème du même type

$$\left[\frac{\partial}{\partial t} - a(t, \varrho)\left(1 + \frac{\partial}{\partial \varrho}\right)\right]^p \varphi(t, \varrho) = b(t, \varrho), \frac{\partial^j \varphi}{\partial t^j}(0, \varrho) = 0$$

$$pour \ j = 0, \ldots, p-1,$$

où

$$a(t, 0) = 0.$$

Supposons:

$$0 \ll \left(\frac{\partial}{\partial t}\right)^j \lambda A(t, \varrho) \ll \left(\frac{\partial}{\partial t}\right)^j a(t, \varrho) \quad pour \quad j = 0, \ldots, p-1;$$

$$0 \ll \lambda B(t, \varrho) \ll b(t, \varrho).$$

Alors

$$\left(\frac{\partial}{\partial t}\right)^j \lambda \Phi(t, \varrho) \ll \left(\frac{\partial}{\partial t}\right)^j \varphi(t, \varrho) \quad pour \quad j = 0, \ldots, p.$$

Preuve pour $p = 1$. Les formules du produit et de la dérivée donnent

(12.4) $\lambda[A(t, \varrho)\left(1 + \frac{\partial}{\partial \varrho}\right)\Phi(t, \varrho)] \ll a(t, \varrho)\left(1 + \frac{\partial}{\partial \varrho}\right)\lambda\Phi(t, \varrho)$.

Donc

$$\left[\frac{\partial}{\partial t} - a(t, \varrho)\left(1 + \frac{\partial}{\partial \varrho}\right)\right]\lambda\Phi(t, \varrho) \ll \lambda B(t, \varrho) \ll b(t, \varrho);$$

donc, vu le lemme 9.2 (croissance):

$$\lambda\Phi(t, \varrho) \ll \varphi(t, \varrho), \quad \frac{\partial}{\partial t}\lambda\Phi \ll \frac{\partial\varphi}{\partial t}.$$

Preuve pour $p > 1$. Notons

$$\varphi_j = \left[\frac{\partial}{\partial t} - a(t, \varrho)\left(1 + \frac{\partial}{\partial \varrho}\right)\right]^{p-j}\varphi, \quad \varphi_0 = b, \quad \varphi_p = \varphi;$$

nous avons les formules analogues à $(9.4)_j$:

$(12.5)_j$ $\left[\frac{\partial}{\partial t} - a(t, \varrho)\left(1 + \frac{\partial}{\partial \varrho}\right)\right]\varphi_j(t, \varrho) = \varphi_{j-1}(t, \varrho), \quad \varphi_j(0, \varrho) = 0.$

Puisque le lemme vaut pour $p = 1$, $(9.4)_1$ et $(12.5)_1$ donnent

$$\lambda\Phi_1 \ll \varphi_1, \quad \frac{\partial}{\partial t}\lambda\Phi_1 \ll \frac{\partial}{\partial t}\varphi_1;$$

$(9.4)_2$ et $(12.5)_2$ donnent alors:

$$\lambda\Phi_2 \ll \varphi_2, \quad \frac{\partial}{\partial t}\lambda\Phi_2 \ll \frac{\partial}{\partial t}\varphi_2;$$

d'où, en appliquant $\dfrac{\partial}{\partial t}\,\lambda$, puis (12.4), à (9.4)$_2$

$$\frac{\partial^2}{\partial t^2}\,\lambda\Phi_2 \ll \frac{\partial^2}{\partial t^2}\,\varphi_2 .$$

Le raisonnement se poursuit de façon évidente et donne

$$\left(\frac{\partial}{\partial t}\right)^j \lambda\Phi_i \ll \left(\frac{\partial}{\partial t}\right)^j \varphi_i \quad \text{pour} \quad 0 \leqq j \leqq i \leqq p\,;$$

en particulier, puisque $\Phi_p = \Phi$ et $\varphi_p = \varphi$, on a les inégalités énoncées.

13. Classes de Gevrey formelles

Définition. — Etant donné un entier $p \geqq 0$ et un nombre $\alpha \geqq 1$, nous nommons classe de Gevrey formelle $\Gamma^{p,(\alpha)}$ l'ensemble des séries formelles

$$\Phi(t,\varrho) = \sum_{s=0}^{\infty} \frac{\varrho^s}{s!}\,\Phi_s(t), \quad F\,[\tau,\varrho,\theta] = \sum_{s,\gamma} \frac{\varrho^s}{s!}\,\frac{\theta^\gamma}{\gamma!}\,F_{s\gamma}(\tau)$$

vérifiant la condition suivante pour t ou τ petits :

$$\left(\frac{\partial}{\partial t}\right)^j \lambda\Phi(t,\varrho) = \sum_s \frac{\varrho^s}{(s!)^\alpha}\,\frac{\partial^j\Phi_s}{\partial t^j},$$

$$\left(\frac{\partial}{\partial \tau}\right)^j \lambda F\,[\tau,\varrho,\theta] = \sum_{s,\gamma} \frac{1}{[(s+|\gamma|)!]^{\alpha-1}}\,\frac{\varrho^s}{s!}\,\frac{\theta^\gamma}{\gamma!}\,F_{s\gamma}(\tau) \quad (|j| \leqq p)$$

sont des fonctions de ϱ ou de (ϱ,θ) holomorphes à l'origine, uniformément par rapport à t ou τ; c'est-à-dire : il existe un voisinage de l'origine, indépendant de t ou τ, où elles ont une borne, indépendante de t ou τ.

Cette condition peut s'énoncer :

$$\sup_{s,t} \frac{1}{[1+s]^\alpha}\left|\frac{d^j\Phi_s}{dt^j}\right|^{\frac{1}{1+s}} < \infty$$

ou

$$\sup_{s,\gamma,\tau} \frac{1}{[1+s+|\gamma|]^\alpha}\left|\frac{\partial^j F_{s\gamma}(\tau)}{\partial\tau^j}\right|^{\frac{1}{1+s+|\gamma|}} < \infty$$

Propriétés. Les propriétés de λ montrent que l'addition, le produit, la dérivation en ϱ et la composition transforment des éléments de $\Gamma^{p,(\alpha)}$ en éléments de $\Gamma^{p,(\alpha)}$.

Note. Si $\Phi(t,\varrho)$ et $\Psi(t,\varrho)$ sont des séries formelles en ϱ, alors la série formelle composée $\Psi(t,\Phi(t,\varrho))$ *est définie quand* $\Phi(t,0) = 0$ et seulement dans ce cas [à moins que $\Phi(t,\varrho)$ ne soit fonction holomorphe de ϱ].

14. Résolution du problème de Cauchy (10.1)

L'opérateur λ permet de déduire du lemme 11 la propriété suivante, qu'emploiera le § 4.

Théorème d'existence, pour le problème de Cauchy formel, non linéaire

Complétons les hypothèses (10.3) *par les suivantes :*

(14.1) $$F_0 \in \Gamma^{p,(\alpha)}, \quad F_q \in \Gamma^{0,(\alpha)}, \quad \text{où } 1 \leq \alpha \leq \frac{p}{q}.$$

Alors le problème de Cauchy formel (10.1) *possède, pour*

$$0 \leq t \leq T \quad (T \text{ petit}, \ T > 0)$$

au moins une solution $\Phi(t, \varrho)$ *vérifiant* (10.2) *et*

(14.2) $$\Phi \in \Gamma^{p,(\alpha)}.$$

Cette solution Φ va être construite par approximations successives.
Définition d'approximations successives $\Phi_K(t, \varrho)$ $(K = 0, 1, \ldots)$.

$$\Phi_0(t, \varrho) = 0 \, ;$$

quand la série formelle en ϱ, fonction de $t \geq 0$, $\Phi_K(t, \varrho)$ a été définie, $\Phi_{K+1}(t, \varrho)$ l'est par le problème de Cauchy suivant :

$(14.3)_K$ $$\left[\frac{\partial}{\partial t} - F_0(\Phi_K)\left(1 + \frac{\partial}{\partial \varrho}\right)\right]^p \Phi_{K+1} = F_q(D^q \Phi_K), \quad \frac{\partial^i \Phi_{K+1}}{\partial t^j}(0, \varrho) = 0$$
$$(j = 0, \ldots, p-1).$$

Rappelons que ce problème $(14.3)_K$ s'intègre par quadratures (lemme 9.1).

Positivité des approximations successives. Prouvons l'inégalité, évidente pour $K = 0$:

$(14.4)_K$ $$0 \ll \left(\frac{\partial}{\partial t}\right)^j \Phi_K(t, \varrho) \qquad \text{pour } j = 0, \ldots, p.$$

Puisque F_0 et $F_q \gg 0$, $(14.4)_K$, $(14.3)_K$ et (9.3) impliquent $(14.4)_{K+1}$.

Croissance des approximations successives. Prouvons l'inégalité, évidente d'après la précédente quand $K = 0$:

$(14.5)_K$ $$\left(\frac{\partial}{\partial t}\right)^j \Phi_K(t, \varrho) \ll \left(\frac{\partial}{\partial t}\right)^j \Phi_{K+1}(t, \varrho) \qquad \text{pour } j = 0, \ldots, p.$$

En appliquant le lemme de croissance 9.2 aux problèmes de Cauchy $(14.3)_K$ et $(14.3)_{K+1}$, on voit que $(14.5)_K$ implique $(14.5)_{K+1}$.

Définition d'une série formelle $\varphi(t, \varrho)$, qui servira à majorer les approximations successives. — Les hypothèses (14.1) signifient ceci : il existe des fonctions $f_0(\tau, \varrho, \theta)$ et $f_q(\tau, \varrho, \theta)$, holomorphes au point $(0, 0, 0)$ et à coefficients de Taylor ≥ 0, telles que :

$$\left(\frac{\partial}{\partial \tau}\right)^j \lambda F_0 \ll \left(\frac{\partial}{\partial \tau}\right)^j f_0 \quad \text{pour } j = 0, \ldots, p, \ |\tau| \leq T_0$$

$$\lambda F_q \ll f_q \quad \text{pour } |\tau| \leq T_q,$$

quand τ est à composantes ≥ 0. Comme (10.3) le permet, nous choisissons

$$f_0(\iota, 0, 0) = 0.$$

Considérons le problème de Cauchy

$$(14.6) \quad \begin{cases} \left[\dfrac{\partial}{\partial t} - f_0(\varphi) \left(1 + \dfrac{\partial}{\partial \varrho} \right) \right]^p \varphi = f_q \left(D^q \left(1 + \varrho \dfrac{\partial}{\partial \varrho} \right)^{p-q} \varphi \right) \\[3mm] \dfrac{\partial^j \varphi}{\partial t^j}(0, \varrho) = 0 \quad \text{pour} \quad j = 0, \ldots, p-1. \end{cases}$$

D'après le lemme 11, ce problème (14.6) possède une solution $\varphi(t, \varrho)$, définie pour

$$0 \leq t \leq T \quad (T \text{ petit, } T > 0)$$

telle que

$$\left(\frac{\partial}{\partial t} \right)^j \varphi(t, \varrho) \gg 0 \quad \text{pour} \quad j = 0, \ldots, p-1.$$

Nous choisissons T assez petit pour que

$$\varphi(t, 0) \leq T_0 \, ; \quad \left| D^q \left(1 + \varrho \frac{\partial}{\partial \varrho} \right)^{p-q} \varphi \right| \leq T_q.$$

Majoration des approximations successives. Prouvons l'inégalité, évidente pour $K = 0$:

$$(14.7)_K \qquad \left(\frac{\partial}{\partial t} \right)^j \lambda \Phi_K \ll \left(\frac{\partial}{\partial t} \right)^j \varphi \quad \text{pour} \quad j = 0, \ldots, p, 0 \leq t \leq T.$$

Vu les propriétés de λ (n° 12), $(14.7)_K$ implique

$$\left(\frac{\partial}{\partial t} \right)^j \lambda F_0(\Phi_K) \ll \left(\frac{\partial}{\partial t} \right)^j f_0(\lambda \Phi_K) \ll \left(\frac{\partial}{\partial t} \right)^j f_0(\varphi)$$

$$\lambda F_q(D^q \Phi_K) \ll f_q(\lambda D^q \Phi_K) \ll f_q \left(D^q \left(1 + \varrho \frac{\partial}{\partial \varrho} \right)^{p-q} \lambda \Phi_K \right) \ll f_q \left(D^q \left(1 + \varrho \frac{\partial}{\partial \varrho} \right)^{p-q} \varphi \right),$$

car $\alpha \leq p/q$. D'où $(14.5)_{K+1}$, en appliquant le lemme 12 aux problèmes de Cauchy $(14.3)_{K+1}$ et (14.6).

Fin de la preuve du théorème. Pour $0 \leq t \leq T$, la suite

$$\frac{\partial^j \Phi_0}{\partial t^j}, \ldots, \frac{\partial^j \Phi_K}{\partial t^j}, \ldots \quad (0 \leq j \leq p)$$

est croissante d'après (14.5) et bornée d'après (14.7); elle possède donc une limite $\Phi(t, \varrho)$, qui vérifie (10.1) d'après (14.3), (10.2) d'après (14.4) et appartient à $\Gamma^{p,(\alpha)}$ d'après (14.7).   C.Q.F.D.

Nous aurons besoin du résultat suivant, que fournit la démonstration précédente:

Théorème de convergence

Donnons-nous deux séries formelles en ϱ, fonctions de $t(0 \leqq t \leqq T)$: $A(t, \varrho)$ et $B(t, \varrho)$ telles que

$$A(t, 0) = 0, \quad \left(\frac{\partial}{\partial t}\right)^j A(t, \varrho) \gg 0 \ (j = 0, \ldots, p), \quad A \in \Gamma^{p, (\alpha)}$$

$$B(t, \varrho) \gg 0, \quad B \in \Gamma^{0, (\alpha)}.$$

Donnons-nous un opérateur différentiel, d'ordre $q \leqq p$, ne contenant pas $\left(\frac{\partial}{\partial t}\right)^p$ si $q = p$:

$$L_q\left(\varrho, \frac{\partial}{\partial t}, \frac{\partial}{\partial \varrho}\right),$$

ayant pour coefficients des séries formelles en ϱ, fonction de t, appartenant à $\Gamma^{0, (\alpha)}$ et $\gg 0$.
Supposons

$$1 \leqq \alpha \leqq p/q.$$

Définissons, pour $0 \leqq t \leqq T$, des séries formelles en ϱ, fonctions de t, $\varphi_K(t, \varrho)$, $(K = 1, 2, \ldots)$ par les problèmes de Cauchy suivants:

$$\left[\frac{\partial}{\partial t} - A(t, \varrho)\left(1 + \frac{\partial}{\partial \varrho}\right)\right]^p \varphi_1(t, \varrho) = B(t, \varrho), \quad \frac{\partial^j \varphi_1}{\partial t^j}(0, \varrho) = 0 \quad (j = 0, \ldots, p-1)$$

$$\left[\frac{\partial}{\partial t} - A(t, \varrho)\left(1 + \frac{\partial}{\partial \varrho}\right)\right]^p \varphi_{K+1}(t, \varrho) = L_q\left(\varrho, \frac{\partial}{\partial t}, \frac{\partial}{\partial \varrho}\right)\varphi_K(t, \varrho),$$

$$\frac{\partial^j \varphi_{K+1}}{\partial t^j}(0, \varrho) = 0, \quad\quad\quad (j = 0, \ldots, p-1)$$

(Rappelons que ces problèmes s'intègrent par quadratures.) Alors

$$\left(\frac{\partial}{\partial t}\right)^j \varphi_K(t, \varrho) \gg 0, \quad \varphi_K \in \Gamma^{p, (\alpha)}, \quad \sum_K \left(\frac{\partial}{\partial t}\right)^j \varphi_K(t, \varrho) \quad converge \quad (j = 0, \ldots, p),$$

$$\sum_K \varphi_K \in \Gamma^{p, (\alpha)}, \quad pour \quad 0 \leqq t \leqq T' \quad (\text{où } 0 < T' \leqq T).$$

Preuve. Les approximations successives Φ_K qu'emploie la preuve du théorème précédent ont pour expression:

$$\Phi_0 = 0, \quad \Phi_K = \varphi_1 + \cdots + \varphi_K \quad \text{si} \quad K > 0.$$

Or nous avons vu que $\left(\frac{\partial}{\partial t}\right)^j \Phi_K$ $(j = 0, \ldots, p)$ est $\gg 0$, croît avec K et tend vers une série formelle dont chaque coefficient est une fonction bornée de t.

Note. [10] prouve (§ 4) et emploie (§ 5 et § 6) un résultat plus précis: on peut prendre $T' = T$ si $\alpha < p/q$.

§ 4. Etude d'une application non-linéaire: $v \to u$

Cette étude permettra au § 5 de résoudre l'équation non linéaire par approximations successives.

15. Classes de Gevrey

Définitions. Soit une fonction $f: X \to \mathbf{C}$; nous disons que

$$f \in \gamma_2^{n,(\alpha)}(X) \quad \text{si} \quad |D^{n,\infty}f, S_t, \varrho| \in \Gamma^{0,(\alpha)} \quad \text{pour} \quad 0 \leq t \leq |X|;$$

c'est-à-dire si

$$\sup_{\beta,\sigma,t} \frac{1}{[1+|\sigma|]^\alpha} [|D_x^{\beta+\sigma}f, S_t|]^{\frac{1}{1+|\sigma|}} < \infty, \quad \text{pour} \quad |\beta| \leq n, \; \sigma_0 = 0, \; 0 \leq t \leq |X|.$$

De même, soit une fonction $F: X \times Y \to \mathbf{C}$; nous disons que

$$F \in \gamma_2^{n,(\alpha)}(X \times Y) \quad \text{si} \quad |D^{n,\infty}F, S_t \times Y, \varrho, \eta, \nu| \in \Gamma^{0,(\alpha)} \quad \text{pour} \quad 0 \leq t \leq |X|;$$

c'est-à-dire si

$$\sup_{\beta,\sigma,\tau,t} \frac{1}{[1+|\sigma|+|\tau|]^\alpha} [|D_x^{\beta+\sigma}D_y^\tau F, S_t \times Y, \nu|]^{\frac{1}{1+|\sigma|+|\tau|}} < \infty$$

$$\text{pour} \quad |\beta| \leq n, \; \sigma_0 = 0, \; 0 \leq t \leq |X|;$$

ν est fixe et son choix n'altère pas la condition ci-dessus.

En remplaçant $|\ldots|$ par $\|\ldots\|$ dans les définitions précédentes, on obtient celles de $\gamma_{[2]}^{n,\alpha}$.

Propriétés. Les propriétés des quasi-normes formelles (n° 4) et de $\Gamma^{0,(\alpha)}$ (n° 13) ont pour conséquence évidente ceci:

$$D_x^\beta : \gamma_{[2]}^{n,(\alpha)}(X) \to \gamma_{[2]}^{n-\beta_0,(\alpha)}(X), \quad \text{si} \quad \beta_0 \leq n;$$

si $n > l/2$, alors $\gamma_{[2]}^{n,(\alpha)}(X)$ est une *algèbre*; si $n > l/2 + 1$, $f = (f_1, f_2, \ldots): X \to Y$, $f_j \in \gamma_2^{n,(\alpha)}$ et $F \in \gamma_{[2]}^{n,(\alpha)}(X \times Y)$, alors $F(x, f(x)) \in \gamma_{[2]}^{n,(\alpha)}(X)$.

Dans toutes ces propriétés, [2] peut être remplacé par 2.

Note. En particulier, si $n > l/2 + 1$, si $f \in \gamma_{[2]}^{n,(\alpha)}(X)$ et si $1/f$ est borné, alors

$$1/f \in \gamma_{[2]}^{n,(\alpha)}(X).$$

Cette propriété permet, si $n > l/2 + 1$, de diviser chaque opérateur différentiel $\in \gamma_{[2]}^{n,(\alpha)}(X)$ par son premier terme, sans qu'il cesse d'être dans $\gamma_{[2]}^{n,(\alpha)}(X)$; autrement dit: l'hypothèse, faite ci-dessus, que ces opérateurs sont *normaux* devient superflue.

16. Définition d'une application $v \to u$

Etant donnée une fonction $v: X \to \mathbf{C}$ telle que $D_0^j v \mid S_0 = 0$ pour $j < m$, nous définirons une fonction

$$u: X' \to \mathbf{C}, \quad \text{où} \quad X': 0 \leq x_0 < |X'| \, (X' \subset X),$$

par le problème de Cauchy:

$$(16.1) \qquad \begin{cases} a(x, D^{m-1}v, D) u = b(x, D^m v), \\ D_0^j u \mid S_0 = 0 \quad \text{pour} \quad j < m. \end{cases}$$

Nous supposons que $a(x, y, D)$ et $b(x, y)$ ont les propriétés qu'énonce le n° 20: (20.2) ... (20.7); les dérivées de v qu'on substitue aux composantes de y s'annulent donc sur S_0; Y est donc un voisinage de l'origine.

Nous supposons en outre

(16.2) $v \in \gamma_2^{m+n,(\alpha)}(X)$, $\quad D_0^j v \mid S_0 = 0 \quad$ pour $\quad j < m+n$,

(16.3) $D_0^j b(x, 0) \mid S_0 = 0 \quad$ pour $\quad j < n$.

Enfin, χ sera le caractère de régularité de l'ensemble des a_j.

Nous allons voir que, sous ces hypothèses, (16.1) définit une application $v \to u$; nous allons la majorer et majorer son module de continuité; il suffira d'appliquer la formule de composition (4.4) et le lemme 7.

17. Existence et majoration de l'application $v \to u$

D'après l'hypothèse (16.2), il existe des séries formelles en ϱ, fonctions de $t(0 \leq t \leq |X|)$, $\Psi_k(t, \varrho)$ telles que:

$$|D^{m+n-p+k,\infty} v, S_t, \varrho| \ll \Psi_k(t, \varrho) \qquad\qquad (k = 0, \ldots, p);$$

(17.1) $\Psi_k \in \Gamma^{0,(\alpha)}$; $\Psi_0 \in \Gamma^{p,(\alpha)}$; $\left(\dfrac{\partial}{\partial t}\right)^j \Psi_0(t, \varrho) \geqslant 0$ pour $j \leq p$, $0 \leq t \leq |X|$;

Notons $\Psi_0(0, 0) = 0$.

(17.2) $\psi(t) = \Psi_0(t, 0)$, ce qui implique $\psi(0) = 0$.

La formule de composition (4.4) et les hypothèses (20.2) ... (20.7) permettent de construire, à partir de

$$\| D^{m_j+n-j,\infty} a_{j+1}(x, y, D), S_t \times Y, \varrho, \eta, v \|$$
$$\| D^{n,\infty} a(x, y, D), S_t \times Y, \varrho, \eta, v \|$$
$$| D^{n,\infty} b(x, y), S_t \times Y, \varrho, \eta, v |$$

des séries formelles en deux variables (ϱ, θ), à coefficients fonctions de τ:

$$C[\tau, \varrho, \theta] \in \Gamma^{p,(\alpha)}, \quad C_j[\tau, \varrho, \theta] \in \Gamma^{0,(\alpha)}, \quad B[\tau, \varrho, \theta] \in \Gamma^{0,(\alpha)}$$

telles que[5]:

$$\| D^{m_j+n-j,\infty} a_{j+1}(x, D^{m-m_j-p+j} v, D), S_t, \varrho \| \ll C(\Psi_0),$$
$$\| D^{n-p+k,\infty} a(x, D^{m-1} v, D), S_t, \varrho \| \ll C_k(\Psi_{k-1}) \qquad (k = 1, \ldots, p),$$
$$| D^{n,\infty} b(x, D^m v), S_t, \varrho | \ll B(\Psi_q),$$
$$\ll B\left(\Psi_{p-1} + \frac{\partial}{\partial \varrho} \Psi_{p-1}\right) \quad \text{si} \quad q = p;$$

$C(\Psi_0)$, $C_k(\Psi_{k-1})$, $B(\Psi_q)$ sont des séries formelles en ϱ, fonctions de t; ces séries $\in \Gamma^{0,(\alpha)}$. Leur définition exige $D^m v \in Y$; pour réaliser cette condition, il suffit (Sobolev) de prendre $\psi(t)$ suffisamment petit; donc de prendre:

$$0 \leq t \leq T(\psi)$$

---

[5] Rappelons que $C(\Psi)$ désigne la série formelle en ϱ, fonction de t:

$$C[\Psi(t, 0), \varrho, \Psi(t, \varrho) - \Psi(t, 0)].$$

où $T(\psi)$ est une fonctionelle de ψ, dont la définition est évidente et qui vérifie:

$$0 < T(\psi) \leqq |X|.$$

Nous choisissons $C[\tau, \varrho, \theta]$ tel que

$$\frac{\partial^j}{\partial \tau^j} C[\tau, \varrho, \theta] \gg 0 \quad \text{pour} \quad j \leqq p.$$

Comme au n° 7, nous considérons la fonction de t

(17.3)
$$A_0(\psi) = c(l, m, \chi, C[\psi(t), 0, 0]),$$

la série formelle en ϱ, fonction de t, définie pour $0 \leqq t \leqq T(\psi)$:

(17.4)
$$A(\psi, \Psi_0) = A_0(\psi) \{ C[\psi(t), \varrho, \Psi_0(t, \varrho) - \psi(t)] - C[\psi(t), 0, 0] \}$$

enfin la série formelle en ϱ, fonction de t, $\Phi(t, \varrho)$ que définit le problème de Cauchy formel

(17.5)
$$\begin{cases} \left[\frac{\partial}{\partial t} - A(\psi, \Psi_0) \left(1 + \frac{\partial}{\partial \varrho} \right) \right]^p \Phi(t, \varrho) = B(\Psi_q) \quad \text{si} \quad q < p, \\ \qquad\qquad\qquad\qquad\qquad = B\left(\left(1 + \frac{\partial}{\partial \varrho} \right) \Psi_{p-1} \right) \quad \text{si} \quad q = p; \\ \frac{\partial^j \Phi}{\partial t^j}(0, \varrho) = 0 \quad \text{pour} \quad j < p. \end{cases}$$

$A, B \in \Gamma^{0,(\alpha)}$; vu le théorème du n° 14 et l'unicité de la solution du problème (17.5),

$$\Phi \in \Gamma^{p,(\alpha)};$$

Φ est défini pour $0 \leqq t \leqq T(\psi)$.

Le lemme 7 montre ceci:

La solution $u(x)$ du problème de Cauchy (16.1) existe et est unique sur la bande

$$X_\psi: 0 \leqq x_0 \leqq T(\psi);$$
$$u \in \gamma_2^{m+n,(\alpha)}(X_\psi);$$

plus précisément on a

(17.6)
$$|D^{m+n-p+k,\infty} u, S_t, \varrho| \ll \Phi_k(t, \varrho) \qquad\qquad (k = 0, \ldots, p)$$

où

(17.7)
$$\begin{cases} \Phi_0 = A_0(\psi) \Phi \\ \Phi_k = c_k'' A_0(\psi) [1 + C_k(\psi_{k-1})]^k \left(1 + \frac{\partial}{\partial t} + \frac{\partial}{\partial \varrho} \right)^k \Phi \qquad (k = 1, \ldots, p). \end{cases}$$

Notons que

$$D_0^j u | S_0 = 0 \quad \text{pour} \quad j < m + n;$$

en effet

$$D^j b(x, D^{m-p+q} v(x)) | S_0 = 0 \quad \text{pour} \quad j < n.$$

Ces résultats vont servir à prouver le lemme que voici:

18. Un sous-ensemble de $\gamma_2^{m+n,(\alpha)}(X')$ que l'application $v \to u$ applique en lui-même

Lemme 18. *Il existe une bande*

$$X' : 0 \leq x_0 \leq |X'|$$

et des séries formelles en ϱ, fonctions de $t(0 \leq t \leq |X'|)$

$$\Phi_k(t, \varrho) \in \Gamma^{0,\alpha} \quad (k = 0, \ldots, p)$$

telles que si

$$|D^{m+n-p+k,\infty}v, S_t, \varrho| \ll \Phi_k(t, \varrho), \, D_0^j v|S_0 = 0$$

sous les hypothèses

$$0 \leq t \leq |X'|, k \leq p, j < m+n,$$

alors on a, sous ces mêmes hypothèses :

$$|D^{m+n-p+k,\infty}u, S_t, \varrho| \ll \Phi_k(t, \varrho), \, D_0^j u|S_0 = 0.$$

Preuve. Il suffit de choisir au n° 17 les $\Psi_k(k = 0, \ldots, p)$ tels que

$$\Psi_k(t, \varrho) = \Phi_k(t, \varrho),$$

c'est-à-dire, vu (18.7), tels que

$$(18.1) \quad \begin{cases} \Psi_0 = A_0(\psi) \, \Phi \\ \Psi_k = c_k'' A_0(\psi) \, [1 + C_k(\Psi_{k-1})]^k \left(1 + \dfrac{\partial}{\partial t} + \dfrac{\partial}{\partial \varrho}\right)^k \Phi \, (k = 1, \ldots, p). \end{cases}$$

Notons

$$\varphi(t) = \Phi(t, 0) ;$$

la définition (17.2) de $\psi(t)$ s'écrit donc :

$$(18.2) \quad \begin{cases} \psi = A_0(\psi) \, \varphi, \\ \text{où } A_0(\psi) \text{ est une fonction de } \psi \text{ que définit (17.3); elle vérifie} \\ A_0(0) > 0, \dfrac{d^j A_0(\psi)}{d\psi^j} \geq 0 \quad \text{pour} \quad \psi \geq 0, j = 0, \ldots, p. \end{cases}$$

Nous montrerons (fin de ce n° 18) que, pour ψ petit, (18.2) équivaut à une relation

$$(18.3) \quad \begin{cases} \psi = f(\varphi) \quad (\varphi \text{ petit}), \\ \text{où } f \text{ est une fonction vérifiant} \\ f(0) = 0, \dfrac{d^j f}{d\varphi^j} \geq 0 \quad \text{pour} \quad \varphi \quad \text{petit} \geq 0, j = 0, \ldots, p. \end{cases}$$

Les relations (18.3) et (18.1) permettent d'exprimer ψ et $\Psi_k(k = 0, \ldots, p)$ en fonction des dérivées de Φ d'ordres $\leq k$; on peut donc éliminer ψ, Ψ_0 et Ψ_q(ou Ψ_{p-1}) de (17.5), qui s'écrit avec les notations du n° 10 :

$$(18.4) \quad \begin{cases} \left[\dfrac{\partial}{\partial t} - F_0(\Phi) \left(1 + \dfrac{\partial}{\partial \varrho}\right)\right]^p \Phi = F_q(D^q\Phi) \\ \dfrac{\partial^j \Phi}{\partial t^j} (0, \varrho) = 0 \quad \text{pour} \quad j < p ; \end{cases}$$

(18.4) est un problème de Cauchy formel d'inconnue Φ; $F_0[\tau, \varrho, \theta]$ et $F_q[\tau, \varrho, \theta]$ sont des séries formelles en (ϱ, θ), fonctions de τ, vérifiant:

$$F_0 \in \Gamma^{p,(\alpha)}, \quad F_q \in \Gamma^{0,(\alpha)},$$

$$F_0[\tau, 0, 0] = 0, \frac{\partial^j F_0[\tau, \varrho, \theta]}{\partial \tau^j} \gg 0 \quad \text{pour} \quad j \leqq p, F_q[\tau, \varrho, \theta] \gg 0;$$

si $q = p$, alors $\dfrac{\partial^p \Phi}{\partial t^p}$ ne figure pas dans $F_q(D^p \Phi)$.

Pour satisfaire (17.1), il suffit qu'on ait

(18.5) $$\Phi \in \Gamma^{p,(\alpha)}, \frac{\partial^j \Phi(t, \varrho)}{\partial t^j} \gg 0 \quad \text{pour} \quad j \leqq p.$$

D'après le théorème d'existence du n° 14, le problème de Cauchy formel (18.4) possède une solution Φ vérifiant (18.5). La preuve du lemme est achevée.

Preuve de (18.3). Faisons croître ψ de 0 à un nombre $\tilde{\psi}$ suffisamment petit pour que $\psi/A_0(\psi)$ soit croissant, c'est-à-dire pour que

(18.6) $$\frac{\psi}{A_0} \frac{dA_0}{d\psi} < 1.$$

Alors φ croît de 0 à $\tilde{\varphi}$ et la relation $\psi = A_0(\psi)\,\varphi$ équivaut à une relation

$$\psi = f(\varphi) \quad (0 \leqq \varphi \leqq \tilde{\varphi}),$$

où f est une fonction croissante telle que $f(0) = 0$, $f(\varphi) \geqq 0$. Supposons prouvé que

$$f, \dots, \frac{d^{j-1} f}{d\varphi^{j-1}} \geqq 0 \quad (j \leqq p).$$

Alors l'application de $\dfrac{d^j}{d\varphi^j}$ à la relation $f(\varphi) = A_0(f(\varphi))\,\varphi$ donne

$$\left[1 - \varphi \frac{dA_0}{d\psi}\right] \frac{d^j f}{d\varphi^j} \geqq 0,$$

c'est-à-dire

$$\left[1 - \frac{\psi}{A_0} \frac{dA_0}{d\psi}\right] \frac{d^j f}{d\varphi^j} \geqq 0$$

et, vu (18.6):

$$\frac{d^j f}{d\varphi^j} \geqq 0.$$

Voici prouvé (18.3).

19. Module de continuité de l'application $v \to u$

Notons $\Gamma^{(\alpha)}$ l'ensemble des séries formelles en ϱ, indépendantes de t, appartenant à $\Gamma^{0,(\alpha)}$.

Lemme 19. *Supposons qu'on ait sur X, pour* $h = 0, 1$:

(19.1)
$$\begin{cases} a(x, D^{m-1}v_h, D)\, u_h = b(x, D^m v_h) \\ D_0^j u_h | S_0 = 0 \quad \text{pour} \quad j < m \end{cases}$$

$$|D^{m+n,\infty} v_h, S_t, \varrho| \ll \Theta(\varrho), \quad |D^{m+n,\infty} u_h, S_t, \varrho| \ll \Theta(\varrho), \quad \Theta \in \Gamma^{(\alpha)},$$

$$D^j v_h | S_0 = 0 \quad \text{pour} \quad j < m+n, \quad \text{donc} \quad D^j u_h | S_0 = 0 \quad \text{pour} \quad j < m+n.$$

Il existe alors des séries formelles en ϱ, *appartenant à* $\Gamma^{(\alpha)}$, *dépendant de* a, b, Θ, *mais indépendantes de* u_h *et* v_h,

$$A(\varrho), B(\varrho)$$

vérifiant :

$$A(\varrho) \gg 0, \quad A(0) = 0, \quad B(\varrho) \gg 0,$$

telles qu'on ait, pour $k = 0, \ldots, p$:

$$|D^{m+n-p+k,\infty}(u_1 - u_0), S_t, \varrho| \ll C(\varrho) \left(1 + \frac{\partial}{\partial t} + \frac{\partial}{\partial \varrho}\right)^k \varphi(t, \varrho),$$

si l'on a, pour $k = 0, \ldots, p$:

$$|D^{m+n-k,\infty}(v_1 - v_0), S_t, \varrho| \ll C(\varrho) \left(1 + \frac{\partial}{\partial t} + \frac{\partial}{\partial \varrho}\right)^k \varphi(t, \varrho)$$

si $C(\varrho) \gg 0$, $C \in \Gamma^{(\alpha)}$ *et si* $\varphi(t, \varrho)$ *est la solution du problème de Cauchy formel*

$$\begin{cases} \left[\frac{\partial}{\partial t} - A(\varrho)\left(1 + \frac{\partial}{\partial \varrho}\right)\right]^p \varphi(t, \varrho) = B(\varrho)\, C(\varrho) \left(1 + \frac{\partial}{\partial t} + \frac{\partial}{\partial \varrho}\right)^q \psi(t, \varrho) \\ \qquad\qquad\qquad\qquad\qquad\qquad\qquad\qquad\qquad\qquad\qquad\qquad\qquad\qquad\quad \text{quand } q < p, \\ \qquad\qquad\qquad\qquad = B(\varrho)\, C(\varrho) \left(1 + \frac{\partial}{\partial t} + \frac{\partial}{\partial \varrho}\right)^{p-1} \frac{\partial \psi}{\partial \varrho} \\ \qquad\qquad\qquad\qquad\qquad\qquad\qquad\qquad\qquad\qquad\qquad\qquad\qquad\qquad\quad \text{quand } q = p, \\ \frac{\partial^j \varphi}{\partial t^j}(0, \varrho) = 0 \quad \text{pour} \quad j < p. \end{cases}$$

Preuve. Nous avons

$$a(x, D^{m-1}v_0, D)(u_0 - u_1) = b(x, D^m v_0) - b(x, D^m v_1) - $$
$$ - [a(x, D^{m-1}v_0, D) - a(x, D^{m-1}v_1, D)]\, u_1 ;$$

autrement dit, en notant

$$v_h = (1-h)\, v_0 + h v_1, \ h \text{ variant maintenant de } 0 \text{ à } 1,$$

nous avons

(19.2)
$$a(x, D^{m-1}v_0, D)(u_0 - u_1) = \sum_\beta D^\beta(v_0 - v_1) \cdot \int_0^1 b_\beta(x, D^m v_h)\, dh$$

$$ - \sum_\beta D^\beta(v_0 - v_1) \cdot \int_0^1 a_\beta(x, D^{m-1}v_h, D)\, u_1\, dh$$

où

$$|\beta| \leqq m - p + q, \quad D^\beta + D_0^m \quad \text{si} \quad \mu = q,$$

$$b_\beta(x, y) = \frac{\partial b(x, y)}{\partial y_\beta} \quad \text{et} \quad a_\beta(x, y, \xi) = \frac{\partial a(x, y, \xi)}{\partial y_\beta}.$$

Or, par hypothèse:

$$|D^{m+n,\infty} v_h, S_t, \varrho| \ll \Theta(\varrho) \quad \text{pour} \quad 0 \leqq h \leqq 1.$$

Donc, vu la formule de composition (4.4) on peut construire, en fonction de Θ et des normes formelles de a et b, une série formelle $B(\varrho)$, indépendante de v_h et u_h, telle que

$$|D^{n,\infty} b_\beta(x, D^m v_h), S_t, \varrho| + \|D^{n,\infty} a_\beta(x, D^{m-1} v_h, D) u_1, S_t, \varrho\| \ll B(\varrho);$$

vu les propriétés des classes de Gevrey formelles, on peut choisir

$$B \in \Gamma^{(\alpha)}.$$

Donc (19.2) donne, vu la formule du produit (4.1) et la formule de la dérivée (4.2)

$$|D^{n,\infty} a(x, D^{m-p+q} v_0, D)(u_0 - u_1), S_t, \varrho| \ll B(\varrho)|D^{m+n-p+q}(v_0 - v_1), S_t, \varrho|$$

$$\text{quand } q < p,$$

$$\ll B(\varrho)\left(1 + \frac{\partial}{\partial \varrho}\right)|D^{m+n-1}(v_0 - v_1), S_t, \varrho| \quad \text{quand } q = p.$$

Il suffit d'appliquer le lemme 7 à cette inégalité pour obtenir le lemme 19.

§ 5. L'équation quasi-linéaire

20. Enoncé des résultats

Donnons-nous sur une bande de $\mathbf{R}^{l+1}$

$$X : 0 \leqq x_0 < |X|, \quad \text{de bord} \quad S_0 : x_0 = 0,$$

le problème de Cauchy

$$(20.1) \qquad \begin{cases} a(x, D^{m-1} u, D) u = b(x, D^m u) \\ D_0^j u | S_0 \quad \text{donné} \ \in \gamma^{(\alpha)}(S_0) \quad (j < m); \end{cases}$$

son inconnue est la fonction numérique complexe $u(x)$.

Nous faisons les hypothèses suivantes:

$$(20.2) \qquad a(x, y, D) \in \gamma_{[2]}^{n,(\alpha)}(X \times Y) \quad \text{et} \quad b(x, y) \in \gamma_?^{n,(\alpha)}(X \times Y)$$

sont respectivement un opérateur différentiel d'ordre m et une fonction, donnés sur X, dépendant d'un paramètre $y \in Y$; Y est un ouvert de l'espace vectoriel complexe de dimension égale au nombre des dérivées de u d'ordres $\leqq m$; Y contient l'adhérence des valeurs prises par les données de Cauchy $D_0^j u | S_0$; quand on substitue à y, dans $a(x, y, D)$ et $b(x, y)$, les dérivées d'une fonction $v(x)$, on obtient $a(x, D^{m-1} v, D)$ qui ne dépend que des dérivées de v d'ordres $\leqq m - 1$, et $b(x, D^m v)$, que nous supposons indépendant de $D_0^m v$; nous supposons

$$(20.3) \qquad a(x, D^{m-1} v, D) = a_1 \ \dots \ a_{j+1}(x, D^{m-m_j-p+j} v, D) \dots a_p,$$

où

(20.4) $a_{j+1}(x, y, D) \in \gamma_{[2]}^{m_j + n - j, (\alpha)}(X \times Y)$

est un opérateur *régulièrement hyperbolique* sur $X \times Y$, $\forall j$; on a noté:

(20.5) $m_j = \text{ordre}(a_1) + \cdots + \text{ordre}(a_j), m_p = m, m_0 = 0$.

Soit q le plus petit entier tel que

(20.6) $\begin{cases} 0 \leqq q \leqq p \\ a(x, D^{m-1}v, D) = a(x, D^{m-p+q}v, D) \\ b(x, D^m v) = b(x, D^{m-p+q}v) \, ; \end{cases}$

le sens de cette dernière relation est, bien entendu, le suivant: $b(x, D^m v)$ ne dépend que des dérivées de v d'ordres $\leqq m - p + q$.

Nous supposons enfin

(20.7) $1 \leqq \alpha \leqq \dfrac{p}{q}, \ \dfrac{l}{2} + p < n$.

Voici les théorèmes que nous allons prouver:

Théorèmes d'existence et d'unicité

Il existe une bande

$$X' : 0 \leqq x_0 < |X'| \qquad\qquad (X' \subset X)$$

sur laquelle le problème de Cauchy (20.1) *possède une solution*

$$u \in \gamma_2^{m+n,(\alpha)}(X').$$

Sur aucune bande plus petite

$$X'' : 0 \leqq x_0 < |X''| \qquad\qquad (X'' \subset X)$$

il ne possède de solution $\in \gamma_2^{m+n,(\alpha)}(X'')$ *autre que u.*

Note. Si $q = 0$, on peut prendre $\alpha = \infty$, c'est-à-dire employer comme dans [3] des espaces de Sobolev au lieu de classes de Gevrey: on est dans le cas strictement hyperbolique: voir P. DIONNE [3].

Note. Un exemple de GIORGI [6] montre que ces théorèmes d'existence et d'unicité sont faux si $\dfrac{p}{q} < \alpha$.

Théorème local d'unicité (domaine d'influence)

Supposons

$$1 \leqq \alpha < p/q.$$

Soient deux fonctions $u_h \in \gamma_2^{m+n,(\alpha)}(X')$ $(h = 0, 1)$ *qui, sur un domaine D' de X', soient solution du problème de Cauchy* (20.1). *Supposons que D' possède la propriété suivante, relativement au cône caractéristique de l'opérateur*

$$a(x, D^{m-1}u_0, D):$$

l'émission rétrograde[6] dans X' de tout point de D' appartient à $D' \setminus S_0$. Alors

$$u_0 = u_1 \quad \text{sur} \quad D'.$$

Note. Il suffirait de supposer $u_h \in \gamma_2^{m+n,(\alpha)}(D')$, moyennant diverses complications, dont la première serait de définir $\gamma_2^{m+n,(\alpha)}(D')$.

Prouvons d'abord le théorème d'existence.

21. Réduction à des données de Cauchy nulles

Il est aisé de déduire de (20.1) les valeurs que doit avoir $D_0^j u | S_0$ pour $j = m, \ldots, m+n-1$; ces valeurs $\in \gamma_2^{(\alpha)}(S_0)$. Construisons sur X une fonction $w \in \gamma_2^{m+n,(\alpha)}(X)$ telle que $D_0^j w | S_0$ $(j \leq m+n-1)$ ait ces valeurs; prenons pour nouvelle inconnue $u - w$.

Nous voici ramenés au cas suivant: les données de Cauchy sont nulles, c'est-à-dire:

$$(21.1) \qquad\qquad D^{m-1} u | S_0 = 0 ;$$

de plus le problème (20.1) implique

$$D^{m+n-1} u | S_0 = 0 .$$

D'où, en appliquant D^{n-1} à $au = b$:

$$D^{n-1} b(x, D^m u) | S_0 = 0 ;$$

c'est-à-dire:

$$(21.2) \qquad\qquad D_0^j b(x, 0) | S_0 = 0 \quad \text{pour} \quad j < n .$$

Voici donc réalisées les hypothèses (16.3) qu'emploient les lemmes 18 et 19.

22. Définition d'approximations successives

Notons u_K $(K = 0, 1, \ldots)$ ces approximations successives de u. Nous choisissons

$$u_0 = 0 ;$$

nous définissons u_{K+1} à partir de u_K par le problème de Cauchy

$$(22.1) \qquad \begin{cases} a(x, D^{m-p+q} u_K, D) u_{K+1} = b(x, D^{m-p+q} u_K), \\ D_0^j u_{K+1} | S_0 = 0 \quad \text{pour} \quad j < m . \end{cases}$$

23. Majoration des approximations successives

Le lemme 18, où l'on remplace les $\Phi_k(t, \varrho)$ par une série formelle, $\Theta(\varrho)$, qui les majore et est indépendante de t, a pour conséquence immédiate ceci: il existe une série formelle en ϱ, indépendante de K, $\Theta \in \Gamma^{(\alpha)}$, et une bande indépendante de K:

$$X' : 0 \leq x_0 \leq |X'|$$

$$\cdots$$

---

[6] L'émission rétrograde d'un point x de X' est la réunion des arcs de X' d'extrémité x, à tangente dans le cône caractéristique (cône convexe) de l'opérateur linéaire $a(x, D^{m-p+q} u_0, D)$; ces arcs sont orientés dans le sens où x_0 croît.

sur laquelle tous les $u_K(x)$ sont définis et vérifient

$$(23.1) \qquad |D^{m+n,\infty} u_K, S_t, \varrho| \ll \Theta(\varrho).$$

24. Convergence des approximations successives

Le lemme 19 a pour conséquence évidente ceci : il existe des séries formelles en ϱ, indépendantes de K, $A(\varrho)$, $B(\varrho)$ appartenant à $\Gamma^{(\alpha)}$ et vérifiant

$$A(\varrho) \gg 0, \quad A(0) = 0, \quad B(\varrho) \gg 0$$

telles qu'on ait pour $0 \leq t \leq |X'|$ et pour $k = 0, \ldots, p$:

$$(24.1) \quad |D^{m+n-p+k,\infty}(u_{K+1} - u_K), S_t, \varrho| \ll C(\varrho) \left(1 + \frac{\partial}{\partial t} + \frac{\partial}{\partial \varrho}\right)^k \varphi_{K+1}(t, \varrho),$$

quand on choisit $C \in \Gamma^{(\alpha)}$ tel qu'on ait (24.1) pour $K = 0$ et quand φ_{K+1} est défini, pour $K > 0$, par le problème de Cauchy formel

$$(24.2) \quad \begin{cases} \left[\dfrac{\partial}{\partial t} - A(\varrho)\left(1 + \dfrac{\partial}{\partial \varrho}\right)\right]^p \varphi_{K+1}(t, \varrho) = B(\varrho)C(\varrho)\left(1 + \dfrac{\partial}{\partial t} + \dfrac{\partial}{\partial \varrho}\right)^q \varphi_K(t, \varrho) \\ \qquad\qquad\qquad\qquad\qquad\qquad\qquad\qquad\qquad\qquad\qquad\text{quand } q < p, \\ \qquad = B(\varrho)C(\varrho)\left(1 + \dfrac{\partial}{\partial t} + \dfrac{\partial}{\partial \varrho}\right)^{p-1}\dfrac{\partial \varphi_K}{\partial \varrho} \qquad\quad \text{quand } q = p, \\ \dfrac{\partial^j \varphi}{\partial t^j}(0, \varrho) = 0 \quad \text{ pour } \ j < p. \end{cases}$$

D'après le théorème de convergence du n° 14, la série

$$\sum_K \left(\frac{\partial}{\partial t}\right)^j \varphi_K(t, \varrho) \qquad\qquad\qquad (j < p)$$

converge pour $0 \leq t < |X''|$; donc

$$\lim_{K \to \infty} \left(\frac{\partial}{\partial t}\right)^j \varphi_K(t, \varrho) = 0 \quad \text{ pour } \quad 0 \leq t < |X''|.$$

Par suite u_K converge vers une limite u sur la bande

$$X'' : 0 \leq x_0 < |X''|.$$

Plus précisément, vu (24.1), $D^{m+n} u_K$ converge vers $D^{m+n} u$ sur X'' ;

$$u \in \gamma_2^{m+n,(\alpha)}(X'').$$

Les théorèmes de Sobolev permettent de préciser que $D^m u_K$ converge uniformément ; vu (22.1), u est donc solution du problème (20.1).

Voici prouvé le théorème d'existence qu'énonce le n° 20.

25. Preuve de premier théorème d'unicité (énoncé n° 20)

Supposons que

$$(25.1) \qquad\qquad u_h \in \gamma_2^{m+n,(\alpha)}(X) \qquad\qquad (h = 0, 1)$$

soient deux solutions distinctes du même problème de Cauchy (20.1). Notons

$$X^+ : 0 \leqq x_0 < |X^*|$$

la plus grande bande semi-ouverte où elles sont identiques. En remplaçant X par $X - X^*$, nous obtenons deux solutions u_1, u_2 d'un même problème de Cauchy (20.1), qui sont distinctes sur toute bande

$$X' : 0 \leqq x_0 < |X'| \quad (X' \subset X).$$

Montrons que c'est incompatible avec l'hypothèse (25.1).

Réalisons les conditions (21.1) et (21.2); appliquons le lemme 19, en y faisant $u_h = v_h$; nous obtenons ceci: l'inégalité

$$(25.2)_K \quad |D^{m+n-p+k,\infty}(u_1 - u_0), S_t, \varrho| \ll C(\varrho) \left(1 + \frac{\partial}{\partial t} + \frac{\partial}{\partial \varrho} \right)^k \varphi_K(t, \varrho), \quad (k = 0, \ldots, p)$$

implique l'inégalité $(25.2)_{K+1}$, si φ_{K+1} est défini par le problème de Cauchy formel:

$$\begin{cases} \left[\frac{\partial}{\partial t} - A(\varrho) \left(1 + \frac{\partial}{\partial \varrho} \right) \right]^p \varphi_{K+1}(t, \varrho) = B(\varrho) C(\varrho) \left(1 + \frac{\partial}{\partial t} + \frac{\partial}{\partial \varrho} \right)^q \varphi_K(t, \varrho) \\ \hspace{9cm} \text{quand } q < p, \\ \qquad = B(\varrho) C(\varrho) \left(1 + \frac{\partial}{\partial t} + \frac{\partial}{\partial \varrho} \right)^{q-1} \frac{\partial \varphi_K}{\partial \varrho} \hspace{1.5cm} \text{quand } q = p, \\ \frac{\partial^j \varphi_{K+1}}{\partial t^j}(0, \varrho) = 0 \quad \text{pour} \quad j < p; \end{cases}$$

ce problème est indépendant de K; $A(0) = 0$.

Choisissons, ce qui est possible par hypothèse, φ_1 tel que $(25.2)_1$ soit vrai et que

$$\varphi_1 \in \Gamma^{p,(\alpha)}, \quad \frac{\partial^j \varphi_1}{\partial t^j}(t, \varrho) \geqslant 0 \quad \text{pour} \quad j \leqq p.$$

D'après le théorème de convergence du n° 14,

$$\sum_K \left(\frac{\partial}{\partial t} \right)^j \varphi_K(t, \varrho) \hspace{5cm} (j \leqq p)$$

converge sur un intervalle $0 \leqq t < |X'|$; donc

$$\lim_{K \to \infty} \left(\frac{\partial}{\partial t} \right)^j \varphi_K(t, \varrho) = 0 \quad \text{pour} \quad 0 \leqq t < |X'|, \quad j \leqq p;$$

donc

$$u_0 = u_1 \quad \text{sur la bande} \quad X' : 0 \leqq x_0 < |X'|;$$

cette conclusion contredit les hypothèses.

Voici prouvé le premier théorème d'unicité. Son seul intérêt est de ne pas exiger $\alpha < p/q$, ce que va supposer le théorème d'unicité locale, dont les conclusions sont plus fortes.

26. Preuve du théorème d'unicité locale (énoncé n° 20)

Soient deux fonctions

$$u_h \in \gamma_2^{m+n,(\alpha)}(X') \qquad (h = 0, 1),$$

solutions sur D' du problème de Cauchy (20.1). Sur D', nous avons donc (19.1), avec $v_h = u_h$. Donc $u_0 - u_1$ vérifie une équations hyperbolique non stricte, linéaire et homogène; ses coefficients vérifient les hypothèses qu'énonce le n° 23 de [10]; $D^{m-1}(u_0 - u_1)|S_0 = 0$. D'après le théorème d'unicité qu'énonce le n° 24 de [10] et la note qui suit ce théorème, nous avons donc

$$u_0 = u_1 \quad \text{sur} \quad D';$$

le théorème est prouvé.

§ 6. Systèmes quasi-linéaires diagonaux

L'extension des théorèmes du n° 20 aux systèmes quasi-linéaires diagonaux est aisée; nous ne donnerons pas le détail des preuves; mais nous expliciterons les résultats, que A. Lichnérowicz [12] et Mme. Y. Choquet-Bruhat [2] appliquent à la magnéto-hydrodynamique relativiste.

27. Énoncé des résultats

Donnons-nous sur une bande de $\mathbf{R}^{l+1}$

$$X : 0 \leqq x_0 < |X|, \quad \text{de bord} \quad S_0 : x_0 = 0$$

le problème de Cauchy[7]:

$$(27.1) \qquad \begin{cases} a^\nu(x, D^{m^\mu - n^\nu - 1} u^\mu, D) u^\nu = b^\nu(x, D^{m^\mu - n^\nu} u^\mu), \\ D_0^j u^\nu | S_0 \quad \text{donné} \quad \in \gamma^{(\alpha)}(S_0), \quad (j < m^\nu - n^\nu), \end{cases}$$

où μ, ν valent $1, \ldots, N$; les inconnues sont les N fonctions numériques complexes $u^\nu(x)$.

Nous faisons les hypothèses suivantes:

$$(27.2) \qquad a^\nu(x, y, D) \in \gamma_{[2]}^{n^\nu,(\alpha)}(X \times Y) \quad \text{et} \quad b^\nu(x, y) \in \gamma_2^{n^\nu,(\alpha)}(X \times Y)$$

sont respectivement N opérateurs différentiels d'ordres $m^\nu - n^\nu$ et N fonctions, donnés sur X, dépendant d'un paramètre $y \in Y$; Y est un ouvert de l'espace vectoriel complexe de dimension égal au nombre des dérivées des u^ν ($\nu = 1, \ldots, N$) d'ordres $\leqq \sup_\mu m^\mu - n^\nu$; Y contient l'adhérence des valeurs prises par les données de Cauchy $D_0^j u^\nu | S_0$; quand on substitue dans $a^\nu(x, y, D)$ et $b^\nu(x, y)$ à y les dérivées de fonctions $v^\mu(x)$, on obtient

$$a^\nu(x, D^{m^\mu - n^\nu - 1} v^\mu, D) \quad \text{et} \quad b^\nu(x, D^{m^\mu - n^\nu} v^\mu),$$

que nous supposons indépendant des $D_0^{m^\mu - n^\nu} v^\mu$; nous supposons

$$(27.3) \qquad a^\nu(x, D^{m^\mu - n^\nu - 1} v^\mu, D) = a_1^\nu \ldots a_{j+1}^\nu(x, D^{m^\mu - m_j^\nu - p^\mu + j} v^\mu, D) \ldots a_{p^\nu}^\nu$$

---

[7] Bien entendu, si $m^\mu < n^\nu$, alors ni u^μ ni aucune de ses dérivées ne figure dans $b^\nu(x, D^{m^\mu - n^\nu} u^\mu)$.

où

(27.4)
$$a_{j+1}^{\nu}(x, y, D) \in \gamma_{[2]}^{m_j^{\nu} - j, (\alpha)}(X \times Y)$$

est un opérateur *régulièrement hyperbolique* sur $X \times Y$; on a noté

(27.5) $m_j^{\nu} = n^{\nu} + \text{ordre}(a_1^{\nu}) + \cdots + \text{ordre}(a_j^{\nu}), \quad m_{p^{\nu}}^{\nu} = m^{\nu}, \quad m_0^{\nu} = n^{\nu}.$

Soient q^{μ} *les plus petits entiers* tels que

$$0 \le q^{\mu} \le p^{\mu}$$

(27.6)
$$\begin{cases} a^{\nu}(x, D^{m^{\mu} - n^{\nu} - 1} v^{\mu}, D) = a^{\nu}(x, D^{m^{\mu} - n^{\nu} - p^{\mu} + q^{\mu}} v^{\mu}, D) \\ b^{\nu}(x, D^{m^{\mu} - n^{\nu}} v^{\mu}) = b^{\nu}(x, D^{m^{\mu} - n^{\nu} - p^{\mu} + q^{\mu}} v^{\mu}). \end{cases}$$

Nous supposons enfin

(27.7)
$$1 \le \alpha \le \frac{p^{\nu}}{q^{\nu}} \; ; \; \frac{l}{2} + p^{\nu} < n^{\nu}, \forall \nu.$$

Théorèmes d'existence et d'unicité

Il existe une bande

$$X': 0 \le x_0 < |X'| \quad (X' \subset X)$$

sur laquelle le problème de Cauchy (27.1) *possède une solution*

$$u^{\nu} \in \gamma_2^{m^{\nu}, (\alpha)}(X').$$

Sur aucune bande plus petite X'' il ne possède de solution $\in \gamma_2^{m^{\nu}, (\alpha)}(X'')$, autre que u^{ν}.

Note. Si $q^{\nu} = 0, \forall \nu,$ on peut prendre $\alpha = \infty$, c'est-à-dire employer des espaces de Sobolev au lieu de classes de Gevrey; on est dans le cas strictement hyperbolique.

Théorème local d'unicité (domaine d'influence)

Supposons

$$1 \le \alpha < p_{\nu}/q_{\nu}, \forall \nu.$$

Soient, sur un domaine D' de X', deux solutions

$$u_h^{\nu} \in \gamma_2^{m^{\nu}, (\alpha)}(X') \quad (h = 0, 1)$$

du problème de Cauchy (27.1). *Supposons que D' possède la propriété suivante, relativement au cône caractéristique de l'opérateur $\prod_{\nu} a^{\nu}(x, D^{m^{\mu} - n^{\nu} - 1} u_0^{\mu}, D)$:*

l'émission rétrograde dans X' de tout point de D' appartient à $D' \cup S_0$. Alors

$$u_0^{\nu} = u_1^{\nu} \quad \text{sur} \quad D'.$$

28. Preuve sommaire

On opère, comme au § 5, par approximations successives, après s'être ramené au cas:

$$D_0^j u^{\nu}|S_0 = 0 \quad \text{pour} \quad j < m^{\nu} - n^{\nu}; \quad D_0^j b^{\nu}(x, 0)|S_0 = 0 \quad \text{pour} \quad j < n^{\nu}.$$

Il faut d'abord avoir étudié, comme au §4, l'application $v \to u$ que définit le problème de Cauchy

(28.1)
$$\begin{cases} a^\nu(x, D^{m^\mu - n^\nu - 1} v^\mu, D)\, u^\nu = b(x, D^{m^\mu - n^\nu} v^\mu) \\ D_0^j u^\nu | S_0 = 0 \quad \text{pour} \quad j < m^\nu - n^\nu. \end{cases}$$

Majoration de l'application $v \to u$. — Supposons, comme au n° 17,

$$|D^{m^\mu - p^\mu + k, \infty} v^\mu, S_t, \varrho| \ll \Psi_k^\mu(t, \varrho), \qquad (k = 0, \ldots, p^\mu)$$

les Ψ vérifiant (17.1); on pose

(28.2)
$$\psi(t) = \sum_\mu \Psi_0^\mu(t, 0);$$

on obtient, sur une bande X_ψ:

$$|D^{m^\nu - p^\nu + k, \infty} u^\nu, S_t, \varrho| \ll \Phi_k^\nu(t, \varrho) \qquad (k = 0, \ldots, p^\nu),$$

en posant

(28.3)
$$\begin{cases} \Phi_0^\nu = A_0(\psi)\, \Phi^\nu \\ \Phi_{p^\nu - j}^\nu = c''\, A_0(\psi)\, [1 + C(\Psi_{r_j^\mu}^\mu)]^{p^\nu - j} \left(1 + \dfrac{\partial}{\partial t} + \dfrac{\partial}{\partial \varrho}\right)^{p^\nu - j} \Phi^\nu, \end{cases}$$

où

$$0 \leq j \leq p^\nu,\; r_j^\mu = \inf(p^\mu - j - 1, q^\mu - j),\; \Psi_r^\mu = \Psi_0^\mu \quad \text{pour} \quad r \leq 0,$$

et en définissant les Φ^ν par le système de Cauchy formel:

(28.4)
$$\begin{cases} \left[\dfrac{\partial}{\partial t} - A\left(\psi, \sum_\mu \Psi_0^\mu\right) \left(1 + \dfrac{\partial}{\partial \varrho}\right)\right]^{p^\nu} \Phi^\nu(t, \varrho) = B^\nu(\Psi_{q^\mu}^\mu) \\ \dfrac{\partial^j \Phi^\nu}{\partial t^j}(0, \varrho) = 0 \quad \text{pour} \quad j < p^\nu; \end{cases}$$

dans B^ν, on remplace $\Psi_{q^\mu}^\mu$ par $\left(1 + \dfrac{\partial}{\partial \varrho}\right) \Psi_{p^\mu - 1}^\mu$, quand $q^\mu = p^\mu$.

Un sous-ensemble de $\sum_\nu \gamma_2^{m^\nu, (\alpha)}(X')$ que $\{v^\nu\} \to \{u^\nu\}$ applique en lui-même s'obtient alors, comme au n° 18, en montrant qu'on peut choisir

(28.5)
$$\Psi_k^\nu = \Phi_k^\nu.$$

On note

$$\varphi(t) = \sum_\nu \Phi^\nu(t, 0);$$

la définition (28.2) de ψ s'écrit donc

$$\psi = A_0(\psi)\, \varphi;$$

on met, comme au n° 18, cette relation sous la forme

(28.6)
$$\varphi = f(\psi).$$

En éliminant les Φ_k^ν entre (28.5) et (28.3), on obtient

$$(28.7) \quad \begin{cases} \Psi_0^\nu = A_0(\psi)\, \Phi^\nu \\ \Psi_{p^\nu - j}^\nu = c'' A_0(\psi)\, [1 + C(\Psi_{r_j^\mu}^\mu)]^{p^\nu - j} \left(1 + \dfrac{\partial}{\partial t} + \dfrac{\partial}{\partial \varrho}\right)^{p^\nu - j} \Phi^\nu. \end{cases}$$

Puisque $r_j^\mu = p^\mu - i$, où $i > j$, les équations (28.7) se résolvent par un nombre fini d'itérations; on obtient, en employant une généralisation évidente de la notation du n° 10:

$$\Psi_{p^\nu - j}^\nu = G_j^\nu(\psi, D^{p^\nu - j}\Phi^\nu, D^{r_i^\mu}\Phi^\mu), \quad \text{où} \quad i \geqq j.$$

Prenons $j > 0$, ce qui implique $i > 0$, donc $r_i^\mu < q^\mu$; il vient:

$$\Psi_{p^\nu - j}^\nu = G_j^\nu(\psi, D^{p^\nu - j}\Phi^\nu, D^{q^\mu - 1}\Phi^\mu) \quad \text{pour} \quad j > 0;$$

d'où, en faisant $j = p^\nu - q^\nu$ quand $q^\nu < p^\nu$, puis $j = 1$ quand $q^\nu = p^\nu$:

$$(28.8) \quad \begin{cases} \Psi_{q^\nu}^\nu - G^\nu(\psi, D^{q^\nu}\Phi^\nu, D^{q^\mu - 1}\Phi^\mu) \quad \text{pour} \quad q^\nu \neq p^\nu, \\ \Psi_{p^\nu - 1}^\nu = G^\nu(\psi, D^{q^\mu - 1}\Phi^\mu) \quad \text{pour} \quad q^\nu = p^\nu. \end{cases}$$

En portant (28.8) dans (28.4), nous voyons que $\{\Phi^\nu\}$ doit être une solution du problème de Cauchy formel

$$(28.9) \quad \begin{cases} \left[\dfrac{\partial}{\partial t} - F_0^\nu(\Phi^\mu)\left(1 + \dfrac{\partial}{\partial \varrho}\right)\right]^{p^\nu} \Phi^\nu = F^\nu(D^{q^\mu}\Phi^\mu) \\ \dfrac{\partial^j \Phi^\nu}{\partial t^j}(0, \varrho) = 0 \quad \text{pour} \quad j < p^\nu; \end{cases}$$

ce problème a des propriétés analogues à celles du problème (18.4); par exemple:

$$\frac{\partial^{p^\nu}\Phi^\nu}{\partial t^{p^\nu}} \text{ ne figure pas dans } F^\nu(D^{q^\mu}\Phi^\mu).$$

Il s'agit de trouver une solution du problème (28.9) telle que

$$(28.10) \quad \Phi^\nu \in \Gamma^{p^\nu, (\alpha)}, \frac{\partial^j \Phi^\nu}{\partial t^j}(t, \varrho) \geqslant 0 \quad \text{pour} \quad j \leqq p^\nu.$$

Une telle solution existe, car le théorème d'existence du n° 14 s'étend aisément à des systèmes formels du type (28.9).

Voici achevée la construction de l'ensemble que l'application $v \to u$ applique en lui-même.

La majoration des approximations successives en résulte, comme au n° 23.

Le module de continuité de l'application $v \to u$ est donné par un lemme analogue au lemme 19: on suppose

$$|D^{m^\mu, \infty} v_h^\mu, S_t, \varrho| \ll \Theta(\varrho), |D^{m^\mu, \infty} u_h^\mu, S_t, \varrho| \ll \Theta(\varrho), \quad \text{où} \quad \Theta \in \Gamma^{(\alpha)}, h = 0, 1;$$

on a

$$(28.11) \quad |D^{m^\mu - p^\mu + k, \infty}(u_1^\mu - u_0^\mu), S_t, \varrho| \ll C(\varrho)\left(1 + \frac{\partial}{\partial t} + \frac{\partial}{\partial \varrho}\right)^k \varphi^\mu(t, \varrho)$$

$$(k = 0, \ldots, p^\mu)$$

si l'on a

$$(28.12) \quad |D^{m^\mu - p^\mu + k, \infty}(v_1^\mu - v_0^\mu), S_t, \varrho| \ll C(\varrho) \left(1 + \frac{\partial}{\partial t} + \frac{\partial}{\partial \varrho}\right)^k \psi^\mu(t, \varrho)$$

$$(k = 0, ..., p^\mu)$$

et si les φ^μ sont la solution du problème de Cauchy formel:

$$(28.13) \quad \begin{cases} \left[\dfrac{\partial}{\partial t} - A(\varrho) \left(1 + \dfrac{\partial}{\partial \varrho}\right)\right]^{p^\nu} \varphi^\nu = \sum_\mu B_\mu^\nu(\varrho) \, C(\varrho) \left(1 + \dfrac{\partial}{\partial t} + \dfrac{\partial}{\partial \varrho}\right)^{q^\mu} \psi^\mu \\ \dfrac{\partial^j \varphi^\nu}{\partial t^j}(0, \varrho) = 0 \quad \text{pour} \quad j < p^\nu; \end{cases}$$

dans (28.13), $\left(1 + \dfrac{\partial}{\partial t} + \dfrac{\partial}{\partial \varrho}\right)^{q^\mu} \psi^\mu$ est remplacé par $\left(1 + \dfrac{\partial}{\partial t} + \dfrac{\partial}{\partial \varrho}\right)^{p^\mu - 1} \dfrac{\partial \psi^\mu}{\partial \varrho}$ quand $q^\mu = p^\mu$; A, B_μ^ν dépendent de a^ν, b^ν, Θ, sans dépendre de u_h^μ ni de v_h^μ; $A(0) = 0$; A, B_μ^ν sont $\geqslant 0$ et $\in \Gamma^{(\alpha)}$.

La convergence des approximations successives en résulte, comme au n° 24, en employant une extension facile, aux systèmes formels, du théorème de convergence du n° 14.

Les théorèmes d'unicité se prouvent, comme aux n° 25 et 26.

§ 7. Systèmes quasi-linéaires ou non-linéaires

29. Un tel système peut être transformé en un système à partie principale diagonale, c'est-à-dire du type qui vient d'être étudié au § 6: voir Mme. Y. Choquet-Bruhat [2].

Bibliographie

[1] Chern, S. S., et H. Lewy: Plongement d'une multiplicité riemannienne dans un espace euclidien (en préparation).

[2] Choquet-Bruhat, Y.: Diagonalisation des systèmes quasi-linéaires et hyperbolicité non stricte. Journal de Math. **45**, 371—386 (1966); — Etude des équations des fluides chargés relativistes inductifs et conducteurs. Commun. math. Physics **3**, 334—357 (1966).

[3] Dionne, P.: Sur les problèmes de Cauchy hyperboliques bien posés. J. d'Analyse Math., **10**, 1—90 (1962).

[4] Gårding, L.: Cauchy's problem for hyperbolic equations. Lecture Notes, University of Chicago, 1957; — Energy inequalities for hyperbolic systems. Colloque international de Bombay, 1964.

[5] Gevrey, M.: Sur la nature analytique des solutions des équations aux dérivées partielles. Ann. sci. école norm. super. **35**, 129—189 (1917).

[6] de Giorgi, E.: Un teorema di unicità per il problema di Cauchy relativo ad equazioni differenziali lineari a derivate parziali di tipo parabolico. Annali di Mat. **40**, 371—377 (1955); — Un esempio di non-unicita della soluzione del problema di Cauchy; Università di Roma, Rendiconti di Matematica, **14**, 382—387 (1955); — J. Leray, Equations hyperboliques non strictes: contre-exemples du type de Giorgi, aux théorèmes d'existence et d'unicité. Math. Ann. **162**, 228—236 (1966).

[7] Hörmander, L.: Linear partial differential operators. Berlin-Göttingen-Heidelberg: Springer 1963.

[8] LEDNEV, N. A.: Nouvelle méthode pour résoudre les équations aux dérivées partielles, Mat. Sb. **22**, 205—259 (1948) (en russe), voir: L. GÅRDING, Une variante de la méthode de majoration de Cauchy. Acta math. **116**, 143—158 (1965).

[9] LERAY, J.: Hyperbolic differential equations. Institute for adv. study, Princeton 1953, Notes miméographiées; — La théorie de Gårding des équations hyperboliques linéaires. CIME, Varenna, 1956, Notes miméographiées.

[10] —, et Y. OHYA: Systèmes linéaires, hyperboliques non stricts. Colloque de Liège, 1964, C. N. R. B.

[11] —, et L. WAELBROECK: Normes des fonctions composées. Colloque de Liège, 1964, C. N. R. B.

[12] LICHNEROWICZ, A.: Etude mathématique des équations de la magnétohydrodynamique relativiste. C. R. Acad. Sci. **260**, 4.449—4.453 (1965).

[13] OHYA, Y.: Le problème de Cauchy pour les équations hyperboliques à caractéristiques multiples. J. Math. Soc. Japan **16**, 268—286 (1964).

[14] PUCCI, C.: Nuove ricerche sul problema di Cauchy. Mem. Acc. Sci. Torino, serie 3 , **1**, 45—67 (1955).

[15] TALENTI, G.: Sur le problème de Cauchy pour les équations aux dérivées partielles. C. R. Acad. Sci. **259**, 1932—1933 (1964).

Professor JEAN LERAY
Collège de France
11 Place Marcelin Bethelot
Paris 5ième/France

Professeur YUJIRO OHYA
Institut de Mathématiques Faculté des Ingénieurs
Université de Kioto
Kioto, Japon

(Received July 2, 1965)

Lectures in Applied Mathematics, Volume 8, p. 170. *Reprinted with Permission of the Publisher The American Mathematical Society.*

Relativistic Hydrodynamics

A. H. Taub

1. **Introduction.** We begin our discussion of relativistic hydro-dynamics with the special relativity formulation of the subject. We shall consider a fluid as a collection of a number of particles, each with a rest mass m, which are subject to an external force, and which may interact through collisions. If we write

$$(1.1) \quad \xi_i = \frac{v_i}{(1 - v^2/c^2)^{1/2}}, \qquad v_i = \frac{\xi_i}{(1 - \xi^2/c^2)^{1/2}}, \qquad i = 1, 2, 3,$$

where

$$v^2 = \sum_{i=1}^{3} v_i^2, \qquad \xi^2 = \sum_{i=1}^{3} \xi_i^2,$$

then, if the v_i are the components of the velocity of a particle relative to an inertial coordinate system in Minkowski space-time, the ξ_i are the components of the momentum per unit rest mass of this particle relative to the same coordinate system. Let $f(t, x^i, \xi_i)$ be the number of particles in the region x^i to $x^i + dx^i$ with values of ξ_i between ξ_i and $\xi_i + d\xi_i$ at time t in the chosen coordinate system. The Boltzman equation for f is then

$$(1.2) \quad Df = \frac{\partial f}{\partial t} + \frac{\xi_i}{(1 + \xi^2/c^2)^{1/2}} \frac{\partial f}{\partial x^i} + F_i \frac{\partial f}{\partial \xi_i} = \Delta_c (f)$$

where F_i is the external force per unit mass and $\Delta_c(f)$ is the time rate of change of f due to collisions between the particles.

Let x stand for the four coordinates x^1, x^2, x^3, and t.

If $f(x, \xi_i)$, $f'(x', \xi_i')$ are the distribution functions of a gas with respect to two different inertial coordinate systems and if x, x' and ξ_i, ξ_i' belong to the same event and the same four-momentum, respectively, then $f(x, \xi_i) = f'(x', \xi_i')$; in this sense the distribution function is a scalar. In fact, the equation

$$\frac{d_3\xi}{(1 + \xi^2/c^2)^{1/2}} = \frac{d_3\xi'}{(1 + \xi'^2/c^2)^{1/2}}$$

expresses that the parameter-intervals $d_3\xi$, $d_3\xi'$ describe the same infinitesimal domain of four-momenta, and

$$(1 + \xi^2/c^2)^{1/2} d_3x = (1 + \xi'^2/c^2)^{1/2} d_3x'$$

holds if d_3x, d_3x' are the volumes of two cross-sections $t = \text{const.}$, $t' = \text{const.}$, respectively, of a world tube swept out in space-time by a cloud of particles with four-momenta close to $m(\xi_i, (1 + \xi^2/c^2)^{1/2})$ (Lorentz-contraction). Hence, $d_3x d_3\xi = d_3x' d_3\xi'$ holds under conditions where the same particles "belong to" $d_3x d_3\xi$ and to $d_3x' d_3\xi'$, which proves our assertion, cf. [13].

In terms of the function f we may define the following functions of x_i and t:

$$(1.3) \quad U^\mu(x) = \int V^\mu(\xi) f(x, \xi_i) \frac{d_3\xi}{(1 + \xi^2/c^2)^{1/2}} = \int V^\mu(\xi) \, d\mu(\xi)$$

and

$$T^{\mu\nu}(x) = mc^2 \int V^\mu(\xi) V^\nu(\xi) f(x, \xi) \frac{d_3\xi}{(1 + \xi^2/c^2)^{1/2}} = T^{\nu\mu}(x),$$

$$(1.4) \qquad\qquad\qquad\qquad\qquad\qquad\qquad \mu, \nu = 1, 2, 3, 4,$$

$$= mc^2 \int V^\mu(\xi) V^\nu(\xi) \, d\mu(\xi),$$

where

$$V^i = \xi_i/c, \qquad V^4 = (1 + \xi^2/c^2)^{1/2},$$

and hence

$$(1.5) \qquad\qquad g_{\mu\nu} V^\mu V^\nu = (V^4)^2 - \sum_{i=1}^{3} (V^i)^2 = 1,$$

and the integrations are carried out over the entire volume of
ξ_1, ξ_2, ξ_3-space $(- \infty \leqq \xi_i \leqq \infty)$.

Since, as remarked above, $d_3 \xi / (1 + \xi^2/c^2)^{1/2}$ is Lorentz-invariant,
and f is a scalar and $V^\mu(\xi)$ are four-vectors, it follows from these
definitions that $U^\mu(x)$ is a four-vector field and that $T^{\mu\nu}(x)$ is a tensor
field in space-time.

We next assume that the laws of collision are such that

$$(1.6) \qquad \int \phi^\alpha \Delta_c(f) \, d_3 \xi = 0, \qquad \alpha = 0, 1, 2, 3, 4,$$

where

$$\phi^0 = m, \qquad \phi^i = m \xi_i, \qquad \phi^4 = m(1 + \xi^2/c^2)^{1/2}.$$

This assumption is the special relativistic equivalent of the assumption that the total rest mass, momentum and energy of the
collection of the particles in unit volume at x is conserved in the
collisions.

In case $\alpha = 0$ in Equation (1.6) it follows that we must have

$$m \int Df \, d_3 \xi = m \int \left(\frac{\partial f}{\partial t} + \frac{\xi_i}{(1 + \xi^2/c^2)^{1/2}} \frac{\partial f}{\partial x^i} + F_i \frac{\partial f}{\partial \xi_i} \right) d_3 \xi = 0,$$

or

$$(1.7) \qquad m(\partial U^4/\partial t + c U^i \partial f/\partial x^i) = mc(U^\alpha)_{,\alpha} = 0,$$

where

$$f_{,i} = \frac{\partial f}{\partial x^i}, \qquad f_{,4} = \frac{1}{c} \frac{\partial f}{\partial t}.$$

We shall define the scalar rest mass density of the fluid by the
equation

$$\rho^2 = m^2 U^\alpha U_\alpha$$

and write

$$U^\alpha = \frac{\rho}{m} u^\alpha.$$

Then

$$u^\alpha u_\alpha = 1,$$

and Equation (1.7) becomes

(1.8)
$$(\rho u^{\mu})_{,\mu} = 0.$$

This equation is known as the equation of conservation of mass in classical hydrodynamics. It is often referred to as the equation of conservation of particle number.

In case $\alpha = i$ in Equation (1.6) it follows that we must have

(1.9)
$$\int m\,\xi_i\,Df\,d_3\xi = m\int \left(\xi_i\frac{\partial f}{\partial t} + \frac{\xi_i\xi_j}{(1+\xi^2/c^2)^{1/2}}\frac{\partial f}{\partial x^j} + \xi_i F_j\frac{\partial f}{\partial \xi^j}\right)d_3\xi$$
$$= T^{i\mu}_{\ \ ,\mu} - m\,U^4(x)\,F_i = 0.$$

The last term in the last equation arises from the corresponding term in the next to last equation by an integration by parts. Equation (1.9) and the equation arising from Equation (1.6) by setting $\alpha = 4$ may be written as

(1.10)
$$T^{\mu\nu}_{\ \ ,\nu} = \rho\,\mathscr{F}^{\mu},$$

where

$$\mathscr{F}^i = \frac{F_i}{(1-u^2/c^2)^{1/2}}, \qquad \mathscr{F}^4 = F_i u^i/c$$

is the four-dimensional force vector per unit rest mass.

We shall consider Equations (1.8) and (1.10) as the laws of hydrodynamics in special relativity. In an arbitrary coordinate system in Minkowski space-time and when there are no external forces present they may be written as

(1.11)
$$(\rho u^{\mu})_{;\mu} = 0$$

and

(1.12)
$$T^{\mu\nu}_{\ \ ;\nu} = 0.$$

There are five equations, corresponding to the five conservation conditions described by Equations (1.6).

2. **Specific internal energy.** The tensor $T^{\mu\nu}$ may be expressed in terms of the time-like unit vector u^{μ} as

(2.1)
$$T^{\mu\nu} = wu^{\mu}u^{\nu} + W^{\mu}u^{\nu} + W^{\nu}u^{\mu} + W^{\mu\nu},$$

where

(2.2)
$$w = u^{\mu}u^{\nu}T_{\mu\nu},$$
$$W^{\mu} = T_{\rho\sigma}u^{\nu}(g^{\rho\mu} - u^{\rho}u^{\mu}),$$
$$W^{\mu\nu} = T_{\rho\sigma}(g^{\rho\mu} - u^{\rho}u^{\mu})(g^{\sigma\nu} - u^{\sigma}u^{\nu}),$$

and hence

(2.3) $W^\mu u_\mu = W^{\mu\nu} u_\nu = 0.$

The vector W^μ is known as the heat flow-vector and the tensor $W^{\mu\nu}$ is known as the stress-tensor.

It follows from Equation (2.1) that

(2.4) $T = g_{\mu\nu} T^{\mu\nu} = w - 3p,$

where

(2.5) $-3p = W^{\mu\nu} g_{\mu\nu}.$

The scalar function p is called the hydrostatic pressure.

We define the rest specific internal energy of the fluid by the equation

$$w = \rho(c^2 + \epsilon).$$

It is then a consequence of Equations (2.2), (1.3), (1.4), (1.8) and Schwarz's inequality that ϵ when considered as a function of ρ and p must satisfy the inequality

(2.6) $\epsilon \geq \dfrac{3}{2} p/\rho + c^2 \left[\left(1 + \dfrac{9}{4} \left(\dfrac{p}{\rho c^2} \right)^2 \right)^{1/2} - 1 \right].$

The proof of the inequality (2.6) is as follows: Schwarz's inequality states that

$$\left(\int g(\xi) \, d\mu(\xi) \right)^2 \leq \left(\int g^2(\xi) \, d\mu(\xi) \right) \left(\int d\mu(\xi) \right)$$

for functions $g(\xi)$ for which the integrals exist. Now define $g(\xi)$ by the equation

$$g(\xi) = V^\mu(\xi) \, U_\mu.$$

Hence

$$\int g(\xi) \, d\mu(\xi) = U^\mu U_\mu = \frac{\rho^2}{m^2}.$$

Further, it follows from Equations (1.4) and (1.5) that

$$w - 3p = T = mc^2 \int g_{\mu\nu} V^\mu V^\nu d\mu = mc^2 \int d\mu$$

and that

$$T_{\mu\nu}U^\mu U^\nu = mc^2 \int g^2(\xi)\, d\mu(\xi) = \frac{\rho^2}{m^2}\, w.$$

Hence we must have

$$\rho^2 c^4 \leq (w - 3p)w,$$

that is, the inequality (2.6) must hold.

In classical hydrodynamics, the specific internal energy ϵ is written as

$$(2.7) \qquad \epsilon = \frac{1}{\gamma - 1}\, p/\rho.$$

If we consider Equation (2.7) as defining γ, no longer a constant, we may write (2.6) as

$$(2.8) \qquad \gamma \leq 1 + \frac{2}{3}\, \frac{x}{x + \frac{2}{3}\left(\left(1 + \frac{9}{4}x^2\right)^{1/2} - 1\right)},$$

where

$$(2.9) \qquad x = \frac{p}{\rho c^2} \geq 0.$$

For x small compared to one the right-hand side of the inequality (2.8) is approximately 5/3 whereas for x large compared to one it is approximately 4/3.

3. **Perfect fluids.** Such fluids are those for which the distribution function $f(x, \xi_i)$ gives

$$W^\mu = 0$$

and

$$W^{\mu\nu} = -p(g^{\mu\nu} - u^\mu u^\nu),$$

i.e.,

$$(3.1) \qquad T^{\mu\nu} = \rho(c^2 + i)u^\mu u^\nu - pg^{\mu\nu}.$$

Here

$$(3.2) \qquad i = \epsilon + p/\rho$$

is called the rest specific enthalpy. It and ϵ are functions of p and ρ.

It follows from Equations (3.1) and (1.9) that

$$T^\mu{}_\nu u^\nu = \rho(c^2 + \epsilon)u^\mu$$

and that

$$T^{\mu}{}_{\nu} X^{\nu} = -pX^{\nu}$$

if

$$u_{\nu} X^{\nu} = 0.$$

That is, the stress energy tensor of a perfect fluid may be characterized algebraically by the following properties: (1) it has a single time-like vector as a proper vector with a positive proper value $\rho(c^2 + \epsilon)$, and it has three space-like proper vectors each with the same proper value $-p$. Such stress energy tensors satisfy the equation

(3.3)
$$(T^{\mu}{}_{\nu} + p\delta^{\mu}{}_{\nu})(T^{\nu}{}_{\lambda} - \rho(c^2 + \epsilon)\,\delta^{\nu}{}_{\lambda}) = 0,$$

where

(3.4)
$$p = -T/4 + (12S - 3T^2)^{1/2}/12,$$
$$\rho(c^2 + \epsilon) = T/4 + (12S - 3T^2)^{1/2}/4$$
$$T = T^{\mu}{}_{\mu},$$
$$S = T^{\mu}{}_{\rho} T^{\rho}{}_{\mu},$$

and are characterized by this equation. Hence if we are given $T^{\mu}{}_{\nu}$ we may compute T and S (p and $\rho(c^2 + \epsilon)$) and verify if Equation (3.3) is satisfied. If it is we may then be assured that it is the stress energy tensor of a perfect fluid and determine its velocity field by determining the time-like proper vector of $T^{\mu}{}_{\nu}$. From this point of view there is no reason for imposing a restriction such as that given by the inequality (2.6) on ϵ. We shall see, however, that unless an equality such as (2.6) is imposed we will not be able to insure that the velocity of sound in the fluid is less than the velocity of light as it must be in special relativity.

4. The equations $T^{\mu\nu}{}_{;\nu} = 0$. Before discussing these equations we introduce the specific entropy S and temperature Θ of the fluid as in classical thermodynamics. We define $S(p, \rho)$ and $\Theta(p, \rho)$ by the requirement that Θ be an integrating factor so that

(4.1)
$$dS = \frac{1}{\Theta}\left(d\epsilon + pd\left(\frac{1}{\rho}\right)\right)$$

is a perfect differential. We may also write this equation as

(4.2)
$$\Theta \, dS = di - \frac{1}{\rho} \, dp.$$

It follows from Equation (3.1) that

$$T^{\mu\nu}{}_{;\nu} = (\rho(c^2 + i) u^\nu)_{;\nu} u^\mu + \rho(c^2 + i) u^\mu{}_{;\nu} u^\nu - p_{,\nu} g^{\nu\mu}.$$

Hence if the right-hand side of this equation is set equal to zero and the resulting equation is multiplied by u_μ and summed we obtain, taking into account $u_\mu u^\nu = 1$,

(4.3)
$$(\rho(c^2 + i) u^\nu)_{;\nu} - p_{,\nu} u^\nu = 0$$

and

(4.4)
$$\rho(c^2 + i) u^\mu{}_{;\nu} u^\nu = p_{,\nu}(g^{\mu\nu} - u^\mu u^\nu)$$

as equations equivalent to $T^{\mu\nu}{}_{;\nu} = 0$. Equations (4.4) are the relativistic analogues to the equations of conservation of momentum. We shall now see that Equations (4.3) alone do not correspond to a classical conservation theorem. This equation may be written as

$$(c^2 + i)(\rho u^\nu)_{;\nu} + \rho \left(\epsilon_{,\mu} + p \left(\frac{1}{\rho} \right)_{,\mu} \right) u^\mu = 0$$

or as

(4.5)
$$(c^2 + i)(\rho u^\nu)_{;\nu} + \rho \Theta S_{,\mu} u^\mu = 0.$$

In classical hydrodynamics, it can be shown that when shock waves are absent then the equation describing the conservation of energy and the equations of conservation of momentum imply that the entropy is conserved along the stream-lines, that is along the world lines of the elements of the fluid. The relativistic form of this statement is

(4.6)
$$S_{,\mu} u^\mu = 0.$$

Equation (4.6) follows from the conservation law $T^{\mu\nu}{}_{;\nu} = 0$ if and only if the law of conservation of mass, the equation

(4.7)
$$(\rho u^\mu)_{;\mu} = 0,$$

also holds. Equation (4.6) does not imply that $S = $ constant. If however S is constant on some initial hypersurface and if no shocks are present it will be constant throughout the fluid.

We may write Equation (4.3) as

(4.8) $\rho(c^2 + i)\, u'_{;\nu} + [\rho(c^2 + \epsilon)]_{,\nu} u' = 0.$

If the motion of the fluid is such that $\rho(c^2 + \epsilon)$ is a function of $\rho(c^2 + i) = \rho(c^2 + \epsilon) + p$ alone, then we may define a quantity α by the equation

$$da/\alpha = d[\rho(c^2 + \epsilon)]/(\rho(c^2 + \epsilon) + p),$$

and Equation (4.8) becomes

$$(\alpha u^\mu)_{;\mu} = 0.$$

We shall have

$$\alpha = \rho$$

if and only if

$$dS = 0.$$

5. **The Rankine-Hugoniot equations.** From a study of Equations (1.11) and (1.12) in Minkowski space-time for the case of one-dimensional motion of a perfect fluid (where $u^2 = u^3 = 0$ and all quantities are functions of x^1 and x^4 alone) it can be shown [1] that the velocity of sound is given by

(5.1) $a^2 = \dfrac{\rho}{(1 + i/c^2)} \left(\dfrac{di}{d\rho}\right)_S$

where the entropy S is kept constant in evaluating $di/d\rho$.

In case the enthalpy is a function of $x = p/\rho c^2$ alone, we may write Equation (5.1) as

$$a^2 = i'x/(i' - c^2)(1 + i/c^2)$$

where

$$i' = di/dx$$

and

$$\frac{\rho di}{c^2 d\rho} = \frac{\rho i'}{c^2}\frac{dx}{d\rho} = \rho\frac{dx}{d\rho} + x;$$

the last equation is a consequence of the fact that $dS/d\rho = 0$.

A fluid for which the equality holds in (2.6) will be called a limiting fluid. For such a fluid

$$1 + i/c^2 = 5x/2 + (1 + 9x^2/4)^{1/2}$$

and

(5.2)
$$\lim_{x \to \infty} a^2 = \frac{1}{3},$$

(5.3)
$$\lim_{x \to \infty} \rho(c^2 + \epsilon) = 3p.$$

This limit for the velocity of sound in an extreme relativistic gas was obtained by Curtis [2] and de Hoffman and Teller [3]. Taub [1] and Israel [4] have shown by considering sound waves as infinitesimal shock waves the relation between the requirement that the speed of sound be less than the speed of light and certain conditions that must obtain between the various thermodynamic variables that characterize the fluid.

The discussion of one-dimensional motion of a fluid in Minkowski space given in [1] shows that shocks form from compressive motions just as in the classical theory. When this occurs, Equations (1.11) and (1.12) cannot be applied at the shocks and must be replaced by statements that relate the variables describing the flow of matter, momentum and energy across the shocks. These are the relativistic Rankine-Hugoniot equations. They are derived as follows: Equations (1.11) and (1.12) are equivalent to the statements that

(5.4)
$$(\rho u^\mu f)_{;\mu} = \rho u^\mu f_{,\mu}$$

and

(5.5)
$$(T^{\mu\nu} \lambda_\nu)_{;\mu} = T^{\mu\nu} \lambda_{\mu;\nu}$$

for arbitrary functions f and vector fields λ_μ which have continuous first derivatives.

In view of Gauss' theorem we may write these equations as

(5.6)
$$\int \rho u^\mu f n_\mu \, dv = \int \rho u^\mu f_{,\mu} (-g)^{1/2} d^4x$$

and

(5.7)
$$\int T^{\mu\nu} \lambda_\nu n_\mu \, dv = \int T^{\mu\nu} \lambda_{\mu;\nu} (-g)^{1/2} d^4x.$$

The integrals on the right in Equations (5.6) and (5.7) refer to an arbitrary four dimensional volume and those on the left refer

25*

to a closed hypersurface enclosing this volume. Equations (5.6) and (5.7) are meaningful even when the integrands are discontinuous. We assume they hold in case ρ, u^μ and $T^{\mu\nu}$ are discontinuous across a hypersurface Σ in Minkowski space which is the history of a two-dimensional surface in the spaces $t = \text{constant}$.

By enclosing an arbitrary portion Σ' of the hypersurface in a thin four-dimensional volume and taking the limit as this volume shrinks to zero we may show that

(5.8)
$$\int_{\Sigma'} [\rho u^\mu] f n_\mu \, dv = 0$$

and

(5.9)
$$\int_{\Sigma'} [T^{\mu\nu}] \lambda_\nu n_\mu \, dv = 0,$$

where we have used the notation

$$[F] = F_+ - F_-$$

where F_+, F_- are the boundary values of F on the two sides of Σ.

Since the integrals in Equations (5.8) and (5.9) refer to arbitrary portions of Σ and since f and λ_μ are arbitrary, we must have

(5.10) $[\rho u^\mu] n_\mu = 0,$

(5.11) $[T^{\mu\nu}] n_\mu = 0.$

In case the $T^{\mu\nu}$ entering into Equations (5.11) is given by Equations (8.1), Equations (5.10) and (5.11) are called the relativistic Rankine-Hugoniot equations, the equations that express the conservation of mass, energy and momentum across a singular hypersurface in space-time, a hypersurface on which the flow variables may be discontinuous. The derivation given above holds in a general space-time.

If in Minkowski space-time in an inertial coordinate system the parametric equations of Σ are given by

(5.12) $x^\mu = X^\mu(u, V, t),$

that is,

(5.13)
$$t = t,$$
$$x^i = X^i(u, V, t),$$

then the three-dimensional velocity of a point labelled by u, V on the hypersurface is given by

$$V^i = \partial X^i / \partial t,$$

and the three-dimensional normal to the surfaces in three-space given by the last three of Equations (5.13) has components proportional to

$$\Lambda_i = \epsilon_{ijk} \frac{\partial X^j}{\partial u} \frac{\partial X^k}{\partial V}.$$

We normalize the Λ_i by requiring

$$\sum_{i=1}^{3} \Lambda_i^2 = 1.$$

Then the quantity

$$V = \Lambda_i V^i$$

is the velocity of these surfaces in three-space in the direction of the normal. The four-vector

$$n_\mu = \left(\Lambda_1, \Lambda_2, \Lambda_3, \frac{V}{c} \right)$$

is normal to the hypersurface Σ. We may verify that

$$n_\mu n^\mu = \frac{V^2}{c^2} - \sum_i \Lambda_i^2 \leq \left(\sum \Lambda_i^2 \right) \left(\frac{\sum V_i^2}{c^2} - 1 \right),$$

and hence

$$n_\mu n^\mu \leq 0$$

is equivalent to

(5.14)
$$\sum_{i=1}^{3} V_i^2 \leq c^2.$$

We define a unit normal vector

$$N_\mu = \alpha n_\mu$$

such that

$$N_\mu N^\mu = -1$$

when Equation (5.14) holds. Note that if u^μ is the four-velocity of the fluid then

$$u^\mu = (1 - u^2/c^2)^{-1/2}(u_1/c, u_2/c, u_3/c, 1)$$

where u_i are the components of the three-dimensional velocity of the fluid and

$$u^\mu N_\mu = (1 - V^2/c^2)^{-1/2}(1 - u^2/c^2)^{-1/2}(V - U)/c$$

where

$$U = \sum \Lambda_i u_i$$

is the three-dimensional velocity of the fluid in the direction of the normal to the surface. Thus $u^\mu N_\mu$ is proportional to the normal component of the velocity of the surface relative to the fluid.

When Equations (3.1) hold, Equations (5.10) and (5.11) may be written as

(5.15) $$M = \rho_+ u_+^\mu N_\mu = \rho_- u_-^\mu N_\mu$$

and

(5.16) $$M(\mu_+ u_+^\mu - \mu_- u_-^\mu) = (p_+ - p_-) N^\mu$$

with

(5.17) $$\mu = c^2 + i = c^2 + \epsilon + p/\rho$$

respectively. If V_i are the components of a unit three-vector in the surface, then the four-vector

$$Y^\mu = (V_1, V_2, V_3, 0)$$

satisfies the relation

$$Y^\mu N_\mu = 0,$$

and

$$Y^\mu u_\mu = (1 - u^2/c^2)^{-1/2}(V_i u_i/c)$$

is proportional to the velocity of the fluid in the direction V_i. Multiplying Equations (5.16) by Y_μ and summing, we obtain

(5.18) $$M(\mu_+ u_+^\mu Y_\mu - \mu_- u_-^\mu Y_\mu) = 0.$$

Hence, either

(5.19) $$M = 0,$$

or

(5.20) $$\mu_+ u_+{}^\mu Y_\mu = \mu_- u_-{}^\mu Y_\mu,$$

or both (5.19) and (5.20) hold. We shall say that the case $M = 0$ represents a slip-stream discontinuity or a density discontinuity and say that the case $M \neq 0$ represents a shock wave. In the former case it follows from Equations (5.15) and the interpretation of $u^\mu N_\mu$ that no matter crosses the surface of discontinuity. That is, this surface is made up of stream-lines of the fluid. Equation (5.20) need not hold when $M = 0$. If it does in this case, we shall say that there is a density discontinuity present.

If ρ_- and $T_-{}^{\mu\nu}$ both vanish, we are dealing with a particular case for which $M = 0$. And then Equations (5.16) and (5.15) become

$$p_+ = 0$$

and

$$\rho_+ u_+{}^\mu N_\mu = 0.$$

It is evident that ρ_+ need not vanish if

$$u_+{}^\mu N_\mu = 0,$$

that is, if the stream-lines of the flow do not cross the surface of discontinuity.

6. **Shock waves.** As was stated earlier for such waves we have $M \neq 0$. It is convenient to rewrite Equations (5.15) and (5.16) in terms of the vector

(6.1) $$V_\lambda = (\mu/c^2) u_\lambda$$

and the quantity

(6.2) $$\tau = \rho/\mu.$$

Then

(6.3) $$V^\mu V_\mu = \mu^2/c^4 = (1 + i/c^2)^2,$$

and we have instead of Equations (5.15) and (5.16) the equations

(6.4) $$M = c^2\tau_+ V_+{}^\mu N_\mu = c^2\tau_- V_-{}^\mu N_\mu,$$

(6.5) $$M(V_+{}^\mu - V_-{}^\mu) = (p_+ - p_-) N^\mu.$$

Equations (5.20), which must hold for shock waves, become

(6.6) $$V_+{}^\mu Y_\mu = V_-{}^\mu Y_\mu$$

for any Y_μ satisfying

$$Y^\mu N_\mu = 0,$$

$$Y^\mu Y_\mu = -1.$$

Equations (6.6) are two of the four equations contained in Equations (6.5). The remaining two will be derived below.

Multiply Equations (6.5) by $V_{+\mu}$ and by $V_{-\mu}$ and sum and obtain

$$\left(\frac{\mu_+^2}{c^4} - V_-{}^\mu V_{+\mu} \right) = (p_+ - p_-)\frac{1}{c^2 \tau_+},$$

$$\left(V_+{}^\mu V_-{}^\mu - \frac{\mu_-^2}{c^4} \right) = (p_+ - p_-)\frac{1}{c^2 \tau_-}.$$

Hence

(6.7)
$$\frac{\mu_+^2}{c^4} - \frac{\mu_-^2}{c^4} = \frac{1}{c^2}(p_+ - p_-)\left(\frac{1}{\tau_+} + \frac{1}{\tau_-} \right).$$

When ϵ is a known function of p and ρ, i is determined as a function of p and ρ and so is τ. Then for fixed p_-, ρ_-, the p_+, ρ_+ satisfying Equation (6.7) lie on a curve. This curve connects those (p, ρ)-states that can be obtained from (p_-, ρ_-) by passing through a shock. The curve is known as the Hugoniot curve. We must select that part of the curve passing through p_-, ρ_- corresponding to states p_+, ρ_+ whose entropy $S_+ > S_-$. Otherwise, the second law of thermodynamics would be violated.

Multiply Equations (6.5) by N_μ and sum. We obtain

(6.8)
$$\frac{m^2}{c^2}\left(\frac{1}{\tau_+} - \frac{1}{\tau_-} \right) = -(p_+ - p_-)(-N_\mu N^\mu)$$

$$= -(p_+ - p_-).$$

Equations (6.4), (6.6), (6.7), and (6.8) are equivalent to Equations (6.4) and (6.5).

Since shock waves must travel with a velocity less than that of light and since M must be real, we must have

$$\frac{p_+ - p_-}{1/\tau_- - 1/\tau_+} > 0,$$

that is, when

$$p_+ > p_-,$$

we must have

$$\tau_+ > \tau_-.$$

In other words an increase (decrease) in pressure across a shock wave must correspond to an increase (decrease) in τ.

Israel has shown [4] that if $\epsilon(p,\rho)$ is such that

$$(\partial p/\partial\rho)_S > 0,$$

$$(\partial S/\partial p)_\rho > 0,$$

$$(\partial p/\partial\rho)_S < (1 + i/c^2),$$

then the speed of shock waves is always less than the speed of light. The last of these conditions is the requirement that the speed of sound is less than the speed of light, and we have seen that it is satisfied for the limiting fluid as are the first two conditions.

7. **Lagrangian and co-moving coordinates.** Classical hydrodynamics deals with a fluid moving in three-space in which a coordinate system is introduced in which every point of the three-space is labelled by coordinates x^i. Relative to this coordinate system, the velocity vector of the fluid is described by its components $v^i(x,t)$. The v^i are called the Eulerian velocities. The stream-lines of the fluid are then the curves given by the solutions of the ordinary differential equations

(7.1) $$dx^i/dt = v^i(x,t), \qquad i = 1, 2, 3.$$

These solutions may be written as

(7.2) $$x^i = x^i(\xi,t),$$

where the ξ^i are the initial values of the x^i. That is,

(7.3) $$x^i = x^i(\xi,0) = \xi^i.$$

The ξ^i are called the Lagrange coordinates and may be used to describe the motion of the fluid. Equations (7.2) supplemented by the equation $t = \tau$ may be regarded as a set of equations describing transformation between the variables t, x^i and τ, ξ^i, and the equations of conservation of mass, momentum, and energy may be rewritten in terms of the variables τ, ξ^i.

The congruence of curves given by Equation (7.2) may be pictured in space-time. In this interpretation, they represent the world lines of an element of the fluid which is labelled by the three parameters ξ^i. These world lines may be described in terms of their four-dimensional tangent vector

(7.4) $dx^\mu/dt = w^\mu(x)$, $\mu = 1, 2, 3, 4$,

where

$$w^i = v^i, \qquad v^4 = c,$$

since

$$x^4 = ct.$$

We define a hypersurface Σ in space-time by the equation

(7.5) $t = 0$.

In this hypersurface we label points by the coordinates ξ^i. That is, we describe the hypersurface Σ by the parametric equations

(7.6) $x^\mu = X^\mu(\xi^1, \xi^2, \xi^3)$

instead of the functional equation

$$F(x^\mu) \equiv t = 0.$$

Then the congruence of world lines given by Equations (7.2) may be described as the set of solutions of Equations (7.4) which intersect the hypersurface Σ in the points ξ^i of Σ.

Equations (7.4) may be written as

$$dx^\mu/dt = \rho u^\mu,$$

where u^μ is the unit four-dimensional velocity vector of the fluid and ρ is given by the equation

$$\rho^2 = \frac{ds^2}{dt^2} = g_{\mu\nu} \frac{dx^\mu}{dt} \frac{dx^\nu}{dt} = g_{\mu\nu} w^\mu w^\nu.$$

If the parameter which labeis points on the world line determined by ξ^i is changed to any other parameter, say ω, Equations (7.4) may still be written as

(7.7) $dx^\mu/d\omega = \rho u^\mu = \overline{w}^\mu$,

where now ω is related to ρ by a similar equation to the one given above. If ω is taken to be equal to s, the proper time along the world line, we have

$$dx^\mu/ds = u^\mu.$$

Now let Σ be an arbitrary space-like hypersurface in space-time described by the parametric Equations (7.6), and let us write the solutions of Equations (7.7) as

$$x^\mu = x^\mu(\eta; \omega),$$

where

$$x^\mu = x^\mu(\eta; 0) = \eta^\mu.$$

That is, the η^μ are the initial values of the x^μ. We further require that the η^μ lie on the hypersurface Σ. Then

$$\eta^\mu \equiv X^\mu(\xi),$$

and we may write the solutions of Equations (7.7) as

(7.8) $$x^\mu = x^\mu(\xi; \omega).$$

These equations may be regarded as a transformation of coordinates from the coordinates x^μ to the coordinates ξ^1, ξ^2, ξ^3, and ω. The latter coordinates are called co-moving coordinates in space-time. They differ from Lagrangian coordinates in that the hypersurface Σ used in their definition need not be given by the Equation (7.5) in the Eulerian coordinate system.

The functions x^μ and $\bar{w}^\mu$ entering into Equations (7.7) may be regarded as functions of ξ and ω in view of Equations (7.8). If they are so regarded, then they should be written as

(7.9) $$\partial x^\mu / \partial \omega = \bar{w}^\mu(\xi, \omega),$$

and it should be understood that the ξ^i are kept constant in the differentiation involved on the left-hand side of these equations.

From the transformation law of vectors, it follows that in the co-moving coordinate system, the components of the vector $\bar{w}^\mu$ are given by

$$\bar{w}^{\mu*} = \bar{w}^\rho \, \partial x^{\mu*}/\partial x^\rho,$$

where

$$x^{i*} = \xi^i, \qquad x^{4*} = \omega.$$

In view of Equation (7.9), we have

$$\bar{w}^{\mu*} = \frac{\partial x^{\mu*}}{\partial x^\rho} \frac{\partial x^\rho}{\partial w} = \frac{\partial x^{\mu*}}{\partial x^{4*}} = \delta_4^\mu.$$

Hence, in a co-moving coordinate system, the four-velocity vector of the fluid satisfies the equations

(7.10) $$u^\mu = \delta^\mu_4 g_{44}^{-1/2}$$

since we must have

$$g_{\mu\nu} u^\mu u^\nu = 1.$$

8. The equations of hydrodynamics in co-moving coordinates. Co-moving coordinates may be introduced both in special relativity and in general relativity. In the former theory, they complicate the metric in that we no longer have

$$g_{\mu\nu} = \eta_{\mu\nu}.$$

However, Equations (4.4), (4.6), and (4.7) have relatively simple forms in co-moving coordinates.

Thus if Equation (7.10) holds, Equation (4.6) becomes

$$S_{,4} = 0,$$

that is

(8.1) $$S = S(\xi).$$

Equation (4.7) may be written as

$$(g^{1/2} g_{44}^{-1/2} \rho)_{,4} = 0$$

or as

(8.2) $$\rho = g_{44}^{1/2} (-g)^{-1/2} \rho_0(\xi),$$

where g is the determinant of the $g_{\mu\nu}$ and

(8.3) $$g_{\mu\nu} = \eta_{\alpha\beta} \frac{\partial x^\alpha}{\partial x^{*\mu}} \frac{\partial x^\beta}{\partial x^{*\nu}},$$

$$ds^2 = g_{44} d\omega^2 + 2g_{4i} d\omega d\xi^i + g_{ij} d\xi^i d\xi^j$$

in the case of special relativity. They are such that the Riemann-Christoffel tensor vanishes:

(8.4) $$R^\mu_{\nu\sigma\tau} = 0.$$

Since

$$u^\mu_{;\nu} = \frac{\partial u^\mu}{\partial x^{*\nu}} + u^\sigma \Gamma^\mu_{\sigma\nu},$$

where the $\Gamma^\mu_{\sigma\rho}$ are the Christoffel symbols evaluated in this coordinate system, we have in this coordinate system:

$$u^{\mu}_{;\nu} = \frac{\partial}{\partial x^{*\nu}}\left(\frac{1}{g_{44}^{1/2}}\right)\delta^{\mu}_{4} + \frac{1}{g_{44}^{1/2}}\Gamma^{\mu}_{4\nu},$$

and

$$u^{\nu}u^{\mu}_{;\nu} = \frac{1}{g_{44}^{1/2}}\frac{\partial}{\partial\omega}\left(\frac{1}{g_{44}^{1/2}}\right)\delta^{\mu}_{4} + \frac{1}{g_{44}}\Gamma^{\mu}_{44},$$

where

$$\Gamma^{\mu}_{44} = g^{\mu\nu}\left(\frac{\partial g_{4\nu}}{\partial\omega} - \frac{1}{2}\frac{\partial g_{44}}{\partial x^{*\nu}}\right).$$

If we define

$$(8.5) \qquad\qquad F = \log g_{44}^{1/2},$$

we may write

$$u^{\nu}u^{\mu}_{;\nu} = \left(g^{\mu\nu} - \frac{\delta^{\mu}_{4}\delta^{\nu}_{4}}{g_{44}}\right)\frac{\partial F}{\partial x^{*\nu}} + g^{\mu\nu}\frac{\partial}{\partial\omega}\left(\frac{g_{\nu4}}{g_{44}}\right).$$

Equations (4.4) then become

$$(8.6) \quad \left(g^{\mu\nu} - \frac{\delta^{\mu}_{4}\delta^{\nu}_{4}}{g_{44}}\right)\left(\frac{\partial p}{\partial x^{*\nu}} + \rho c^{2}\left(1 + \frac{i}{c^{2}}\right)\frac{\partial F}{\partial x^{*\nu}}\right) = g^{\mu\nu}\frac{\partial}{\partial\omega}\left(\frac{g_{4\nu}}{g_{44}}\right).$$

It follows from this equation that if the flow is stationary in co-moving coordinates, that is if all quantities are independent of ω, the time variable in co-moving coordinates, then we must have[†]

$$(8.7) \qquad\qquad \frac{\partial p}{\partial\xi^{i}} + \rho c^{2}\left(1 + \frac{i}{c^{2}}\right)\frac{\partial F}{\partial\xi^{i}} = 0.$$

If we recall Equation (8.5) and interpret g_{44} in terms of the Newtonian potential, then Equation (8.7) is recognized as the equation for hydrostatic equilibrium.

Equations (8.7) must hold whenever the right-hand sides of Equations (8.6) vanish. In particular, they must hold when we have a co-moving coordinate system in which $g_{4i} = 0$. It can be shown that this implies that the flow is irrotational.

Time does not permit a further discussion of the equations of hydrodynamics. I shall close this section by referring you to the

---

[†] This result was communicated to me verbally by Dr. J. R. WRIGHT.

discussion of circulation, Bernoulli's theorem and other extensions of classical results that may be found in reference [5].

9. **General relativistic hydrodynamics.** Problems in this area are of two sorts: (I) Those in which the gravitational field of the fluid can be neglected, but the gravitational field due to some other matter or energy must be taken into account, and (II) those in which the only matter and gravitational fields which are present are the fluid and its own gravitational field.

For problems of type I, the basic equations are Equations (1.11), (1.12), (5.10) and (5.11), but now the covariant differentiation is done with respect to a metric tensor $g^{\mu\nu}$ which is determined by the stress-energy tensor $\theta^{\mu\nu}$, the source of the exterior gravitational field and its potentials, through the equations

$$(9.1) \qquad R^{\mu\nu} - \frac{1}{2} g^{\mu\nu} R = -kc^2\theta^{\mu\nu},$$

where $R^{\mu\nu}$ is the Ricci curvature tensor computed from the $g^{\mu\nu}$, R is the scalar curvature, and

$$k = 8\pi G/c^2$$

with G the Newtonian constant of gravitation. Thus for this type of problem the only difference between the special theory of relativity and the general theory is that the underlying space-time is changed from a Minkowski one to a general one whose metric tensor satisfies Equation (9.1).

In problems of type II Equations (9.1) with $\theta^{\mu\nu} \doteq T^{\mu\nu}$ hold. Thus we must determine both $T^{\mu\nu}$ and $g^{\mu\nu}$ so that

$$(9.2) \qquad R^{\mu\nu} - \frac{1}{2} g^{\mu\nu} R = -kc^2 T^{\mu\nu}$$

and Equations (3.1), (1.11), (1.12), (5.10) and (5.11) obtain. Equations (1.12) are a consequence of Equations (9.2). The former may be considered as a first integral of equations (9.2), and this observation has led McVittie [6] to a method for solving problems in prerelativity hydrodynamics.

Equations (3.1), (1.8), and (9.2) are sometimes said to describe the interior problem for the Einstein field equations. The "solution" of this problem requires the determination of the ten components of the tensor $g^{\mu\nu}$, three independent components of the four-velocity vector u^μ and the two scalar functions p and ρ. The scalar function ϵ is supposed to be known as a function of p and ρ when the nature of the fluid medium is given. Thus one is required to "solve" the ten Equations (9.2) when neither the right-hand side nor the left-hand side of these equations is given. Surprising as it may seem this can be done at least in a variety of cases. In the following we shall see how this can be done by using the algebraic properties of the $T^{\mu\nu}$ satisfying Equations (3.1) or by imposing the conservation of mass (Equation (1.11)) condition and the condition of isentropy and irrotationality.

By substituting for $T^\mu{}_\nu$ from Equations (9.2), Equations (3.3) and (3.4) become a set of partial differential equations for the $g_{\mu\nu}$ alone. The solutions of these equations may be substituted into Equations (9.2) to determine $T^\mu{}_\nu$. The equations for the $g_{\mu\nu}$ obtained by the method described above have been called the consistency equations by McVittie [6]. He dealt with an approximate form of these equations and thus obtained various solutions for problems in pre-relativity hydrodynamics. Rainich [7] and Misner and Wheeler [8] used a similar technique for formulating the Einstein-Maxwell field equations for the case where the only stress-energy tensor creating the gravitational field is that due to the electromagnetic field and no electromagnetic sources are present.

The vorticity vector v^μ is defined by the equation

$$(9.3) \qquad v^\mu = (-g)^{-1/2} \epsilon^{\mu\nu\sigma\tau} u_{\sigma;\tau} u_\nu,$$

where g is the determinant of the metric tensor and $\epsilon^{\mu\nu\sigma\tau}$ is the alternating tensor density whose components are equal to plus or minus one depending on whether $(\mu\nu\sigma\tau)$ is an even or odd permutation of (1234). A flow will be said to be irrotational if

$$v^\mu = 0.$$

If we define the vector

$$V^\mu = (1 + i/c^2) u^\mu,$$

it may be shown [5] as a consequence of Equations (2.6) that

$$\Omega_{\mu\nu} = (1 + i/c^2)(-g)^{1/2} v^\sigma u^\tau \epsilon_{\sigma\tau\mu\nu} + TS_{;\nu} u_\mu - TS_{;\mu} u_\nu = V_{\mu,\nu} - V_{\nu,\mu}$$

where T is the temperature and S is the entropy.

Hence for isentropic $(S = \text{constant})$, irrotational $(v^\mu = 0)$ flow we have

$$\Omega_{\mu\nu} = 0.$$

As a consequence

$$u_\mu = \frac{1}{1 + i/c^2} \frac{\partial \theta}{\partial x^\mu},$$

where θ is   arbitrary scalar function. We may adapt our co-ordinate system so that $\theta = x^4$, and then in this comoving co-ordinate system

$$u_\mu = \frac{1}{(1 + i/c^2)} \delta^4{}_\mu.$$

It then follows from Equation (1.9) that

$$g^{44} = (1 + i/c^2)^2.$$

Hence for irrotational isentropic flow we may choose a co-moving coordinate system in which the stress-energy tensor depends on g^{44} alone, since in this case the enthalpy, i, is a function of pressure alone as is the rest density. If these functions of g^{44} and the above expression for u_μ are used in the evaluation of the tensor $T^{\mu\nu}$ we may write the field Equations (9.2) wholly in terms of the tensor $g_{\mu\nu}$. Thus the Einstein equations for the case of relativistic hydrodynamics involve a single set of dependent variables and one has a not impossible problem.

Various exact and approximate solutions of these equations have been discussed. In particular, the problem of the one-dimensional motion of a gas analogous to the problem in classical hydrodynamics of the motion generated by pushing a piston into a tube filled with gas has been treated approximately [9], [10]. The procedure starts with the special relativistic solution to the problem and modifies this solution in accordance with the Einstein field equations. It can be seen that the nonlinearities in the problem reside in the special relativistic approximations and that the gravitational corrections satisfy linear equations. Thus, in the zero approximation shocks occur. It is unlikely that the gravitational corrections satisfying linear equations will remove them. We must therefore be prepared in general relativity to treat with and allow for coordinate transformations involving discontinuous derivatives. The metric tensor thus becomes singular [11] on certain hypersurfaces. Although such singularities are associated with coordinate transformations they are essential ones since the coordinates involved have intrinsic and physical significance.

References

1. A. H. Taub, *Relativistic Rankine-Hugoniot equations*, Phys. Rev. 74 (1948), 328-334.

2. A. R. Curtis, Proc. Roy. Soc. Ser. A 200 (1950), 248.

3. F. deHoffmann and E. Teller, Phys. Rev. 80 (1950), 692.

4. W. Israel, Proc. Roy. Soc. Ser. A 259 (1960), 129.

5. A. H. Taub, *On circulation in relativistic hydrodynamics*, Arch. Rational Mech. Anal. 3 (1959), 312-324.

6. G. C. McVittie, *General relativity and cosmology*, Chapters VI and VII, Wiley, New York, 1956.

7. G. Y. Rainich, *Electrodynamics in general relativity theory*, Trans. Amer. Math. Soc. 27 (1925), 106-136.

8. C. W. Misner and J. A. Wheeler, *Geometrodynamics*, Ann. Physics 2 (1957), 525-603.

9. _____, *Isentropic hydrodynamics in plane symmetric space-time*, Phys. Rev. 103 (1956), 454-467.

10. _____, *Approximate solutions of the Einstein equations for isentropic motions of plane symmetric distributions of perfect fluids*, Phys. Rev. 107 (1957), 884-900.

11. _____, *Singular hypersurfaces in general relativity*. III, J. Math. 1 (1957), 370-388.

General References

12. A. Lichnerowicz, *Theories relativistes de la gravitation et de l'ectromagnetisme*, Masson, Paris, 1955.

13. J. L. Synge, *The relativistic gas*, North-Holland, Amsterdam, 1957.

14. _____, *Relativity, the general theory*, North-Holland, Amsterdam, 1960.

UNIVERSITY OF CALIFORNIA, BERKELEY
BERKELEY, CALIFORNIA

Offsetdruck: Werk- und Feindruckerei Dr. Alexander Krebs, Weinheim u. Hemsbach (Bergstr.) und Bad Homburg v. d. H.